Utilization, Treatment, and Disposal of Waste on Land

Utilization, Treatment, and Disposal of Waste on Land

Proceedings of a workshop held in Chicago, IL, 6–7 Dec. 1985.

Organizing Committee

K.W. Brown*
Texas A&M University, College Station, TX
B.L. Carlile*
Texas A&M University, College Station, TX
Cecil Lue-Hing
Chicago Metropolitan Sanitary District, Chicago, IL
I.P. Murarka
Electric Power Research Institute, Palo Alto, CA
R.H. Miller*
North Carolina State University, Raleigh, NC
A.L. Page
University of California-Riverside, Riverside, CA
E.M. Rutledge*
University of Arkansas, Fayetteville, AR
R.E. Thomas
Environmental Protection Agency, Washington, DC
E.C.A. Runge,* chair
Texas A&M University, College Station, TX

* Indicates service on the Editorial Committee

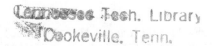

Soil Science Society of America, Inc.
Madison, Wisconsin, USA

1986
Reprinted 1988

CIP

ISBN 0-89118-781-2

Printed in the United States of America

Table of Contents

Foreword

These proceedings result from a two-day workshop devoted to the discussion, review, and update of information available on the "Utilization, Treatment, and Disposal of Waste on Land." The workshop was held immediately following the annual meeting of the Soil Science Society of America in Chicago. The objectives of the workshop were: (1) to bring together experts involved with various aspects of utilization, treatment, and disposal of waste on land; (2) to update interested individuals in this important use of soils in solving society's increasing problem with waste management; (3) to publish these papers in a proceedings so the information can be made widely available in one publication; and (4) to draw attention to the role soil science disciplines have in helping solve societal problems in an economic way.

The workshop consisted of invited papers by well-known experts from various disciplines under four general areas:

1. Land treatment of sewage effluents with minimum pretreatment--large flows--land application,
2. Small flows--onsite clustered systems,
3. Land application of municipal sludges, and
4. Hazardous waste in the soil environment.

Papers include the latest information available on these subjects as well as papers on case histories of various projects that have been underway for various periods of time.

The participants and the organizing committee wish to express their gratitude for the support of the six sponsoring organizations, the financial support from four of the sponsors, and the technical and logistical support of Mr. David M. Kral and his staff of the Headquarters office of the Soil Science Society of America. The editorial assistance of Mrs. Sherri Mickelson from the Headquarters office is specifically acknowledged.

On behalf of the participants, I would like to thank the invited speakers for their efforts in focusing attention on their assigned topic and for providing manuscripts in a timely manner. Finally, I hope readers of this book continue to develop information that will enhance our understanding of soil science and how we can improve the "Utilization, Treatment, and Disposal of Waste on Land."

E. C. A. Runge, President, 1985
Soil Science Society of America

Acknowledgment

The sponsors of this symposium were: Soil Science Society of America, American Society of Agricultural Engineers, Soil Conservation Society of America, Electric Power Research Institute, Environmental Protection Agency, and Southern California Edison.

Financial assistance was provided by the Electric Power Research Institute, Southern California Edison, Soil Science Society of America, and Waste Management Inc.

Utilization, treatment, and disposal of waste on land

A. L. Page, Department of Soil and Environmental Science, University of California, Riverside, CA 92521
Cecil Lue-Hing, Metropolitan Sanitary District of Greater Chicago, Chicago, IL 60611
A. C. Chang, Department of Soil and Environmental Science, University of California, Riverside, CA 92521

Although land application of wastes has been practiced since the ancient time and still represents an integral part of the waste management scheme around the world, the federal mandate on water pollution control and waste disposal marked the beginning of an era that the technology of land treatment was systematically studied and widely accepted in the U.S. (Jewell and Seabrook, 1979).

The nation's attention on water pollution control began with the enactment of the Federal Water Pollution Control Act of 1956 (PL 84-660). Since then, the legislation has undergone numerous changes and been extended through a series of amendments. The most significant change of the law-making process occurred with the 1972 amendment (PL 92-500) which defined the ultimate goal of water pollution control and called for "zero discharge" to restore and to maintain the integrity of nation's water. The law further stipulated that close circuit recycling and land treatment of wastes should be employed to satisfy the discharge requirements. Today, there are more than 3000 land application systems throughout the nation to treat wastes from a diverse origin. In this presentation, we attempted to review concepts of present day land application and waste utilization systems.

Soil as a Waste-Receiving Medium

Soils are a chemically and biologically complex porous medium consisting of weathered mineral fragments, organic matter, microorganisms, water, and air. Throughout the years, fates of organic chemicals and inorganic constituents in soils have been extensively studied. For the purpose of describing the behavior of various constituents in soils, the physical features of the soil frequently are divided into four compartments: atmospheric, aqueous, solid and biotic. The physical, chemical, and biological processes that partition a constituent in the soil among these four compartments are fundamental to the land application of wastes. Land treatment is an engineered delivery of wastes to a designated site with prescribed objectives.

The mechanisms responsible for immobilizing and attenuating anthropogenic constituents in soils are essentially the same as those in the other segments of the environment. The reaction kinetics that determine the mass transfer, however, may be different from those in the aqueous or the atmospheric mediums. Experience indicates that rate constants for the transport process in the soil are influenced by both the properties of the soil and the nature of the chemical constituent. Among the processes that are most likely to limit the ability of the soil to receive wastes are cation exchange, precipitation, adsorption, volatilization, degradation, and bio-absorption.

A porous medium such as the soil is capable of trapping particulates while permitting the solution to pass through. Depending on the particle sizes, solids suspended in the water may be deposited at the soil water interface or incorporated into the soil matrix upon the water's contact with the soil. Even particles whose sizes are considerably smaller than the soil pores may be effectively retained (Ives, 1973). These constituents will be immobile unless they are chemically or biologically altered and enter into the aqueous phase. The movement of solutes in the soil is governed to a large extent by the movement of water. Reactions with the soil, however, will retard the movement of solutes. Ideally, water leaving a land treatment zone shall be free of undesirable constituents. In land treatment, it is essential to match the reaction rates with the hydraulic loading to ensure desirable results.

Cation Exchange

During the course of weathering, soil particles frequently acquire negative charges through amorphous substitution in silicate minerals, from proton dissociation by iron and aluminum hydrous oxides, silicates, alumino-silicates, and organic functional groups. The deficit of positive charge of the soil is balanced by adsorbed cations which may exchange with those present in solution phase of the soil matrix. Because of differences in weathering processes and the soil's mineral composition and organic matter content, soils exhibit distinctly different capacities to adsorb cations. Depending on the concentrations and the charge characteristics, soils also exhibit preferential adsorption of specific cations under given conditions. Chemically, the capacity of soil to exchange cations and the chemical equilibrium between cations of different charges may be described by the following mathematical expression (Vanselow, 1932):

$$\frac{(M^{+})\ [D^{++}\text{-soil}][D^{++}\text{-soil} + M^{+}\text{-soil}]}{(D^{++})\ [M^{+}\text{-soil}]} = K_{v} \qquad (1)$$

where (M^{+}) and (D^{++}) are ionic activities of the monovalent and the divalent ions in soil solution $(\text{mmol}\cdot L^{-1})$, respectively; $[M^{+}\text{-soil}]$ and $[D^{++}\text{-soil}]$ are concentrations of the exchangeable monovalent and the divalent cations in the solid phase $(\text{mmol}\cdot\text{unit weight}^{-1})$, repectively; and K_v is a selectivity coefficient characteristic of the adsorption affinity of the two cations for a given soil. In the presence of an electrolyte solution, cation exchange reactions proceed until the equilibrium described by eq. (1) is attained. Since their concentrations in the soil are far greater than other cations, Ca^{++},

Mg^{++}, K^+, and Na^+ are more commonly involved in the ion exchange reactions. Metallic cations such as Cd^{++}, Zn^{++}, Cu^{++}, Ni^{++}, are not frequently found in soil solution in large concentrations. But they may be introduced into the soil through land application of wastes. Their concentrations in soil solutions, however, are orders of magnitude lower than Ca^{++}, Mg^{++}, and Na^+. For this reason, cation exchange reactions are not expected to play a major role in the transport of trace metal elements (Cd^{++}, Zn^{++}, Ni^{++} ...).

Precipitation and Adsorption

Precipitation identifies the process during which a finely divided solid appears in a solution. This occurs when concentrations of the reacting chemical constituents exceed a saturated condition. In this system, the undissolved solid is in equilibrium with the dissolved solute. Adsorption denotes the chemical and physical adhesion of dissolved substances on the surface of a solid medium in a thin layer. In soil, inorganic as well as organic substances may be adsorbed by soil particles. A variety of intermolecular electrostatic forces are known to be responsible for the adsorption, and under different circumstances, different mechanisms may be operative and different components of the soil may be involved. Regardless of the reactions, they will limit the solution concentration of sparingly soluble elements and retard their movement through the soil.

Although their mechanisms are not the same, it is difficult to experimentally distinguish between precipitation and adsorption reactions in the soil. Quantitatively, they often are described collectively by adsorption equations (e.g., Langmuir or Freundlich). A number of studies have shown that concentrations of most trace metals in the soil solution were considerably lower than those predicted by the solubility products of hydroxides or carbonates, indicating either the formation of less soluble precipitates or a situation where equilibrium concentrations of these metals in the soil were controlled by specific adsorption (Garcia-Miragaya and Page, 1978). Soils, in general, exhibited extraordinary ability to adsorb metals from the soil solution (Doner, 1978).

For organic solutes, adsorption undoubtedly is the most important soil reaction in retarding their mobilities in the soil. Considering the computational requirements in solving the solute transport equation numerically, the non-linear adsorption isotherm for organic chemicals in low concentrations are customarily fitted into the following proportionality equation (Karickhoff et al., 1979):

$$\frac{X}{M_s} = K_d \times C \qquad (2)$$

where X is the amount of solute adsorbed (μg), M_s is the amount of soil (g), C is the equilibrium solution concentration ($\mu g \cdot mL^{-1}$), and K_d is the soil distributed coefficient (mL/g). For non-polar hydrophobic organic substances, their adsorption can be positively related to the organic matter content of the soil (Karickhoff et al., 1979; Means et al., 1982). Therefore, eq. (2) may be written to reflect the influence of soil organic matter so that:

$$\frac{X}{M_{oc}} = \frac{K_d}{f_{oc}} \times C = K_{oc} \times C \tag{3}$$

where f_{oc} is the soil organic carbon expressed in weight fraction, M_{oc} is the amount of soil organic matter (g) that equals M_s multiplied by f_{oc}, and K_{oc} is the soil adsorption constant coefficient in terms of soil organic carbon. The value of K_{oc} for given organic substance remains fairly constant and is independent of the soil type. The normalization against the soil organic matter content greatly reduced the variability of partitioning coefficients across various types of soil.

When measured adsorption data are not available, K_{oc} of organic chemicals may be estimated from the octanol-water partition coefficient (K_{ow}) of the compound. Several equations may be used to empirically estimate K_{oc} (Karickhoff et al, 1979; Schwarzenbach and Westall, 1981; Kenaga and Goring, 1978; Rao and Davidson, 1980). The K_{ow} of organic compounds are well defined and extensively tabulated (Leo et al., 1971).

Volatilization

In a moist soil, volatilization from dilute aqueous solution is determined by the partitioning coefficient between the water and the air (K_w). The K_w of an organic chemical is defined as the ratio of the equilibrium concentration in water (g/mL) to the corresponding concentration in the air (g/mL) at a given temperature (25°C). This parameter provides a comparative measurement of the ability of a chemical to partition between the water and the air. Volatilization of organic chemicals (mostly pesticides) from soils has been reviewed by Spencer et al. (1982).

A low value of K_w indicates a greater tendency for the compound to become volatilized. Since a steady-state equilibrium is seldom reached in the soil, the partitioning coefficient of a compound between the water and the air is best determined experimentally. If experimental data are not available, the K_w may be estimated. Mathematically, the value of K_w for an organic chemical is the inverse of the Henry's Law constant and can be approximated by dividing the water solubility of the compound by its saturated vapor concentration. Using the ideal gas law to evaluate the saturated vapor concentration, the K_w may then be expressed as (Laskowski et al., 1982):

$$K_w = [S \times 0.062366 \times (273.16 + T)]/[V_T \times M] \quad (4)$$

where S is the water solubility of the chemical (μg/mL), T is the temperature (^{o}C) at which water solubility was measured, V_T is the saturated vapor pressure at T (mm Hg), and M is the molecular weight of the chemical.

Although volatilization is dependent on the rate constant of the mass transfer across the air and water interface, the actual amount of chemical volatilized is influenced by many environmental factors as well as the behavior of the chemical at the soil-air-water interphase.

Degradation

In the soil, hazardous organic chemicals may be adsorbed by soil particles or become volatilized, but the chemical structure of the compounds are not altered in the processes. Experimental data have demonstrated that adsorbed organic chemicals may be rapidly desorbed when the soil is leached with a water free of the given chemical (Schwarzenbach and Westall, 1981). This reversal of the adsorption process releases the potentially hazardous chemicals to the aqueous solution.

Degradation refers to the combined biological, chemical, and photochemical transformation of organic chemicals. Among them, the photochemical oxidation is important only when the chemical is deposited on the soil surface and plays a minimal role once the chemical has been incorporated into the soil. Chemical reactions are widespread in soils and they may be oxidative transformation, hydrolytic nucleophilic transformation, or non-hydrolytic nucleophilic displacement reactions. Biodegradations are enzyme activated biochemical reactions whereby organic substances are decomposed. Bacteria, actinomycetes, and fungi are microorganisms most important in the breakdown of organic substances in the soils. Besides the naturally occurring organic matter, soil microorganisms may also be acclimated to attack xenobiotic organic compounds. Although the biochemical pathways of microbial decomposition processes are not always known, microbial metabolisms are well understood. Biochemical reactions such as β-oxidation, cleavage of ester linkage, ring hydroxylation, ring cleavage, ester hydrolysis, dehalogenation, and n-dealkylation are all common to biodegradation processes in the soil (Alexander, 1981). Degradation processes are important in land treatment of hazardous organic wastes because they offer the means for detoxifying waste constituents in the soil.

Despite our knowledge on the organic matter decomposition, only limited information is available about the reaction rates (Overcash and Sims, 1981). Because of the difficulty in making measurements under natural conditions, widely scattered values are often found.

For practical purposes, degradation of organic matter has frequently been represented by the difference between final and initial quantities that were present in the soil. Assuming a first order reaction, the degradation characteristics of various organic constituents may be represented by the degradation half-life, which indicates the time it takes for 1/2 of the introduced organic compounds to disappear via decomposition. Rate constants for many organic pesticides and other hazardous organic chemicals are summarized by Rao and Davidson (1980) and Callahan et al. (1978). Because of its high microbial density, the soil is capable of receiving large amounts of readily decomposable organic matter. Measured in terms of the pollutional strength of the wastes, it is not uncommon for a soil to receive 1000 x 10^3 kg BOD per hectare per year in the form of municipal or industrial wastewater, oil,or organic solid wastes.

Bioabsorption and Waste Utilization

Besides organic matter, most wastes also contain plant nutrient elements. During the course of microbial decomposition, nutrient cycling in the soil is also affected. Depending upon the ratios of amounts of organic carbon to amounts of nutrients, nitrogen, phosphorus and sulfur may be assimilated into the microbial mass or become mineralized. It is conceivable that land application of wastes, under appropriate circumstances may serve the purpose of supplying and/or regulating plant available nutrients in the soil. In this manner, wastes introduced into the soil may be utilized beneficially. Among them, nitrogen transformation has attracted the most attention.

The end product of nitrogen mineralization, nitrate, is not only an essential plant nutrient but also a potential water pollutant. Under the proper environmental conditions, nitrate may be reduced biologically to N_2 or N_2O. But uncontrolled land application of wastes and/or high hydraulic loading inadvertently result in nitrate escaping the zone of root extraction. Properly operated land application systems can minimize the threat of nitrate pollution.

The possibility of plants absorbing potentially hazardous chemical constituents from waste-treated soils should not be overlooked. Despite the favorable soil environment for adsorption and precipitation, there is ample evidence that plants grown on waste treated soils accumulate potentially harmful metal elements (Logan and Chaney, 1983). The extent of metal absorption by plants is influenced by a variety of waste, soil, and plant associated factors and is difficult to describe quantitatively.

Examples from pesticide residues and polynuclear aromatic hydrocarbon studied in soil showed some degree of plant absorption of organic chemicals occurs (Brown and Weiss, 1978). Since there is no known physiological mechanism for plants to absorb organic chemicals, the most likely pathway appeared to be mass flow. Once they are adsorbed by the soil, amounts of the organic chemicals available for plants become limited. Because of their hydrophobic nature, organic compounds with high K_{ow} are expected, upon entering the plants, to be deposited primarily in the plant roots (Iwata and Gunther, 1976; Fuchsbichlev et al., 1978). Low molecular weight organic compounds that are gradually assimilated and metabolized by plants should not exhibit symptoms of bioaccumulation.

In most land treatment systems, both plant nutrients and potentially harmful chemicals are present in the wastes. It is necessary for one to maximize the utilization of plant nutrients and, at the same time, minimize the absorption of potentially harmful chemicals.

Design of Land Application Systems

Unit processes are the building blocks of conventional waste treatment systems. Depending on the treatment objectives, several unit processes may be selected, arranged in a logic sequence, and designed to accommodate the hydraulic and pollutant loads. In the course of designing, each unit process is optimized to achieve a single goal. The overall treatment objective is accomplished stepwise as the wastes travel through the treatment train. Unlike a conventional treatment system, land applications entail a single stage, multiple processes waste treatment system which needs to be operated to accomplish several goals at the same time.

Conceptually, the design of land treatment systems requires the following considerations:

- Identification of environmental constraints.
- Determination of principal limiting constituents of land application.
- Determination of rate and capacity limits of each principal limiting constituent.

Discussions on designing specific land treatment systems are available (Overcash and Pal, 1979).

Environmental Constraints

For land application systems, the treatment is not confined in specially designed and configurated reaction vessels. Instead, a portion of the terrestrial environment is utilized as the medium for waste treatment. It is essential that the integrity of the environment is not breached and the receiving soil is not irreversibly removed from normal land uses. The non-degradation environmental constraint is often the criteria for designing and operating land treatment systems (Overcash, 1976). Under given conditions, the non-degradation constraints may be transcribed into specific design objectives.

Limiting Constituents

The transport process in soils involves many pathways. The rates of transfer and the partition of waste constituents between various compartments in soils are dependent on many factors. Any one of the pathways and any one of the waste constituents may limit the land application. To design and operate a land treatment system properly, it is essential to identify which of the pathways and which of the waste constituents are most limiting.

Land limiting constituent analysis is often used to design land treatment systems (Overcash and Pal, 1977; Loehr, 1979). It is understood that soil has a definitive assimilative capacity for waste constituents. Based on waste characteristics and the assimilative capacities assigned to the site, land requirements for accommodating various constituents in the waste may be determined. Obviously, the constituent that requires the largest land area is the limiting constituent.

Jury et al. (1983a, b, 1984) developed a simulation model using the one dimensional mass conservation equation for homogeneous porous medium to assess the relative extent of solute transport. If rate constants are known, the model may be used to screen various chemical constituents of the waste and to identify the most reactive waste constituent for each pathway. Unlike the land limiting analysis, the screening model based on the transport process takes into account interactions among various pathways in a dynamic manner. According to the outcome of the screening, appropriate application rates of wastes and total capacities of the site may be determined.

The capacity of soil for disposal utilization or treatment of waste is highly variable and depends on physical, chemical, and biological properties of the soil and the water. Although general information on the transport process is known and the environmental fate of many chemicals have been studied, the development of a land treatment system must rely on site and waste specific cause-effect relationships to calibrate reaction coefficients of the transport process.

Application Rate and Total Capacity

In a land application, the most limiting constituent of the waste will determine the rate of application. The upper limit is necessary to prevent waste applications from breaching the environmental constraints, which could be leaching that resulted in groundwater pollution, volatilization that emitted toxic chemicals, phytotoxicity that inhibited plant growth, etc. If a degradable constituent is limiting, each waste application will bring the amount of this constituent in soil to the maximum tolerable level. As the degradation or other transport mechanisms lower levels to near the background, waste application may be resumed. Levels of non-degradable, cumulative constituents in soils, however, rise with each waste input. It is conceivable that there will be a threshold beyond which land application of waste no longer is acceptable. These constituents will determine the total waste capability of the soil. The application rate of wastes coupled with the total capacity of the soil will define the life span of a land application site.

Conclusions

- Mechanisms controlling the transport process in soils are the basis of any land application system for disposal, utilization, and treatment of wastes.

- Pathways of the transport process most important to land application of wastes are adsorption/precipitation, volatilization, degradation, and bioabsorption.

- To design a land application system, it is essential that environmental constraints are identified, principal waste limiting constituents are established, and the rate of application and capacities of the site are calculated.

- Site and waste specific cause-effect relationships are needed to calibrate rate coefficients of the transport process equation.

References

1. Alexander, M. 1981. Biodegradation of chemicals of environmental concern. Science 211:132-138.

2. Brown, R. A., and F. T. Weiss. 1978. Fate and effects of polynuclear aromatic hydrocarbons in the aquatic enviroment. Environ. Affairs Dept. Pub. No. 4297. Amer. Petroleum Institute, Washington, D.C.

3. Callahan, M. A., M. W. Slimak, N. W. Gabel, I. P. May, C. F. Fowler, J. R. Freed, P. Jenning, R. L. Dufee, F. C. Whitemore, B. Maestri, W. R. Mabey, B. R. Holt, and C. Gould. 1979. Water-related environmental fate of 129 priority pollutants (Volume I and II). EPA-440/4-79-029A,B. U.S. Environmental Protection Agency, Washington, D.C.

4. Doner, H. E. 1978. Chloride as a factor in mobilities of Ni(II), Cu(II) and Cd(II) in soil. Soil Sci. Soc. Am. J. 42:882-885.

5. Fuchsbichler, G. A. Suss, and P. Wallnofer. 1978. Uptake of 4-chloro and 3,4 dichloroanilline by cultivated plants. Z. Pflanzenkr. Pflanzenchutz 85:298-307. (Ger.)

6. Garcia-Miragaya, J., and A. L. Page. 1978. Sorption of trace quantities of cadmium by soils with different chemical and mineralogical composition. Water Air Soil Pollut. 9:289-299.

7. Iwata, Y., and F. A. Gunther. 1976. Translocation of the polychlorinated biphenyl Arclor 1254 from soil into carrots under field conditions. Arch. Environ. Contam. Toxicol. 4:44-59.

8. Ives, K. J. 1971. Filtration of water and wastewater. Crit. Rev. Environ. Control 2(2):292-335.

9. Jewell, W. J., and B. L. Seabrook. 1979. A history of land application as a treatment alternative. U.S. Environmental Protection Agency Technical Report EPA 430/9-79-012.

10. Jury, W. A., W. F. Spencer, and W. J. Farmer. 1983a. Use of models for assessing relative volatility, mobility and persistence of pesticides and other trace organics in soil systems. In: Hazard Assessment of Chemicals: Current Development, Vol. 2, p. 11-43. Academic Press, New York.

11. Jury, W. A., W. F. Spencer, and W. J. Farmer. 1983b. Behavior assessment model for trace organics in soil: I. Model description. J. Environ. Qual. 12:558-564.

12. Jury, W. A., W. F. Spencer, and W. J. Farmer. 1984. Behavior assessment model for trace organics in soil: II. Chemical classification and parameter sensitivity. J. Environ. Qual. 13:567-572.

13. Karickhoff, S. W., D. S. Brown, and T. A. Scott. 1979. Sorption of hydrophobic pollutants on natural sediments. Water Research 13:241-248.

14. Kenaga, E. E., and C.A.I. Goring. 1978. Relationship between water solubility, soil sorption, octonal-water partitioning and bioconcentration of chemicals in biota. ASTM 3rd Aquatic Toxicity Symposium, New Orleans, Louisiana.

15. Laskowski, D. A., C.A.I. Goring, P. J. McCall, and R. L. Swann. 1982. Terrestrial environment. p. 198-240. In: R. A. Conway (ed.) Environmental risk analysis for chemicals. Van Nostrand Reinhold Co., New York.

16. Leo, A., C. Hansch, and D. Elkins. 1971. Partitioning coefficients and their uses. Chem. Rev. 71:525-616.

17. Loehr, R. C., W. J. Jewell, J. D. Novak, W. W. Clarkson, and G. S. Friedman. 1979 Land Application of Waste. Van Nostrand Reinhold. New York.

18. Logan, T. J., and R. Chaney. 1983. Metals. p. 235-328. In: A. L. Page, T. L. Gleason III, J. E. Smith, Jr., I. K. Iskander, and L. E. Sommers (eds.) Proc. 1983 Workshop on Utilization of Municipal Wastewater and Sludge on Land. University of California, Riverside, CA 92521.

19. Lyman, W. J., W. F. Reehl, and D. H. Rosenblatt. 1982. Handbook of Chemical Property Estimation Models — Environmental Behavior of Organic Compounds.

20. Mattigod, S. V., A. Gibali, and A. L. Page. 1979. Effect of ionic strength and ion pair formation on the adsorption of nickel by kaolinite. Clay Minerals 27:411-416.

21. Means, J. C., S. G. Wood, J. J. Hassett, and W. L. Banwart. 1982. Sorption of amino- and carboxy-substituted polynuclear aromatic hydrocarbons by sediments and soils. Environ. Sci. Tech. 16:93-98.

22. Overcash, M. R., and R. C. Sims. 1981. TERRETOX: Catalog of terrestrial response to organic chemicals literature. Dept. of Chemical Engineering, North Carolina State University, Raleigh, NC.

23. Rao, P.S.C., and J. M. Davidson. 1980. Estimation of pesticide retention and transformation parameters required in nonpoint source pollution models. p. 23-67. In: M. R. Overcash (ed.) Environmental impact of nonpoint source pollution. Ann Arbor Sci. Publ. Ann Arbor, MI.

24. Schwarzenbach, R. P., and J. Westall. 1981. Transport of non-polar organic compounds from surface water to groundwater: Laboratory sorption studies. Environ. Sci. Tech. 15:1360-1367.

25. Spencer, W. F., W. J. Farmer, and W. A. Jury. 1982. Review: Behavior of organic chemicals at soil, air water interphases as related to predicting the transport and volatilization of organic pollutants. Environ. Toxicol. Chem. 1:17-26.

26. Vanslow, A. P. 1932. Equilibria of the base-exchange reactions of bentonites, permutites, soil colloids and zeolites. Soil Sci. :301-309.

Land treatment of wastewaters: Some perspectives

Richard E. Thomas, U.S. Environmental Protection Agency, Washington, DC 20460

Land treatment is once again emerging as one of many popular alternatives to treating municipal wastewater in a conventional technology plant and then discharging it to surface waters. History has a way of repeating itself, and the record shows that our national concept of wastewater management is no exception. This rebirth of land treatment as an alternative technology that recycles nutrients while reclaiming wastewater is a classic example of history repeating itself.

Contrary to the impression fostered by some recent discourse on the topic, applying wastewater to the land is neither a new nor a novel approach to wastewater management. Land disposal of collected urban wastewater dates back at least four centuries and has experienced several cycles of popularity. Historically, the purpose of land application approaches has emphasized wastewater disposal, whereas the current trend is away from the concept of disposal and toward the concept of treatment and reuse with recycle of nutrients. This change in concept has been accompanied by coinage of many new phrases to describe methods of applying wastewater to the land.

It is desirable to group these identifying phrases into categories to avoid confusion about terminology. One convenient grouping is based on a combination of terms involving the treatment achieved and the fate of the applied wastewater. This grouping identifies systems as slow-rate, rapid-infiltration. and overland-flow.

Slow-rate systems achieve the greatest degree of treatment; they may or may not control surface runoff; low loading rates of 0.5 to 2.5 m/yr allow much of the applied wastewater to be lost through evapotranspiration. The contribution to groundwater is largely dependent on these evapotranspirative losses. Overall, evapotranspiration and groundwater contributions are of comparable magnitude and account for most of the applied wastewater. The slow-rate category includes systems which emphasize recycle of nutrients while reusing wastewater to grow crops. Some people prefer to use terms such as crop irrigation systems or the living filter to describe these systems.

Rapid-infiltration systems also achieve a high level of treatment, but they have limitations for removal of nitrogen compared to slow rate systems. These systems are usually designed to prevent surface runoff; high loading rates (up to 90 m/yr) make evaporative losses relatively insignificant; and up to 99 percent of the applied wastewater may be contributed to ground water as recharge. Systems designated as percolation ponds and recharge basins fall in this category although many of them are designed more for disposal than for treatment of the wastewater.

Overland flow systems provide a lesser level of advanced wastewater treatment. They are designed to return 50 percent or more of the applied wastewater as direct surface runoff. Intermediate loading rates of about 3 to 7 m/yr cause evapotranspiration losses to be relevant. Selection of sites with impermeable soils restricts the contribution to groundwater. The term overland flow avoids specifying a mode of application which is inherent in the frequently used spray-runoff term.

This first session includes four presentations on applying partially stabilized municipal wastewater to the land. The first paper presents an overview of design and management options which are available. These cover a broad range of choices involving the combined objectives of reuse and treatment. For example, the design and management of a project to irrigate a golf course differs substantially from design of a project to grow a forage crop under the high moisture conditions of overland-flow. The second paper gives an overview of health considerations covering biological factors, inorganic chemicals and organic chemicals. The last two presentations address energy and cost savings for actual cases involving overland flow and crop irrigation. The theme of the session highlights the role of soil science professionals in the implementation of individual projects as well as in the advancement of scientific knowledge.

Design and management considerations in applying minimally treated wastewater to land

Larry D. King, Department of Soil Science, North Carolina State University, Raleigh, NC 27650

Land has a capacity to treat wastewater just as a primary clarifier, aeration basin, sand filter or any other component of a sewage treatment plant has a capacity to treat wastewater. The treatment capacity of a particular area is more or less fixed since it is determined by the soil type, topography, cropping system, etc. Therefore, the design should start with the land and work *backward* to determine what, if any, pretreatment is needed. Too often the opposite approach has been used. As a result, systems were designed with excessive pretreatment and were more expensive and energy intensive than they should have been.

A number of good publications are available for use in designing land application systems (see Appendix). Consequently, this paper is not intended as a design manual; it is a brief overview of the factors that should be considered in designing and managing land application systems.

Slow-rate, high-rate, and overland flow are the land application systems currently used in the United States. In slow-rate systems, wastewater is applied by irrigation at a rate of from 0.3 to 3 m per year using irrigation equipment normally used in conventional agriculture. Vegetation may be forages, row crops, or forest. High-rate or rapid infiltration systems apply from 3 to 150 m of wastewater per year and thus require very permeable soils. In areas where soils are impermeable, overland flow systems receiving from 1.5 to 7.5 m per year may be used. With these systems, treatment is effected by filtering the wastewater via surface flow through vegetation. The slow-rate system provides the highest level of treatment and is more widely adapted than are the other two systems. Consequently, it will be emphasized in this paper.

The general design procedure for land application systems is to characterize the wastewater (rate of flow, nutrients, organics, toxic materials), determine the treatment capacity of the proposed site and then determine the land area and pretreatment required (Fig. 1). Another component must be added in the design stage. The treatment capacity of a site (hereafter called the site assimilative capacity) is affected by weather conditions to a greater extent than is the treatment capacity of a unit process in a conventional treatment plant. Consequently, the design must include a facility to store the wastewater during periods when weather conditions reduce site assimilative capacity (e.g., rainfall or low temperatures).

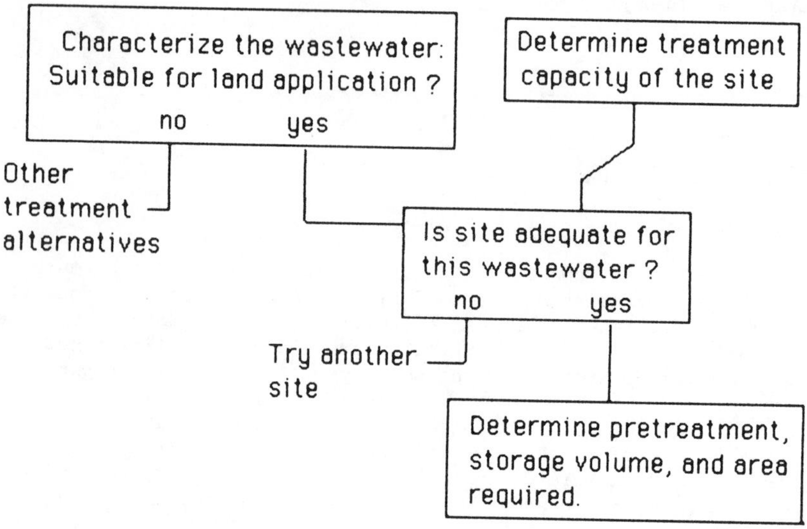

Figure 1. Flow chart of design procedure.

Wastewater Characterization

Wastewater considered for land application usually is from municipal or industrial treatment plants. From a chemical standpoint, wastewater from municipal plants is usually innocuous if the raw sewage is domestic but the potential for increased concentrations of toxic metals, organics, etc., increases as industrial input to the sewer system increases. Industrial wastewater ranges from relatively clean by-products (e.g., from food processing and fermentation) to those which contain toxic substances and require special attention.

Generation Rate

The annual quantity of wastewater generated must be determined so annual generation of constituents can be calculated. Variability in rate must be considered because it will affect storage capacity, cropping systems and land requirements. Most municipal flows are constant but recreational areas will have large seasonal fluctuations. Industrial flows may be constant but will vary during plant shutdown periods or if batch processes are used.

Composition

Analysis of wastewater should include:

Suspended solids	Cu	Na
Total N	Ni	SO_4
NH_4-N	Zn	Cl
NO_3-N	Pb	B
P	Cr	COD
K	Hg	Conductivity
Ca	As	Fecal Coliforms
Mg	pH	Temperature upon application
	PCB's	Any suspected toxic substance

The solids content of municipal wastewater is low but some industrial wastewaters may contain appreciable solids. The suspended solids content and the size of the particles will determine the type of irrigation system by dictating such factors as nozzle size, flow velocity, and pressure requirements. Pretreatment will be required to remove solids too large to be handled by the irrigation system and flushing may be required to remove solids from the irrigation lines after each irrigation. Another limiting factor will be the potential for soil sealing by the solids which accumulate on the soil surface.

Plant nutrient content of wastewater is required so one can select loading rates which do not exceed the nutrient retention capacity of the soil and uptake capacity of the crop. Nitrogen and P are the important nutrients from an environmental standpoint but information on the other nutrients is necessary to determine if supplemental fertilizer is needed or if nutrient imbalance may be a problem. Salt content is important from two standpoints. High concentrations of total dissolved salts (usually measured by electrical conductivity) cause crop injury due to the osmotic effect. High concentrations of Na in the absence of adequate Ca and Mg causes clay dispersion and subsequent reduction of soil permeability.

Heavy metal concentrations are important because of the effect of metals on crop production and animal or human health. High rates of Cu, Zn, and Ni will cause crop injury before concentrations in the crop are high enough to be toxic to consumers of the crop. In contrast, Cd can accumulate to concentrations which are not harmful to the crop but may pose a hazard to consumers of the crop. Lead and Cr are very unavailable to crops so entry into the food chain through crop uptake is slight. If they did enter the food chain, it would be directly by ingestion of contaminated plant material.

In land treatment systems, COD rather than BOD is used to estimate the organic content of wastewater. COD is more appropriate because the microbial population in soil is much more diverse than that in water and this population decomposes organic matter at a greater rate than would occur in water. A measure of organic matter content is necessary so rates can be selected that prevent anaerobic conditions from developing in the soil. Anaerobic conditions cause odor problems but more importantly they can cause a reduction in infiltration capacity due to the sealing effect of gels and slimes secreted by anaerobic microorganisms.

Wastewater may contain synthetic organics that are resistant to decomposition or may be toxic or carcinogenic. Examples are chlorinated hydrocarbons, some halogenated insecticides, and PCB's. Although these materials are not taken up by plants, they may enter the food chain by ingestion of contaminated plant material.

Although experience over many years has shown the health hazard associated with land application is very low, analysis for fecal coliforms is generally required by regulatory agencies. Normal sanitation practices by workers (i.e., like practices used in sewage treatment plants) and prohibition of growing crops for direct human consumption help alleviate concerns about health hazards.

Sampling

Wastewater analyses are valuable only if the samples analyzed are representative of the entire flow; thus, a scheme of periodic or composite sampling must be used. This sampling scheme must continue after the land application system is in operation so the system can be modified if wastewater composition changes significantly.

Pretreatment and Source Reduction

The objective of pretreatment is to reduce or remove constituent that limit the acceptibility of wastewater for land application. Generally, pretreatment is used to remove large solids that may plug the irrigation system, remove toxic substances, reduce pathogens, and control odor. After analyzing the wastewater and doing a preliminary design, one may find that one or two constituents reduce the suitability of wastewater for land application. Industries may find it economical to make changes in their process to reduce the concentration of the limiting constituent. Examples are substitution of KCl for NaCl for recharging water softeners for boilers or the substitution of KOH for NaOH for cleaning equipment. Municipalities can use ordinances to limit certain types of industrial inputs into the sewer system.

Site Evaluation and Selection

Once the wastewater has been characterized, a rough estimate of the land area required can be made using the procedures to be outlined in the next section. Then using maps of soils, topography, geology, hydrology, and land use, etc., one can begin to look for potential sites. Consideration of the factors influencing suitability--current and planned land use, topography, soils, potential groundwater pollution, flood hazard, and size of the site--will narrow the choices for potential sites. Once acceptable sites have been located, on-site investigation will be necessary to verify the information on the maps.

A variety of people should be involved in site selection. Since "wastewater" has negative connotations, a representative of the local government should be involved early in the planning stages so he will be informed and can communicate with concerned residents near the site. The county extension agent can assist in agronomic recommendations and can help with an educational program for public meetings. The county health department should be involved so they can have input on any potential health hazard they deem significant. The local Soil Conservation Service office can provide information on soils on the site and can develop a soil map of the area if one is not available. A representative of the state agency regulating land application must be involved because that agency will be responsible for issuing a permit to develop and operate the system.

Site Assimilative Capacity

Each potential site for wastewater application has a capacity to accept wastewater, bring about decomposition of the organic constituents, and channel the decomposition products into environmentally sound sinks. This site assimilative capacity has three phases (Fig. 2):

1. Aerial removal of decomposition products from the site. This includes nutrient removal by crop uptake and subsequent harvest, CO_2 and NH_3 volatilization, N_2 or NO_x loss from denitrification and loss of applied water by evapotranspiration.

2. Permanent storage in the soil. The most important examples are P fixation by reaction with Al and Fe and heavy metal fixation by reaction with organic matter and oxides of Al, Fe, and Mn.

3. Removal from the site by drainage. A portion of the applied water leaves the site by draining into the groundwater. Both anions (NO_3^-, Cl^-, $SO_4^=$) and cations (Na^+, K^+, Ca^{++}, Mg^{++}) move down with drainage water, but the anions move more rapidly since they are not attracted to the negatively charged soil particles.

These three phases must be considered in detail for each system design since the importance of each phase is a function of the site and the wastewater being applied.

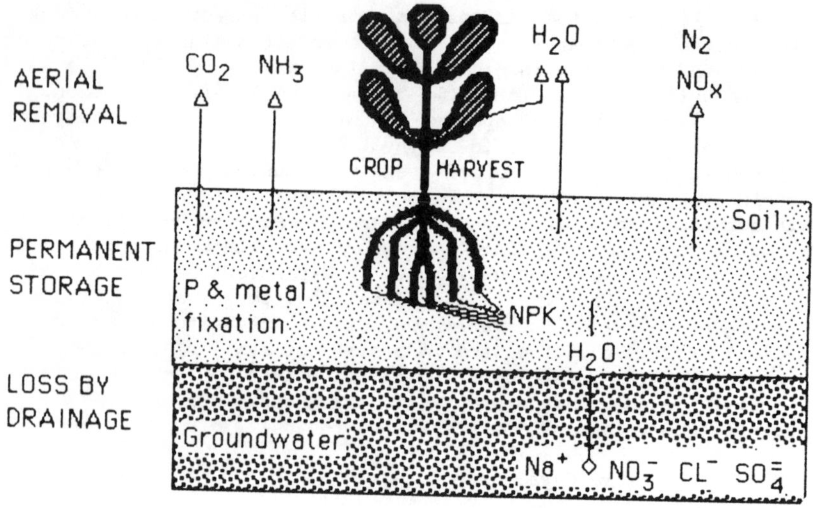

Figure 2. The three phases of site assimilative capacity.

Aerial Removal

Removal of nutrients from the site is accomplished via crop uptake and subsequent crop harvest. This is the main method of removal of N from the site if the N loading rate does not exceed the uptake capacity of the crop. Uptake and harvest removes some of the P applied by the wastewater but if the P-fixing capacity of the soil is high, fixation may be more important than removal by crops.

An oversight has crept into land application design because of a misconception of nutrient uptake by crops. As a crop is supplied with increasing amounts of a nutrient, the crop will use that nutrient with decreasing efficiency. Thus, although a crop may take up 400 kg N/ha per year, this uptake may have been attainable only when 1200 kg N/ha per year was applied. Therefore, it is meaningless to present tables of nutrient uptake by various crops and to use these for design purposes. These tables seldom contain information on quantity of nutrient applied, length of growing season (e. g., North Carolina vs. Florida for Coastal bermudagrass), whether the crop was irrigated, etc. Unfortunately, the tables imply that if a crop takes up 400 kg N/ha per year, one can apply this quantity of N and it will be taken up by the crop. In reality, uptake will be some fraction of that amount.

A more reasonable approach is to use loading rates that supply available nutrients at a rate that has been shown or can be estimated logically to pose no pollution hazard. Current fertilizer recommendations for the crops to be used on the site are a good baseline for setting wastewater-applied nutrients for most crops. Fertilizer requirements for trees are not well established but research and full-scale forested systems have been in operation for some time so data is available to help establish nutrient loading rates. Adjustments in rates can be made based on NH_3 volatilization losses, increased denitrification potential, P fixing potential of the particular soil, increased crop uptake due to irrigation, and, if available, data from land application systems in the area.

Gaseous loss of waste constituents is another mechanism of aerial removal. Evapotranspiration of applied wastewater is determined by climate and soil moisture content. Data on potential evapotranspiration are available and can be used to calculate water loss by this mechanism.

Release of CO_2 by microorganisms decomposing the organic matter in wastewater is the mechanism of C removal from the site. Thus, the soil must be kept aerobic to facilitate decomposition and to prevent odors and soil sealing that occur when soils become anaerobic. The assimilative capacity of soils for organic matter can be estimated using the O_2 diffusion rate in soils as a function of moisture content.

Organic loading rate is normally not the limiting factor with municipal wastewater but some industrial wastewaters have high organic matter contents. With these wastewaters, organic loading can be critical because the soil O_2 content is decreased not only by reduced diffusion due to high soil moisture content, but also by the high O_2 demand exerted by the microorganisms decomposing the organic matter in the wastewater.

Certain industrial wastes contain synthetic organic compounds that may be resistant to decomposition and/or toxic to vegetation or animals. Potential toxicity and rate of decomposition of these compounds must be determined from the literature or from actual tests so the site assimilative capacity for these substances can be determined.

If wastewater contains appreciable NH_4^+, then NH_3 will be volatilized at a rate determined by air temperature, wind speed, wastewater pH, and droplet size. Nitrogen losses also occur when NO_3^- is reduced to N_2 or oxides of N via microbial activity (denitrification). Although land application sites must remain aerobic to properly treat the wastewater, some microsites in the soil will be anaerobic and denitrification will occur if available C and NO_3^- diffuse into these microsites. In some cases the design can take advantage of off-site denitrification. If the site is located near a poorly drained area (marsh, swamp, etc.), appreciable denitrification can occur as subsurface drainage from the site moves through this poorly drained area.

Permanent Storage in Soil

Many soils have high P-fixing capacities because of their high Fe and Al content. Reaction products of P with Fe and Al are relatively insoluble so P does not move appreciably when applied to soil. The P-fixation capacity of soil is mainly a function of texture (Al and Fe generally increase with clay content), drainage class and extent of previous P applications. In very sandy soils where Al and Fe contents are low, P movement may be significant.

A greater potential hazard than P leaching to groundwater is enrichment of surface water by P carried in eroded soil or runoff from land application sites. The P/N ratio in municipal wastewaters is much higher than the ratio required by crops. Therefore, if wastewater is applied at rates to supply the N needs of the crop, excess P will be applied and soil P concentration will increase rapidly. To prevent movement of this soil into surface water, conservation measures must be used to minimize erosion and adequate buffer strips must be maintained between the site and surface water so sediment in runoff will be redeposited prior to reaching the watercourse.

Heavy metals react with soil organic matter and oxides of Fe, Al, and Mn to form relatively insoluble compounds. Although soil cation exchange capacity (CEC) is not directly involved in heavy metal fixation, it is currently used as a general measure of the ability of a soil to fix heavy metals. Current EPA guidelines limit metal loading rates based on soil CEC and pH.

Since soils have finite capacities to retain P and heavy metals, each site has a finite useful life. The concept of site life must not be interpreted to mean that the site is no longer useful for agriculture. Sufficient safety factors have been incorporated into the guidelines so normal agricultural production can continue after land application of wastewater ceases.

Removal by Drainage

The water not lost from the site by evapotranspiration moves down through the soil profile to the groundwater. To adequately estimate an irrigation rate which will not result in site failure from hydraulic overloading, one must estimate the drainage rate by considering the entire pathway from the soil surface to the outflow of groundwater through natural discharge areas or artificial drains. For analysis this pathway can be divided into three sections (Fig. 3):

1. Infiltration and storage in the A horizon.
2. Downward movement through the B horizon and parent material to the groundwater.
3. Groundwater movement to an outlet.

The A horizon must have adequate infiltration capacity to allow wastewater to be applied relatively rapidly and adequate capacity to store the water until it slowly moves through the less permeable B horizon. The B horizon and parent material must have adequate permeability to allow the A horizon to drain rapidly enough to prevent anaerobic conditions from developing. Groundwater movement must be rapid enough to prevent excessive mounding of the watertable under the site since this would reduce the soil depth in which the wastewater is being treated.

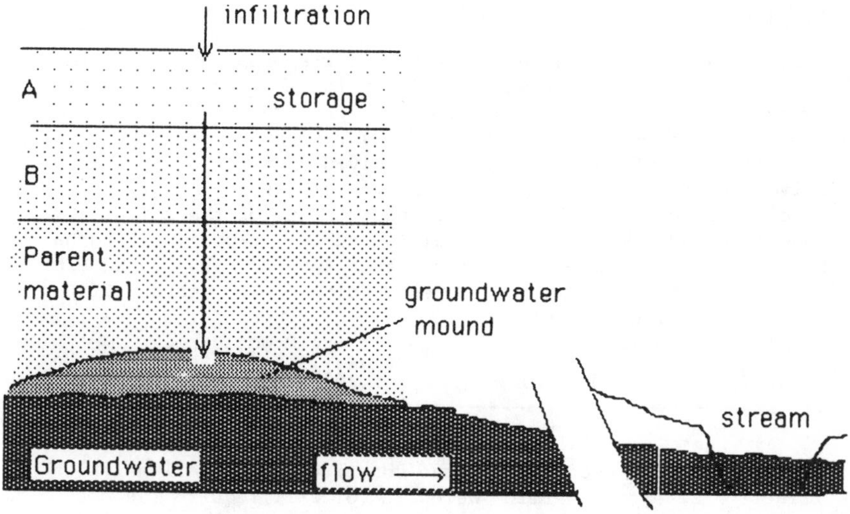

Figure 3. Schematic of drainage pattern in a land application site.

Once the drainage rate is estimated, it is used with rainfall and evapotranspiration data to calculate a monthly water balance. From this balance the design irrigation rate and required storage capacity can be determined.

Drainage water moving to the groundwater carries with it waste constituents not removed by storage in the soil or aerial removal. A mass balance can be used to determine loading rates which will prevent excessive concentrations of constituents (e.g., NO_3^-, Cl^-, $SO_4^=$) from entering the groundwater.

Sodium is one constituent that must be removed from the soil to prevent clay dispersion and loss of permeability. The dispersion hazard increases as the clay content increases and as the ratio of Na to Ca+Mg in the wastewater increases. This ratio is called the sodium adsorption ratio and is expressed as:

$$SAR = \frac{Na/23}{[0.5\ (Ca/20 + Mg/12)]^{0.5}}$$ where concentrations are in mg/L

If the SAR exceeds 15 then clay dispersion is likely and preventive measures must be taken. Gypsum ($CaSO_4 \cdot 2H_2O$) is normally added to application sites to provide a relatively soluble source of Ca. The Ca addition reduces the SAR of the soil solution and thus reduces the quantity of Na retained by the soil. Gypsum rates must be limited to those which prevent an excessive amount of $SO_4^=$ from entering the groundwater.

Land Requirement

After the wastewater has been analyzed for appropriate constituents and the site assimilative capacity for those constituents determined, the amount of land for environmentally sound wastewater application can be calculated. The quantity of each wastewater constituent generated per year divided by the site assimilative capacity for that constituent yields the amount of land required. For example, if the P-fixing capacity of the soil is found to be 4000 kg/ha, the annual P uptake by crops is 35 kg/ha, and a 20-year site life is desired, then 4000/20 + 35=235 kg P/ha would be the annual P assimilative capacity for 20 years. If wastewater flow was 50 million L/day and P concentration was 8 mg/L then the land required to handle the P would be:

$$\frac{50\text{x}10^{6}\text{L/day x 365 days/yr x 8 mg P/L x }10^{-6}\text{kg/mg}}{235\text{ kg P/ha yr}} = 621\text{ ha}$$

This procedure is used for all the constituents as well as the irrigation rate to determine which factor will require the most land. The system is then designed for that land area. Additional land will be necessary for buffer areas. With municipal wastewater, the irrigation rate (hydraulic loading rate) is generally the limiting factor.

System Management

Land application systems are complex physical-chemical-biological systems that require proper management. One would not design and build a coventional wastewater treatment plant and turn the operation over to a haphazard manager. Similarly, competent managers will be needed for land application systems. Lack of proper management can result in environmental degradation and possibly irreversible damage to the land. The manager must be knowledgeable in both engineering and agronomy and/or forestry. Engineering skills are needed to manage the pretreatment, storage and application facilities. Agronomic and/or forestry skills will be needed to manage the vegetation on the site.

The pretreatment system will be operated according to standard procedures for conventional wastewater unit processes. Storage facilities should be kept at as low a level as possible to assure that adequate storage is available in case of inclement weather. The design storage volume is based on some return frequency of weather (e.g., a 10-year return frequency) and the possibility exists that more severe weather could occur during the life of the site. Consequently, contingency plans should be made to handle the wastewater if inclement weather occurs when the storage volume is full. For example, in an emergency, irrigating a very wet soil and allowing some runoff may be the most environmentally feasible alternative. If sludge accumulates in the storage lagoon, some provision must be made to remove it periodically or to re-suspend it so it can be applied via irrigation.

Wastewater will be applied either through a stationary system (e.g., buried pipe sprinkler system), a traveling irrigator (e.g., hose or cable tow gun types or center pivot systems) or in some cases by furrow or flood irrigation. Stationary systems have a higher initial cost but the labor requirement for operation is less than with traveling types. Stationary systems are well-suited for forested systems since little vehicle traffic is on the site. For sites with agronomic crops traveling systems make crop planting and harvesting much easier because the system can be removed from the field during these operations.

The cropping system will have been determined during the design phase but scheduling irrigation and cropping cultural practices will be a major task for the site manager. In forested systems, scheduling problems are minimal except during periods of tree harvest. In agronomic systems, irrigation must be coordinated with planting, weed and insect control, harvesting, hay baling, etc., so scheduling is more complex. Since wastewater usually does not supply all the nutrient needs of crops, soil and plant analysis will be required so a program of supplemental fertilization, liming, and, in cases where Na is a problem, the application of gypsum can be developed.

In addition to scheduling irrigation around cultural practices, the scheduling must also be based on soil moisture conditions so the time of application or rate of application can be adjusted to prevent runoff or ponding and to allow adequate reaeration of the A horizon for proper crop growth and wastewater treatment. Therefore, some type of soil moisture measurements will be necessary to help make irrigation decisions.

In addition to soil and plant monitoring, some groundwater, surface water, and wastewater monitoring will be required to assure the system is operating as designed. The groundwater and surface water monitoring should begin before startup of the land treatment system so background data can be obtained.

Record-keeping is the key to proper site management. Unless the manager knows what is being applied and what has been applied to the site, how the crops are responding, the effect the system is having on soil properties and surface and groundwater quality, he will not be able to properly manage the system. A computer program would be the best approach to record-keeping. A program could be developed or one of the commercial software packages could be used.

Summary

Land has a capacity to treat wastewater and that capacity should be considered first in system design so excessive pretreatment can be avoided. The general design process consists of determining the flow rate of the wastewater and the concentration of solids, nutrients, heavy metals, COD, salts, toxic organics, etc., to determine if it is suitable for land application. The treatment capacity of the proposed application site is then determined by estimating its assimilative capacity for various constituents of the wastewater. Using the wastewater characteristics and the site assimilative capacity, the limiting constituent can be determined and the land area and required storage volume calculated.

Land application systems are complex physical-chemical-biological systems and thus require good management to assure proper performance. The manager(s) must have engineering skills to operate the pretreatment, storage, and application systems and agronomic/forestry skills to manage the vegetation and soil on the site. Monitoring of wastewater, groundwater, surface water, soil, and crops and having a good record-keeping system are essential in a good management program.

Appendix

The following references are recommended for use in land application design:

EPA. 1981. Process design manual, land treatment of municipal wastewater. EPA 625/1-81-013.

Loehr, R. C., et al. 1979. Land application of wastes, Vols. 1 and 2. Van Nostrand Reinhold Co., New York.

McKim, H. L. (ed.) 1978. State of knowledge in land treatment of wastewater, Vols. 1 and 2. U.S. Army Corp of Eng., Cold Regions Research and Engineering Laboratory, Hanover, New Hampshire.

Overcash, M. R., and D. Pal. 1979. Design of land treatment systems for industrial wastes -- theory and practice. Ann Arbor Science Publishers, Inc., Ann Arbor, Michigan.

Health considerations in applying minimum treated wastewater to land

N. E. Kowal, Health Effects Research Laboratory, U.S. Environmental Protection Agency, Cincinnati, OH 45268

There are three types of land treatment systems in general use: slow rate (or "irrigation"), rapid infiltration (or "infiltration-percolation"), and overland flow. Slow rate is the most commonly used land treatment system. Wastewater, usually pretreated by some process, is applied by sprinklers, surface flooding, or ridge-and-furrow irrigation, at a rate of 2-20 feet (0.6-6 m) per year. Soils are usually medium to fine textured with moderate permeability, and percolated water is either collected by drainage tile or reaches the groundwater. Surface vegetation has included lawns and golf courses for highly treated wastewater, pastures, and forests, but is most commonly crops, usually for animal consumption. Climatic constraints often require some winter storage of wastewater (Reed, 1979). Recycling benefits include moderate groundwater recharge and the utilization of wastewater nutrients in crop production.

In rapid infiltration, wastewater is flooded, usually intermittently, into shallow basins at a rate of 20-600 feet (6-183 m) per year. Soils are usually coarse textured with high permeability, and percolated water moves to groundwater, recovery wells, or underdrains. Surface vegetation is usually absent, and climatic constraints not of concern (Reed, 1979). Groundwater recharge is a recycling benefit.

Overland flow is the least commonly used land treatment system. Wastewater is applied to the top of gently sloping (2-4 percent) fields at a rate of 10-70 feet (3-21 m) per year, and moves by sheet flow down the slope to collection ditches at the base. Soils are usually fine textured with very low permeability. Surface vegetation consists of water-tolerant grasses, e.g., reed canary grass. Climatic constraints may require some winter storage of wastewater (Reed, 1979). Recycling benefits may include the utilization of wastewater nutrients in harvested grass for animal forage.

Many of the examples of land application systems in the USA utilize wastewater treated by conventional means up to tertiary level (secondary in the case of overland flow). The objectives of these systems are usually to produce clean irrigation water (e.g., for golf course application) or highly treated water for groundwater recharge. From a wastewater treatment point of view, land application in these systems is a form of tertiary treatment or effluent "polishing," rather than true land treatment. In land treatment systems, the subject of this report, raw wastewater is given a minimum preapplication treatment or "pretreatment," e.g., by a stabilization pond, before being applied to the land, and the land itself is the site of the major portion of the wastewater treatment.

With the application to land of large volumes of minimally pretreated wastewater, it is evident that considerable potential for adverse health effects exists. These potentials have been briefly summarized by Lance and Gerba (1978) in Table 1. They identified the greatest health risks as arising from aerosols in slow rate, groundwater pollution in rapid infiltration, and surface water pollution in overland flow.

Table 1. Potential Land Treatment Health Effects.

Type of land treatment system	Food contamination	Groundwater pollution	Surface Water pollution	Aerosols
Slow rate	+	+	+	++
Rapid infilitration	-	++	-	-
Overland flow	-	-	++	+

-Little or no potential problem
+Moderate potential
++Considerable potential

The agents, or pollutants, of concern from a health effects viewpoint can be divided into the two broad categories of pathogens and toxic substances. The pathogens include bacteria (e.g., Salmonella and Shigella), viruses (i.e., enteroviruses, hepatitis virus, adenoviruses, rotaviruses, and Norwalk-like agents), protozoa (e.g., Entamoeba and Giardia), and helminths (or worms, e.g., Ascaris, Trichuris, and Toxocara). The protozoa and helminths are often grouped together under the term, "parasites." The toxic substances include organics, trace elements (or heavy metals, e.g., cadmium and lead), nitrates, and sodium. Nitrates and sodium are usually not viewed as "toxic" substances, but are here so considered because of their potential hematological and long-term cardiovascular effects when present in water supplies at high levels. The major health effects of these agents are listed below:

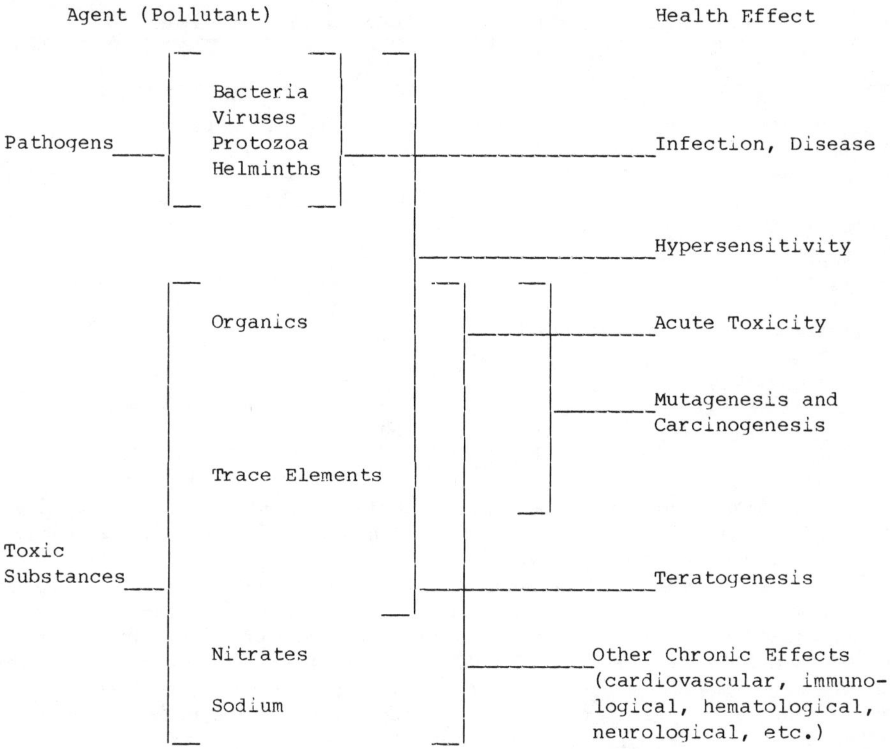

This paper is primarily a summary of two previously published detailed reports on the health effects of land treatment (Kowal, 1982, 1985), and the reader is encouraged to seek there the supporting data and arguments for the generalizations made here. In those reports the types and levels commonly found in municipal wastewater and the efficiency of preapplication treatment (usually stabilization pond) are briefly reviewed for each agent of concern. A discussion of the levels, behavior, and survival of the agent in the medium or route of potential human exposure, i.e., aerosols, surface soil and plants, subsurface soil and groundwater, and animals, follows as appropriate. For the pathogens, infective dose, risk of infection, and epidemiology are then briefly reviewed. Finally, conclusions and research needs are presented.

Bacteria

The pathogenic bacteria of major concern in wastewater are listed in Table 2. All have symptomless infections and human carrier states, and many have important nonhuman reservoirs as well.

Table 2. Pathogenic Bacteria of Major Concern in Wastewater.

Name	Nonhuman reservoir
Campylobacter jejuni	Cattle, dogs, cats, poultry
Escherichia coli (pathogenic strains)	-
Leptospira spp.	Domestic and wild mammals, rats
Salmonella paratyphi (A, B, C)	-
Salmonella typhi	-
Salmonella spp.	Domestic and wild mammals, birds, turtles
Shigella sonnei, *S. flexneri*, *S. boydii*, *S. dysenteriae*	-
Vibrio cholerae	-
Yersinia enterocolitica, *Y. pseudotuberculosis*	Wild and domestic birds and mammals

The human exposure to aerosol bacteria at land treatment sites can be roughly estimated from the data at Kibbutz Tzora, Israel, where raw wastewater was sprayed, thus yielding higher bacterial levels than those found at Deer Creek, Ft. Huachuca, or Pleasanton, where treated wastewater was sprayed (see Kowal, 1982). Thus, an adult male, engaged in light work, breathing at a rate of 1.2 m^3/hr, and exposed to 34 coliforms/m^3 (the Kibbutz Tzora average) at 100 m downwind from a sprinkler, would inhale approximately 41 coliforms per hour. Since the ratio of aerosolized *Salmonella* to coliforms is 1:10^5 (Grunnet and Tramsen, 1974) the rate of inhalation of *Salmonella* would be about 10^5-fold less, an extremely low rate of bacterial exposure. More recent data from Kibbutz Tzora allows a more accurate estimate of human exposure (Teltsch et al., 1980). During a period of time in 1977-78, when the wastewater total coliforms were 2.4 x 10^6 to 1.4 x 10^7/100 ml and *Salmonella* was 0-60/100 ml, the density of aerosol *Salmonella* at 40 m, the maximum distance found, was 0-0.054/m^3, with a mean of 0.014/m^3. This would result in an inhalation rate of 0.017/hr at 40 m, higher than the previous estimate, but still an extremely low rate of bacterial exposure.

The surface soil and plants of an active land treatment site are constantly heavily laden with enteric bacteria; these are the specific locations where the actual treatment of the wastewater and inactivation of the bacteria occur. (In some situations bacteria may be deposited on plants in the environs of a land treatment site, due to aerosol drift). The survival time of bacteria in surface soil and on plants is only of concern when decisions must be made on how long a period of time must be allowed after last application before permitting access to people or animals, or harvesting crops.

In view of the large number of environmental factors affecting bacterial survival in soil, it is understandable that the values found in the literature vary widely. Two useful summaries of this literature are those of Bryan (1977) and Feachem et al. (1978). The ranges given in Table 3 are extracted from these summaries, as well as other literature. "Survival" as used in this table, and throughout this report, denotes days of detection. It should be noted that inactivation is a rate process and, therefore, detection depends upon the initial level of organisms, sensitivity of detection methodology, and other factors. If kept frozen, most of these bacteria would survive longer than indicated in Table 3, but this would not be a realistic soil situation.

Table 3. Survival Times of Bacteria in Soil.

Coliform	4-77 days
Fecal coliform	8-55 days
Fecal streptococci	8->70 days
Leptospira	<15 days
Mycobacterium	10 days - 15 months
Salmonella paratyphi	>259 days
Salmonella typhi	1-120 days
Salmonella spp.	11->280 days
Streptococcus faecalis	26-77 days

Perhaps the largest epidemiological study of the health effects of land treatment was a retrospective study of 77 kibbutzim (agricultural cooperative settlements) in Israel practicing slow-rate land treatment with nondisinfected oxidation pond effluent, and 130 control kibbutzim (Katzenelson et al., 1976). The incidence of typhoid fever, salmonellosis, shigellosis, and infectious hepatitis was 2-4 times higher in the land-treatment kibbutzim than the controls. The study, however, did not rule out a number of pathways of infection other than aerosols, e.g., direct contact via clothing or bodies of sewage irrigation workers, and there were problems with the data reporting methods. Consequently, it is generally felt that no conclusive findings may be based on the report, and the study is currently being repeated, correcting for the deficits of the original study (Shuval and Fattal, 1980). Preliminary results suggest little effect of land treatment on disease incidence.

Viruses

The human enteric viruses that may be present in wastewater are listed in Table 4 (Melnick et al., 1978; Holmes, 1979). These are referred to as the enteric viruses and new members are constantly being identified. Since no viruses are normal inhabitants of the gastrointestinal tract and none of these have a major reservoir other than man (with the likely exception of rotaviruses), all may be regarded as pathogens, although most can produce asymptomatic infections.

Table 4. Human Wastewater Viruses.

Enteroviruses
Poliovirus
Coxsackievirus A
Coxsackievirus B
Echovirus
New Enterovirus
Hepatitis A Virus
Rotavirus ("Duovirus," "Reovirus-like Agent")
Norwalk-Like Agents (Norwalk, Hawaii, Montgomery County, etc.)
Adenovirus
Reovirus
Papovavirus
Astrovirus
Calicivirus
Coronavirus-Like Particles

From Pleasanton, California data (Johnson et al., 1980) it can be calculated that an adult male, engaged in light work, breathing at a rate of 1.2 m^3/hour and exposed to 0.014 PFU/m^3 at 50 m downwind from a sprayer, would inhale approximately 0.13 PFU of enterovirus during an 8-hour work day. This is probably an insignificant level of exposure. However, since the recovery of enteric viruses form environmental samples is not perfectly efficient, isolation of viruses increases as more cell culture types are used, and some enteric viruses cannot yet be isolated in cell cultures, the actual exposure to enteric viruses may be as much as ten to a hundred times the reported level (Teltsch et al., 1980). Thus, it might be prudent to recommend a 100 m or 200 m minimum exposure distance of the general public to a land treatment spray source.

As is the case with bacteria, the surface soil and plants of an active land treatment site are constantly receiving enteric viruses. The survival time of viruses is primarily of concern when decisions must be made on how long a period of time must be allowed after last application before permitting access to people or animals, or harvesting crops. Another concern is that the longer viruses survive at the surface the greater opportunity they have for being desorbed and moving into the soil toward the groundwater. This is not a problem with overland flow systems, which, although 68 to 85% of the enteric viruses are deposited at the surface, little virus penetrates into the soil profile (Schaub et al., 1978, 1980).

Much of the recent literature on survival time of enteric viruses in soil is summarized in Table 5. Approximately one hundred days appears to be the maximum survival time of enteric viruses in soil, unless subject to very low temperatures, which prolong survival beyond this time. Exposure to sunlight, high temperatures, and drying greatly reduce survival times.

Table 5. Survival Time of Enteric Viruses in Soil.

Virus	Soil	Moisture and temperature	Survival (days)	Reference
Enterovirus	Sandy or loamy podzol	10-20% 3-10°C	70-170	Bagdasaryan, 1964
		10-20% 18-23°C	25-110	
		Air dry, 18-23°C	15-25	
Poliovirus	Sand	Moist Dry	91 <77	Lefler and Kott, 1974
Poliovirus	Loamy fine sand	Moist, 4°C	84 (<90% reduction)	Duboise et al., 1976
		Moist, 20°C	84 (99.999% reduction)	
Coxsackie-virus	Clay	300 mm rain-fall, -12-26°C	<161	Damgaard-Larsen et al., 1977
Poliovirus	-	-14-27°C 15-33°C	89-96 <11	Tierney et al., 1977
Poliovirus	Sugarcane field	Open, direct sunlight	7-9	Lau et al., 1975
		Mature sugarcane	≤60	
Poliovirus and coxsac-kievirus	Sandy loam	Saturated, 37°C	12	Yeager and O'Brien, 1979
		Saturated, 4°C	≥180	
		Dried, 37°C and 4°C	<3-<30	

While viruses near the soil surface are rapidly inactivated due to the combined effects of sunlight, drying, and the antagonism of aerobic soil microorganisms, those that penetrate the aerobic zone can be expected to survive over a more prolonged period of time. The longer they survive, the greater the chance that an event will occur to promote their penetration into groundwater (Gerba and Lance, 1980).

Viruses do not regrow on foods or other environmental media, as bacteria sometimes do. Therefore, the risk of infection is completely dependent upon being exposed to an infective dose (which may be very low) in the material applied. In any event, as is the case with bacteria, it would seem prudent for humans to maintain a minimum amount of contact with an active tretament site, and to rely on the viral survival data discussed earlier for limiting the hazard from crops grown for human consumption on wastewater-amended soils.

Parasites

The protozoa and helminths (or worms) are often grouped together under the term, "parasites," although in reality all the pathogens are biologically parasites. Because of the large size of protozoan cysts and helminth eggs, compared with bacteria and viruses it is extremely unlikely that they will find their way into either aerosols or groundwater at land treatment sites, and, thus, these routes of exposure are not further considered. Little attention has been given to the presence of parasites in wastewater, and their potential for contaminating food crops in the United States, probably because of the popular impression that the prevalence of parasite infection in the USA is minimal (Larkin et al., 1978). However, because of the increasing recognition of parasite infections in the USA, the return of military personnel and travelers from abroad, the level of recent immigration and food imports from countries with a high parasitic disease prevalence, and the existence of resistant stages of the organisms, a consideration of parasites is warranted.

The most common protozoa which may be found in wastewater are listed in Table 6. Of these, only three species are of major significance for transmission of disease to humans through wastewater: Entamoeba histolytica, Giardia lamblia, and Balantidium coli. Toxoplasma gondiii also causes significant human disease, but the wastewater route is probably not of importance. Eimeria spp. are often identified in human fecal samples, but are considered to be spurious parasites, entering the gastrointestinal tract from ingested fish.

Table 6. Types of Protozoa in Wastewater.

Name	Protozoan Class	Nonhuman Reservoir
HUMAN PATHOGENS		
Entamoeba histolytica	Ameba	Domestic and wild mammals
Giardia lamblia	Flagellate	Beavers, dogs, sheep
Balantidium coli	Ciliate	Pigs, other mammals
Toxoplasma gondii	Sporozoan (Coccidia)	Cats
Dientamoeba fragilis	Ameba	
Isospora belli	Sporozoan (Coccidia)	
I. hominis	Sporozoan (Coccidia)	
HUMAN COMMENSALS		
Endolimax nana	Ameba	
Entamoeba coli	Ameba	
Iodamoeba butschlii	Ameba	
ANIMAL PATHOGENS		
Eimeria spp.	Sporozoan (Coccidia)	Fish, birds, mammals
Entamoeba spp.	Ameba	Rodents, etc.
Giardia spp.	Flagellate	Dogs, cats, wild mammals
Isospora spp.	Sporozoan (Coccidia)	Dogs, cats

The pathogenic helminths whose eggs are of major concern in wastewater are listed in Table 7. They are taxomomically divided into the nematodes, or roundworms, and cestodes, or tapeworms. The trematodes, or flukes, are not included since they require aquatic conditions and intermediate hosts, usually snails, to complete their life cycles, and thus are unlikely to be of concern at land treatment sites. Several of the human pathogens listed in Table 7, e.g., *Toxocara* spp., are actually animal parasites, rather than human parasites, infesting man only incidentally, and not completing their life cycle in man.

Table 7. Pathogenic Helminths of Major Concern in Wastewater.

Pathogen	Common Name	Disease	Nonhuman Reservoir
NEMATODES (Roundworms)			
Enterobius vermicularis	Pinworm	Enterobiasis	
Ascaris lumbricoides	Roundworm	Ascariasis	
A. suum	Swine roundworm	Ascariasis	Pig*
Trichuris trichiura	Whipworm	Trichuriasis	
Necator americanus	Hookworm	Necatoriasis	
Ancylostoma duodenale	Hookworm	Ancylostomiasis	
A. braziliense	Cat hookworm	Cutaneous larva migrans	Cat, dog*
A. caninum	Dog hookworm	Cutaneous larva migrans	Dog*
Strongyloides stercoralis	Threadworm	Strongyloidiasis	Dog
Toxocara canis	Dog roundworm	Visceral larva migrans	Dog*
T. cati	Cat roundworm	Visceral larva migrans	Cat*
CESTODES (Tapeworms)			
*Taenia saginata***	Beef tapeworm	Taeniasis	
T. solium	Pork tapeworm	Taeniasis, Cysticercosis	
Hymenolepis nana	Dwarf tapeworm	Taeniasis	Rat, mouse
Echinococcus granulosus	Dog tapeworm	Unilocular hydatid disease	Dog*
E. multilocularis		Alveolar hydatid disease	Dog, fox, cat*

*Definitive host; man only incidentally infested.
**Eggs not infective for man.

Protozoa cysts are highly sensitive to drying. Rudolfs et al. (1951) have reported survival times for Entamoeba histolytica of 18-24 hours in dry soil and 42-72 hours in moist soil. Somewhat longer times, i.e., 8-10 days, have been reported by Beaver and Deschamps (1949) in damp loam and sand at 28-34°C.

Helminth eggs and larvae, in contrast to protozoan cysts, live for long periods of time when applied to the land, probably because soil is the transmission medium for which they have evolved, while protozoa have evolved toward water transmission. Thus, under favorable conditions of moisture, temperature, and sunlight, Ascaris, Trichuris, and Toxocara can remain viable and infective for several years (Little, 1980). Hookworms can survive up to 6 months (Feachem et al., 1978), and Taenia a few days to seven months (Babayeva, 1966); other helminths survive for shorter periods.

Single eggs of helminths are infectious to man, although, since the symptoms of helminth infections are dose-related, many light infections are asymptomatic. However, Ascaris infection may sensitize individuals so that the passage of a single larval stage through the lungs may result in allergic symptoms, i.e., asthma and urticaria (Muller, 1953).

Because of the low infective doses of helminth eggs, and their longevity, it would be prudent for humans to maintain a minimum amount of contact with an active or inactive land treatment site, unless the wastewater has been pretreated to remove helminths.

Organics

The potential health effects of toxic organic compounds are myriad. Systems affected range from the dermatological to the nervous to the subcellular, and effects produced range from rash to motor dysfunction to cancer. The degree of toxicity of organic compounds varies widely, from essentially harmless (e.g., most carbohydrates) to moderately toxic (e.g., most alcohols) to extremely toxic (e.g., aflatoxins).

A glance at the current edition of The Merck Index will reveal that the number of organic compounds described thus far is almost unfathomable. Nearly any of these may appear in wastewater, depending upon its sources. Thus, the discussion below must be perforce rather general, and the presence of any particular toxic organic in high concentration in the wastewater may require a site-specific evaluation of potential health effects.

The major contributors of toxic organics to municipal wastewaters are usually assumed to be industrial discharges. However, household wastewater discharge may represent an important contributor since many consumer products in daily use contain toxic substances. A recent study (Hathaway, 1980) identified consumer products containing toxic compounds on EPA's list of 129 priority pollutants, which may eventually end up in wastewater. The most frequently used products are cleaning agents and cosmetics, containing solvents and heavy metals as main ingredients. Next are deodorizers and disinfectants, containing naphthalene, phenol, and chlorophenols. Discarded into wastewater infrequently, but in large volumes, are pesticides, laundry products, paint products, polishes, and preservatives. The organic priority pollutants most frequently used and discharged into domestic wastewater were predicted to be the following:

benzene
phenol
2,4,6-trichlorophenol
2-chlorophenol
1,2-dichlorobenzene
1,4-dichlorobenzene
1,1,1-trichloroethane
naphthalene
toluene
diethylphthalate
dimethylphthalate
trichloroethylene
aldrin
dieldrin

Because of the difficulty of analysis of complex mixtures, it has only recently been possible to measure the actual levels of organics in wastewater using advanced methods of extraction, gas and other chromatography, mass spectrometry, and computer analysis. The U. S. Environmental Protection Agency has sponsored two extensive surveys of the types and levels of priority pollutants in municipal wastewaters, which, of course, result from both domestic and industrial discharges. The first (DeWalle et al., 1981), supported by the Municipal Environmental Research Laboratory in Cincinnati, covered 25 cities located throughout the United States, and the second (Burns & Roe, 1982), supported by the Effluent Guidelines Division in Washington, D. C., covered 40 cities.

In the 25-city survey (DeWalle et al., 1981) most of the 24-hour composite samples of raw wastewaters contained a total of less than 1 mg/l of priority organics, and the numbers of compounds detected clustered between 20 and 50. In the 40-city survey (Burns & Roe, 1982) six days of 24-hour sampling was completed, and the samples from 20 cities were analyzed. The priority organics detected in at least 50% of the samples analyzed in either survey are listed in Table 8.

Comparison of the results of the two surveys with the list of organic priority pollutants most likely to be discharged into domestic wastewater, reveals considerable overlap, and gives one some confidence that these two studies have yielded a reasonable characterization of the priority organics in municipal wastewater, at least of those identifiable by modern methods. The broad range of concentrations detected among the samples, however, suggests that wastewater applied to land should be regularly monitored for toxic organics. This measure is emphasized by the occasional discharge of toxic substances into municipal wastewater systems with resulting medical effects in treatment plant workers, such as the recent hexachlorocyclopentadiene episode in Louisville, Kentucky (Kominsky et al.,1980).

Table 8. Most Frequently Detected Priority Organics in Raw Municipal Wastewater (detection frequency, %).

Compound	DeWalle et al., 1981	Burns & Roe, 1982
Phenol	94	79
1,1,1-Trichloroethane	94	85
Trichloroethylene	94	90
Tetrachloroethylene	94	95
Ethylbenzene	94	80
Trichloromethane (Chloroform)	94	91
Diethylphthalate	91	53
Di-n-butylphthalate	91	64
Toluene	90	96
Dichloromethane	90	92
Bis(2-ethylhexyl)phthalate	89	92
Naphthalene	86	49
1,4-Dichlorobenzene	83	17
Phenanthrene	83	20
Benzene	79	61
Heptachlor	77	5
Butylbenzylphthalate	77	57
BHC-G (Lindane)	71	26
1,2-Dichlorobenzene	69	23
Dimethylphthalate	66	11
BHC-D	63	3
Dieldrin	63	1
1,3-Dichlorobenzene	60	7
BHC-A	60	8
DDT	60	<1
Di-n-octylphthalate	57	7
1,1-Dichloroethane	55	31
1,2-Dichloroethane	55	15
DDD	54	1
Anthracene	51	18
Aldrin	51	1
Endosulfan-B	51	-
1,2-Trans-dichloroethylene	20	62

In view of the multitudinous variety of organic compounds existing, it is difficult to generalize about their biodegradation in soil. It appears, however, that most organics do become microbially decomposed in the soil, at least to some extent. This is especially true of naturally occurring compounds, or those resembling them, because of the eons of evolution that have developed microbial enzyme systems to do the job. The more structually complex the molecule is, e.g., condensed rings or dense branching, and more halogenated it is, the more difficult is biodegradation.

From existing data it is evident that toxic organics can be transported to groundwater below land treatment sites, although the degree can be controlled by the level of preapplication treatment and application rate as well as choice of effluent and site characteristics. This certainly is cause for concern, but it should be kept in mind that groundwater is not the pristine substance it was once thought to be (Burmaster, 1982). The synthetic organic compounds most commonly found in groundwater in the USA, deriving primarily from industrial wastes, are (Environmental Health Letter 21(6):7, 1982):

trichloroethylene	benzene
tetrachloroethylene	chlorobenzene
carbon tetrachloride	dichlorobenzene
1,1,1-trichloroethane	trichlorobenzene
1,2-dichloroethane	1,1-dichloroethylene
vinyl chloride	cis-1,2-dichloroethylene
methylene chloride	trans-1,2-dichloroethylene

Trace Elements

The trace elements (including the "heavy metals") in wastewater of public health concern, i.e., those for which primary drinking water standards (USEPA 1977) exist (but excluding silver since its effect is largely cosmetic), are:

	Primary Drinking Water Standard (mg/l)
Arsenic (As)	0.05
Barium (Ba)	1.0
Cadmium (Cd)	0.010
Chromium (Cr)	0.05
Lead (Pb)	0.05
Mercury (Hg)	0.002
Selenium (Se)	0.01

Of these, cadmium, lead, and mercury are usually regarded as of most concern, and barium of minor concern. Chromium and selenium are essential elements in man; arsenic and cadmium have been shown to be essential to experimental animals and, thus, may be essential to man as well (National Research Council,1980). Secondary drinking water standards (USEPA,1979), i.e., those related to aesthetic quality, also exist for copper, iron, manganese, and zinc. These latter elements, as well as all other trace elements, are toxic if ingested or inhaled at high levels for long periods (Underwood,1977), but this fact does not warrant considering them in the land treatment context, where low levels are expected.

Limits for the maximum cumulative application of trace elements to agricultural land have been recommended by various governmental agencies, for the protection of public health and the prevention of phytotoxicity. These have almost invariably been proposed in the context of the land application of sludge, but should be just as valid for land treatment of wastewater, at least in the slow rate mode. Limits for rapid infiltration and overland flow could be less restrictive because of the lack of production of crops for human consumption and the greater depth of soil involved in treatment (in the former case). These limits have been used, together with typical wastewater levels of trace elements, by Page and Chang (1981) to predict the useful life of a typical land treatment site where crops are grown for human consumption. The results appear in Table 9. It is evident that cadmium, with a 17-67 year limit, is the element most likely to restrict the use of wastewater for irrigation of crops for human consumption. In the case of crops not for human consumption, other elements may be limiting--in particular molybdenum because of its toxicity to livestock, and nickel because of its phytotoxicity. These latter elements yield limits of 47-48 years in Page and Chang's analysis.

Table 9. Annual Input of Trace Elements and Years of Land Treatment Required to Exceed Recommended Cumulative Input Limits (modified from Page and Chang, 1981).

Element	Concentration in Wastewater (mg/l)	Annual Input[1] (kg/ha)	Recommended Cumulative Input Limits (kg/ha)		Years Required to Exceed Limits	
			USA[2]	UK[3]	USA	UK
Arsenic	0.005	0.075	-	10	-	133
Cadmium	0.02	0.3	5/10/20[4]	5	17/33/67[4]	17
Chromium	0.05	0.75	-	1000	-	1333
Lead	0.2[5]	3.0	800	1000	267	333
Mercury	0.0009	0.014	-	2	-	143
Selenium	0.005	0.075	-	5	-	67

[1]Assuming an annual application rate of 1.5 m.
[2]USEPA, USFDA, and USDA (1981).
[3]National Water Council, 1977.
[4]For soils with cation exchange capacities of <5, 5-15, and >15 meq/100g, respectively, and soil pH ≥6.5. If soil pH <6.5, first figure holds.
[5]Raised over the value presented in Page and Chang (1981), to reflect recent data.

Nitrates

Inorganic nitrogen is normally quite innocuous from a human health point of view, although high ammonia levels can present an aesthetic problem. The major health concern is that infants, less than about three months of age and consuming large quantities of high-nitrate drinking water through prepared formula, have a high risk of developing methemoglobinemia. The incompletely developed capacity to secrete gastric acid in the infant allows the gastric pH to rise sufficiently to encourage the growth of bacteria which reduce nitrate to nitrite in the upper gastrointestinal tract. The nitrite is absorbed into the bloodstream, and oxidizes the ferrous iron in hemoglobin to the ferric state, in which form it is incapable of carrying oxygen. Fetal hemoglobin (Hb F), 50-89% of total hemoglobin at birth, is particularly susceptible to this transformation. Methemoglobin is normally present in the erythrocytes of adults, at a concentration of about 1% of total hemoglobin, being formed by numerous agents, but kept to a low level by the methemoglobin reductase enzyme system. This enzyme system is normally not completely developed in young infants. At a methemoglobin concentration of about 5-10% of total hemoglobin the body's oxygen deficit results in clinically detectable cyanosis. As a result of epidemiological and clinical studies (Shuval & Gruener, 1977, Craun et al., 1981, Fraser and Chilvers, 1981) a primary drinking water standard of 10 mg/l of nitrate-nitrogen (i.e., nitrate expressed as N) has been established (USEPA, 1977) to prevent this condition from developing.

Besides methemoglobinemia, there is also some concern about nitrates resulting in the formation of carcinogenic N-nitroso compounds in the gut, but this phenomenon probably involves higher concentrations than the 10 mg/l water standard (Fraser et al., 1980, Fraser and Chilvers, 1981).

The relevance of land treatment, of course, centers on the possibility of highly soluble nitrates reaching groundwater which may be used as a potable water supply.

Comparison with Conventional Systems

The comparison of the potential health effects of land treatment, caused by both pathogens and toxic substances, with those of conventional treatment is necessarily highly subjective. Nevertheless, there are suggestions that land treatment is at least equally protective of public health as conventional treatment.

A comparison of bacterial aerosol levels at conventional activated sludge plants and a spray irrigation land treatment site has been performed by Clarke et al. (1978), to evaluate relative human exposure levels at these two types of facilities. They concluded that airborne bacterial levels, as measured by fecal coliforms, appear to be higher at the activated sludge plants than at the spray irrigation facility.

A broad comparison of health risks between activated sludge treatment and slow-rate land treatment has been performed by Crites and Uiga (1979, Uiga et al., 1978). The comparison assumed: (1) A flow of 3 M gal/day of domestic wastewater. (2) Activated sludge treatment is followed by disinfection and surface water discharge. (3) Land treatment is preceded by aerated lagoon preapplication treatment and storage, and followed by percolate water recovery using underdrains and surface water discharge, with no disinfection. They arrived at the following conclusions comparing the two systems:

1. If maintained and operated properly, both conventional and land treatment systems provide a large measure of safety for public health. Slow-rate land treatment offers greater protection against parasites, viruses, trace organics, halogenated organics, trace elements, and nitrate.

2. Since adequate removal of parasite eggs and cysts require such measures as filtration or long detention times in ponds or storage lagoons, land treatment offers greater protection from health risks.

3. Land treatment systems, especially those with ponds and storage lagoons, remove viruses to a higher degree than do conventional treatment and disinfection systems.

4. Land treatment systems are less susceptible to failure or upsets than conventional systems, especially for small systems.

In an overview of existing land treatment systems, Iskandar (1978) concluded that "the potential health hazards from land treatment are no greater and probably less than those from conventional treatment. . . . Although land treatment, like all known waste treatment systems, has potential health hazards associated with it, these risks can be kept to a minimum." The risks associated with alternative systems of wastewater treatment and disposal, conventional as well as land treatment, should be defined and compared. He considered it to be rather odd that land treatment instills so much more public fear than conventional treatment.

Recent Epidemiological Studies

Two extensive prospective epidemiological studies of the effect of land application of wastewater on infectious disease in the exposed population have recently been completed in Lubbock, Texas (Camann et al. 1985) and in Israel (Shuval et al., 1985). This section is extracted from the cited draft final reports for these two projects, and thus is subject to future (probably minimal) modification.

The Lubbock Infection Surveillance Study (LISS) was conducted to monitor infections and acute illness in the primarily rural community surrounding the Lubbock Land Treatment (Demonstration) System (LLTS) at the Hancock farm near Wilson, Texas. The LISS objective was to identify possible adverse effects on human health from slow-rate (sprinkler) land application of wastewater which contained potentially pathogenic microorganisms.

An epidemiological analytic prospective cohort study of 478 area residents and Hancock farm workers was maintained during the first 20 months of operation of the LLTS (February 1982-October 1983) and during the 20-month period immediately preceding LLTS operation (June 1980-January 1982). Blood samples collected semiannually were analyzed for antibody titers to 14 enteroviruses, 3 adenoviruses, 2 reoviruses, rotavirus, Norwalk virus, hepatitis A, Legionella, Entamoeba histolytica, and influenza A. Routine fecal specimens were collected regularly to isolate enteric viruses and overt and opportunistic bacterial pathogens. Electron microscopic examination was performed to detect a variety of other virus-like particles. Tuberculin skin tests were administered annually to detect non-tuberculosis mycobacterial infections. Illness information was provided by study participants on a weekly basis. Concentrations of microorganisms also were measured in the wastewater, wastewater aerosol, and drinking water. Dispersion modeling, participant activity diaries, and a weekly log of extensive wastewater contact were used to calculate an aerosol exposure index of relative cummulative exposure of each participant to the wastewater aerosol within each of the four major irrigation seasons.

Very high levels of bacteria and enteric viruses were present in the sprayed wastewater obtained via pipeline directly from the Lubbock sewage treatment plant. Enteroviruses were consistently found in the wastewater aerosol in 1982.

Participants in the high and low exposure groups were generally well balanced with regard to age, gender, previous titer, and time spent in Lubbock. However, aerosol exposure was largely confounded with patronage of a local restaurant and use of evaporative cooler air conditioners.

Disease surveillance did not disclose any obvious connection between the self-reporting of acute illness and degree of aerosol exposure.

Whenever a sufficient number of infections was observed during an irrigation season, this infection episode was analyzed by four different methods: confirmatory statistical analysis, exploratory logistic regression analysis, confidence intervals of incidence density ratios, and risk ratio scoring. The association of infection status with wastewater aerosol exposure and other relevant factors was investigated.

Comparisons of crude seroconversion incidence densities indicated that some excess risk of viral infection (risk ratio of 1.5 to 1.8) appeared to be associated with level of aerosol exposure. A symmetric risk ratio scoring approach provided evidence of a dose-related stable assocation ($\underline{P}$=0.002) between the infection events in the observed episodes of infection and aerosol exposure. More than the expected number of statistically significant associations of the presence of infection with wastewater aerosol exposure were found in the confirmatory analysis of independent infection episodes using Fisher's exact test. Thus, three different statistical approaches provided similar evidence that the rate of viral infections was slightly higher among members of the study population who had a high degree of aerosol exposure.

In the episode of poliovirus 1 seroconversion in spring 1982, some of the infections were probably caused by wastewater aerosol exposure because a strong association existed and no alternative explanation could be identified. Three distinct factors (poliovirus immunization in spring 1982, low polio 1 antibody titer in January 1982, and a high degree of aerosol exposure) were independently associated with the poliovirus 1 seroconversion and each appears to have been responsible for some of the poliovirus 1 infections. Weak evidence of association was found between aerosol exposure and infection by other enteric viruses (specific coxsackie B viruses and echoviruses) which were simultaneously recovered from the wastewater during the summer irrigation season of 1982. However, it could not be determined whether aerosol exposure or identified alternative explanations were the actual risk factor(s) in these enteric viral infections. The association of viral infections with aerosol exposure shows a dose effect, since the study population was exposed to more enteroviruses via the wastewater aerosol in 1982 than in 1983.

In summary, a general association existed between exposure to irrigation wastewater and new infections. A viral dose-response relationship was observed over the four irrigation seasons, since the aerosol exposure-associated episodes of viral infection occurred primarily in 1982 during the irrigation seasons of greater enterovirus aerosol exposure. Some poliovirus 1 seroconversions during the spring of 1982 were probably related to wastewater aerosol exposure. However, even during 1982, the strength of association remained weak and frequently was not stable. Wastewater of poor quality comprised much of the irrigation water in 1982. Of the many infection episodes observed in the study population, few appear to have been associated with wastewater aerosol exposure, and none resulted in serious illness.

In Israel, a prospective epidemiological study of possible disease transmission associated with land application of wastewater was carried out in kibbutzim (collective agricultural settlements) between March 1980 and February 1982 (Shuval et al., 1985). Medical data was collected directly from the patients' files and daily logs of physicians and nurses at each kibbutz clinic. The morbidity data was collected in 29 kibbutzim. The study population was 15,605, and 104,298 clinic visits were recorded in both nurses and physicians records.

No significant excess of enteric disease rates was found in any age group, including the 0-5 years old age group, in kibbutzim exposed to wastewater aerosols (category A) as compared with kibbutzim using wastewater - but not exposed to aerosols (category B), or in kibbutzim of the control group - not exposed to wastewater in any form (category C).

Children of the 0-5 years old age group of all water contact workers (including wastewater contact) in category A had a statistically significant excess of enteric disease as compared with children of the same age group in categories B and C. This excess was not found among children of wastewater contact workers; however their number was too small for effective analysis. Notwithstanding the suggestive evidence supporting direct person to person contact infection of children of water contact workers, it is difficult to draw any firm conclusions from these findings at this stage.

The serological portion of the study proved to be a particularly sensitive technique, since it surmounted problems of nonuniform clinic reporting and allowed the detection of numerous subclinical cases that did not appear in the morbidity data. This study, which encompassed a sample of 1810 persons in 29 kibbutzim, confirmed, in general, the findings of the morbidity study. Of 8 enteroviruses tested, including infectious hepatitis A, no excess in antibody levels was found in the population of 7 kibbutzim exposed to wastewater as compared to controls. However, an excess of ECHO 4 antibodies in the 0-5 age group was detected in kibbutz subcategory A1, exposed to sprinkler irrigation with effluent from neighboring towns, for 1980. this virus was not present in the population for many years prior to the 1980 epidemic, resulting in a particularly high number of susceptibles in younger age groups. This finding suggests that under unusual conditions, such as in an epidemic of an enteric virus for which there is low immunity among the population, transmission and infection by aerosolized wastewater may occur.

In the study, little or no health risks (i.e., excess disease) associated with wastewater irrigation were detected, despite the poor quality of effluent used. Apparently, the dominant mode of transmission of enteric disease in the study population was by direct person to person contact or by food-borne contact. The possibility of aerosol transmission of viruses under unusual circumstances is supported by the circumstantial evidence provided by the serological study.

Recent International Views

In July of 1985, a meeting was convened by The World Bank and the World Health Organization in Engelberg, Switzerland, to discuss the health aspects of excreta and wastewater use in agriculture and aquaculture (IRCWD, 1985). The meeting was hosted by the International Reference Centre for Wastes Disposal, and included environmental engineers, epidemiologists, and social scientists.

The meeting reviewed the progress that had been made in understanding the health effects of human waste reuse since the publication of the World Health Organization's widely accepted report published in 1973, and entitled "Reuse of Effluents: Methods of Wastewater Treatment and Health Safeguards," WHO Technical Report Series No. 517. The concensus of the meeting was that many standards for human waste reuse, including those recommended in WHO Technical Report No. 517, are unjustifiably restrictive and not supported by currently available epidemiologic evidence. Thus, the meeting recommended that WHO initiate revision of its Technical Report No. 517 in the nearest possible future. It recommended that other interested international agencies such as The World Bank, FAO, and UNEP participate in this revision or be otherwise consulted in an appropriate manner.

The meeting's recommendations for the microbiological quality of treated wastewaters to be used for agricultural irrigation are reproduced in Table 10. These recommendations were considered to be technically feasible and in accord with the best currently available epidemiologic evidence. They introduce for the first time a guideline for the helminthic quality of treated wastewater. The quality guideline for restricted irrigation (trees, industrial and fodder crops, fruit trees, and pasture) implies a high removal (>99 per cent) of helminth eggs, and its purpose is to protect the health of agricultural laborers. It was thought to be readily achievable through a variety of treatment technologies but, in many cases, the most appropriate treatment method will be a two-cell waste stabilization pond system (either a 1-day anaerobic pond followed by a 5-day facultative pond, or two 5-day facultative ponds).

The guidelines for unrestricted irrigation (edible crops, sports fields, and public parks) comprise the same requirement for helminth eggs and a maximum geometric mean concentration of 1000 fecal coliforms per 100 ml. The latter recommendation implies a very high level of removal of fecal bacteria (5-6 log units or >99.999 per cent). Its purpose is to protect the health of the consumers of crops (principally vegetables). This was thought to be readily achievable in a properly designed series of waste stabilization ponds. For the range of temperatures normally encountered in tropical and subtropical areas, a series of four 5-day ponds will normally produce an effluent of this required quality. Such a series of ponds would also produce a stable and aesthetically acceptable effluent. The irrigation of sports fields and public parks, especially hotel lawns, may require a more stringent standard as the health risks may be greater to those who come into contact with recently irrigated grass.

Table 10. Tentative Microbiological Quality Guidelines for Treated Wastewater Reuse in Agricultural Irrigation[1]

Reuse Process	Intestinal nematodes[2] (geometric mean no. of viable eggs per litre)	Fecal coliforms (geometric mean no. per 100 ml)
Restricted irrigation[3]		
Irrigation of trees, industrial crops, fodder crops, fruit trees[4], and pasture[5]	$\leqslant$ 1	not applicable
Unrestricted irrigation		
Irrigation of edible crops, sports fields, and public parks[6]	$\leqslant$ 1	$\leqslant$ 1000[7]

[1]In specific cases, local epidemiological, sociocultural, and hydrogeological factors should be taken into account, and these guidelines modified accordingly.
[2]Ascaris, Trichuris, and hookworms
[3]A minimum degree of treatment equivalent to at least a 1-day anaerobic pond followed by a 5-day facultative pond or its equivalent is required in all cases.
[4]Irrigation should cease two weeks before fruit is picked, and no fruit should be picked off the ground.
[5]Irrigation should cease two weeks before animals are allowed to graze.
[6]Local epidemiological factors may require a more stringent standard for public lawns, especially hotel lawns in tourist areas.
[7]When edible crops are always consumed well-cooked, this recommendation may be less stringent.

Conclusions

Types and Levels in Wastewater

The types and levels in wastewater of most pathogens are fairly well understood, with the exception of viruses. Since only a small fraction of the total viruses in wastewater and other environmental samples may actually be detected, the development of methods to recover and detect viruses continues to be a research need. The occurrence of viruses in an environmental setting should probably be based on viral tests rather than bacterial indicators since failures in this indicator system have been reported.

The tremendous number of organic compounds possibly present in wastewater, together with their myriad health effects and poorly understood behavior in the environment, represent a considerable potential for adverse health effects. Most of these can probably be prevented by simple design and monitoring measures; this, of course, would not be true in the case of discharges containing high levels of particular chemicals.

It seems reasonable to conclude that cadmium is the only trace element likely to be of health concern to humans as a result of land treatment of wastewater, with the exposure being through food plants or organ meats.

Preapplication Treatment

The level of preapplicaton treatment required for the protection of public health may be as little as properly designed sedimentation at land treatment sites with limited public access, where crops are protected by appropriate crop choice and waiting periods, and groundwater is protected by appropriate hydrological studies and application rate selection. Where protection of groundwater cannot be assured, wastewater stabilization ponds should be considered for virus removal, but further investigations into the survival of viruses in these ponds is an important research need, as is that of protozoan cysts. Because of potential contamination of crops and infection of animals, slow-rate and overland-flow systems should have high removal rate of helminth eggs. These relatively simple pretreatment requirements would be appropriate for many land treatment systems in the USA, e.g., for many slow-rate sites where crops for animal feed are grown.

Preapplication treatment by storage lagoons may remove considerable quantities of organics, but cannot be relied upon to efficiently remove all toxic organics, particularly since most pretreatment design questions center on inactivation of pathogens.

The recommendations made in the paragraphs below assume a minimum level of preapplication treatment, i.e., properly designed sedimentation. In situations with greater public access (e.g., renovated water reuse on golf courses), shorter waiting periods before grazing or harvest of crops (e.g., agriculture in arid areas), or threat of groundwater contamination (e.g., shallow water table used as a drinking water source), more extensive preapplication treatment may be required. This treatment may consist of wastewater stabilization ponds, conventional treatment unit processes, or even disinfection. The exact degree of pretreatment required for these situations is site-specific, and recommendations should be determined separately for each system (Lance and Gerba, 1978).

Aerosols

Because of the potential exposure to aerosolized viruses at land treatment sites, it would be prudent to limit public access to 100-200 m from a spray source, unless the effluent has been disinfected. At this distance bacteria are also unlikely to pose significant risk. Human exposure to pathogenic protozoa or helminth eggs through aerosols is extremely unlikely.

Suppression of aerosol formation by the use of downward-directed, low-pressure nozzles, ridge-and-furrow irrigation, or drip irrigation is recommended where these application techniques are feasible.

Although removal rates of organics from wastewater by aerosolization and volatilization are high, exposure through this route is unlikely to present any significant health effect.

Surface Soil and Plants

The survival times of pathogens on soil and plants are summarized in Table 11 (after Feachem et al., 1978). Since pathogens survive for a much longer time on soil than plants, the recommended waiting periods before harvest are based upon probable contamination with soil.

Aerial crops with little chance for contact with soil should not be harvested for human consumption for at least one month after the last wastewater application; subsurface and low-growing crops for human consumption should not be grown at a land treatment site for at least six months after last application. These waiting periods need not apply to the growth of crops for animal feed, however.

An important research need is the effect of drying of the soil between wastewater applications on the survival of surface-soil viruses.

The levels of toxic organics likely to be present in soils at land treatment sites will probably result in extremely low levels in above-ground portions of plants, but levels in roots, tubers, and bulbs may present a health hazard.

The potential increase in cadmium levels in human food due to land treatment or irrigation is still an unsettled question. It is clear, however, that increased cadmium in the soil results in increased cadmium in the plants grown in that soil, the degree of increase being a function of cadmium amendment, plant species and cultivar, soil pH, organic matter, and time since application (Ryan et al., 1982). The degree of risk to man, of course, is dependent upon the amount of the food supply affected and the diet selection of the individual. Present levels of total dietary intake of cadmium for most people appear to be fairly safe. However, in view of human variability in sensitivity and the variability in food supply, these levels probably should not be allowed to rise greatly.

TABLE 11. Survival times of Pathogens on Soil and Plants.

	Soil		Plants	
Pathogen	Absolute Maximum	Common Maximum	Absolute Maximum	Common Maximum
Bacteria	1 year	2 months	6 months	1 month
Viruses	6 months	3 months	2 months	1 month
Protozoa	10 days	2 days	5 days	2 days
Helminths	7 years	2 years	5 months	1 month

The most significant research need in the area of trace elements probably continues to be the development of an understanding of the factors controlling the uptake of trace elements by plant crops at land treatment sites, and their entry into the human food supply.

Movement in Soil and Groundwater

Properly designed slow-rate land treatment systems pose little threat of bacterial or viral contamination of groundwater. Considerable threat of bacterial contamination exists, however, at rapid-infiltration sites where the water table is shallow, particularly if the soil is porous. The survival of bacteria in groundwater, once they get there, is poorly understood, and is an important research need.

Likewise, considerable potential for viral contamination of groundwater exists at rapid-infiltration sites, and appropriate preapplication treatment or management techniques should be instituted, e.g., intermittent application of wastewater. Until then, groundwater drawn for use as portable water supplies should be disinfected. The factors controlling the migration of viruses in soils, and the survival of viruses in groundwater, are poorly understood, and are significant research needs.

Human exposure to pathogenic protozoa or helminths through groundwater is extremely unlikely.

Toxic organics can enter the groundwater, particularly at rapid infiltration sites, and the application and soil factors controlling this transport, together with the factors governing their movement and decomposition within groundwater, are significant research needs.

Groundwater is unlikely to represent a significant trace element threat except at rapid infiltration sites.

Land treatment systems, particularly rapid-infiltration, threaten to raise the nitrate concentration in their underlying groundwater above the drinking water standard of 10 mg/l as N. This can be prevented, however, by proper siting and management practice, e.g., using high C:N ratio wastewater, matching loading rate to crop uptake (for slow-rate systems), and optimizing the flooding-drying regime. These management practices and the agronomic factors controlling the entrance of nitrates into groundwater are important research needs.

Increased groundwater sodium concentrations beneath land treatment sites should be kept in mind as a possible future health concern.

Animals

There appears to be little danger of bacterial, viral, or protozoan disease to animals grazing at land treatment sites if grazing does not resume until four weeks after last application. However, the role of animals in transmitting human viral diseases at land application sites is poorly known, and is a research need. Removal of helminth eggs during preapplication treatment should eliminate the potential of disease from those long-lived parasites. The feeding of land-treatment-site-grown plants to animals is unlikely to pose a health problem, but grazing animals may accumulate significant levels of toxic organics. The issue of accumulation of organics from the soil by plants and animals (particularly into milk) is poorly understood, and more research is required.

Infective Dose, Risk of Infection, Epidemiology

Because of the possibility of picking up an infection, it would be wise for humans to maintain a minimum amount of contact with an active land treatment site. The comparison of the respiratory infective dose of enteric viruses with the oral infective dose is a significant research need.

Epidemiological studies to date suggest little effect of land treatment on disease incidence. However, they do indicate that nonsymptomatic viral infections can be transmitted by aerosols. Nevertheless, there appears to be a consensus developing to make guidelines for the application of wastewater to land less restrictive.

References

Babayeva, R.I. 1966. Survival of beef tapeworm oncospheres on the surface of the soil in Samarkand. Med. Parazitiol. Parazit. Bolezn. 35:557-560.

Bagdasaryan, G.A. 1964. Survival of viruses of the enterovirus group (poliomyelitis, ECHO, coxsackie) in soil and on vegetables. J. Hyg. Epidemiol. Microbiol. Immunol. 8:497-505.

Beaver, P.C., and G. Deschamps. 1949. The viability of E. histolytica cysts in soil. Amer. J. Trop. Med. 29:189-191. (Cited in Feachem et al. 1978).

Burmaster, D.E. 1982. The new pollution: Groundwater contamination. Environment 24(2):6-13,33-36.

Burns and Roe Industrial Services Corporation. 1982. Fate of Priority Pollutants in Publicly Owned Treatment Works. EPA-440/1-82-303. U.S. Environmental Protection Agency, Washington, D.C.

Bryan, F.L. 1977. Diseases transmitted by foods contaminated by wastewater. J. Food Protection 40:45-56.

Camann, D.E., et al. 1985. Health Effects Study for the Lubbock Land Treatment Project. USEPA, Cincinnati, Ohio.

Clark, C.S., et al. 1978. A seroepidemiologic study of workers engaged in wastewater collection and treatment. In: "State of Knowledge in Land Treatment of Wastewater" (H.L. McKim, ed.), Vol. 2, 263-271. U.S. Army Corps of Engineers, CRREL, Hanover, New Hampshire.

Craun, G.F., D.G. Greathouse, and D.H. Gunderson. 1981. Methaemoglobin levels in young children consuming high nitrate well water in the United States. Int. J. Epidem. 10:309-317.

Crites, R.W., and A. Uiga. 1979. An Approach for Comparing Health Risks of Wastewater Treatment Alternatives: A Limited Comparison of Health Risks Between Slow Rate Land Treatment and Activated Sludge Treatment and Discharge. EPA 430/9-79-009. USEPA, Washington, D.C. [cf.: Uiga, A., and R.W. Crites. 1980. Relative health risks of activated sludge treatment and slow-rate land treatment. J. Water Poll. Control Fed. 52:2865-2874.]

Damgaard-Larsen, S., K.O. Jensen, E. Lund, and B. Nissen. 1977. Survival and movement of enterovirus in connection with land disposal of sludges. Water Research 11:503-508.

DeWalle, F.B., E.S.K. Chian, et al. 1981. Presence of Priority Pollutants in Sewage and Their Removal in Sewage Treatment Plants. Unpublished report. USEPA, Cincinnati, Ohio.

Duboise, S.M., B.E. Moore, and B.P. Sagik. 1976. Poliovirus survival and movement in a sandy forest soil. Appl. Environ. Microbiol. 31:536-543.

Feachem, R.G., et al. 1978. Health Aspects of Excreta and Wastewater Management. The World Bank, Washington, D.C. [cf.: R.G. Feachem, D.J. Bradley, H. Garelick, and D.D. Mara. Sanitation and Disease: Health Aspects of Excreta and Wastewater Management. World Bank Studies in Water Supply and Sanitation, No. 3. John Wiley & Sons.]

Fraser, P., and C. Chilvers. 1981. Health aspects of nitrate in drinking water. Sci. Total Environ. 18:103-116.

Fraser, P., C. Chilvers, V. Beral, and M.J. Hill. 1980. Nitrate and human cancer: A review of the evidence. Int. J. Epidem. 9:3-11.

Gerba, C.P., and J.C. Lance. 1980. Pathogen removal from wastewater during groundwater recharge. In: "Wastewater Reuse for Groundwater Recharge" (T. Asano and P.V. Roberts, eds.), 137-144. Office of Water Recycling, California State Water Resources Control Board.

Grunnet, K., and C. Tramsen. 1974. Emission of airborne bacteria from a sewage treatment plant. Rev. Intern. Oceangr. Med. 34:177. (Cited in Katsenelson et al. 1977).

Hathaway, S.W. 1980. Sources of Toxic Compounds in Household Wastewater. USEPA, Cincinnati, Ohio.

Holmes, I.H. 1979. Viral gastroenteritis. Progr. Med. Virol. 25:1-36.

IRCWD. 1985. The Engelberg Report: Health Aspects of Wastewater and Excreta Use in Agriculture and Aquaculture. Sponsored by The World Bank and World Health Organization. International Reference Centre for Wastes Disposal, Dubendorf, Switzerland.

Iskandar, I.K. 1978. Overview of existing land treatment systems. In: "State of Knowledge in Land Treatment of Wastewater" (H.L. McKim, ed.), Vol. 1, 193-200. U.S. Army Corps of Engineers, CRREL, Hanover, New Hampshire.

Johnson, D.E., et al. 1980. The Evaluation of Microbiological Aerosols Associated with the Application of Wastewater to Land: Pleasanton, California. EPA-600/1-80-015, USEPA, Cincinnati, Ohio.

Katsenelson, E., et al. 1976. Risk of communicable disease infection associated with wastewater irrigation in agricultural settlements. Science 195:944-946.

Kominsky, J.R., C.L. Wisseman, and D.L. Morse. 1980. Hexachlorocylopentadiene contamination of a municipal wastewater treatment plant. Amer. Ind. Hyg. Assoc. J. 41:552-556.

Kowal, N.E. 1982. Health Effects of Land Treatment: Microbiological. EPA-600/1-82-007. USEPA, Cincinnati, Ohio.

Kowal, N.E. 1985. Health Effects of Land Treatment: Toxicological. EPA-600/1-84-030. USEPA, Cincinnati, Ohio.

Lance, J.C., and C.P. Gerba. 1978. Pretreatment requirements before land application of municipal wastewater. In: "State of Knowledge in Land Treatment of Wastewater" (H.L. McKim, ed.), Vol. 1, 293-304. U.S. Army Corps of Engineers, CRREL, Hanover, New Hampshire.

Larkin, E.P., J.T. Tierney, J. Lovett, D. Van Donsel, D.W. Francis, and G.J. Jackson. 1978. Land application of sewage wastes: Potential for contamination of foodstuffs and agricultural soils by viruses, bacterial pathogens and parasites. In: "State of Knowledge in Land

Treatment of Wastewater" (H.L. McKim, ed.), Vol. 2, 215-223. U.S. Army Corps of Engineers, CRREL, Hanover, New Hampshire.
Lau, L.S., et al. 1975. Water recycling of sewage effluent by irrigation: A field study on Oahu. Technical report No. 94. Water Resources Research Center, University of Hawaii, Honolulu.
Lefler, E., and Y. Kott. 1974. Virus retention and survival in sand. In: "Virus Survival in Water and Wastewater Systems" (J.F. Malina and B.P. Sagik, eds.), 84-94. Center for Research in Water Resources, Austin, Texas.
Little, M.D. 1980. Agents of health significance: Parasites. In: "Sludge -- Health Risks of Land Application" (G. Bitton, B.L. Damron, G.T. Edds, and J.M. Davidson, eds.), 47-58, Ann Arbor Science Publishers, Ann Arbor, Michigan.
Melnick, J.L., C.P. Gerba, and C. Wallis. 1978. Viruses in water. Bull. World Health Org. 56:499-508.
Muller, G. 1953. Investigations on the lifespan of Ascaris eggs in garden soil. Zentralbl. Bakteriol, 159:377-379.
National Research Council. 1980. Recommended Dietary Allowances. Ninth revised edition. National Academy of Sciences, Washington, D.C.
National Water Council. 1977. Report of the Working Party on the Disposal of Sewage Sludge to Land. National Water Council, London.
Page, A.L., and A.C. Chang. 1981. Trace metal in soils and plants receiving municipal wastewater irrigation. In: "Municipal Wastewater in Agriculture" (F.M. D'Itri et al., ed.), 351-372. Academic Press, New York.
Reed, S.C. 1979. Health Aspects of Land Treatment. U.S. Army Corps of Engineers, CRREL, Hanover, New Hampshire.
Rudolfs, W., et al. 1951. Contamination of vegetables grown in polluted soil. II. Field and laboratory studies on Endamoeba cysts. Sewage Ind. Wastes 23:478-485.
Ryan, J.A., H.R. Pahren, and J.B. Lucas. 1982. Controlling cadmium in the human food chain: A review and rationale based on health effects. Environ. Res. 28:251-302.
Schaub, S.A., et al. 1978. Evaluation of the overland runoff mode of land wastewater application for virus removal. In: "State of Knowledge in Land Treatment of Wastewater" (H.L. McKim, ed.), Vol. 2, 245-252. U.S. Army Corps of Engineers, CRREL, Hanover, New Hampshire.
Schaub, S.A., et al. 1980. Evaluation of the overload runoff mode of land wastewater tretament for virus removal. Appl. Environ. Microbiol. 39:127-134.
Shuval, H.I., and B. Fattal. 1980. Epidemiological study of wastewater irrigation in kibbutzim in Israel. In: "Wastewater Aerosols and Disease" (H.R. Pahren and W. Jakubowski, ed.), EPA-600/9-80-028. USEPA, Cincinnati, Ohio.
Shuval, H.I., and N. Gruener. 1977. Health Effects of Nitrates in Water. EPA-600/1-77-030. USEPA, Cincinnati, Ohio.
Shuval, H.I., et al. 1985. Disease Transmission Associated with Land Application of Wastewater. USEPA, Cincinnati, Ohio.
Teltsch, B., et al. 1980. Isolation and identification of pathogenic microorganisms at wastewater-irrigated fields: Ratios in air and wastewater. Appl. Environ. Microbiol. 39:1183-1190.
Tierney, J.T., R. Sullivan, and E.P. Larkin. 1977. Persistence of poliovirus I in soil and on vegetables grown in soil previously flooded

with inoculated sewage sludge or effluent. Appl. Environ. Microbiol. 33: 109-113.

Uiga, A., et al. 1978. Relative health factors comparing activated sludge systems to land application systems. In: "State of Knowledge in Land Treatment of Wastewater" (H.L. McKim, ed.), Vol. 2, 253-261. U.S. Army Corps of Engineers, CRREL, Hanover, New Hampshire.

Underwood, E.J. 1977. Trace Elements in Human and Animal Nutrition. 4th ed. Academic Press, New York.

USEPA. 1977. National Interim Primary Drinking Water Regulations. EPA-570/9-76-003. USEPA, Washington, D.C.

USEPA. 1979. National Secondary Drinking Water Regulations. Federal Register 44(140):42195-42202.

USEPA, USFDA, and USDA. 1981. Land Application of Municipal Sewage Sludge for the Production of Fruits and Vegetables: A Statement of Federal Policy and Guidance. Joint Policy Statement, SW-905. USEPA, Washington, D.C.

Yeager, J.G., and R.T. O'Brien. 1979. Enterovirus inactivation in soil. App. Environ. Microbiol. 38:694-701.

Energy and cost savings in applying minimum treated wastewater to overland flow systems

D. Donald Deemer, ERM-Southeast, Inc., Marietta, GA 30066

The following is a case history of how a particular small community should be able to save $1-million by avoiding the use of excessive preapplication treatment prior to an overland flow land treatment system. Overland flow is a site-specific land treatment process, suitable principally for sites which contain low permeability soils. Site development consists of re-shaping the site into a network of terraces and drainage channels, and establishing a water tolerant grass cover crop. Most, if not all, overland flow systems are dedicated land treatment sites with controlled public access. These features make the process very suitable for application of minimum treated wastewater.

A fair number of research, pilot-scale, and full-scale projects have demonstrated the ability of overland flow systems to produce a very high quality of effluent with only minimal levels of preapplication treatment. Nevertheless, many state land treatment guidelines insist on relatively high levels of preapplication treatment prior to overland flow. It is not unusual to find secondary treatment required and it is very common to encounter requirements for reducing the BOD and suspended solids concentrations to levels of at least 60 to 70 mg/L prior to overland flow application.

Thus, while many regulators take a very conservative stance on preapplication treatment requirements, they often unjustifiably raise the cost of a project. In some cases, the cost increases can be quite substantial, thereby converting an economical treatment process (land treatment) into an alternative which is economically less competitive than other alternatives.

Performance Capabilities

The overland flow process is capable of producing a treated effluent quality which is equivalent to many advanced wastewater treatment processes. EPA's Process Design Manual for Land Treatment of Municipal Wastewater (EPA, 1981) lists the effluent quality which can be expected from an overland flow system being used to treat municipal wastewater. This quality is shown on Table 1.

Table 1. Expected Treated Effluent Quality from Overland Flow Systems.

Constituent	Average	Upper Range
BOD, mg/L	10	< 15
Suspended solids, mg/L	10	< 20
Ammonia nitrogen, mg/L as N	< 4	< 8
Total nitrogen, mg/L as N	5	< 10
Total phosphorus, mg/L as P	4	< 6
Fecal coliform, No./100 mL	200	< 2,000

The effluent quality shown on Table 1 is not significantly affected by the level of preapplication treatment provided. Higher levels of preapplication treatment may allow slightly higher hydraulic loading rates and, consequently, less land area than low levels of preapplication treatment. However, the difference is usually not as significant as the cost to provide the higher level of treatment.

Preapplication Treatment Alternatives

Overland flow systems can be combined with numerous methods of preapplication treatment processes to produce a satisfactory effluent quality. These alternatives range from a minimum of screening, to primary treatment, lagoons, and secondary treatment. An attractive feature of most land treatment systems, including overland flow, is that they are compatible with most existing municipal wastewater treatment systems. Therefore, they can be used very effectively as an "add-on" process step to most existing municipal treatment systems, particularly where those existing systems are not capable of meeting current or proposed effluent discharge limitations.

Case Study

The following is a case study of a 0.67 MGD municipal wastewater treatment system which has frequently been in violation of its effluent discharge limitations. Moreover, a new NPDES permit will contain even more stringent effluent limits. A study was conducted to upgrade the existing aerated lagoon system with the use of overland flow. Presented herein is a comparison of cost controlled upgrading of the existing treatment plant to provide preapplication treatment versus considerably more upgrading as recommended by the town's consultant following the state guidelines and requirements.

Effluent Limitations

The Facilities Plan determined the design flow of the upgraded treatment facility to be 0.67 MGD. The receiving stream has been classified by the state as a water quality limited stream. Consequently, the effluent limitations established are quite stringent. These are listed on Table 2.

Table 2. Effluent Discharge Limitations.

Constituent	Discharge Limits
BOD, mg/L	3
Suspended solids, mg/L	3
Ammonia nitrogen, mg/L (April - October)	1.5
Dissolved oxygen, mg/L	6.5

Alternatives Considered

Upgrading the existing sewage treatment plant to produce effluent of secondary quality will not be sufficient to meet the new effluent discharge limitations. Therefore, it was necessary to consider advanced wastewater treatment technologies. In addition, both slow rate and overland flow land treatment were considered. While the slow rate process could be used, the soil and hydrogeologic conditions of the project area are not at all suitable for slow rate and would severely restrict the hydraulic loading rate. Overland flow, on the other hand, was found to be very suitable for the site conditions. Furthermore, the addition of an in situ sand filtration process to polish the overland flow runoff was tested on a pilot scale and found to have the potential for meeting the effluent limitations.

Proposed Upgrading - Option 1

The town's consultant, using guidance provided by the state sewage treatment regulations, proposed the utilization of the existing sewage treatment plant, but with considerable upgrading, including the following:

1. A new grit chamber at the head of the plant.

2. A new influent flow meter.

3. Additional aerators for the aerated lagoons.

4. A new final clarifier.

5. Two new aerobic sludge digesters, and new sludge drying beds.

6. A new chlorine contact chamber.

7. A new control building.

When added to the various items in the overland flow system, including the cost of land and the engineering, legal, and administrative fees, the preliminary cost estimate for the total project amounted to $3.66 million.

Proposed Upgrading - Option 2

The proposed Option 1 upgrading of the existing treatment plant, while consistent with state guidelines, will basically create a complete secondary treatment system prior to the overland flow system. However, this degree of preapplication treatment will have little or no influence on the performance of the overland flow system and its ability to meet the proposed effluent discharge limitations. Therefore, it appears to make no economic sense to propose major upgrading of the existing system when such upgrading will not improve final effluent quality. Option 2 represents an attempt to minimize the upgrading of the existing treatment plant without interfering with the ability of the overland flow system to achieve the required final effluent quality.

The proposed grit chamber was omitted entirely since grit has never been a problem in the plant operation and is not likely to have an adverse impact on the overland flow system. The proposed additional aeration capacity was retained to assure good mixing in the aerated lagoons. A new final clarifier was omitted because the overland flow system can readily accept the completely mixed effluent from an aerated lagoon. By eliminating the final clarifier, the need for sludge digesters and drying beds is also eliminated. The location of the new chlorination system was changed from ahead of the overland flow system to after the overland flow system, just prior to stream discharge. All other components remained the same between Options 1 and 2, although a strong case could probably be made for eliminating, or at least reducing, the size of the control building.

Table 3 shows the cost comparison between the two overland flow options. As seen on Table 3, the cost estimates show a potential savings of over one-million dollars by providing minimal upgrading of the existing treatment plant. This is a savings of over 28 percent of the estimated cost of Option 1.

Table 3. Cost Comparison between Options 1 and 2.

	Option 1 Extensive Upgrading	Option 2 Minimal Upgrading
Preapplication Treatment		
Grit chamber	$ 175,000	-0-
Flow meter	60,000	$ 60,000
Aerators	60,000	60,000
Clarifier	350,000	-0-
Chlorination	130,000	-0-
Sludge treatment	270,000	-0-
Standby generator	115,000	115,000
Control building	160,000	160,000
Sub-total	$1,320,000	$ 395,000
Overland Flow System		
Pumping to site	$ 50,000	$ 50,000
Piping to site	80,000	80,000
Pumping at site	80,000	80,000
Storage basin	300,000	300,000
Site clearing	50,000	50,000
Terrace construction	200,000	200,000
Distribution piping	125,000	125,000
Agriculture	60,000	60,000
Chlorination	-0-	130,000
Effluent discharge pipeline	110,000	110,000
Sub-total	$1,110,000	$1,240,000
TOTAL CONSTRUCTION COST	$2,430,000	$1,635,000
Land		500,000
500,000		
Engineering/Legal Administrative Construction Management @30%	730,000	490,000
TOTAL PROJECT COST	$3,660,000	$2,625,000

Summary

Numerous opportunities exist today where overland flow and other land application systems can be used very effectively in treating municipal and various industrial wastewaters. However, there are also many examples of land treatment processes loosing their cost-effectiveness because of unnecessarily stringent preapplication treatment requirements or selections. This situation can only be overcome by recognizing the high treatment potential that land application systems offer and allowing them to be designed and operated with cost-effective levels of preapplication treatment.

Muskegon, Michigan--A case study of project costs at an existing slow rate land treatment system

W. Henry Waggy, Metcalf & Eddy Inc., Arlington Heights, IL 60005
Douglas A. Griffes, CH2M HILL, Emeryville, CA

One of the major benefits of land treatment is its flexibility to adapt to a wide range of site conditions and system objectives. Contrary to the impression given by some design guidelines and regulatory contraints, there is no universal land design appropriate to every condition. Typically, a system consists of several components, which may include: preapplication treatment, storage, wastewater distribution, crop management, and drainage. Each of these can be used in a variety of ways to suit individual requirements. Because of this flexibility, the costs of land treatment are highly variable and difficult to generalize.

Experiences at the Muskegon County Wastewater Management System offer a good opportunity to view a range of typical costs. The facility was placed in operation in 1973 and was one of the first of the current generation of innovative land treatment systems. It has operated successfully for the past twelve years, although there is currently a need for upgrading and expansion. To meet these current needs, a facility plan was recently completed in which several variations of land treatment were evaluated (Metcalf & Eddy, 1982). This paper presents some of the cost comparisons from the facilities plan. The variability of cost components and the need to adapt to local conditions are highlighted.

All cost information presented in this paper is from the 1982 report. Due to grant funding constraints and regulatory requirements, the planning and design process for Muskegon is still ongoing. Consequently, some of the information presented here is no longer current. Design flows and several aspects of the recommended plan have been modified; however, the major concepts and the relationships between alternatives remain unchanged. For purposes of clarity, only the original alternatives and costs are considered here.

THE FIRST TWELVE YEARS AT MUSKEGON

The Wastewater Management System was formed to provide a regional solution to wastewater treatment problems in Muskegon County, Michigan. Prior to the formation of the Wastewater Management System, area industries and municipalities handled wastewater disposal on an individual basis. Disposal practices included discharge of untreated or partially treated wastewater into the surface streams and lakes of the area. Water quality had deteriorated and water pollution was described as "severe."

The facilities, located in Figure 1, were developed to serve area communities and industries. Land application of partially treated wastewater was selected as the treatment method. Land treatment had been previously used in this country, but not to the scale proposed for this project. The main facility serving the metropolitan area of Muskegon was designed to treat 42 million gallons per day (mgd) by irrigating 5,500 acres. A smaller but similar system, designed to treat 1.36 mgd, was constructed to serve the Whitehall-Montague area. Both systems were placed in operation in 1973. Additional pretreatment facilities were added to the Whitehall site in 1978 to more adequately handle the high-strength wastewater influent.

This paper will focus predominately upon the Metro system although some cost information will, of necessity, include the Whitehall facility.

METRO PROCESS DESCRIPTION

The existing treatment facility at Metro consists of aerated lagoons, storage lagoons, irrigation system, and renovated wastewater collection system as shown schematically in Figure 2. Advanced wastewater treatment including nitrification and phosphorus removal is provided through land treatment. Facilities at the Whitehall site are generally similar to those at Metro.

The preapplication treatment system is intended to provided removal of biodegradable organics and suspended solids to a degree that loadings to the storage lagoons do not create nuisance conditions. Recommended loadings to the storage lagoons are less than 20 pounds of BOD per acre per day (lb BOD/acre/day). Adequate treatment performance has been provided by the aerated lagoons with only one or two of the three lagoons in use. Reduced mixing of the lagoons has allowed higher solids removal and reduced energy consumption. Solids accumulation is permitted in the aerated lagoons with periodic cleaning by the County staff. The settling lagoon has been abandoned as unnecessary and ineffective.

The storage lagoons provide storage for wastewater during non-irrigation periods. Climatic conditions necessitate 4 1/2 months of storage. The two lagoons have a combined volume of 5.1 billion gallons, which is equivalent to 37.5 mgd in raw wastewater flow over the 4 1/2 month storage season. (Note that this is less capacity than the original design intent of 42 mgd.) Seepage from the lagoons is controlled by two interception ditches. At present, lagoon seepage is mixed with plant effluent and directly discharged to surface waters.

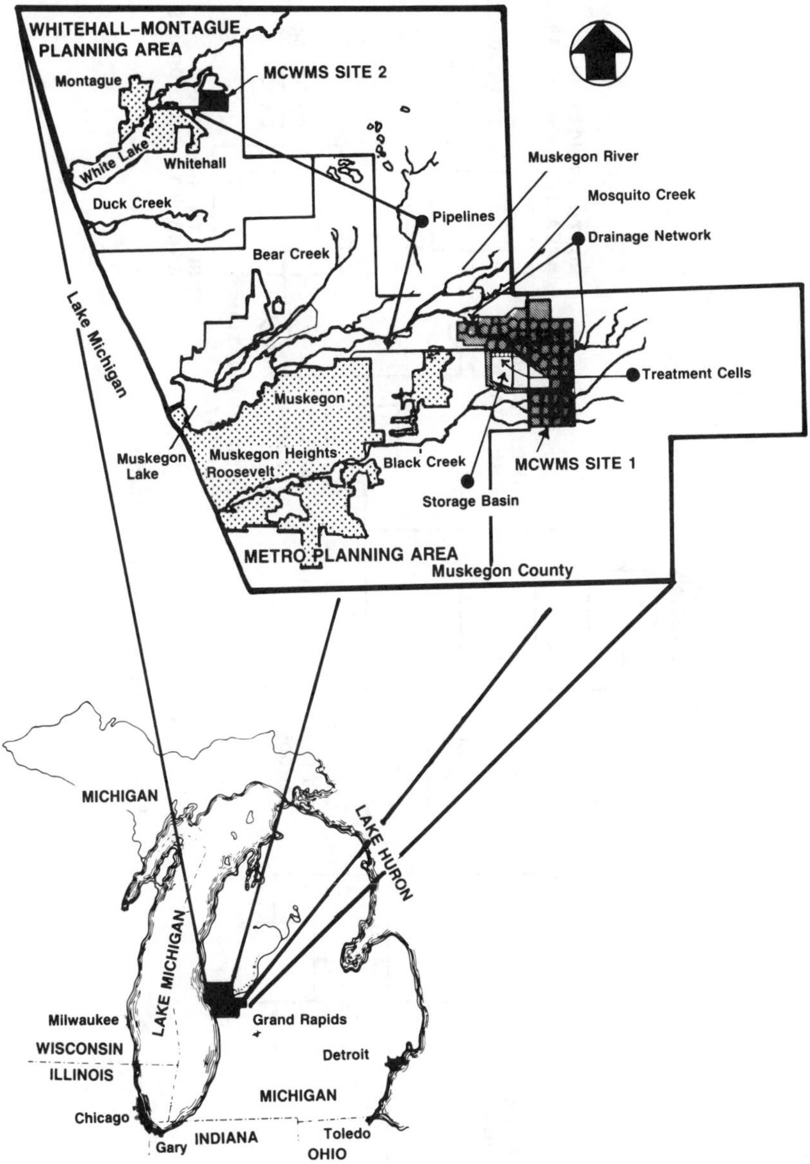

Fig. 1. Muskegon County Wastewater Management System.

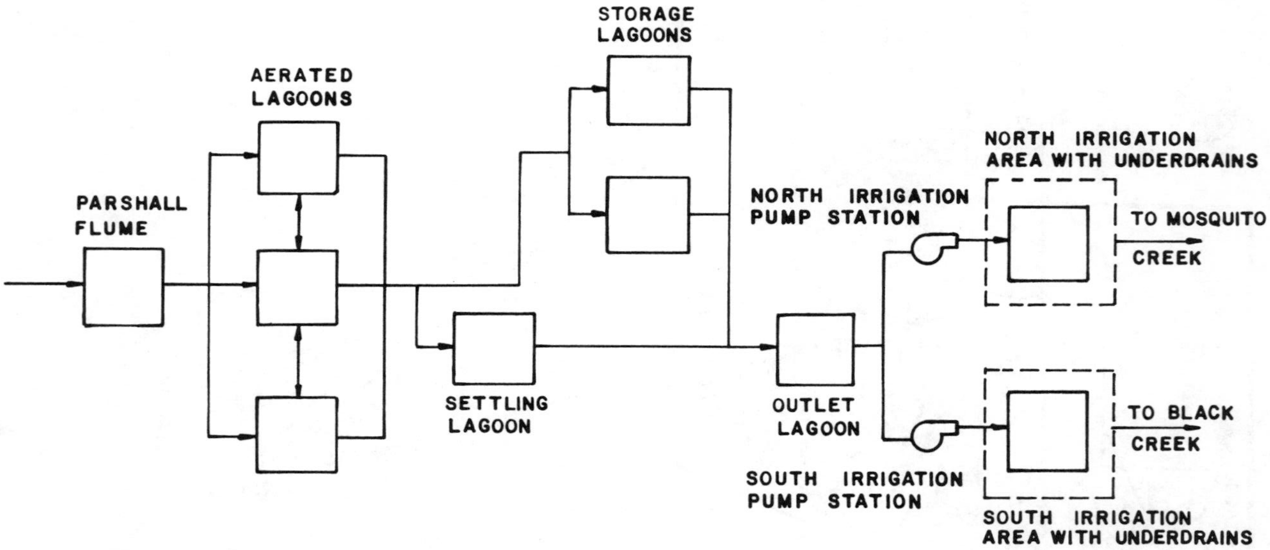

Fig. 2. Schematic of metro facility.

After storage, the wastewater is utilized for irrigation. The irrigation distribution system, which is shown in Figure 3, consists of two pumping stations, a network of distribution forcemains, and center pivot irrigation rigs. At the present time, approximately 5,300 acres (mostly of corn) are irrigated. In addition to the normal irrigation season, a significant amount of wastewater must be applied both before planting and after harvest to meet the water balance constraints at the Metro site. Wastewater application rates range from 30 to 100 inches per year on the irrigated fields. The application rate to a given circle is governed by the soil type found in that circle and the effectiveness of the underdrains in that circle.

An extensive drainage system is provided at Metro to collect renovated wastewater from the irrigation areas and lagoon seepage for surface discharge. This system consists of a combination of underdrains, wells and interception ditches discharging into two main collector drains. The underdrains and interception ditches accomplish two important objectives: they lower the groundwater from its normally high levels and they recapture the applied wastewater for surface discharge. The first objective is necessary to increase the disposal capacity of the site. The second objective is necessary to comply with Michigan's increasingly stringent groundwater protection laws.

TREATMENT COSTS OF THE EXISTING SYSTEM

Neither the capital costs from the construction of original facilties nor the annual operating costs are segregated into separate costs for Metro and Whitehall. Therefore, it is necessary to consider both facilities together when discussing existing costs.

CAPITAL COSTS OF THE EXISTING SYSTEM - Table 1 presents a cost breakdown from the construction of the original facilities. The total construction cost was approximately $43.5 million through the end of 1975. This translates into a unit cost of approximately one million dollars in construction cost per million gallons per day of design capacity. Since the actual capacity of the two systems is closer to 34.3 mgd rather than the design level of 43.7 mgd, the actual unit cost was $1.27 million per mgd capacity in 1975 dollars, or approximately $2.56 million in 1982 dollars (The base date for comparison of expansion alternatives).

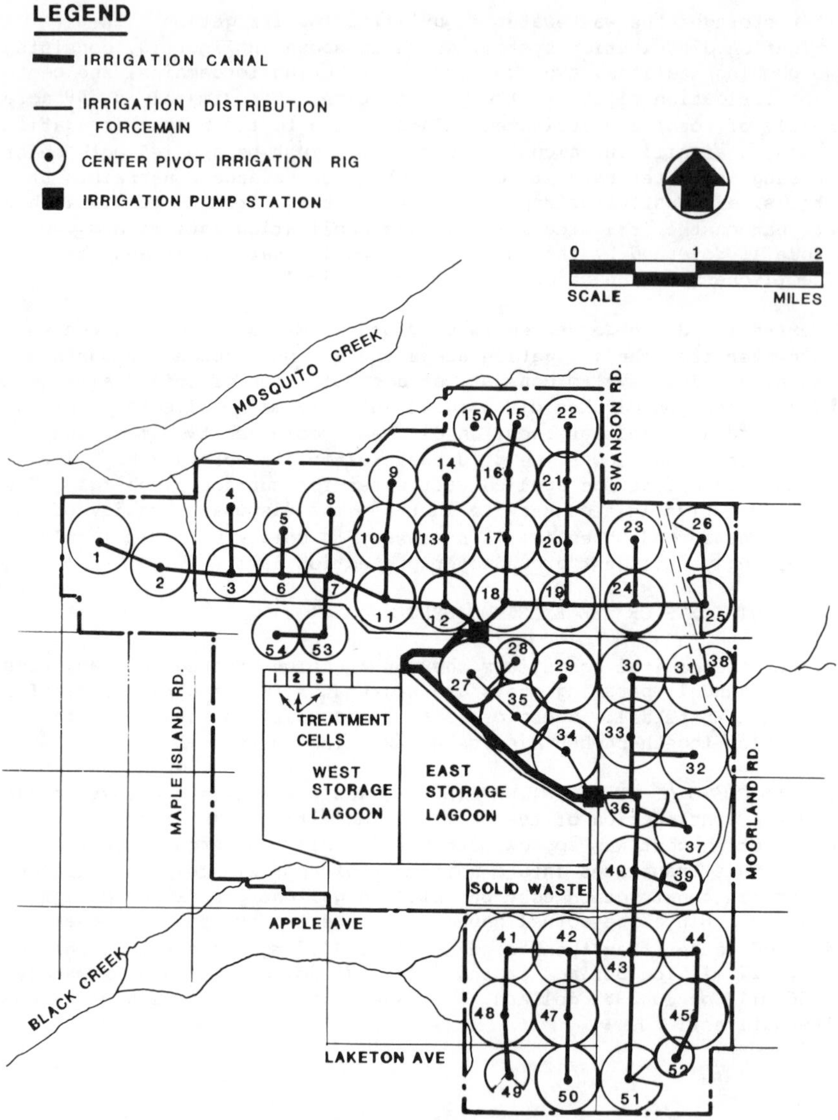

Fig. 3. General layout metro irrigation system.

Table 1. Muskegon County wastewater management system costs - Dec. 31, 1985 (Demirjian, et al., 1980).

Line Item	Dollars (X1000)
Land	6,648
Site Improvements	12,073
Pipes, Force Mains, Ditches, Etc.	9,951
Mechanical and Electrical	3,842
Machinery and Equipment	6,735
Engineering, Interest, Loan Costs	3,403
	42,652
Non - Capitalized Costs	1,038
	43,690

* Demirjian, et al. 1980

OPERATING COSTS OF THE EXISTING SYSTEM - Table 2 presents the operating costs for the combined Metro and Whitehall systems over the period from 1981 through 1983. These costs cover annual operations and maintenance expenses only and exclude all repayment of capitalized debt. Note that the costs for both the laboratory and the farm operations are net costs (the difference between expenses and revenue). The laboratory generates revenue by performing laboratory analysis work for outside agencies and externally funded research projects. The farm operation shows a net profit which helps defer the costs of wastewater treatment. Note that the farm income fluctuates greatly from one year to another.

Table 2. MCWMS operating costs, 1981 - 1983.

	Dollars (X1000)		
	1981	1982	1983
Operations & Maintenance	2,838	3,140	3,133
Laboratory	122	166	106
Farm	(472)	87	(667)
Net Operating Cost	2,488	3,393	2,572
Treated Flow (MGD)	29.8	34.8	34.4
Net Cost Per MGD	83.5	97.5	74.8

Flow treated over this three year period averaged 33 mgd. The net annual cost of treatment per million gallons of flow averaged $234. The 1983 user charge in Muskegon was $309 per million gallons of wastewater treated, which covers the cost of both treatment and collection. This rate compares favorably with the $650 per million gallons average charge estimated from Analysis of Operations and Maintenance Costs for Municipal Wastewater Treatment Systems (Dames & Moore, 1978) for advanced wastewater treatment plants.

Power costs represent approximately a third of the total operating costs. Consequently, energy conservation is an import consideration at the Muskegon County Wastewater Management System. Table 3 shows the pattern of electricity consumption by the System over a five year period.

Table 3. Power consumption at MCWMS.

	Flow, mgals	Electricity Consumed, Kw-hr	Consumption Rate, Kw/hr/gals
1979	10,400	27,900,000	2,680
1980	10,600	32,400,000	3,060
1981	10,900	30,400,000	2,790
1982	12,700	39,000,000	3,070
1983	12,600	38,400,000	3,050

TREATMENT PERFORMANCE

Treatment performance at the Metro plant has been excellent. Historically the plant has bettered its stringent effluent requirements. Table 4 shows the treatment performance at Metro for 1982 and 1983. Problems are developing for the future, however, because of increasing strength of the wastewater (60 percent of the flow is industrial) and tightening effluent standards. At the time of the original evaluation, a phosphorus standard of 0.3 mg/l was proposed for the Muskegon River, which is the major receiving water body. The standard currently proposed is 0.13 mg/l during the growing season.

Table 4. Summary of Treatment Performance - Metro.

		BOD, mg/l	SS, mg/l	P, mg/l	NH_3,[1] mg/l
1982					
	Influent	265.14	299.92	2.67	11.64
	Effluent				
	001	2.89	5.60	0.13	0.72
	002	1.65	16.25	0.10	0.38
1983					
	Influent	293.40	364.92	3.84	13.45
	Effluent				
	001	2.59	5.25	0.12	0.68
	002	1.40	13.83	0.09	0.38
Permit Effluent					
001[3]		4.0/10.0[2]	15.0	0.4/0.3[2]	3.0/2.0/5.0[2]
002[4]		4.0/17.0[2]	25.0	0.20	2.0/6.0[2]

1. Expressed as Nitrogen
2. Seasonal Standard
3. Effluent point 001 is Mosquito Creek
4. Effluent point 002 is Black Creek

OPERATIONAL LIMITATIONS

The Metro system has a number of limitations or problems that either impair its present ability to operate or threaten to do so in the near future. Many of these limitations are related to capacity and will have to be alleviated if waste loads or flows are to be increased in the future. Others are related to treatment performance and control. The following are the major operational limitations which are addressed in the alternative plans for improvement of the system.

- Agricultural drainage - subsurface drainage pipe network and drainage wells are not performing as designed in many areas. Applicaton rates have been limited and the capacity of the system has been correspondingly reduced.

- Lagoon seepage - intercepted seepage from the lagoons is currently of adequate quality for direct surface discharge. Long-term phosphorus removal capability of the soils in the flow path may be limited, however. Consequently, with direct discharge of seepage, the ability of the system to meet future effluent limitations is in question.

- Groundwater control - the groundwater flow pattern in the northwest corner of the site is uncertain. More positive control of the groundwater is required.

- Preapplication treatment - the existing aerated lagoon system provides adequate treatment but has a number of operational problems including: release of odors, formation of foam, limited mixing and aeration capabilities, and sludge accumulation.

DEVELOPMENT OF ALTERNATIVES

Based on the evaluation of the existing system, it became apparent that the following improvements would be required:

- Expansion of capacity to 49 mgd
- Drainage/agricultural improvements to existing system
- Improved phosphorus removal from lagoon seepage
- Improved trace organics control
- Improved groundwater control at two locations
- Improvements in existing preapplication system

The most obvious method for achieving these requirements is to make necessary modifications to the existing system, and then expand the irrigation area in order to provide the additional capacity. Although this approach has several important advantages, it was apparent from initial investigations that the incremental cost of the new capacity would be considerably more expensive than for the original system. Several new ideas and alternatives were then introduced. The three major alternatives evaluated in the study are described below.

EXPANSION OF IRRIGATION AREA

The irrrigation expansion alternatives (M-1A, M-1B, M-1C) are shown in Figure 4. They entail a continuation of the same basic processes that are currently employed at Muskegon. Improvements would be made to the existing system to correct deficiencies and to increase capacity to the maximum extent possible within the present boundaries. Approximately 500 to 800 acres of additional irrigation land would still be required, however. As shown in Figure 4, three sites were identified as possible expansion areas.

Irrigation expansion would also require a corresponding expansion of storage capacity. Approximately 480 acres of existing irrigation in an area of poor soils would be taken out of service for the construction of a new lagoon.

In addition to the main system of wastewater treatment by irrigation, a small area of rapid infiltration would be provided for final treatment (primarily phosphorus removal) of collected seepage from the storage lagoons.

IRRIGATION/RAPID INFILTRATION

This alternative (M-2) is similar to the previous set of alternatives except that there would be no acquisition of new land for irrigation. As indicated in Figure 5, a rapid infiltration system would be designed

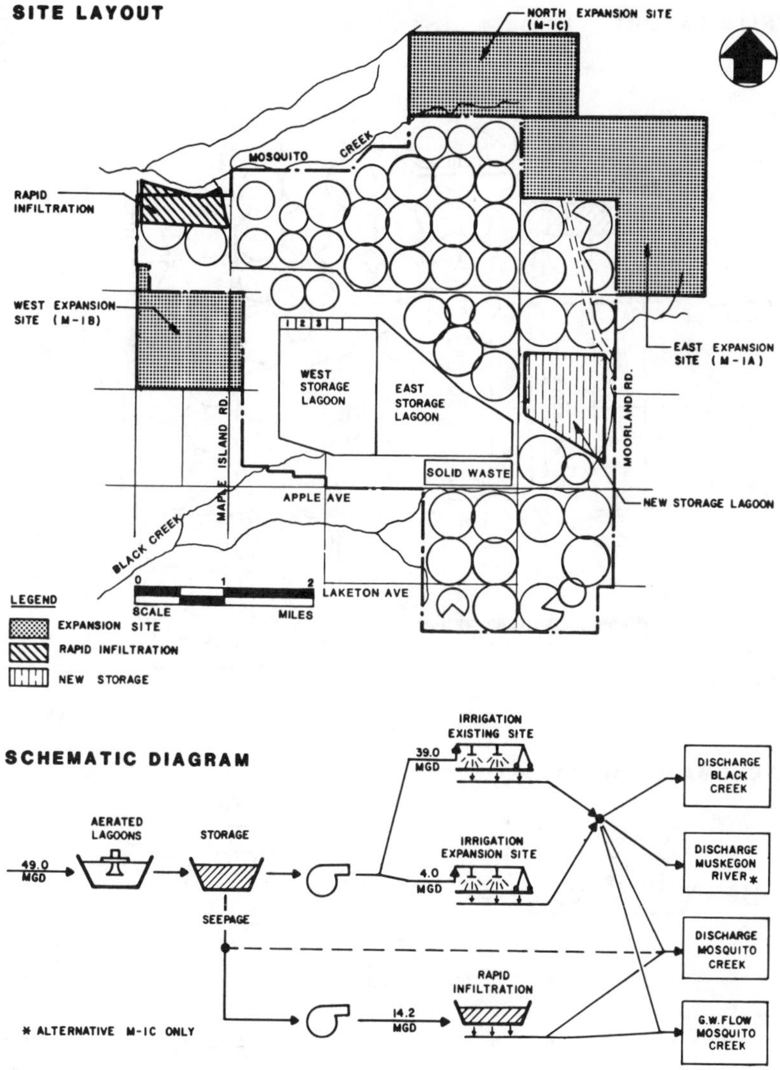

Fig. 4. Metro irrigation expansion. Alternatives (M-1A, M-1B, M-1C).

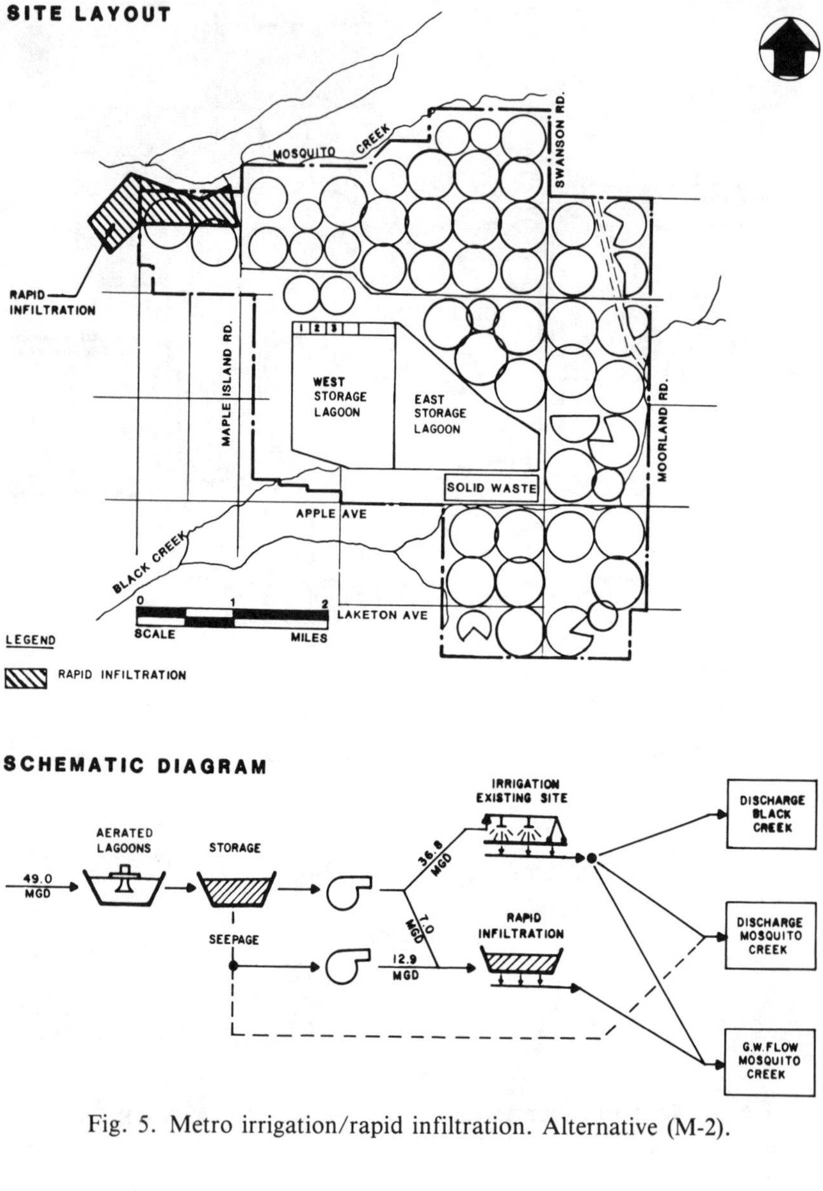

Fig. 5. Metro irrigation/rapid infiltration. Alternative (M-2).

to handle all of the expansion flows as well as seepage flows from the storage lagoon. An area of well-drained soils on the existing site would be taken out of irrigation and used for this purpose. The system would consist of approximately 155 acres of infiltration basins, distribution pipelines, and a network of interception ditches and wells for groundwater control. An annual application rate of 145 feet is planned. The system would be operated year-round, so that no additional storage capacity would be required.

IRRIGATION/CONVENTIONAL TREATMENT

Alternative M-3 is parallel to the previous one except that conventional treatment is substituted for rapid infiltration. As indicated in Figure 6, the expansion flows would be nitrified by means of rotating biological contactors. Following nitrification, the flow would then be combined with the lagoon seepage for phosphorus removal and disinfection. The effluent would be discarged to the north interception ditch, which is the same location as in the previous alternatives.

EVALUATION OF ALTERNATIVES

In order to select the best overall solution, the alternatives previously described were compared on the basis of present worth costs and other non-monetary factors. Some of the highlights of the cost comparison are presented here together with a discussion of the major cost trade-offs considered.

The present worth costs include capital cost of the new facilities, present worth of 20 years of annual operations and maintenance costs, present worth of the replacement of facilities over the 20 year design period, and a deduction for salvage values at the end of 20 years. Table 5 summarizes the costs for the six basic alternatives considered. It is obvious from Table 5 that Alternative M-2, enhancement of the existing irrigation system plus construction of a rapid infiltration system to treat expansion flows and lagoon seepage, is the most cost effective alternative.

Table 5. Alternative cost summary (present worth cost in millions of dollars).

Alternative	Capital Cost	Annual O&M	Replacement Cost	Salvage Value	Total Cost
M-1A	42.7	23.0	3.1	2.5	66.3
M-1B	37.4	22.8	3.0	2.2	61.0
M-1C	41.1	23.1	3.0	2.5	64.7
M-2	30.2	23.3	2.9	1.3	55.0
M-3	36.1	23.8	3.0	1.5	61.4

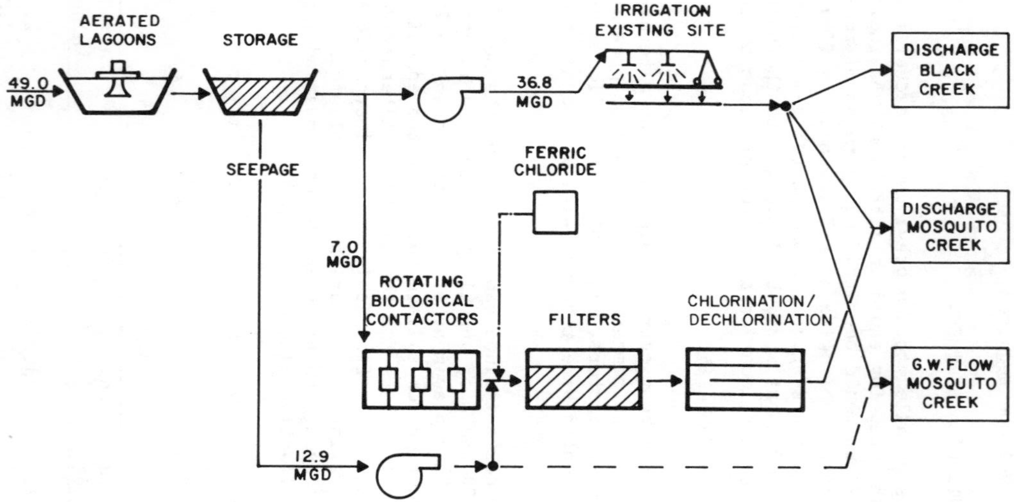

Fig. 6. Metro irrigation/conventional treatment. Alternative (M-3).

The principal advantages Alternative M-2 has over Alternative M-1B, its nearest competitor, are in the avoidance of major land purchases and new storage construction.

COMPARISON OF MAJOR COST FACTORS

An inherent part of the evaluation process was an optimization of treatment system components and design variables. At Muskegon, this involved consideration of several different cost trade-offs in the following areas:

- Land acquisition for irrigation expansion
- Agricultural drainage
- Treatment level

IRRIGATION EXPANSION

As indicated above, three alternative sites were identified for expansion of the irrrigation system. Each had its own combination of favorable and unfavorable characteristics with offsetting effects on cost. The most important of these characteristics are: existing land uses, land cost, ease of acquisition, proximity, and drainage. The major factors involved in evaluating the irrigation expansion alternatives are summarized in Table 6. The costs shown on Table 6 are solely those relating to development of an off-site parcel of new land. Facilities constructed within the bounds of the existing system and common to all three of the M-1 variations (i.e., an additional storage lagoon) are not considered in these comparative costs.

Table 6. Irrigation expansion costs factors.

FACTOR	A	B	C
LAND ($ / ACRE	2900	6200	1900
APPLICATION (IN / YR)	70	100	100
GROUNDWATER	SHALLOW	DEEP	DEEP
PROXIMITY	ADJACENT	ADJACENT	1-2 MI
COST ($ MILLION / MGD)	2.1	1.9	3.0

Alternative M-1A has the advantage of a low land cost due to the lightly populated nature of the site. Land costs are estimated at $2,900 per acre to purchase the land and relocate current residents. The M-1A site is adjacent to the existing facility, so the cost of the distribution system would be relatively low. The major disadvantage is that much of the site contains soils that are moderately to poorly drained. Based on experience with similar soils at the existing system, application rates would be limited to approximately 70 in./yr, thus increasing the land area required. The total capital cost for development of this site was estimated to be approximately $2.1 million per mgd of additional capacity provided.

Both of the other alternatives (M-1B and M-1C) are situated in areas with well-drained soils, similar to the best soils of the existing system. Consequently, allowable application rates would be higher (approximately 100 in./yr), so less land would be required.

With Alternative M-1B, the major disadvantage is the higher land cost (estimated at $6,200 per acre) and relatively high degree of development. In spite of the higher costs for purchase of land and relocation of residents, this alternative has the lowest estimated overall capital cost of $1.9 million per mgd of additional capacity provided.

Alternative M-1C has two major disadvantages. The first is the extra cost for transmission to its relatively distant location. The second disadvantage is the need for extensive groundwater control measures. Although land costs are low (estimated at $1,900 per acre), the estimated overall capital cost is $3 million per mgd.

It is interesting to note that before cost estimates were fully developed, Alternative M-1A was the favored alternative due to its lower land cost and greater ease of development.

AGRICULTURAL DRAINAGE

The installation of underdrainage represented a significant portion of the original construction cost at Muskegon. As previously noted, much of this has not performed up to original expectations, thus limiting application rates over approximately 50% of the site. Because of this limitation, one important consideration in the present project was the comparison of costs between drainage improvements and the development of new irrigation land area.

Proposed drainage improvements consisted of several deep interception ditches and the installation of new underdrains in the soils classified as having moderate limitations (approximately 2,000 acres). The new underdrains are to be installed at 100-foot spacing between the existing drains which are mostly at 500 feet. Based on operational experience and recent field investigations, it is expected that application rates in these soils can be increased by approximately 50%. Performance could also be improved in the 650 acres of soils with severe drainage limitations, but only marginally, and at a greater cost. Consequently, no improvements are recommended in these areas. Overall, the recommended improvements would provide about about 3.8 mgd of capacity at a unit capital cost of $836,000/mgd.

As an alternative, the existing system could continue to be irrigated at existing rates with no drainage improvements. The additonal required capacity would then be obtained by the development of new land. As previously indicated, the least expensive option for irrigation expansion would provide additional capacity at a unit cost of approximately $1.9 million/mgd.

TREATMENT LEVEL

The wastewater constituent of primary concern in the Muskegon River is phosphorus. At the time of the evaluation, a limit of 0.3 mg/l had been proposed. This limitation could readily be met by expansion of the irrigation system (slow rate land treatment), however, it was recognized that this approach would provide a higher quality effluent than required. Alternate processes providing lesser degrees of phosphorus removal were then considered for a portion of the waste stream such that the blended discharge would meet the new phosphorus standard.

As indicated in the description of alternatives, rapid infiltration was found to be a good complement to irrigation for this purpose. In Alternative M-2, the existing irrigation system would be utilized to its maximum, and a rapid infiltration system would be developed to treat the remaining flow. The flow would consist of 7 mgd from expansion and 12.9 mgd from lagoon seepage. The overall capital cost for the portion attributed to expansion was estimated to be $1.9 million per mgd of capacity provided.

For comparison purposes, the equivalent capital cost for irrigation expansion was estimated to be $2.3 million per mgd. This includes the cost of a new storage lagoon which would be required only for the irrigation alternatives.

Conventional treatment was also considered as an alternative treatment method for the expansion flows and lagoon seepage (Alternative M-3). A conventional treatment system could match the rapid infiltration system effluent quality at a unit cost of $2.3 million per mgd, over 20 percent higher in capital cost than the rapid infiltration system.

Note that although the unit capital costs for conventional treatment and irrigation expansion appear to be the same, they are not comparable because the irrigation effluent quality is much higher. Consideration of operating costs would place conventional treatment at an even greater disadvantage.

CONCLUSIONS

Based on the experiences at Muskegon, it is possible to make several general conclusion regarding the design and costs of land treatment systems.

- Appropriate designs must be developed for the specific characteristics of each individual site. Characteristics such as soil types, existing land uses, land costs, drainage, and other factors can have a large impact on the overall cost of the system.

- The design of land treatment systems can be adapted to suit different treatment requirements and system objectives. Overall costs can be substantially reduced if irrigation (slow rate land treatment) is combined with other methods such as rapid infiltration.

- Even when climatic conditions are unfavorable as they are in Michigan, land treatment can be very competitive relative to conventional methods of wastewater treatment.

REFERENCES

Dames &. Moore. Analysis of Operations and Maintenance Costs for Municipal Wastewater Treatment Systems. EPA 430/9-77-015. February, 1978.

Demirjian, Y.A. et al. Muskegon County Wastewater Management System - Progress Report 1968 through 1975. EPA 905/2-80-004. February, 1980.

Metcalf & Eddy, Inc. County of Muskegon, Michigan, Wastewater Management System Facilities Plan Update. August, 1982.

The public's view of waste management

Commissioner Joanne Alter, The Metropolitan Sanitary District of Greater Chicago, 100 East Erie Street, Chicago, IL 60611

The public's view of waste management, particularly sewage sludge, has been an interest of mine for the past 15 years, and the utilization, treatment and disposal of wastes, which you have been discussing here today, is close to my heart.

How the public perceives those of us in this field of waste management is really the greatest challenge in our profession today -- we sewage treatment plant agencies and landfill operators, we government representatives and scientists and technical experts. We face this challenge of explaining our work and of trying to learn what they -- the public -- think and want. But, let's face it -- the fact is: THEY is US!

A decade ago, when there were only 5,000 Americans or so who were members of the Sierra Club, it wasn't such a sin or a crime or a political error to graciously ignore the opinions of the Sierra Club, the Audubon Society, the Citizens for a Better Environment -- those so-called Earth People. But today, there are 4 1/2 million paid members of the National Wildlife Federation, half million members of the Audubon Society, and 365,000 Sierra Club paying members. It is like John Kennedy's "Ich bin ein Berliner" -- "Ich bin ein Environmentalist!" -- everyone is an environmentalist in the United States today, not just a few vegetarians and campers.

We can't easily separate those of us in this room who are scientists and technicians, owners and operators, elected officials and the public. Yes, from time to time, some of us represent these special interests, but the truth is that every single one of us in this room cares desperately about how we manage our waste -- our sewage waste, our solid waste, our toxic waste -- and all of us are politicians in that we all vote.

So, therefore, let us assume that we are the public inside and outside this room, and that we all have the same basic goal--to understand how we can work together in order to achieve our ends of managing wastes properly, using state-of-the-art techniques, in the public interest.

But, first a few words about the Metropolitan Sanitary District which I call the District or MSD. We are a governmental body whose main function is to treat the wastewater generated in the greater Chicago area. The District is located within the boundaries of Cook County and serves an area of approximately 875 square miles, including 125 municipalities, with a population of 5 1/2 million. We own and operate seven sewage treatment facilities with a total capacity of 1.7 billion gallons per day for secondary treatment and 114 million gallons per day for tertiary treatment.

We are the wholesalers of sewage treatment service and have constructed 500 miles of interceptor sewers to service local sewer systems owned by our member municipalities. Each day we produce 500+ dry tons of sewage sludge and have to utilize or dispose of it. We have put it on land in Fulton County, distributed it to the public as Nu Earth, and are currently using this material to cover landfills. Managing our sludge is one of our biggest technical challenges.

The District was created in 1889 by an act of the Illinois Legislature and is governed by an elected nine-member Board of Commissioners, who serve six-year terms. Every two years, three members of the Board are elected.

Our founding and fame came 100 years ago with the reversal of the flow of the Chicago and Calumet River systems. Before the construction of a 70-mile system of canals, the Chicago and Calumet Rivers flowed into Lake Michigan, as all good rivers do. In a major engineering feat using new technology later used to dig the Panama Canal, the direction of the waterways was changed. These rivers now flow into the Illinois River system, and effluent from District treatment facilities ultimately enters the Gulf of Mexico rather than Lake Michigan.

Therefore, the MSD is the Chief environmental agency in the lives of the people in the Chicago metropolitan area. We are responsible for the quality of the water in Lake Michigan. As a local agency, we have dual challenges. First we have our own local authority to collect, treat and dispose of the sewage, and in the process of doing that, we have embarked on more than a few controversial programs. We also have the responsibility of implementing and interpreting federal government regulations and Illinois EPA requirements.

Now, how does the public perceive us? First, we must ask ourselves some serious questions. Does the public perceive us as working in their best interests? Second, does the public think that our technology is advanced? Third, does the public feel they are involved in the decision-making process?

As to the first question, "Are we working in their best interests or are we viewed with suspicion and distrust?" I would ask, "Is our word good? Do others understand what we do? Do they understand the problems of treating sewage? -- of collecting and disposing of solid waste? Do they really understand the serious problems of disposing of toxics? Can we be trusted if there is no knowledge of what we do?"

It is a question of educating, teaching, informing and spreading the word that we are service organizations -- imperfect, but dedicated to performing much-needed services. We must emphasize that there is no basic confrontational battle line between what the environmentalist or special interest groups want and what the sewage treatment agency or landfill operators or soil scientists want. We all must understand what our basic needs are and try to work to achieve some mutual trust as a result of knowledge and education.

The most important thing, of course, is that we talk WITH each other. I don't mean TO or AT each other. I mean in small groups where we can relate to each other as human beings; where those people who are sure that I am poisoning their gardens with cadmium can sit down and talk with me about their concerns, and I can explain to them that we have 500+ dry tons of sludge every day to dispose of; where others can begin to understand the problems of government, and where the leadership -- the scientific leadership, the technical leadership, and the political leadership -- can begin to understand the needs and desires of their fellow taxpayers.

The only way democracy works is when people thoroughly understand and know about their problems. Only then can we begin to trust each other and feel waste is being managed in our best interests.

I am convinced that one of the important things we can do is take the general public on tours of our operations and open our records. When meeting with members of the general public, I always invite them to visit our sewage treatment plants, and more recently, our Deep Tunnel operation. I have taken people out to our laboratory to show them some of the 900,000 tests we do each year, to illustrate how we spend almost $11 million in our Research and Development Department, headed by our highly respected and nationally recognized Dr. Cecil Lue-Hing.

It is important for us to prove we have nothing to hide, for people to have easy access to public records and for them to have as much access as we can possibly provide, at our initiative, to our operations -- and that goes for landfill operations and sewage treatment operators, and most of the activities of management of waste. It is far better for us to show the Sierra Club people the records they want to see than for us to spend three years in court and be forced to give information by court order. There are some legal constraints, but freedom of information laws can work for us in helping the public appreciate the good work we do.

From our point of view, knowing that we can be visited by the public and that we are on display all of the time will keep us on our toes. That's not so bad.

We also must act quickly when incomplete information creates suspicion. Let's discuss a recent example that occurred at the Sanitary District. Recently, the USEPA sponsored a study of air emissions of volatile organic compounds (VOCs) in the Philadelphia area, and though the results were not complete, they gave the media selected data.

Some of the members of the public were very excited about the data that was quoted because the first report said that sewage treatment facilities were the major source of air emissions of volatile organic compounds in the Philadelphia area, and the Sanitary District, by extrapolation, would be one of the greatest air polluters in the country.

When we met with the concerned citizens -- and I was indeed one of those terribly concerned -- it was clear that we had to find out how the study was conducted and whether or not we could do some kind of a cross-check. What happened was that the report to the media about the Philadelphia study caused enough concern at the Illinois EPA that the Department of Energy and Natural Resources commissioned an air emission study, and this study then showed that the volatile emissions from all our sewage treatment plants represented only .017 percent of air emissions in Cook County.

It's very important for the results of studies to be complete -- that's one lesson we learned -- but it's also important for us to be able to say. "Well, we'll look at this report and we'll try to do some kind of impartial survey or double check, and we'll share our information with you." The question is whether the survey will be interpreted as being self-serving or not; and that, of course, depends a lot on the mutual trust that's been developed between the parties.

The second question I would like to ask after "Does the public think we're working in their best interests" is "Does the public think our technology is advanced?"

Many things have changed in the area of sewage treatment and treatment of solid wastes in the past decade, but in general, the way we treat our waste is pretty much the same way they did in Egypt 3,000 years ago and in Greece and Rome two millenniums ago. In this world of space exploration and micro-chips, it's an unconscious assumption that all aspects of our lives have been affected by these technological advances. But the truth is, we can't seem to make that jump in our field. Thus, our procedures, although run by computers, are still basically: add air to the water to clean it up, use sand as a filter -- or bury your waste -- or burn it -- much the same way they did in Greece, Turkey, China and Africa, as we discover from archeological excavations. Things haven't changed a lot, but the public thinks they have.

One of the interesting things about the District's Deep Tunnel is that it is new and old technology. It is a way of collecting our combined sewage that hasn't been done before. But it's really an adaptation of the Roman aqueduct system -- bring it in, use it, then send it as far away as possible. No one is really working to develop a system where we retain and re-use all of our wastewater. We're always sending it away somewhere else. And we are only very slowly achieving usable technology for converting our wastes into energy and recovering resources from our every day solid waste.

Because the public really doesn't understand that we are still operating in a basically primitive fashion, it may be that they expect too much of us and it may be that we, in turn, emphasize the small technical advances and become defensive about systems that are really not yet perfected. Maybe we should say more often. "This is the best we can do with the best available technology." I do remember we used that phrase in our Clean Water legislation.

I believe the future of waste management lies in resource recovery: re-use of our used sewage water, beneficial land application of our good sludges, energy and resources recovery from solid wastes.

Perhaps this approach can encourage the thinking public to prioritize. When the public wants us to cover aeration tanks, for example, we explain it's too expensive and not cost-effective. The same explanation is given for resource recovery of solid waste. Yet rarely do others say that the Sergeant York or MX is too expensive. They say, "We need it!" We need to decide what we really want, and I fear that environmental concerns are not presently in the ascendancy. We need to apply space technology to waste management -- even if it's only getting our solids to the point where we can safely recycle them.

A third question, and one that comes up again and again, is. "Does the public think they are part of the decision-making?" And more important still, do we as managers of waste think it's important for the public to be part of the decision-making?

It has been clear in this 200-year old democracy of ours that we can only be effective in this fragile system by arranging and nurturing participation. It doesn't seem to work any other way. But, oh how fraught with danger that is; and oh, how delicately it has to be handled; and oh, how defensive we get! How can they perceive us in a positive way if we don't reach out to include them?

Of course, this is extremely difficult. For example, when we are making decisions about what heavy metals we are going to test for, what organics we're going to test for, or for what particular contaminants we're looking for, we are sometimes restricted by state and federal regulations. But there is no reason in the world why we should not also include in our testing reasonable requests made by the leaders of the environmental community. We must listen!

You remember those 365,000 Sierra Club members, those 4 1/2 million National Wildlife members, those people who belong to the Clean Water Project and Clean Water Federation? They're not our enemies -- THEY is US!

Besides this reaching out early to include input from all interested parties before decisions are made, we should also think about the format of our meetings and what we are trying to accomplish. None of us can tolerate a rubber stamp situation. It is one thing to testify in front of a congressional committee or in front of a Board sitting behind desks with brass nameplates at the Sanitary District; it is another to meet in small groups around a round table with no head or foot -- to sit around a room in a circle rather than a semi-circle -- and to exchange ideas so that each of us -- technical people, operators and ordinary citizens -- like mothers and fathers worried about their children playing around landfill sites -- can begin to understand each other. Dr. Lue-Hing has arranged these meetings with excellent results.

When a woman comes to my agency and says her babies are dying from cancer because of the nearby landfill, and it's the first time I've heard of it, and of course the story is immediately picked up by the TV news, I know we have been remiss in reaching out to the citizens who are worried about our policies.

I don't mean to suggest that by outreach, input, early consultation, then accommodation, that we will forever be the best friends of the citizens who are most concerned about their health and welfare. All we can hope to accomplish is to convince some of the leaders and some of the members, and a lot of the general public that those of us in elective office and in the scientific and technical community care just as much about our children and their welfare as anybody else.

Sometimes, of course, you really can't win. A good example of this is the location of sewage treatment facilities. The public in general is very wary of any pollution control program that they believe is too close to them geographically or encroaches upon their residences or places of business. This wariness need not even be associated with a perceived health or safety concern, but it is the same concern expressed by citizens when any new land use is proposed for a community -- even a golf course!

Although pollution control facilities have an overall public good, local public concerns about the effects on general community life will often outweigh this perceived public good. Let me give you a concrete example of this phenomenon.

In the mid-1960's, the District determined that projected population growth in the northwest suburbs of Chicago would necessitate the eventual construction of a new wastewater treatment facility in the area referred to as the O'Hare Drainage Basin. With this in mind, the District acquired land in the City of Des Plaines, Illinois, a northwest suburb, to provide a site for the new treatment facility. The O'Hare Water Reclamation Plant (WRP), with a proposed capacity of 72 million gallons per day, would treat the combined sewage and stormwater flows of a 52-square mile service area encompassing seven suburbs with a projected population of 400,000 people.

Although the residents and elected officials in these suburbs felt that additional wastewater treatment facilities would be needed to encourage growth and prevent sewer backups during storms, no one volunteered to locate the facility in their backyard.

So beginning in 1966, when word of the District's plan to build a treatment plant in Des Plaines became known, the City of Des Plaines began an 11-year battle with the District in the state and federal courts in an effort to prevent the construction of the proposed O'Hare WRP. In addition, some of the approximately 1,200 city residents living near or along Oakton Street, an avenue within 2,000 feet of the plant boundary, carried on their own protests with the Illinois Pollution Control Board, the Illinois EPA, the USEPA, and the District's Board of Commissioners.

Although many types of arguments were put forward over the years in an attempt to force relocation of the O'Hare facility, by 1975 the only issue that had not been resolved in the District's favor was the contention that there would be adverse health effects for the people living on Oakton Street, resulting from breathing aerosols which might be emitted from the O'Hare WRP aeration tanks.

The USEPA was particularly concerned about this problem, and made it a condition of the construction grant for the O'Hare WRP that this aerosol situation be addressed. The types of possible solutions ranged from planting some tall trees along the boundary of the plant, thus blocking the aerosols, to completely covering all of the aeration tanks and disinfecting the collected air before release to the atmosphere.

In response to the public's concern, the District, in conjunction with the USEPA, conducted a $1.5 million aerosol suppression study. The study consisted of intensive air and wastewater sampling to measure the physical, chemical and biological characteristics of the aerosols emanating from the aeration tanks, and the extent and pattern of their downwind dispersion.

Based upon the results of the District's study and the results of seven other USEPA-funded studies dealing with the health effects of wastewater aerosols, the USEPA prepared a supplemental Environmental Impact Statement (EIS) dealing with the O'Hare WRP, which concluded that aerosol emissions from the plant would not be a health hazard. This EIS was presented to the residents of Des Plaines at a public hearing held in October 1979, and effectively ended the controversy. The O'Hare WRP began actual operations in May 1980.

It is important to realize that the controversy was in the courts for eight years and then finally resolved by providing sufficient information to the public, proving the safety of the modern sewage treatment facility. This is not to say that if we had done a spectacular job involving the residents early in the decision-making, if we had earlier agreed to aerosol studies, or if we had covered the tanks, they would have been on bended knee asking us to build the plant in their backyard. However, I believe we should do everything we can to stay out of court. It does not do us any good to be in court over matters of waste management.

Another example is the Sanitary District's Fulton County Project. Over a 10-year period, we applied millions of gallons of digested sludge to land that had been stripped bare by strip mining. We grew corn and beans and hay. Our famous $100 million project was innovative, creative, and scientifically productive -- over $5 million in research grants to the University of Illinois alone.

To tell the truth, we were invited to Fulton County, but by the County Board, not by Joe Smith or Mary Jones. And since the public was not involved early, they objected strongly and we had a terrible time -- court suits, property damage and lots of heartache. Eventually, we established the Fulton County Steering Committee and used this committee to educate each other and resolve our differences. But it took a long time to relax those confrontational attitudes developed on both sides.

In sum, it is an uneasy relationship between the public and our profession. We can't live without each other and we have trouble living with each other, but neither side is going to go away. I hope there are some environmental leaders, as well as soil scientists, here today because it's a two-way street -- and we all must make efforts to accommodate.

You may have noted that I didn't say a word about hiring a public relations firm or running ads on national TV. What we have been talking about is attitude. I respect and admire what this distinguished group has been doing -- you are on the cutting edge of the needs of our society. It's important that you even thought of considering this question today. Let us Listen and Accommodate -- and try to stay out of court. Why not, if we have nothing to hide.

Onsite and clustered wastewater renovation by soil treatme[nt]: overview

E. M. Rutledge and D. C. Wolf, Department of Agronomy, University of Arkansas, Fayetteville, AR 72701

When household wastewaters are renovated onsite or near dwellings in clustered systems, soil treatment is the method most frequently used. Soil treatment is chosen because it employs natural treatment processes, gives a high level of treatment, often costs less to construct, and is especially inexpensive to maintain. Economic evaluations are expected to consistently favor onsite and clustered soil treatment systems when soil conditions are favorable and population densities are low or total populations are small.

Soil treatment using onsite and clustered systems generally utilizes a septic tank system which consists of (1) a septic tank, (2) a distribution system, and (3) the soil. The septic tank is a settling chamber and anaerobic digester which yields water containing suspended and dissolved wastes. This wastewater is spread into the soil by the distribution system which normally consists of pipe and gravel. Most of the treatment or renovation of the water occurs within the soil.

To prevent degradation of both ground and surface water quality, adequate renovation of wastewater must occur in the soil. The renovation must remove both biological and chemical contaminants. Biological contaminants of primary concern are pathogenic viruses and bacteria. The public health hazard associated with the biological contaminants is related to the types and numbers, their survival, and their transport in the soil. Pathogens are removed by filtration, sorption, sedimentation, and die-off. Pathogens can be transported over relatively long distances before the wastewater has been fully renovated if rapid rates of water movement occur such as through macropores and sandy textured soils.

Renovation of chemical contaminants must include removal of (1) organic compounds which yield TOC and (2) inorganic compounds such as nitrogen and phosphorus. The most frequently cited concern is nitrate contamination of ground and surface waters. Under aerobic conditions, organic nitrogen can be mineralized to ammonium and subsequently oxidized to nitrate by soil microorganisms. Nitrate is mobile in the soil and can be readily transported in the soil water. Biological denitrification can occur and will result in the reduction of nitrate to dinitrogen and/or nitrous oxide gases. Denitrification requires an absence of oxygen, adequate levels of an available carbon source, and nitrate to serve as a terminal electron acceptor for aerobic bacteria capable of anaerobic growth. Indigenous soil microbes also reduce TOC levels by utilizing organic carbon compounds contained in the wastewater. Under aerobic conditions, the end products of metabolism would be carbon dioxide, water, microbial biomass, and humus. Phosphorus removal from wastewater is also important because contamination of surface waters with phosphorus can result in eutrophication. Many soils are well-suited for removal of phosphorus by sorption and precipitation processes which generally result in effective phosphorus removal in relatively short distances.

Soil evaluation for onsite wastewater renovation consists of assessing the soil's ability to renovate and transmit wastewater. Numerous soil evaluation systems are presently used. Most approaches utilize (1) soil morphology or (2) percolation tests or (3) a combination of the two. Most workers recognize the superiority of a soil morphological approach which is increasing in use. A good evaluation should discriminate among numerous designs and lead to the most appropriate design. Improved distribution systems can enhance treatment by (1) spreading the effluent over the entire bed to reduce the loading rate, (2) introducing the effluent higher in the soil to increase the amount of soil available for treatment, and (3) introducing the effluent in smaller doses to increase the residence time in the soil.

Small communities in some states have utilized onsite and clustered soil treatment systems for several years. Other states are considering these systems because of their economics if installation and maintenance and their environmental soundness. When onsite soil treatment systems are enlarged to clustered systems, more extensive designing is required to insure water flow from and air movement into the soil treatment area. Arrangements should also be made for management and routine maintenance of these systems.

Onsite and clustered wastewater renovation by soil treatment will continue to be a major method for treating wastes from individual households and small communities. If care is exercised not to overburden the capability of these systems, they will allow the soil to provide excellent renovation of wastewater.

Treatment by onsite systems

R. B. Reneau, Jr., J. J. Simon, and M. J. Degen, Virginia Polytechnic Institute and State University, Blacksburg, VA 24061

Treatment of effluent by on-site wastewater disposal systems (OSWDS) is receiving more attention as the number of residences not served by public sewage increases and the amount of land suitable for satisfactory disposal of home sewage with conventional OSWDS decreases. Sewage from 20.9 million residences (24.1% of the total in the USA) was treated via OSWDS using septic tank systems or cesspools as of 1980 (Bureau of Census, 1983b). This represents a 26% increase in the number of residences occupied year-round which are served by OSWDS from 1970 to 1980 (Bureau of Census, 1972, 1983b). Assuming that 170 L of wastewater per capita per day (USEPA, 1980) is generated by 8.25×10^7 people (Bureau of Census, 1983a, b), then 14×10^9 L of wastewater are applied via OSWDS to USA soils each day. This figure is expected to increase as the use of OSWDS expands.

In this paper an OSWDS is viewed as a system in which effluent generated on-site flows into the soil and moves laterally or vertically to ground or surface waters. Examples are subsurface trenches with gravity or low pressure distribution (LPD) or mounds.

This paper examines effluent infiltration and flow through soils and the fate of bacteria, viruses, N, and P associated with this effluent. These biological and chemical contaminants are potential pollutants of ground and surface waters.

WATER FLOW CONCEPTS

Proper performance of OSWDS depends on the ability of the soil or a soil material to transmit and purify wastewater. Failure occurs if either of these functions is not performed. Both processes are directly related to soil hydraulic conductivity characteristics, which are largely controlled by soil pore geometry (USEPA, 1977).

Processes Limiting Effluent Infiltration

Transmission of effluent away from an OSWDS can be visualized as two interdependent flow processes: infiltration of effluent across the gravel-soil interface and percolation of the water through the surrounding regolith. The presence of a restrictive horizon, a shallow water table, or surface and ground water intrusion may limit infiltration and effluent surfacing may occur.

Although the permeability and the hydraulic gradient of a soil indicate that OSWDS effluent transmission should be adequate, pore clogging may result in reduced infiltration rates. Physical factors which may contribute to the development of a clogging zone at the infiltrative surface of an OSWDS include compaction, puddling, or smearing of the trench bottom and sidewalls during construction as well as blockage of soil pores by solids filtered from the effluent. Swelling and dispersion of soil colloids resulting in structural degradation can occur when cations with large hydrated radius are exchanged onto colloids (USEPA, 1977). Also effluent infiltration may be limited by formation of a biological mat or crust (Bouma et al., 1972; Jones & Taylor, 1964; Kristensen, 1981; Magdoff et al., 1974a) or by surface smearing (Daniel & Bouma, 1974). The development of a biological mat or crust plays an important role in effluent treatment. With restricted flow through the biological mat or crust, larger more continuous pores may become desaturated, flow becomes more tortuous, travel times lengthen, and treatment is enhanced.

Mat development is attributed to suspended solids and exudates of microbial organisms and may be accelerated as a result of excessive effluent loads to OSWDS (Winneberger et al., 1960; Bouma et al., 1972; Jones & Taylor, 1964; Mitchell & Nevo, 1964). Kristiansen (1981) concluded that although C:N ratios indicated large quantities of polysaccharides were not likely to be present, they may bind bacterial cells together, thus increasing resistance to flow. Anaerobic conditions in a portion of the OSWDS, inherent in the trickling nature of gravity distribution of effluent, also perpetuate and accelerate mat formation (Avnimelech & Nevo, 1964). Mat formation may eventually seal or partially seal the entire gravel-soil interface and may result in surfacing of effluent.

Effluent Flow Away from OSWDS

A simplified view of the hydraulic effects of crust formation is that a flow restrictive layer overlies a more permeable layer. When infiltration approaches steady state (a reasonable approximation except during major rainfall or water table fluctuation events) a matric potential (Ψ_m) will develop below the crust in the more permeable material with a gradient approaching unity. However, even with effluent ponding in the OSWDS due to flow restriction, Ψ_m were wetter than -9.8 KPa below the crust (Bouma et al., 1972; Simon & Reneau, 1985). Effluent moves primarily in response to gravitational gradients (Hillel and Gardner, 1969) even though soils are not saturated. The infiltration rate through the saturated crust should equal the flux through the underlying unsaturated material (Hillel & Gardner, 1969; Magdoff & Bouma, 1974). Infiltration through the crust is a function of the crust resistance (R_c), the subcrustal matric potential (Ψ_{sc}), the crust thickness (Z_c) and the height that effluent is ponded above the trench bottom (H_o). The flux density (q) for one dimensional flow is

$$q=R_c^{-1}\ (H_o + \Psi_{sc} + Z_c). \qquad [1]$$

Equation [1] allows evaluation of different crust resistances when q, H_o, and Ψ_{sc} are known if Z_c is assumed to be negligible relative to values of Ho and Ψ_{sc} (Magdoff & Bouma, 1974). It also illustrates that if a perched or free water table approaches the infiltrative surface, q may decrease significantly, resulting in OSWDS failure.

Transmission of effluent away from OSWDS may be limited by slow permeability in receiving or underlying soil layers, shallow depth to ground water or perched water tables, or intrusion of surface water and/or through-flow from upslope. In coarse-textured coastal plain soils with shallow water tables and surface or tile drains, water table gradients were horizontal away from OSWDS with limited vertical effluent movement (Reneau, 1978; Stewart & Reneau, 1981a, 1981b, 1984). Effluent transmission was also largely horizontal in soils with plinthite and fragipans when OSWDS were installed above these layers (Reneau & Pettry, 1975). A rapid increase in effluent ponding levels in OSWDS trenches following major precipitation events was reported by Simon & Reneau (1985) and Hargett (1985).

Treatment of effluent can be enhanced by maximizing the distance between an underlying water table or restrictive layer and an OSWDS. Mounds have been utilized for this purpose with some success, but in some cases subsurface flow is not adequate to transmit hydraulic loads (Converse & Tyler, 1985). Transmission is enhanced if OSWDS are long and narrow to minimize water mounding below the OSWDS (Converse & Tyler, 1985; Pask et al., 1985). Stewart & Reneau (1984) reported that long, low basal area mounds placed in a Typic Ochraquult (clayey, mixed, thermic) functioned adequately. Tile drainage has been effective in water table drawdown in permeable, poorly drained soils with shallow seasonally fluctuating water tables (Wilson et al., 1982; Reneau, 1978), but large gradients may result in ground and surface water contamination if effluent distribution is poor.

The LPD system has been successfully used to improve effluent distribution. A full-scale LPD system was installed in the surface horizon of a Typic Ochraquult (fine-loamy, mixed, thermic) soil which had insufficient outfall for tile drainage (Stewart & Reneau, 1984). The site was landscaped to enhance runoff. No effluent surfacing was noted even though the water table rose to within 30 cm of the surface during high water table periods. Hargett (1985) reported that a survey of 12 shallow placed OSWDS with LPD indicated that systems ponded in response to wet soil conditions during months when rainfall exceeded evapotranspiration and shallow water tables were near the surface. Effluent surfacing during pumping was also noted when storage volume in the gravel was inadequate. This study illustrates that proper design of alternative OSWDS is required in marginal soils.

Simon & Reneau (1985) applied effluent at flux densities of 0.4, 1.8, and 3.6 cm day^{-1} (trench bottom area basis) to a Typic Paleudult (clayey, mixed, mesic) with slow subsoil permeability. Increased ponding depths in the OSWDS in response to major rainfall events indicated that the weakly structured horizon underlying the trench restricts downward flow. Results from this study and similar studies on two other soils (unpublished data) indicate that soils with moderately to strongly structured clay horizons will dissipate effluent applied at uniform flux densities as high as 1.8 cm day^{-1}. Biological clogging with continuous ponding occurred at the 3.6 cm day^{-1} flux density.

BIOLOGICAL CONTAMINANTS

Most infectious agents contained in human wastes are potentially waterborne. Cabelli (1978) indicated that waterborne disease outbreaks from non-community water systems were more than double those from community systems and that fecal wastes from warm blooded animals, especially man, are the major source of contamination of recreational waters. Craun (1981) attributed 41% of the outbreaks of waterborne diseases in the USA from 1971-1978 to overflow or seepage of sewage from septic tanks and cesspools.

Pathogenic viruses and bacteria present in OSWDS are the primary biological contaminants of concern. The number of viruses entering an OSWDS depends on the presence of an infected person in the household. Virus levels up to 10^7 to 10^8 infectious units per gram of feces have been reported (Safferman, 1982). In raw sanitary sewage wastewater, 5 to 21,000 plaque-forming units (PFU) $(100\ ml)^{-1}$ have been isolated with an estimated average of 200 to 7,000 PFU $(100\ ml)^{-1}$ (Clark et al., 1964). The number of fecal coliform bacteria entering a drainfield from a septic tank is estimated at 4.2×10^5 $(100\ ml)^{-1}$ (Ziebell et al., 1974). Fecal rather than total coliforms are the preferred indicator for bacterial contamination in soils (Clark & Kabler, 1964).

Processes Affecting Biological Contaminants in OSWDS

Biological contaminants are removed by the filtering action of the soil, adsorption to soil constituents, sedimentation, and natural die-off (Gerba et al., 1975). Unsaturated flow with effluent movement through smaller pores increases the efficiency of both bacterial and viral removal due to slower average pore water velocities and increased surface contact per net distance traveled through the soil system. Conditions which contribute to unsaturated flow include good effluent distribution, development of a surface clogging mat, well drained soils, and moisture deficits (Bouma et al., 1972). Caldwell (1937, 1938a, b) and Caldwell & Parr (1937) apparently demonstrated the effectiveness of mat formation in the late 1930's. They observed that B. coli were initially present at horizontal distances up to 24.4 m but within 7 months the detection distance had retreated to the edge of the study latrine. This reduction in bacteria was attributed to "soil defense" which may correspond to the modern day term "biological clogging". The effectiveness of unsaturated flow and good effluent distribution for removal of bacteria was demonstrated by Stewart & Reneau (1984). They reported no movement beyond 1.5 m vertically or horizontally from a shallow placed OSWDS in a soil with a seasonally high water table. While some filtration of pathogens by surface mats may be anticipated, increased contact time caused by unsaturated flow conditions are more important (Bouma et al., 1972) in limiting organism transport.

The physical straining that is considered causative for bacteria removal is of minimal importance to the removal of viruses (Bitton, 1980), due to the smaller size of the viruses, 18-25 nm as opposed to 750 nm for bacteria. Viral adsorption to soil constituents is the primary immobilization mechanism. Mutual repulsion by the net negative charge present on viruses and most soil colloids is probably overcome by the presence of cations in the soil solution (Carlson et al., 1968). Carlson et al. (1968) measured the effect of varying clay content and type, ion species, and ionic strength on the viral adsorption process of colloids suspended in river waters. Repulsive forces between the negatively charged viruses and colloids were overcome by cation bridging. With increased concentrations of cations, a corresponding increased removal of viruses was observed. They noted that divalent salts were more effective in promoting adsorption than were monovalent salts. Also Al concentrations (pH 6.0) as low as 10^{-6} resulted in 98% virus removal. Ninety percent of the inactivation occurred within <5 min after the solutions were mixed, indicating rapid adsorption. In soils the surface cation exchange capacity (CEC), particle size and shape, and mode of ismorphous substitution are of most importance.

Unsaturated conditions in sand columns were more efficient for virus inactivation than saturated conditions (Lance et al., 1976; Lance & Gerba, 1980, 1984). Macropore flow and higher pore water velocities associated with saturated flow are anticipated to reduce adsorption efficiency as compared to unsaturated conditions. Adsorption may be reduced or desorption may occur as a result of competition between viruses and organic materials (Carlson et al., 1968). Desorption of viruses has been reported or discussed when solutions of much lower ionic strength (rainfall or distilled water) were flushed through a soil or laboratory system (Bitton, 1975; Duboise et al., 1976; Hurst et al., 1980; Wellings, 1980). In most soil systems, such desorption would not be anticipated due to depth of OSWDS placement unless ground waters of low salinity could intrude.

Pathogenic bacteria and viruses in soil are generally reduced to minimal numbers within 2 to 3 months, although enteric bacteria have been known to survive up to 5 yr in some extreme environments (Romero, 1970; Bitton, 1975; Gerba et al., 1975). Laboratory studies indicate that moist soil conditions and low temperatures favor bacterial and viral survival in soils (Gerba et al., 1975; Hurst et al., 1980). Although dry soil conditions of <10% moisture have been reported to result in rapid viral population decline (Bitton, 1975; Bitton et al., 1984), these conditions are atypical for most OSWDS, as soils are usually near saturation. The presence of actinomycetes result in depression of bacterial survival times (Bryanaskaga, 1966 as reported in Gerba et al., 1975) probably due to the release of antibiotics. Virus survival has also been reported to decrease in the presence of unspecified aerobic organisms (Romero, 1970; Hurst et al., 1980). Acidic conditions have a deleterious effect on enteric bacterial survival (Gerba et al., 1975) but may lengthen viral survival due to increased soil adsorption. The higher concentration of Al^{3+} at lower pH values probably promotes adsorption (Hurst et al., 1980) thus offering some protection for the virus (Gerba et al., 1975; Bitton, 1975).

Transport Through Soil

Transport of bacteria and viruses through soils is primarily dependent on their removal by straining and adsorption, respectively, until death of the organism due to environmental conditions can occur in the OSWDS. Bacteria and viruses are transported through soils in the direction of groundwater flow (Romero, 1970). He concluded from several privy and land application studies of pathogen movement that a travel distance of 15 to 30 m should be sufficient for removal of biological contaminants under saturated conditions, with most treatment occurring in <3 m under unsaturated conditions.

A limited number of studies concerned with viral transport away from OSWDS have been conducted. Romero (1970) notes that although many reports of viral travel exceeding 15 m exist, controlled experiments indicate that most viruses die within 3 m of the source. Viral deactivation over short travel distances in column studies (Lance et al., 1976; Lance & Gerba, 1984) support Romero's conclusions. Under unsaturated flow conditions they reported removal within 40 cm of the point of application. Brown et al. (1979) examined coliphage and fecal coliform movement away from OSWDS lines placed in undisturbed sandy loam, sandy clay, and clay subsoils by collecting samples via ceramic cup lysimeters. In all three materials no coliphage movement was reported beyond 120 cm vertically or 60 cm horizontally except in very low concentrations immediately after the effluent was spiked with coliphages. Coliform movement to the 120 cm depth was observed only on a few occasions. Hagedorn et al. (1981) suggested that results obtained with porous cups be viewed with caution.

Numerous field studies using fecal coliforms as indicators of movement of biological contaminants have been reported. Reneau et al. (1975) demonstrated the potential for ground water and surface water contamination from drainfields placed in unsuitable soils. A watershed in the Coastal Plain/Piedmont of Virginia, dominated by Typic Fragiudults, was examined. Only 17% of the OSWDS were in suitable soils, while 41% were in unsuitable soils. Both surface and subsurface water samples indicated a potential health hazard from contaminated runoff and throughflow.

Macropore flow through saturated strongly structured soils or soils of the sandy textural family may result in pathogen travel over relatively long distances with minimal treatment. Romero (1970) cites a number of pit privy studies where the pits intersected or were within close proximity to the water table. Elevated bacterial levels were temporarily detected up to 24.4 m horizontally from the source. Reneau & Pettry (1975) investigated the movement of bacteria in soils considered to be marginally suited for OSWDS disposal due to restricting soil horizons and/or seasonally perched water tables. Vertical movement of the bacteria through a fragipan was limited. Horizontal movement of effluent above the fragipan resulted in significant removals of the bacteria but only after effluent had travelled horizontally a minimal distance of 6.1 to 12 m. The fecal coliform counts in water samples collected at 12 m were only slightly lower than in samples collected at 6.1 m. A possible explanation is that samples collected at 12 m were in a discharge area below the drainfield and may have more effectively sampled the contribution of macropore flow to the fecal coliform count. Hagedorn et al. (1978) isolated bacteria 16.3 m from an inoculated pit. Flushes of bacteria coincided with rainfall events and a water table rise to within 15 cm of the surface, aided in the rapid transport of the bacteria under saturated flow conditions. Studies with resistance-labeled <u>Escherichia coli</u> bacteria transport in a hillslope soil toposequence (Rahe et al., 1978; McCoy & Hagedorn, 1980) indicated that pathogens may move extended distances via "pipe" or channel flow after rainfall events. Convergence of stream lines below a concave slope resulted in surfacing of water and bacteria after precipitation events. Pipe or macropore flow in the saturated zone upslope from the point of surface seepage was through root channels and animal tunnels, resulting in natural short-circuiting of the soil treatment process. In many instances where extended movement of biological contaminants is observed, transport may have occurred before formation of the clogging zone. Reneau (1979a) studied fecal coliform movement from three OSWDS to a tile drain in a Typic and an Aeric Ochraquult. Even though bacteria decreased in a predictable manner, soil treatment effectiveness was reduced during high water table periods as a result of enhanced movement via macropore flow. A limited number of organisms moved from the OSWDS to a tile drain (10.4 m) in the Typic Ochraquult, resulting in contamination of a receiving stream. Extended movement of fecal coliforms at this site was attributed to poor effluent distribution and a strong hydraulic gradient toward the tile drain. In the Aeric Ochraquult fecal coliforms were not transported to the tile drain.

Under unsaturated flow conditions, bacteria can be adequately removed within 0.9 to 1.2 m of effluent flow through soils (U.S.EPA, 1980; Hansel & Machmeier, 1980). Hagedorn et al. (1981) reviewed a report by Bouma et al. (1972) which examined 19 subsurface soil disposal systems. Fecal coliforms were reduced to background levels within 61 cm of the trench bottom. Even in a sandy soil Ziebell et al. (1974) reported a 3000 fold reduction in bacteria levels 38 cm below the trench bottom and 30 cm laterally. Stewart & Reneau (1984) installed a shallow-placed, LPD system to increase the unsaturated zone in a Typic Ochraquult. After 2 yr, fecal coliforms had been detected in only 5% of the 150 samples collected from shallow wells (150 cm deep). Samples that contained fecal coliforms were restricted to periods of high water tables and were confined to the effluent distribution area.

Converse & Tyler (1985) summarized data from 40 mound-type OSWDS installed and monitored in Wisconsin. Mounds were dosed with effluent, resulting in good effluent distribution and unsaturated flow through the sand layer. Although many of the mounds leaked at the toe at least seasonally, low fecal coliform counts in the seepage waters indicated excellent bacterial removal. Stewart & Reneau (1984) installed long, narrow mounds above a slowly permeable Typic Ochraquult (clayey, kaolinitic, thermic), fecal coliform analysis indicated no movement of bacteria out of the base of the mound.

In summarizing 10 yr of research regarding the spatial and temporal variation of fecal coliform movement surrounding OSWDS, Stewart & Reneau (1981a) concluded that placement of conventional OSWDS in soils with seasonally shallow water tables may result in contamination of surface and ground waters due to apparent poor effluent distribution. Alternative OSWDS such as the mounds (Converse & Tyler,1985; Stewart & Reneau, 1984) and shallow placed systems with LPD (Stewart & Reneau, 1981) appear to offer solutions to problems in certain high water table soils because of uniform distribution throughout the OSWDS and creation of a zone of unsaturated flow before effluent flow intersects the water table.

NITROGEN

Nitrogenous components introduced into OSWDS can potentially contaminate ground and surface waters. Nitrate contamination of these waters increases the risks of methemoglobinemia in infants who ingest water containing excessive concentrations of NO_3^- and NO_2^-. Nitrogen also results in accelerated eutrophication of surface waters with subsequent algal blooms and O_2 depletion. When considering the possible contamination of ground and surface waters via OSWDS, N may pose the greatest potential for contamination of these water resources.

Septic tank effluent averages 40 - 80 mg N L^{-1} and is approximately 25% organic N and 75% soluble NH_4^+ (Walker et al., 1973a; Otis et al., 1974). With aerobic treatment, much of the N is present as NO_3^-. For a family of four approximately 14.9 kg N yr^{-1} is applied to soils in an OSWDS.

Processes Affecting Nitrogen in OSWDS

Processes affecting the fate of N in soils include NH_4^+ adsorption, nitrification, denitrification, mineralization, immobilization, NH_3 volatilization, chemical decomposition, and plant uptake (Keeney, 1981). Ammonium adsorption, nitrification, and denitrification are the dominant processes in OSWDS and will be discussed in more detail.

Ammonium adsorption onto cation exchange sites is the dominant process where nitrification is limited by soil environmental conditions. The quantity of NH_4^+ occupying exchange sites depends primarily on the CEC of the soil, the affinity of the exchange sites for NH_4^+, and the activity of NH_4^+ and competing ions in the soil solution.

In soils surrounding OSWDS where NH_4^+ fluxes exceed removal, an equilibrium is reached between adsorbed NH_4^+ and soil solution NH_4^+. Leaching of NH_4^+ to ground and surface waters may then occur. The velocity of the NH_4^+ front moving through the soil varies between soils as illustrated by Brown et al. (1984) who reported average vertical peak velocities of 25 cm yr^{-1} for both sandy clay and clay loam soils and 100 cm yr^{-1} for a sandy loam soil.

Nitrification is the biological formation of NO_2^- and/or NO_3^- from reduced N by obligate chemolithotropic bacteria. In the natural soil environment mineralization of organic N to NH_4^+ is normally the rate limiting step in nitrification. In OSWDS where N in the effluent is primarily present as NH_4^+, O_2 diffusion to the treatment zone is potentially the most rate limiting factor. Optimally operating OSWDS are conducive to nitrification as O_2 diffusion rates are much higher when unsaturated flow conditions prevail. As Sikora & Corey (1976) indicate, coarser textured soils and strongly aggregated soils should favor nitrification, while poorly aggregated, fine-textured soils may limit nitrification. Nitrification rate reduction at low temperatures and pH values have been reported in natural soils, but evidence indicates that the microbial population acclimates to stressful conditions (Martin & Focht, 1977). Temperature effects in cooler regions of the continental USA should be offset in part by the depth of OSWDS placement which is typically >60 cm and often >100 cm and the temperature of the effluent.

Nitrification has been reported below and adjacent to OSWDS located in sandy soils even though effluent was ponded above a crusted layer (Walker et al., 1973a; Whelan & Barrow, 1984a). They concluded that adequate distance to a water table has to exist before conditions are sufficiently unsaturated to allow nitrification. Nitrification has been reported to occur within 0.3 to 0.6 m of the OSWDS (Preul & Schroepfer, 1968; Whelan & Barrow, 1984a). Walker et al. (1973a) reported nitrification in the subcrust portion (<10 cm) of the trench bottom. Eh and N distribution data (Simon & Reneau, unpublished data) indicate that nitrification is not limited in clayey soils prior to ponding. As effluent ponds, conditions become anoxic below the trench, reducing by at least 1/3 the area available for O_2 exchange and nitrification. This observation is supported by the prediction that nitrification would be limited below OSWDS in fine textured soils (Sikora & Corey, 1976). However, in OSWDS that were not ponded, higher Eh values and predominance of NO_3^- indicate active nitrification (Simon & Reneau, 1985).

Denitrification occurs when facultative anaerobic bacteria use N oxides (e.g., NO_3^-) as terminal electron acceptors in the absence of O_2 and in the presence of an adequate energy source (Firestone, 1982). Denitrification is the only significant biological leak in the system and is thus the primary mechanism for reducing the concentration of N from OSWDS. Firestone (1982) noted that denitrification may occur not only when the soil is saturated, but the presence of anaerobic microsites also results in some denitrification in unsaturated soils. The primary factors influencing denitrification rate include availability of a soluble C source, O_2 concentration, and NO_3^- concentration.

In optimally functioning OSWDS located in well-drained soils, minimal denitrification would be expected and then only in anaerobic microsites (Bouma, 1979). Denitrification may be significant in soils with restricted drainage if conditions are also adequate for nitrification of NH_4^+ before it reaches an anaerobic zone (Bouma, 1975; Otis & Boyle, 1976). Availability of an adequate C source has been cited as the most limiting factor for promoting denitrification (Sikora & Keeney, 1974) especially for OSWDS placed in lower soil horizons. Stewart et al. (1979) conducted laboratory column studies using a mixture of a histic epipedon and sand to evaluate the use of soil organic matter as an energy source. Their results indicate that residual soil organic matter is probably not a satisfactory long-term energy source for denitrification. Reneau (1977, 1979a) reported that in soils with seasonally fluctuating water tables, NO_3^- concentrations decreased rapidly with lateral distance from the disposal site. Also, samples collected in a tile-drained Typic Ochraquult showed an average decrease in NO_3^--N1/Cl^- ratios from 0.128 at 4.56 m to 0.076 at 10.4 m (Reneau, 1979a). In a subsequent study Stewart & Reneau (1984) utilized a shallow-placed, LPD system in a Typic Ochraquult and noted that the NO_3^--N1/Cl ratios average 0.70 in the drainfield and 0.015 at 8.4 m laterally from the drainfield. In these OSWDS, NO_3^- that accumulated during low water table periods was transported upward through the soil profile with the rising water table. Apparently denitrification occurred as the water table approached the surface horizon. The energy source may be a combination of soil organic matter and fresh C sources supplied by the grass cover growing over the shallow placed OSWDS. Cogger & Carlile (1984) postulated that a similar decrease in NO_3^- in wet soils resulted from denitrification.

Denitrification has been reported in mound systems where nitrification occurs in the sand fill followed by denitrification in the underlying soil. Magdoff et al. (1974) reported 32% denitrification in laboratory columns that simulated mounds. Harkin et al. (1979) reported that in mounds with a high dosing rate and a low fill uniformity coefficient, denitrification rates ranged from 48 to 86%. Similar mounds with low dosing had an 86% removal rate.

Studies have demonstrated that introduction of methanol into an OSWDS after nitrification as a C source can result in large quantities of NO_3^- being removed (Sikora & Kenney, 1974; Sikora et al., 1977). Also Sikora & Keeney (1976) demonstrated a S-*Thiobacillus denitrificans* NO_3^- removal system. Laak (1981) reported enhanced denitrification after modification of a conventional OSWDS.

Transport Through Soil

Once NO_3^- is below the OSWDS in well-drained soils limited reduction in NO_3^- concentration would be expected and NO_3^- would be transported via convection and dispersion to surface or ground water. Nitrate contamination of ground waters from OSWDS was assumed by Quam et al. (1974) and Miller (1975) based on NO_3^- concentrations in drainage and well waters. Quam et al. (1974) reported that from 30 280 to 37 850 m^3 day^{-1} of effluent is introduced into OSWDS in a 78 km^2 area in East Portland, Oregon. They reported NO_3^--N levels from 5 to 12 mg L^{-1} in shallow ground waters which eventually reached a surface drain from this area. Nitrate levels in deeper aquifers and upgradient shallow ground waters were <1 mg L^{-1}. Miller (1975) reported increased NO_3^--N concentrations in Delaware Coastal Plain ground waters where OSWDS and home water wells were located on the same site. In well-drained soils with a water table at 4.5 to 7.5 m and a population density of 1029 people km^{-2}. Samples collected ranged from 5 to 30 mg NO_3^--N L^{-1}.

Walker et al. (1973b) reported NO_3^--N concentrations as high as 40 mg L^{-1} in the upper 30 cm of the aquifer adjacent to one of their study systems. Wheland & Barrow (1984a) observed NO_3^--N concentrations in soil solution as high as 224 mg L^{-1} at a depth of 5.5 m for a black water soak well in a Karrakatta sand (Inceptisol) of the Swan Coastal Plain in Australia. These data indicate that in well drained soils when nitrification occurs immediately below the trench bottom, the most probable mechanism for reducing the N concentration in ground waters is by dilution. Walker et al. (1973b) estimated that a relatively large area of 0.2 ha down gradient was needed for NO_3^--N concentrations in the top layer of the ground water to be diluted to <10 mg L^{-1}.

PHOSPHORUS

The primary environmental concern with addition of P to soil in OSWDS is the potential for eutrophication of surface waters if adequate soil retention of P does not occur. In many surface waters P is the limiting nutrient for algae and aquatic weeds. Fortunately, P is relatively immobile in most soils.

Groundwaters normally have a low concentration of P. Values range from 0.005 to 0.1 mg P L^{-1} with typical concentrations of <0.05 mg P L^{-1} (Enfield & Bledsoe, 1975; Tofflemire et al., 1973). Reneau & Pettry (1976) and Reneau (1979a) observed that in soils with shallow ground waters not influenced by OSWDS, P concentrations were <0.05 mg P L^{-1}. These studies were conducted in acid soils, and fractionation of P indicated that iron phosphate was probably controlling P levels in solution. The relatively low concentrations of P in groundwaters (10^{-6} to 10^{-5} mol L^{-1}) indicate the strong interactions of soluble P with certain soil constituents.

The P concentration in OSWDS effluent varies widely. Bicki et al. (1984) reported values ranging from 11 to 31 mg L^{-1} with a mean of 16 mg P L^{-1}. The principle sources of P in OSWDS effluent are laundry detergents and human excreta (Sikora & Corey, 1976). Most P in OSWDS effluent (85%) is present in the orthophoshate forms (Magdoff et al., 1974; Reneau & Pettry, 1976). For a family of four, an average of approximately 4.0 kg P yr^{-1} are applied to soils.

Processes Affecting Phosphorus in OSWDS

Phosphorus may be removed in OSWDS by adsorption or precipitation (sorption) processes [physical adsorption, chemisorption, anion exchange, surface precipitation, and precipitation as separate solid phases (Sample et al., 1980)], plant uptake, and biological immobilization. Since carbon-limited OSWDS effluent is normally introduced into subsurface soil horizons below most of the rooting zone, biological immobilization and plant uptake will not be considered further.

Although several soil properties influence the sorption of P in the natural soil system, mineralogy and texture are two of the most important. Phosphorus is adsorbed by amorphous and crystalline oxides (very effectively), by the edge face of the phyllosilacates, and to CA^{2+} on calcite. Finer textured soils generally sorb more P than do coarser textured soils. The total quantity of P sorbed also depends on the exchangeable Al content, pH, ions present in solution, and reaction time.

Phosphorus added to soils undergo at least two distinct P sorption stages. A rapid initial reaction that may be completed within 24 h is followed by a slower reaction that may continue for several wk or even months (Munns & Fox, 1976). Initially or in solutions with low P concentrations (mmol L^{-1} range) nonspecific adsorption and ligand exchange, often described using the Langmuir adsorption isotherm, are believed to dominate. The second slower reaction or in solutions with higher P concentrations (mol L^{-1} range), precipitation is considered the primary process.

Precipitation depends on factors such as pH; concentrations of P, Fe, Al, and Ca; competing anions; and reaction time. Slightly soluble P compounds with Fe and Al form in acid soils, while in calcareous soils various compounds containing Ca and P form. Lindsay & Moreno (1960) have illustrated the relationships between pH, ion concentration and dominant P species with P solubility diagrams. Magdoff et al. (1974) and Lance (1977) showed that P removal from effluent by calcareous sand columns was proportional to the loading rate. Magdoff et al. (1974) used solubility expressions to conclude that octocalcium phosphate was probably forming in a calcareous sandy loam used in their column studies. Formation of octocalcium phosphate may be followed by the slow conversion to more insoluble Ca phosphates. In acid soils used for OSWDS or sewage water renovation most P sorption is associated with Al. Reneau & Pettry (1976) studied an OSWDS that had been functioning for 15 yr in an initially acid Plinthic Paleudult. Phosphorus fractionation of the argillic horizon (maximum P accumulation) at 0.15 m horizontal distance from the OSWDS compared to the argillic horizon in a control profile indicated that the largest increases in P after long-term exposure to OSWDS effluent were the Al-P and Fe-P fractions. These fractions increased 106- and 7.2-fold, respectively. Similar changes (Reneau, 1978) were observed in an Aeric and a Typic Ochraquult. Beek et al. (1977) showed that the distribution patterns for the different P fractions were virtually the same for sandy soils that had been flooded intermittently for 30 and 50 yr with sewage water. Since Al-combined P prevailed, it was concluded that P retention was primarily governed by reaction with Al. Staunes (1984) reported that the only parameter that significantly influenced P - sorption capacity in his study was dithionite-citrate extractable Al. These data indicate that under OSWDS, including installations in marginal soils with respect to fluctuating water tables, P retention will probably be controlled primarily by active Al and Fe present and that Al may contribute more to P fixation than Fe in acid soils used for OSWDS.

Even though many investigators treat P retention by soil as either a precipitation or adsorption reaction, Sample et al. (1980) have suggested viewing P retention as a continuum embodying precipitation, chemisorption, and adsorption. This outlook is particularly appropriate for OSWDS that are designed to function effectively for several decades after installation. Discussion in the literature with respect to the mechanism of P retention in soils is much too extensive to review here, however, Sample et al. (1980) have recently reviewed the modern literature with respect to P retention by soil. Beek & van Riemsdijk (1979) have reviewed orthophosphate interactions with soil.

Phosphorus sorption studies have been used to estimate P retention capacity of soils below and adjacent to OSWDS systems. In these studies P sorption is often estimated by the Langmuir adsorption isotherm or other P equilibrium methods. These data may also be useful in applying kinetics to the system. Most studies indicate that P sorption maximums underestimate soil P sorption. Whelan & Barrow (1984b) reported a strong correlation between the P sorption characteristics determined on reference soil samples and the amount of inorganic P present around an OSWDS in the matching part of the profile. However, the quantity of P accumulated in the soil tended to be higher than that predicted from equilibrium isotherms. Wastewater P moved almost 40-fold less than predicted in a sandy soil surrounding a leaching pit after 12 yr of use

(Sawhney & Hill, 1975). Horizontal P movement was 135 cm compared to the 5 m predicted by laboratory sorption experiments. The P sorption capacity in soil columns is also higher when treated with wastewater compared to pure solutions (van Riemsdijk et al., 1979). This increased P sorption was attributed to formation of a stable P compound with Al, P, and other cations, possibly Ca, in the wastewater.

The underestimation of P immobilization in soil from P sorption studies may result from the slower long-term reaction or regeneration of P adsorption sites or both. Ellis & Erickson (1969), Kao & Blanchar (1973), and Sawhney & Hill (1975) all observed regeneration of P sorption sites. Apparently regeneration occurs as Al, Fe, or Ca and fresh mineral surfaces come into equilibrium with the soil solution. Ellis & Erickson (1969) noted that the B horizons of the soils used in their study recovered 100% of their P adsorption capacity within 3 months. Sawhney & Hill (1975) showed that soils surrounding 12-yr-old drainfields were not completely saturated with P at the 15 to 30 cm distance and retained a portion of their P sorption capacity.

In contrast, Reneau (1978) observed that P sorption maximums underestimated P movement under soil conditions that promote macropore flow. He reported that the total P which had accumulated in argillic horizons adjacent to a subsurface absorption system (0.15 m) and in the direction of flow after 12 yr in a Typic and an Aeric Ochraquult was <50% of that predicted from a Langmuir adsorption maximum. Macropore flow, enhanced by poor effluent distribution, a large hydraulic gradient, and competition between F and P, probably resulted in extensive P movement. The water supply at this site contained 3.6 mg F^{-} L^{-1}.

Transport Through Soil

Even though P can move downward 50 to 100 cm yr^{-1} in clean silica sand, the limited movement of P away from OSWDS through most soils is well documented. Dudley & Stepheson (1973) (as cited in Jones & Lee, 1977) showed P enrichment of groundwater in a few coarse sands, but effective P removal in soils primarily composed of medium and fine sand. Extremely low P concentrations (99% reduction) were measured in wells located 4.6 m down gradient from OSWDS studied. Reneau & Pettry (1976) conducted a study of septic tank effluent renovation in a Plinthic Paleudult and an Aquic Paleudult. In both soils, very slowly permeable horizons resulted in perched water tables and saturated horizontal flow. After 15 yr of operation an 80% reduction in P had occurred in the Plinthic Paleudult within the first 0.15 m and >90% at 6.1 m. In the younger OSWDS (Aquic Paleudult), 92% of the P was removed within 0.15 m. Jones & Lee (1977) found no P above background in monitoring wells placed 15 m down gradient from a subsurface absorption field. Soils in their study were comprised primarily of fine to medium sand overlying a medium to coarse sand with relatively low P sorption capacity. These data indicate that P sorption may also be very effective under saturated flow conditions.

Cogger & Carlile (1984) studied 15 conventional and alternative OSWDS under varying moisture regimes and a wide range in textural groups. In seven of eight seldom or seasonally saturated systems, P was reduced by

90% within 1.5 m horizontally from the OSWDS. Phosphorus had been reduced by >95% within 7.6 m horizontally of the drainfield for all systems regardless of saturation status (seldom, seasonal, or continuous). Uebler (1984) reported that for a Cecil soil (clayey, kaolinitic, thermic Typic Hapludult) soluble P concentrations were unaffected by rate of effluent application (7.5, 11.3, and 15.0 L $m^{-2}day^{-1}$) and that soluble P was reduced by 95% at a depth of 30 cm below the drainfield trenches. Lime and cement amendments in the trench did not affect the P concentrations at 30 cm.

Reneau (1979a) studied the movement of P from conventional drainfields toward a tile drain in a Typic Ochraquult (fine-loamy, mixed, thermic) and in an Aeric Ochraquult (coarse-loamy, mixed, thermic). The tile drain had been installed approximately 12 yr prior to the initiation of the study. In the Aeric Ochraquult P had been reduced by 90% within 4.5 m of the absorption field. In the Typic Ochraquult only an 84% reduction had been achieved in the first 4.5 m of travel. Equations developed for P movement at these sites indicated that approximately 8 m were required for 99% reduction in P in the Aeric Ochraquult and approximately 30 m for the same reduction in a Typic Ochraquult. The extended movement of P in the Typic Ochraquult is attributed to a large hydraulic gradient, poor effluent distribution throughout the OSWDS, and possible competition between F and P for sorption. In a subsequent study in another Typic Ochraquult where a shallow-placed, LPD OSWDS was used to maximize unsaturated flow (Stewart & Reneau, 1984), P concentrations were reduced to background levels (0.01 mg L^{-1}) within 1.5 m (both vertically and horizontally) of the drainfield.

Most field studies indicate that P contamination is limited to shallow groundwaters adjacent to OSWDS and that P sorption continues under saturated conditions. More extensive P transport is observed for coarser textured soils that are low in hydrous oxides or in situations where there is both poor effluent distribution and rapid flow away from the OSWDS.

SUMMARY

Optimization of transmission and treatment of OSWDS effluent through soil or soil material will become more important as increasing numbers of OSWDS are installed, particularly in marginally suited soils. Most OSWDS in the USA include a subsurface absorption system with perforated lines and gravity distribution. Research indicates that in both fine textured soils and permeable soils with shallow groundwaters, transmission and/or treatment are enhanced by use of shallow placed trenches or mounds and LPD. The improved effluent distribution allows primarily unsaturated flow below OSWDS.

When flow is unsaturated, or travel distances through the soil are adequate, pathogen removal and die-off should be adequate. Likewise P removal was adequate in most OSWDS studied. Nitrogen remains the component with the greatest pollution potential due to leaching of NO_3^- and in some situations NH_4^+. The potential for denitrification adjacent to OSWDS is still largely undetermined. However, the apparent large amounts of denitrification observed in several studies suggests the need for additional research in this area.

REFERENCES

Avnimelech, Y. & Z. Nevo, 1964. Biological clogging of sands. Soil Sci. 98:222-226.

Beek, J., F. A. M. de Hann, & W. H. van Reimsdijk, 1977. Phosphates in soils treated with sewage waters: II. Fractionation of accumulated phosphates. J. Environ. Qual. 6:7-12.

Beek, J. & W. H. van Riemsdijk, 1979. Interaction of orthophosphate ions with soil. In Bolt, G. H. (ed.) Soil Chemistry B. Physico-chemical models. Elsevier Scientific Publishing Company, New York. pp. 259-284

Bici, T. J., R. B. Brown, M. E. Collins, R. S. Mansell, & D. F. Rothwell, 1984. Impact of on-site sewage disposal systems on surface and ground water quality. Rep. to Florida Department of Health and Rehabilitative Services. Contract No. LC170.

Bitton, G. 1975. Adsorption of viruses onto surfaces in soil and water. Water Res. 9:473-484.

Bitton, G. 1980. Introduction to Environmental Virology. John Wiley & Sons, New York.

Bitton, G., O. C. Pancorbo, & S. R. Farrah, 1984. Virus transport and survival after land application of sewage sludge. Appl. Envir. Microbiol. 47:905-909.

Bouma, J. 1975. Unsaturated flow during soil treatment of septic tank effluent. J. Environ. Eng. Div. Amer. Soc. Civil Eng. 101:967-983.

Bouma, J. 1979. Subsurface applications of sewage effluent. In Beatty, M. T. et al. (ed.) Planning the uses and management of land. Amer. Soc. of Agron. Madison, WI. pp. 665-703.

Bouma, J., W. A. Ziebell, W. G. Walther, P. G. Olcott, E. McCoy, & F. D. Hole, 1972. Soil absorption of septic tank effluent. Univ. of Wisconsin Ext. Cir. No. 20., Madison, WI.

Brown, K. W., K. C. Donnelly, J. C. Thomas, & J. F. Slowey, 1984. The movement of nitrogen species through three soils below septic fields. J. Environ. Qual. 13:460-465.

Brown, K. W., H. W. Wolf, K. C. Donnelly, & J. F. Slowey, 1979. The movement of fecal coliforms and coliphage below septic lines. J. Environ. Qual. 8:121-125.

Bureau of the Census, 1972. 1970 Census of Housing. Volume 1. Housing Characteristics for States, Cities and Counties. Part 1. United States Summary. U. S. Department of Commerce. pp. S20-1-9.

Bureau of the Census, 1983a. 1980 Census of Housing. Volume 1. Characteristics of Housing Chapter A. General Housing Characteristics. Part 1. United States Summary. U. S. Department of Commerce. pp. 59.

Bureau of the Census, 1983b. 1980 Census of Housing. Volume 1. Characteristics of Housing. Chapter B. Detailed Housing Characteristics. Part 1. United States Summary. U. S. Department of Commerce. pp. 66.

Cabelli, V. 1978. New standards for enteric bacteria. In R. Mitchell (ed.) Water Pollution Microbiology. Vol. 2. John Wiley and Sons, Inc., New York. pp. 231-242.

Caldwell, E. L. 1937. Pollution from pit latrines when an impervious structure closely underlies the flow. J. Inf. Dis. 61:269-288.

Caldwell, E. L. 1938a. Pollution flow from a pit latrine when permeable soils of considerable depth exist below the pit. J. Inf. Dis. 62:225-258.

Caldwell, E. L. 1938b. Studies in subsoil pollution in relation to possible contamination of ground water from human excreta deposited in experimental latrines. J. Inf. Dis. 62:273-292.

Caldwell, E. L. & L. W. Parr, 1937. Ground water pollution and the bored hole latrine. J. Inf. Dis. 61:148-183.

Carlson, G. F., F. E. Woodward, D. F. Wentworth, & O. J. Sproul, 1968. Virus inactivation on clay particles in natural waters. J. Water. Poll. Control. Fed. 40:R89-R106.

Clark, N. A., G. Berg, P. W. Kabler, & S. L. Chang, 1964. Human enteric viruses in water: Source, survival, and removability. In Advances in water pollution research, Vol. 2. Macmillian Co., NY. pp. 523-536.

Clark, H. P. & P. W. Kabler, 1964. Reevaluation of the significance of the coliform bacteria. J. Am. Water Works Assoc. 56:936.

Cogger, C. G. & B. L. Carlile, 1984. Field performance of conventional and alternative systems in wet soil. J. Environ. Qual. 13:137-142.

Converse, J. C. & E. J. Tyler, 1985. Wisconsin mounds for difficult sites. In Proceedings of the fourth national symposium on individual and small community sewage systems. Am. Soc. of Agric. Eng., St. Joseph, MI. pp. 119-130.

Craun, G. F. 1981. Outbreaks of waterborne diseases in the United States: 1971-1978. J. Amer. Water Works Assoc. 73:360-369.

Daniel, T. C. & J. Bouma, 1974. Column studies of soil clogging in a slowly permeable soil as a function of effluent quality. J. Env. Qual. 4:321-326.

Duboise, S. M., B. E. Moore, & B. P. Sagik, 1976. Poliovirus survival and movement in a sandy forest soil. Appl. Environ. Microbiol. 31:536-543.

Ellis, B. G. & A. E. Erickson, 1969. Movement and transportation of various phosphorus compounds in soils. Soil Science Dept., Michigan State Univ. and Michigan Water Res. Comm., East Lansing, MI.

Enfield, C. G. & B. E. Bledsoe, 1975. Kinetics model for orthophosphate reactions in mineral soils. USEPA Agency Rep. No. 660/2-75-022.

Firestone, M. K. 1982. Biological denitrification. In Stevenson, J. (ed.) Nitrogen in agricultural soils. Monograph No. 22. Am. Soc. of Agron., Madison, WI. pp. 289-326.

Gerba, C. P., C. Wallis & J. L. Melnick, 1975. Fate of wastewater bacteria and viruses in soil. J. Irr. Drain Div. Amer. Soc. Civil Eng. 101:157-174.

Hagedorn, C., D. T. Hansen, & G. H. Simonson, 1978. Survival and movement of fecal indicator bacteria in soil under conditions of saturated flow. J. Environ. Qual. 7:55-59.

Hagedorn, C., E. L. McCoy, & T. M. Rahe, 1981. The potential for ground water contamination from septic effluents J. Environ. Qual. 10:1-8.

Hargett, D. L., 1985. Performance assessment of low pressure pipe wastewater injection systems. In Proceedings of the fourth national symposium of individual and small community sewage systems. Am. Soc. of Agric. Eng., St. Joseph, MI. pp. 131-143.

Hansel, M. J. & R. E. Machmeier, 1980. Onsite wastewater treatment on problem soils. J. Water Pollution Control Fed. 52:548-558.

Harkin, J. M., C. P. Duffy & D. G. Kroll, 1979. Evaluation of mound systems for purification of septic tank effluent. Tech. Rep. Wis-WR G79-05. Univ. of Wisconsin Water Resources Center, Madison, WI.

Hillel, D. & W. R. Gardner, 1969. Steady infiltration into crust-topped profiles. Soil Sci. 108:137-142.

Hurst, C. J., C. P. Gerba, & I. Cech, 1980. Effects of environmental variables and soil characteristics on virus survival in soil. Appl. Envir. Microbiol. 40:1067-1079.

Jones, R. A. & G. F. Lee, 1977. Septic tank disposal systems as phosphorus sources for surface waters. USEPA Rep. No. 6013-77-129.

Jones, J. H. & G. S. Taylor, 1964. Septic tank effluent percolation through sands under laboratory conditions. Soil Sci. 99:301-309.

Kao, C. W. & R. W. Blanchar, 1973. Distribution and chemistry of phosphorus in a Albaqualf soil after 82 years of phosphate fertilization. J. Environ. Qual 2:237-240.

Keeney, D. R, 1981. Soil nitrogen chemistry and biochemistry. In Iskandar, I. K. (ed.) Modeling wastewater renovation land treatment. John Wiley & Sons, New York. pp. 259-276.

Kristiansen, R. 1981. Sand-filter trenches for purification of septic tank effluent: I. the clogging mechanisms and soil physical environment. J. Env. Qual. 10:353-357.

Laak, R. 1981. A passive denitrification system for on-site systems. In Proceedings of the third national home sewage treatment symposium. Am. Soc. of Agric. Eng. St. Joseph, Michigan. pp. 108-115.

Lance, J. C. 1977. Phosphate removal from sewage water by soil columns. J. Environ. Qual. 6:279-284.

Lance, J. C. & C. P. Gerba, 1980. Poliovirus movement during high rate land filtration of sewage water. J. Environ. Qual. 9:31-34.

Lance, J. C. & C. P. Gerba, 1984. Virus movement in soil during saturated and unsaturated flow. Appl. Envir. Microbiol. 47:335-337.

Lance, J. C., C. P. Gerba, & J. L. Melnick, 1976. Virus movement in soil columns flooded with secondary sewage effluent. Appl. Envir. Microbio. 32:520-526.

Lance, J. C., C. P. Gerba, & D. S. Wang, 1982. Comparative movement of different enteroviruses in soil columns. J. Environ. Qual. 11:347-351.

Lindsay, W. L. & E. C. Moreno, 1960. Phosphate phase equilibria in soils. Soil Sci. Soc. Am. Proc. 24:177-182.

Magdoff, F. R., & J. Bouma, 1974. The development of soil clogging in sands leached with septic tank effluent. In Proceedings of the national home sewage disposal symposium. Am. Soc. of Agric. Eng., St. Joseph, MI. pp. 37-47.

Magdoff, F. R., D. R. Keeney, J. Bouma, & W. A. Ziebell, 1974. Columns representing mound-type disposal systems for septic tank effluent: II. Nutrient transformations and bacterial populations. J. Environ. Qual. 3:228-234.

Martin, J. P. & D. D. Focht, 1977. Biological properties of soils. In Elliott, L. F. & F. J. Stevenson (ed.) Soils for management of organic wastes and wastewater. SSSA, ASA, CSSA, Madison, WI. pp. 115-169.

McCoy, E. L. & C. Hagedorn, 1980. Transport of resistance-labeled *Escherichea coli* strains through a transition between two soils in a topographic sequence. J. Environ. Qual. 9:686-691.

Miller, J. C. 1975. Nitrate contamination of the water-table aquifer by septic tank systems in the Coastal Plain of Delaware. In Jewell, W. S. & R. Swan (ed.) Water pollution control in low density areas. Proc. Rural Environ. Eng. Conf. University Press of New England., Hanover, NH. pp. 121-133.

Mitchell, R. & Z. Nevo, 1964. Effects of bacterial polysaccharide accumulation on infiltration of water thorugh sand. Appl. Microbiol. 12:219-223.

Munns, D. N. & R. L. Fox, 1976. The slow reaction which continues after phosphate adsorption: Kinetics and equilibrium in some tropical soils. Soil Sci. Soc. Am. Proc. 40:46-51.

Otis, R. J., W. C. Boyle, & D. R. Sauer, 1974. The performance of household wastewater treatment units under field conditions. In Proceedings of the national home sewage disposal symposium. Am. Soc. of Agric. Eng., St. Joseph, MI. pp. 191-201.

Otis, R. J. & W. C. Boyle, 1976. Performance of single household treatment units. J. Environ. Eng. Div. Amer. Soc. Civil. Eng. 102:175-189.

Pask, D. A., P. J. Casey, J. C. Vaughan, & D. Thirumurthi, 1985. Recent research and developments in on-site sewage disposal in Nova Scotia. In Proceedings of the fourth national symposium of individual and small community sewage systems. Am. Soc. Agric. Eng., St. Joseph, MI. pp. 155-164.

Preul, H.C. & G. J. Schroepfer, 1968. Travel of nitrogen in soils. J. Water Poll. Control Fed. 40:30-48.

Quam, E. L., H. R. Swett, & J. R. Illian, 1974. Subsurface sewage disposal and contamination of groundwater in East Portland, Oregon. Ground water 22:356-367.

Rahe, T. M., C. Hagedorn, E. L. McCoy, & G. F. Kling, 1978. Transport of antibiotic-resistant Escherichia coli through western Oregon hillslope soils under conditions of saturated flow. J. Env. Qual. 7:487-494.

Reneau, R. B., Jr. 1977. Changes in inorganic nitrogenous compounds from septic tank effluent in a soil with a fluctuating water table. J. Environ. Qual. 6:173-178.

Reneau, R. B., Jr. 1978. Movement of biological and chemical contaminants through tile drained soil systems. Rep. to Virginia Department of Health.

Reneau, R. B., Jr. 1979a. Changes in concentrations of selected chemical pollutants in wet, tile-drained soil systems as influenced by disposal of septic tank effluents. J. Environ. Qual. 8:189-196.

Reneau, R. B., Jr. 1979b. Influence of artifical drainage on penetration of coliform bacteria from septic tank effluents into wet tile drained soils. J. Environ. Qual. 7:23-30.

Reneau, R. B., Jr. & D. E. Pettry, 1975. Movement of coliform bacteria from septic tank effluent through selected coastal plain soils of Virginia. J. Environ. Qual. 4:41-44.

Reneau, R. B., Jr. & D. E. Pettry, 1976. Phosphorus distribution from septic tank effluent in Coastal Plain soils. J. Environ. Qual. 5:34-39.

Reneau, R. B., Jr., J. H. Elder, Jr., D. E. Pettry, & C. W. Weston, 1975. Influence of soils on bacterial contamination of a watershed from septic sources. J. Environ. Qual. 4:249-252.

Romero, J. C. 1970. The movement of bacteria and viruses through porous media. Groundwater 8:37-48.

Safferman, R. S. 1982. Viruses in wastewaters. Env. Int. 7:15-20.

Sample, E. C., R. J. Soper, & G. D. J. Race, 1980. Reactions of phosphate fertilizers in soils. In Khasawneh, F. E., E. C. Sample, E. J. Kamprath (ed.) The role of phosphorus in agriculture. Am. Soc. of Agrono. and Others., Madison, WI. pp. 263-310.

Sawhney, B. L. & D. E. Hill, 1975. Phosphate sorption characteristics of soils treated with domestic waste water. J. Environ. Qual. 4:342-346.

Sikora, L. J., J. C. Converse, D. R. Keeney, & R. C. Chen, 1977. Field evaluation of a denitrification system. In Proceedings of the second national symposium on individual and small community sewage systems. Am. Soc. of Agric. Eng., St. Joseph, MI. pp. 202-207.

Sikora, L. J. & R. B. Corey, 1976. Fate of nitrogen and phosphorus in soils under septic tank waste disposal fields. Trans. Amer. Soc. Agric. Eng. 19:866-870, 875.

Sikora, L. V. & D. R. Keeney, 1974. Laboratory studies on simulation of biological denitrification. In Proceedings of the national home sewage disposal symposium. Am. Soc. of Agric. Eng., St. Joseph, MI. pp. 64-73.

Sikora, L. V. & D. R. Keeney, 1976. Evaluation of a sulfur Thiobacillus denitrifans nitrate removal system. J. Environ. Qual. 5:298-303.

Simon, J. J. & R. B. Reneau, Jr. 1985. Hydraulic performance of prototype low pressure distribution systems. In Proceedings of the fourth national symposium on individual and small community sewage systems. Am. Soc. of Agric. Eng., St. Joseph, MI. pp. 251-259.

Stewart, L. W., B. L. Carlile, & D. K. Cassell, 1979. An evaluation of alternative simulated treatments of septic tank effluent. J. Environ. Qual. 8:397-402.

Stewart, L. W. & R. B. Reneau, Jr. 1981a. Movement of fecal coliform bacteria from septic tank effluent through coastal plains soils with high seasonal water tables. In Proceedings of the third national symposium on individual and small community sewage systems. Am. Soc. of Agric. Eng., St. Joseph, MI. pp. 319-327.

Stewart, L. W. & R. B. Reneau, Jr., 1981b. Spatial and temporal variation of fecal coliform movement surrounding septic tank--soil absorption systems in two Atlantic Coastal Plain soils. J. Environ. Qual. 10:528-531.

Stewart, L. W. & R. B. Reneau, J. 1984. Septic tank effluent disposal experiments using nonconventional systems in selected coastal plain soils. Final report to Virginia Department of Health.

Stuanes, A. O. 1984. Phosphorus sorption of soils to be used in wastewater renovation. J. Environ. Qual. 13:220-224.

Tofflemire, J. J., M. Chen, F. E. Van Alstyne, L. J. Aetling, & D. B. Aulenbach, 1973. Phosphate removal by sands and soils. N.Y. State Dept. Environ. Conserv. Tech. Paper No. 31.

Uebler, R. L 1984. Effect of loading rate and soil amendments on inorganic nitrogen and phosphorus leached from a wastewater soil absorption system. J. Environ. Qual. 13:475-479.

U.S.EPA, 1980. Design manual--onsite wastewater treatment and disposal systems. EPA 625/1-80-012. U.S.EPA, Washington, DC.

U.S.EPA, 1977. Alternatives for small wastewater treatment systems. EPA-625/4-77-011. U.S.EPA, Washington, DC.

Van Riemsdijk, W. H., J. Beek, & F. A. M. DeHann, 1979. Phosphates in soils treated with sewage water: IV. Bonding of phosphate from sewage water in sand columns containing aluminum hydroxide. J. Environ. Qual. 8:207-210.

Walker, W. G., J. Bóuma, D. R. Keeney, & F. R. Magdoff, 1973a. Nitrogen transformations during subsurface disposal of septic tank effluent in sands: I. Soil transformations. J. Environ. Qual. 2:475-479.

Walker, W. G., J. Bouma, D. R. Keeney, & P. G. Olcott, 1973b. Nitrogen transformations during subsurface disposal of septic tank effluent in sands: II. Ground water quality. J. Environ. Qual. 2:521-525.

Wellings, F. M. 1980. Virus movement in groundwater. In McClelland, N. I. (ed.) Individual onsite wastewater systems. Proceedings sixth national conference 1979. Ann Arbor Science Publishers, Inc., Ann Arbor, MI. pp. 427-434.

Whelan, B. R. & N. J. Barrow, 1984a. The movement of septic tank effluent through sandy soils near Perth. I. Movement of nitrogen. Aust. J. Soil Res. 22:283-292.

Whelan, B. R. & N. J. Barrow, 1984b. The movement of septic tank effluent through sandy soils near Perth. II. Movement of phosphorus. Aust. J. Soil Res. 22:293-302.

Wilson, S. A., R. C. Paeth, & M. P. Ranayne, 1982. Effect of tile drainage on disposal of system tank effluent in wet soils. J. Environ. Qual. 11:372-375.

Winneberger, J. H., L. Francis, S. A. Klein, & P. H. McCauley, 1960. Biological aspects of failure of septic tank percolation systems. Sanit. Engr. Res., Univ. of Calif., Berkeley.

Ziebell, W. A., D. H. Nero, J. F. Deininger, & E. McCoy, 1974. Use of bacteria in assessing waste treatment and soil disposal systems. In Proceedings of the national home sewage disposal symposium. Am. Soc. of Agric. Eng., St. Joseph, MI. pp. 58-63.

Soil and site criteria for on-site systems

H. J. Kleiss and M. T. Hoover, Soil Science Department, North Carolina State University, Raleigh, NC 27650

Utilization of the soil system for disposal and treatment of wastes has the potential for taking advantage of the soil's unique physical, chemical, and biological processes. The natural soil system offers a dynamic media for not only absorbing waste but for treating and utilizing waste constituents. The challenge is clear - to effectively design waste application systems that optimize a soil's treatment potential.

The porous nature of soil can provide an ideal media for absorbing and transmitting liquid effluent. A tortuous flow path through soil pores and voids that is neither too rapid nor too slow allows for a variety of natural treatment processes. Purification occurs through physical filtration, chemical treatment through ion exchange and transformation, biological decomposition by a myriad of micro-organisms as well as nutrient uptake by plants.

Clearly the ultimate goal in utilizing the complex array of dynamic soil processes is the protection of health and the environment. Herein lies a distinct difference between approaches that address only disposal and those that place a major emphasis on treatment. Regulatory programs challenged with controlling on-site systems for health and water quality reasons must certainly place priority on the goal of adequate treatment.

DEFINING THE SYSTEM

As the knowledge and technology for on-site waste disposal have expanded, the criteria for soil and site characteristics have also become more complex. A generic discussion of soil and site criteria becomes very difficult. It is important that any detailed review of siting criteria be based on a specific waste disposal system. A clear understanding of system parameters will improve the precision that can be attained in defining soil and site limitations.

A conventional system for purposes of this paper consists of a two-compartment septic tank with a sanitary tee outlet structure. The tank is connected by solid pipe to a distribution box that will feed the effluent drainlines via gravity flow. The drain lines (often called nitrification lines) are placed in trenches 60 to 90 cm wide and dug to a similar depth. Effluent drain lines will be placed in a minimium of 30 cm of gravel or stone (2 - 6 cm diameter) with at least 15 cm below the pipe and 5 cm over the pipe. Trenches should be dug such that the bottom has a slope no greater than 0.3 percent.

Variations of the above-defined system are many. Modifications including wider or deeper trenches, seepage beds or pits, multiple tanks, grease traps, dosing systems, mound systems, and other alternatives will influence specific soil and site criteria. Furthermore, the type and quantity of waste must be characterized to assure optimum site suitability.

Detailed refinement of siting criteria will be climate specific as well as system specific. Rainfall and temperature variations that influence evapotranspiration, biological activity and the total amount of hydraulic load to be handled by the soil system must be accounted for in siting criteria. As the type of system varies individual siting parameters will change to reflect system differences. While the basic principles and potentially limiting factors will remain the same the magnitude of certain parameters may change. Irrespective of the specific waste disposal system a thorough understanding of its function is a requisite for defining soil and site limitations.

Integral with defining the general type of system is the underlying and often unstated decision on the allowable pollution risk. This question is most strongly related to existing surface or groundwater quality, proximity to a water body and planned water use. In other words, what are the consequences of inadequate treatment? How will failure be defined? Will failure of a system only be recognized when effluent is visible on the soil surface? These are basic questions that should underlie regulatory requirements and certainly will influence the selection of critical limits for site suitability.

SITING CONSIDERATIONS

After documenting the nature of the waste and the anticipated type of system as well as defining the acceptible risks or performance level a soil-site assessment can be undertaken. The assessment procedure must be sufficiently comprehensive so as to reflect the dynamic and complex nature of the soil system upon which treatment and disposal is based. A truly comprehensive approach is needed.

Factors or conditions that affect on-site waste disposal include those that might influence absorption, movement, treatment and ultimate disposal of the waste effluent. Both internal soil properties and external site factors must be integrated in a land suitability evaluation. For purposes of discussion soil considerations and site factors will be treated separately. It is, however, a major intent of this paper to emphasize the need for full integration of these two components.

The following discussion of site-soil criteria is intended as an overview for individuals who have not received formal soils training.

This group may include sanitarians or others with environmental interests who have a concern with on-site waste disposal but whose background has not included soil science.

SITE FACTORS

A proper land evaluation entails not only a soil investigation but includes a comprehensive 3-dimensional assessment of site conditions. Understanding the continuum of soils across the landscape requires the establishment of relationships of various landform components with relevant soil characteristics. Understanding the broader site conditions before becoming involved in detailed soil characteristics can facilitate the total assessment process.

Geologic Setting

A preliminary step in understanding the present landscape-soil system is to review the geology. Significant insight and predictions of soil conditions can result from a knowledge of soil parent materials. By applying principles of soil genesis, clues on texture, type of clay minerals, thickness of soil, water table depth or presence of restrictive layers can be extrapolated from geologic data. Likelihood of bedrock fractures, solution cavities or other highly porous media that could reduce soil treatment potential can be anticipated. The potential for dense restrictive layers in fluvial and marine sediments should be identified.While a review of general rock or sediment conditions will not answer specific on-site questions it will provide an initial framework for the soil-landscape-hydrologic model to which we are considering waste application. The subsequent site specific investigation may reveal that a geologic condition below the soil is the limiting factor on the acceptability of the site or on the loading rate.

Landscape Position

Location on the landscape is important in its influence on surface and subsurface water flow. Convexity or concavity in surface landform will affect runoff or runon of surface water. Elimination of all excess water is one goal in siting. In a related manner landscape position will be important in subsurface water flow. The lower the site on the slope the greater the likelihood of lateral downslope movement and accumulation of subsurface water. Restrictive soil horizons or rock layers that limit downward percolation force downslope flow along the restrictive feature. Potential interception of this lateral flow by the disposal trench must be recognized and any hydraulic overload eliminated.

Landscape limitations such as floodplains and colluvial footslope positions where excess surface or subsurface water may occur are generally obvious. However, in many upland locations slight topographic changes act in conjunction with subsurface soil or parent material conditions to create subtle but serious constraints to on-site disposal. Understanding geologic, hydrologic, and geomorphic factors are vital in the assessment process.

Landform, soil, and hydrologic relationships can change dramatically within a local area let alone across regional boundaries. In the southeastern USA, a comparison of conditions in the lower Coastal Plain with the upper Coastal Plain and in turn with the Piedmont and Mountain regions will reveal unique water table and water movement constraints for on-site waste disposal. A broader comparison encompassing the northern glacial till plains, the more arid Great Plains or aeolian landscapes would illustrate additional conditions requiring special consideration.

Slope

Apart from the above landscape concerns the slope gradient is itself a major factor. It relates not only to surface and subsurface water flow but to overall site stability including erodibility and complications in system installation. Depth of trenches and ability to achieve uniform effluent distribution by gravity flow are impacted by the steepness of the site. Increasing the downslope lateral component of effluent flow as opposed to the vertical flow can affect trench spacing and waste treatment. The precise degree of slope that could impose a site limitation will vary with internal soil conditions as they influence effluent movement.

Land Use

A fully comprehensive site assessment will include a variety of non-soil factors. These include space limitations, type of adjacent land use, well locations, setback requirements and vegetation just to mention a few. Such site constraints can be critical and in some cases may interact with specific soil considerations. The past management on a site that would suggest compaction, cut and fill or other serious land modification and soil disturbance must be documented. In some cases accelerated erosion resulting from previous mismanagement will have resulted in a complex landscape with limited potential for on-site waste disposal. Such conditions can not be overlooked and will vary with state and local regulations.

SOIL FACTORS

A comprehensive review of all the specific soil factors that interact to influence on-site waste treatment is beyond the scope of this paper. Such a discussion would entail a review of nearly all facets of soil science. The complexity of the soil system as it functions to provide waste absorption and treatment requires the contribution of most sub-disciplines of soil science. These would include soil physics, soil chemistry, soil microbiology, soil morphology, soil and water management and soil mineralogy. Each has a vital and integrated role in understanding the absorption, movement and treatment of waste effluent. The major factor and the usual focus of initial concern in siting criteria concerns the acceptance and movement of liquid effluent through the soil system. Soil properties influencing water movement and waste treatment are many. Several will be reviewed in this paper.

Water Movement

Liquid movement through soil media has been and is the subject of considerable study but is generally categorized into saturated or unsaturated flow. While the forces controlling moisture movement form a continuum between the saturated state and the dry state the two-category approach is convenient. Under saturated conditions or zero moisture tension flow is primarily through large pores and cracks and hydraulic conductivity is at its maximum for a given soil material. In this moisture state the vertical conductivity will decrease in the order of sand to loam to clay. As the moisture content decreases and the large pores have drained the cohesion and adhesion tension forces act in the small pores to reduce the conductivity in this unsaturated state. With further drying the finer pores exerting greater tensions become the dominant force and ultimately result in the soils with fine pores (clays) having greater conductivity than coarser soils.

The effects of moisture content (saturated vs. unsaturated) on hydraulic conductivity is critical since optimal waste treatment will occur in the unsaturated condition. While most data on hydraulic conductivity of soils is for the saturated state, this data must be adjusted when making loading rate inferences about specific soils. The unsaturated flow state facilitates the aerobic conditions needed for various biological and chemical processes of waste treatment and maximizes the physical filtration attributes of soil media. Significant differences in waste treatment are observed when comparing these flow conditions. Attenuation of bacteria and virus movement is improved markedly in the unsaturated vs. the saturated state.

Soil Properties Affecting Water Movement and Treatment

It is clear that any soil property that would affect the quantity, size, shape, or continuity of pores will influence hydraulic conductivity. This suggests a complete morphological evaluation of the full soil column (profile) to a depth sufficient to reveal any layers limiting effluent movement. This should be at least 2 meters.

Texture

Particle-size distribution (proportions of sand, silt, and clay) is the dominant factor in water movement. Sandy soils with a dominance of macro-pores may transmit effluent too fast for adequate treatment. Very clayey soils with many fine pores exerting high tension forces will limit not only waste movement but also reduce oxygen diffusion and slow biological treatment processes. In general, loamy textures provide the most desirable condition not only from a hydraulic conductivity standpoint but from the treatment potential as well.

While texture itself is crucial the sequence or arrangement of soil horizons of varying texture is also important as is the nature of the transition or boundary between the layers. Abrupt textural changes are generally detrimental in water flow continuity. Furthermore due to the increased affinity of small pores (in clayey soils) for water, downward flow from a clay to a sand can be restricted. Such phenomena support the requirements for a textural analysis that encompasses the complete soil column. Texture alone however may be insufficient. Restrictive fragipan horizons that are typically loamy in texture illustrate the problem with the single factor approach.

Structure

The aggregation of the primary sand, silt, and clay size particles into secondary aggregates or structural units (peds) of various shapes and sizes is a soil forming process of major consequence in absorption and movement of liquids. The size, shape, and stability of the structural units will determine the amount and continuity of macro-voids between the units. Structural condition is commonly a clue to other characteristics. Angular blocky and prismatic structure may indicate the presence of expandable clay minerals. This is illustrative of the integrative nature of a soil evaluation. Conclusions based on one soil factor in the absence of aggregating all morphological data can lead to an erroneous assessment.

Soil structure while playing a vital role in water and air movement has the unfortunate characteristic of being easily altered. The effects of compaction or excavation when the soil is very wet can drastically reduce the conductivity afforded by the original in-situ structural development. This delicate character of natural soil structure is also supportive evidence for concern over the use of disturbed soil fill material for use with a ground absorption system.

Consistence

The degree of firmness or friability and the range of plasticity and stickiness recognized in soil profile descriptions provide indirect evidence for restrictive densities and presence of active (expandable) clay minerals. Combining consistence with texture and structure a soil scientist begins to formulate a dynamic model for how a particular soil can handle water and how it will react under varying moisture conditions. How stable are the structural voids. Will the hydraulic conductivity be seriously reduced due to clay mineral expansion?

Restrictive Horizons

The study of soil genesis has provided valuable insight into the characteristics of several natural genetic soil horizons that can be limiting to water movement. These include fragipans, spodic horizons, duripans, petrocalcic horizons and plinthite. Along with rock-like parent materials these layers can be restrictive. The depth, degree of restriction, vertical thickness and horizontal continuity of such features must be determined. Effects of moisture change on these horizons must be understood. The importance of a 3-dimensional site assessment becomes clear when considering the potential impact of these special soil features.

Soil Color

Color and color patterns are perhaps the most obvious soil characteristics but are also the most indirect measures of relevant properties. Brightness versus dullness are clues to moisture movement or general aeration status. Certain genetic horizons such as fragipans or plinthite that may be restrictive have characteristic color patterns. Dark surface layers and gray (low chroma) matrix or mottle colors suggest wetness. More subtle moisture relationships are suggested by reddish versus yellowish brown colors indicative of varying iron oxide mineralogy. Color patterns on the faces of structural units (peds) as opposed to color within a ped are further hints in unraveling the site history as well as the present conditions on a site.

Pores and Roots

Morphological features often overlooked in a soil evaluation and very difficult to assess in an auger boring are the presence of roots and the abundance and size or pores. These features may provide the most direct evidence for the degree of restriction in a given horizon. An abrupt decrease in roots or the sudden concentration of roots at a particular position in a profile should be investigated more fully. While many external factors will influence the nature of root distribution they can provide a valuable piece to the soil puzzle. With the complexity of the soil system no feature should go unnoticed in a soil assessment.

Clogging Mat

The foregoing overview of soil properties that influence the absorption, movement and treatment of effluent must be couched in the understanding of how a ground absorption system will function over time. Research has shown that a biochemical crust or mat forms at and in the soil interface with the absorption trench. This biological clogging layer a few centimeters in thickness may then become the hydraulically limiting feature. The precise nature of this mat will be affected by the soil, the loading rate, the quantity of solids reaching the soil-trench interface and the degree of anaerobism that is produced. Understanding this phenomenon becomes very important in predicting the performance of a given soil. It is clear that in the long term a greater emphasis must be placed on the rates of unsaturated flow in our soil media. If the biological crust provides an initial restriction to effluent movment, the rates of movement through the soil system will reflect unsaturated conditions. Loading rates must reflect this characteristic of ground absorption systems.

The variety of factors influencing water movement and the desire to maintain an aerobic system along with the concern for the clogging mat formation must all be integrated into the formulation of the "long-term acceptance rate" for the soil system of concern. This loading rate should reflect the capability of the mature system after an equilibrium condition is achieved. Such an application rate would assure not only the optimal length of time for system utilization but also for optimum treatment.

APPROACHES TO LAND SUITABILITY ASSESSMENT

The preceding overview of some of the site and soil factors influencing on-site systems illustrates the need for a comprehensive approach to evaluating a soil-site and determining a proper application rate. Clearly there is more we need to know to improve our utilization of the soil system for waste treatment. Notwithstanding our continuing research needs, on-site systems are important and are being installed in great numbers. Several general approaches are used in determining the acceptability of land for disposal. The following discussion will not attempt to review the many ways specific regulatory programs handle land permitting but will overview approaches with respect to their relative utilization of soil information.

Percolation Testing

A long-standing method for evaluating a soil for waste absorption is the percolation test. This simple test to measure the rate of fall of water in a borehole continues to be a major determining factor in site evaluation in many areas. It has been criticized but is still used by many sanitarians. Used alone it provides a relatively simple technique that is designed to integrate many of the factors affecting conductivity into an observed rate of water movement. It does not require intensive knowledge of soil properties and can be performed by most technicians.

The procedure, however, is subject to many errors by the testor. Measured results on a given soil may vary 100%. It is strongly affected by soil moisture status, presence of large roots and the actual digging technique. The statement has often been made that with very slight variations in the recommended field procedure the test could provide any result desired.

The main criticism of the test is that it is a measure of saturated flow. With the knowledge that a system develops a crust or clogging mat at the soil interface, it is the unsaturated flow rate of the soil that becomes important. The unsaturated flow below the mat might be less than 5% of the rate measured in a percolation test. An equally serious practical concern is that where the perc test is used it oftentimes overshadows or even negates an evaluation or observation of the site-soil system. "Perc" values, since they do provide a quantitative measurement, can easily override important observation and interpretations of the soil system. Soil and site conditions are generally too complex for this simplistic measurement if used alone.

More sophisticated in-situ tests for hydraulic conductivity are available and can overcome many of the technical difficulties of the standard percolation test. Such techniques generally measure saturated hydraulic conductivity although some are used to determine unsaturated conductivity over a range of soil moisture. These tests are generally very time consuming and in themselves reflect conditions of a small body of soil. Natural soil variability must be considered in determining the necessary number of tests if they are to provide the sole source of site determining information.

Soil Map Interpretation

Soil surveys as presented in a typical county report provide an inventory of soil resources that can be valuable in making decisions for waste disposal. Interpretative ratings of slight, moderate or severe limitation are given for each soil map unit. The three-category rating scheme indicating the degree of soil limitation is based on clearly established quantitative criteria through the National Cooperative Soil Survey Program. Factors considered include permeability, slope, depth to bedrock, depth to water table, and depth to restrictive horizons along with other considerations of a more localized nature.

The soil survey is, however, designed as a multi-use inventory of soil properties. A soil map shows the distribution of soils on the landscape and like any map it has a limit of scale. In most soil surveys the smallest mapping delineation will be 3-10 acres (1.2 - 4 hectares) in size. Land management decisions on parcels smaller than this must be made with care. It is essential that a user review the map unit descriptions and become familiar with the variability and inclusions of other soils that may be encountered in a given delineation.

A soil map is most appropriately utilized as an initial planning guide. It depicts soil conditions that should be expected. It is not intended to provide site-specific information especially on small parcels of land. As a supporting document and source of preliminary data a soil map is an invaluable tool. The utility of a standard soil survey report can be extended with the development of supplements that would reflect local needs or regulatory criteria. Such a document can be designed to bridge the gap between the criteria used for the soil survey limitation ratings and specific state or local standards. It could also expand the soil ratings to include alternative types of on-site systems (Kleiss, 1981).

Soil Evaluation And Percolation Testing

A third land suitability assessment approach includes an on-site soil evaluation to ascertain the presence of a necessary thickness of unsaturated soil and a percolation test to determine the loading rate for the soil absorption system. The soils evaluation usually is conducted in a backhoe-dug pit or from auger borings. Typically, the depth from the soil surface to a limiting condition is determined. Limiting conditions include the seasonal high water table including a perched water table, bedrock and impervious soil strata. If a limiting condition is sufficiently deep in the soil then a percolation test is conducted. Sites with excessively slow or fast percolation rates are determined to be unsuitable even if no limiting condition was encountered in the soils evaluation. The loading rate and hence the size of the soil absorption system is determined strictly according to the percolation rate.

This assessment approach requires that the site evaluator have substantial knowledge of soils. The evaluator must have an ability to recognize and make inferences concerning the presence of soil mottling, restrictive soil horizons such as fragipans, duripans, plinthite, and orstein, and the soil/bedrock boundary. However, this knowledge is not put to full use since any soils information gathered during the determination of the limiting conditions is not used to determine the system loading rate.

An improvement upon this approach that has been utilized consists of a detailed soil profile description to identify limiting conditions and to compare with the percolation rate later determined at the site. If the magnitude of the percolation rate is not consistent with the described soil properties then the need for further testing is indicated. This may consist of evaluation of soil variability at the site or retesting of the percolation rate. This method requires significant training of soil evaluators to develop their abilities to describe soil morphology and relate it to soil moisture movement.

Soil Evaluation Alone

Another approach to land suitability assessment consists of a detailed soil evaluation to determine suitability and to establish the loading rate or size of the system. The state of Maine has used such an assessment strategy for a number of years as have other states including North Carolina. Sites in Maine are assessed by licensed site evaluators who classify each site into one of the eleven parent material/soil profile categories on the basis of soil morphology (Hoxie and Frick, 1985). These criteria and the depths to bedrock and the ground water table determine the site suitability and the loading rate. Loading rates range from 3.3 cm/day (0.8 gpd/ft^2) for a small seepage bed in sandy soils in stratfied drift to 0.8 cm/day (0.2 gpd/ft^2) for an extra large seepage bed in slowly permeable lacustrine sediments (Hoxie and Frick, 1985).

Another example of this detailed soils evaluation is the approach used by county sanitarians in North Carolina. The on-site investigation consists of evaluation of:

1. Topography and landscape position

2. Soil characteristics

   .texture
   .mineralogy
   .structure
   .organic matter content

3. Soil internal drainage

4. Soil depth

5. Restrictive horizons

6. Available space

Each of these factors is classified as Suitable (S), Provisionally Suitable (PS), or Unsuitable (U). The overall site suitability is determined by the suitability of the most limiting factor. Sites classified as Suitable can be used with little difficulty. Provisionally Suitable sites can be utilized with some modifications and careful planning. Some sites classified as Unsuitable cannot be utilized while others may be reclassified to Provisionally Suitable under certain conditions using acceptable site or system modifications.

Topography and landscape position are classified by slope percentage with slopes greater than 30% generally being considered Unsuitable as to topography. Also, any slopes dissected by gullies and ravines, areas subject to frequent flooding, or containing alluvial soils, or depressions are considered Unsuitable.

Soil morphology is typically evaluated from auger borings taken to depths of 122 cm (48 inches) or less. Texture and mineralogy are estimated from profile characteristics and each site is classified into one of five Soil Groups (I, II, III, IVa, IVb). Soil loading rate and, hence, system size are determined on the basis of these Soil Groups rather than upon a percolation rate (Table 1). The sandy soils (Group I) include sand and loamy sand textural classes. The coarse loamy soils (Group II) include sandy loam and loam textural classes while the fine loamy soils (Group III) include sandy clay loam, silt loam, clay loam, and silty clay loam textural classes. The clayey soils (Group IV) include the sandy clay, silty clay, and clay textural classes and are separated into Soil Groups IVa in which the soils are strongly influenced by the 1:1 type clay minerals and Soil Group IVb in which the soils are strongly influenced by clay minerals that cause excessive shrinking and swelling such as 2:1 type clay minerals. Structure, consistence, color, and the type of parent material provide morphologic indicators of the clay mineral composition. Not only is structure evaluated to determine the type of clay minerals, but platy structure in Groups III and IV and massive structureless conditions in Groups II, III, and IV are designated Unsuitable as to structure. Organic soils are also considered Unsuitable.

Any seasonal high water table, bedrock, or restrictive horizons are specifically evaluated using morphological observations and must be maintained 30 cm (1 ft.) below the bottom of the absorption field. Internal soil drainage is determined by soil color or in some coastal areas with unmottled coarse sands by direct observation or by surface elevation relative to mean sea level. Soil depth is determined to bedrock including saprolite. Restrictive horizons include plinthites, fragipans, and iron pans (cemented spodic horizons).

As can be seen from the previous discussion of soil and site criteria in North Carolina this approach to land suitability assessment requires a high degree of personnel training. In North Carolina this is provided by a few county-hired scientists, by regional state-hired soil scientists, and by the Soil Science Extension at North Carolina State University. Due to the high turnover rate of county sanitarians this training is never completed. In fact, while the better sanitarians do an excellent job of soils evaluation it is questionable whether the average sanitarian has adequate motivation to become a good soils evaluator.

The value of using detailed soil and site criteria rather than the percolation test to site and size systems has been recognized in North Carolina where failure rates have been significantly reduced (Steinbeck, 1982) and in Maine where failure rates remain low even after a ten-year history of soil evaluation programs (Hoxie and Frick, 1985).

Table 1. Loading rates by soil group for conventional systems in North Carolina.

Soil group	Important features	Application rate	
		cm/day	gpd/ft^2
I	Sandy textures	5.0 - 3.3	1.2 - 0.8
II	Coarse loamy textures	3.3 - 2.5	0.8 - 0.6
III	Fine loamy textures with provisionally suitable structure	2.5 - 1.7	0.6 - 0.4
IVa	Clayey textures with 1:1 type clay minerals and provisionally suitable structure	1.7 - 0.8	0.4 - 0.2
IVb	Clayey textures with 2:1 type clay minerals or other clay minerals that swell excessively	Unsuitable	

Soil Evaluation, Classification, And Extrapolation

Another variation of the detailed soil evaluation approach involves a detailed soil description and USDA taxonomic classification of the soils. This would be followed by a more formal extrapolation of application rate data from a number of taxonomically similar sites to the described and classified site (Bouma, 1977 and Bouma, 1979). In this case, specific application rates could be pre-determined in the field using time-consuming unsaturated hydraulic conductivity tests on key soils. After establishment of a data base the results could be used to group soils that behave similarly and the application rates for these groups could be extrapolated to taxonomically similar soils. The opportunity exists here to utilize a two-tiered evaluation system where the extrapolation approach is used for soil series which consistently behave in a similar manner, but those soil series which exhibit wide variation in morphologic or hydraulic properties could be individually tested site-by-site.

The soil and site criteria needed for this approach extend beyond those necessary for the simpler, more empirical North Carolina and Maine approach. In order to use the extrapolation approach, a detailed soil profile description is still necessary as is additional soils information and expertise to properly designate horizons and classify the soil. This seems quite beyond the capabilities of a highly trained sanitarian and is certainly into the realm of the soil scientist. The feasability of having soil scientists conduct the site investigations for all on-site sewage disposal systems in a state depends upon whether 3,000 to 5,000 systems or 40,000 to 50,000 systems per year are installed in the state.

A detailed site evaluation used in conjunction with a site evaluation computer model may provide a workable solution to the dilemma of the need for soil scientists to conduct all site investigations. A site evaluation computer program could be designed to address the capabilities and limitations of the sanitarian at the local level (Fritton et al., 1982). The computer program would have to be user-friendly. This method would require the same level of soils-based training as the better sanitarians in North Carolina have obtained, but less than required for the soils description-classification-extrapolation method previously described.

The soil and site criteria from a site evaluation would be immediately entered into the computer by the sanitarian who would then respond to specific questions concerning setbacks, lot boundaries, housing location preference of the landowner, and location of the site on the published soil survey. The computer would need to be programmed to determine the site suitability relative to state/county regulations and to classify the soil on the basis of the processed soil description, the expected inclusions in the map unit and the sanitarian's response to specific questions. This technique eliminates the requirement of a soil scientist to classify the soil and thereby makes the description-classification-extrapolation technique a more realistic alternative. In fact, the decision-making criteria for matching a described and classified soil to a data base of soil groups with specific wastewater application rates could be included in the computer model. Likewise, as has already been done by Fritton et al., (1982), a user-friendly computer model could be used to mathematically predict the likelihood of a septic tank system working on the specific soil in question. Therefore, a sanitarian could use a very sophisticated mathematical tool that would not normally be available.

By excluding some features with large computer storage requirements (such as the sophisticated mathematical models) one could reduce the computer capacity needed to that available from a microcomputer. Microcomputers are more typically available in county extension offices than in county health departments or soil and water district offices. Likewise, many extension offices have video cassette recorders (VCR's) which could greatly facilitate classification of the soil. Correct classification of the soil profile is more likely if the interactive computer program could not only "ask" questions of the sanitarian concerning the soil characteristics at a site, but could "show" the sanitarian specific examples from which to choose. These could be examples of soil features important to the particular soil's classification or they could be color pictures of two or three soil profiles. Using the VCR, the computer could even access tape segments illustrating specific soil features. For example, these could include the degree of effervesence of a particular soil horizon or the relative expression of structure or of how the soil fell apart upon handling or upon digging with a sharpshooter.

RANGES FOR SOIL AND SITE CRITERIA

It is evident from the previous sections that there exists a great deal of variation in the methodologies utilized for land suitability assessment. Each method of assessment requires measurement of a different set of soil and site parameters ranging simply from the percolation test to determination of detailed soil morphological features as well as assessment of specific soil classification criteria.

The acceptable value for each soil and site criteria is dependent upon a number of factors. These factors vary greatly across the country and include climate, geologic conditions, soil conditions, hydrology, potential for surface and groundwater supply development, past experiences with disease outbreaks, and land development pressures. For example, preservation of a large vertical separation distance from the soil absorption field to the water table may be deemed necessary in a mountainous state; however, the same criteria may be viewed as totally unrealistic in a flat coastal state where entire counties would be deemed unsuitable for any further development.

While no attempt has been made in this paper to list specific criteria used in every state there is value to that type of analysis. Not only does it identify subjects upon which there is a lack of a consensus of opinion, but it can also help one to recognize trends, particularly when previous data are also available. Results of a number of surveys of state regulations were reported by Kreissl (1982) and specific summaries are presented here.

The depth from the bottom of the soil absorption field to groundwater required in 1980 state codes typically was 120 cm (4 ft.) and ranged from 0 to 120 cm (0 to 4 ft.) with an average of 90 cm (3 ft.) separation (Kreissl, 1982). While for adequate purification the most appropriate vertical separation distance to the water table is about 90 cm (3 ft) (Small-Scale Waste Management Project, 1978; Bicki et al., 1984; and USEPA, 1980); 13 of 41 states surveyed in 1980 required less than the 90 cm (3 ft.) separation (Kreissl, 1982).

The separation distance between the bottom of the soil absorption field and bedrock or restrictive horizons ranges at least from 15 to 244 cm (0.5 - 8 ft.) (Ronayne et al., 1982; Shepard, 1981). Since 90 cm (3 ft) of unsaturated soil is necessary for adequate purification (Small-Scale Waste Management Project, 1978) it seems that a 90 cm (3 ft) separation distance from the soil absorption field to bedrock or restrictive horizons would also be adequate. However, mound systems installed with a 120 cm (4 ft) or greater separation distance over restrictive horizons such as fragipans or over impermeable bedrock malfunctioned in Pennsylvania (Hoover et al., 1981). The thickness of natural soil required over restrictive horizons or bedrock for proper functioning of sewage disposal systems also varies with the amount of annual precipitation and the slope according to Ronayne et al. (1982). They also address siting criteria for systems installed in shallow soils with saprolite (weathered rock) close to the soil surface. Kleiss (1985) has discussed criteria for evaluating the suitability of various saprolites in the Piedmont and Mountains of the Southeastern USA.

Maximum slope limits have typically been arbitrarily set at 20-30% for use of on-site systems. Many times these limits are set because of the expected occurrence of shallow soils on steep slopes and also due to the difficulty in operating excavation equipment on slopes greater than 30%. However systems have functioned adequately in deep soils on slopes up to 45% in Oregon (Ronayne et al., 1982). While hand-dug systems are costly, they are an economically feasible system on steep slopes in resort communities in western North Carolina where some undeveloped, 0.2 - 0.4 ha (1/2 - 1 acre), building lots cost $50,000

While fourteen states utilized the percolation test for site evaluation and design in 1947, thirty-six of thirty-nine reporting states in 1975 used the percolation test for system design (Kreissl, 1982). Nineteen of these used the perc test results alone for system sizing. Relatively few states have adopted a detailed soils evaluation as a replacement or as a companion for the percolation test. However, "the blind use of percolation tests for sizing the soil absorption systems does appear to be waining" (Kreissl, 1982).

Design loading (or application) rates for determining system size have decreased from 1947 to 1980 as the size of conventional soil absorption fields has increased, particularly for the sandier soils (Kreissl, 1982). Current application rates range from 5.0 to 3.3 cm (1.2 - 0.8 gpd/ft^2) for sandy soils to 0.8 to 0.0 cm/day (0.2 - 0.0 gpd/ft^2) for clayey soils depending upon shrink-swell potential (North Carolina Administrative Code, 1982; Hoxie and Frick, 1985). These correspond closely to loading rates in Wisconsin (Small-Scale Waste Management Project, 1978; Bouma, 1975) for soils that have a water table greater than 90 cm (3 ft) below the soil absorption field (see Table 2).

The loading rates for the coarser-textured materials are based upon many more in-situ measurements than for the finer-textured soils. There is, then the need for further testing, particularly in the finer-textured soils. Further sources of information on specific soil and site criteria and be identified from a recent bibliography concerning small wastewater flows (Dix and Karolchik, 1984).

Table 2. Recommended maximum loading rates based upon in situ measurements at septic tank soil absorption fields.[1]

Soil type	Texture	Loading rate[2]	
		cm/day	gpd/ft^2
I	Sands	5	1.2
II	Sandy loams	3	0.7
	Loams	2	0.5
III	Silt loams Some silty clay loams	5	1.2[3]
IV	Clays	1	0.2[3]

[1] Assumes the seasonal high water table is >3 ft below the soil absorption

[2] Bottom area only

[3] Should not be applied to soils from expandable clays. These types of soil might be used with lower loading rates.

FROM: Small-Scale Waste Management Project (1978).

SUMMARY

The soil system can provide an excellent medium for waste disposal. Numerous physical, chemical, and biological processes can be utilized in effecting waste treatment. Optimizing the use of these processes requires proper siting, design, installation, and maintenance of the system (Hoover et al., 1981). Many soil and site factors must be considered and their roles integrated to achieve effective treatment. Specific siting criteria will vary regionally as soil and climatic conditions change. It is clear, however, that the complexity of the soil system and its interactions with a given waste and waste disposal system requires a comprehensive and integrated soil evaluation approach to site assessment. This approach can be formulated in various ways but should encompass as many facets of soil science as possible. Care must be taken in single factor approaches that are too simplistic for the variability of the soil system. On the other hand, soil scientists involved in research relevant to this subject must be cognizant of the constraints on local regulatory agencies having the responsibility to implement an on-site program (National Science Foundation, 1983). While tremendous strides have been made in putting "science" into this program a great deal of education, training, and extrapolating remains undone.

REFERENCES

Bicki, Thomas J., Randall B. Brown, Mary E. Collins, Robert S. Mansell and Donald F. Rothwell. 1984. Impact of on-site sewage disposal systems on surface and ground water quality. Report to Florida Department of Health and Rehabilitative Services. Soil Science Department, Institute of Food and Agricultural Sciences, University of Florida, Gainesville, FL.

Bouma, Johannes. 1975. Unsaturated flow phenomena during subsurface disposal of septic tank effluent. J. Environ. Eng. Div., Am. Soc. Civil Eng., Vol 101. no. EE6, Proceedings Paper 11783.

Bouma, Johannes. 1977. Soil survey and the study of water in unsaturated soil. Soil survey papers No. 13. Soil Survey Institute, Wageningen, The Netherlands.

Bouma, Johannes. 1979. Subsurface applications of sewage effluent. In M.T. Beatty et al. (ed.) Planning the uses and management of land. Agronomy 21 : 665-703.

Dix, S.P., and T.L. Karolchik. 1984. Bibliography of Small Wastewater Flows. EPA 600/2-84-183. EPA National Small Flows Clearinghouse, West Virginia University,Morgantown, WV.

Fritton, D.D., J.H. Stahl, and G. Aron. 1982. A site evaluation model for effluent disposal. Proceedings of the Third National Symposium on Individual and Small Community Sewage Treatment. On-site Sewage Treatment. American Society of Agricultural Engineers Publication 1-82. St. Joseph, MI.

Hoover, M.T., G.W. Peterson, and D.D. Fritton. 1981. Utilization of mound systems for sewage disposal in Pennsylvania. In N.I. McClelland and J. L. Evans (ed.) Individual On-Site Wastewater Systems, Proceedings of the Seventh National Conference. National Sanitation Foundation. Ann Arbor, MI.

Hoxie, D.C., and A. Frick. 1985. Subsurface wastewater disposal systems designed in Maine by the site evaluation method: Life expectancy, system design and land use trends. Proceedings of the Fourth National Symposium on Individual Small Community Sewage Systems. On-site wastewater treatment. American Society of Agricultural Engineers Publication 07-85. St. Joseph, MI.

Kleiss, H.J. 1981. Soil ratings for ground absorption sewage disposal systems to be used with the Durham County soil survey. Soil Science Information Series. Soil Science Department, North Carolina State University, Raleigh, NC.

Kleiss, H.J. 1985. Saprolite in sewage treatment and disposal - a working paper. Report to the N.C. Department of Human Resources, Division of Health Services. Committee on Ground Absorption Rules and Regulations. Raleigh, NC.

Kreissl, James F. 1982. Regulations and policy regarding on-site water disposal systems. In A.S. Eikum and R.W. Seabloom (eds.) Alternative Wastewater Treatment. D. Reidel Publishing Co., The Netherlands.

National Science Foundation. 1983. Proceedings from a workshop on research needs relating to soil absorption of wastewater. Ward, R.C. and S.M. Morrison (eds.) Environmental and Water Quality Engineering Program, Division of Civil and Environmental Engineering, National Science Foundation, Washington, D.C.

North Carolina Administrative Code. 1982. Laws and Rules for sanitary sewage collection, treatment and disposal. Section .1900 Title 10, Department of Human Resources, Chapter 10, Health Services, Subchapter 10A, Sanitation.

Ronayne, M.P., R.C. Paeth, and S.A. Wilson, 1982. Final report, Oregon on-site experimental program. State of Oregon Department of Environmental Quality.

Shephard, J.E. 1981. On-site disposal in the granite state. In N.I. McClelland and J.L. Evans (ed.) Individual On-site Wastewater Systems. Proceedings of the Seventh National Conference. National Sanitation Foundation, Ann Arbor, MI.

Small-Scale Waste Management Project 1978. Management of small waste flows. University of Wisconsin. EPA-600/2-78-173 NTIS Report No. PB286-560. September 1978.

Steinbeck, S.J. 1982. Determining the acceptability for individual sewage treatment and disposal systems in North Carolina. Proceedings of 1982 Southeastern On-site Sewage Treatment Conference, North Carolina Division of Health Services, Raleigh, NC.

U.S. Environmental Protection Agency. 1980. Design manual - onsite waste water treatment and disposal systems. Office of Water Program Operation, Washington, D.C.

On-site domestic wastewater renovation system designs to overcome soil limitations

C. Roland Mote, Agricultural Engineering Department, The University of Tennessee, P. O. Box 1071, Knoxville, TN 37901-1071
Carl L. Griffis, Agricultural Engineering Department, University of Arkansas, Fayetteville, AR 72701

In the context of on-site domestic wastewater renovation, soils with limitations are soils that cause renovation systems to have difficulty, or be incapable of, accomplishing one or more of the system objectives. The principal objectives of a wastewater renovation system are:

1. To remove waste from water.

2. To convey water into one of the reservoirs (i.e., ground water, surface water, or water vapor in the atmosphere) of the hydrologic cycle.

Soils which function well as a major component of a wastewater renovation system use inherent biological, chemical, and physical processes to remove waste from water. As wastewater flows through soil, naturally occurring microorganisms oxidize organic matter to carbon dioxide and water vapor. Phosphorous and many other anions in wastewater either combine with aluminum, iron, carbonate, etc. to form relatively insoluble compounds, or they attach to the anion exchange sites on the soil particles. Comparably, many of the cations in wastewater attach to the cation exchange sites in the soil. Fecal bacteria, not being adapted to the soil environment, soon lose their viability. Finally, after having been in contact with the soil for an appropriate amount of time, renovated water enters the ground water reservoir, thus completing the objectives of a wastewater renovation system.

Two characteristics limit the utility of soils for conventionally designed filter fields for on-site domestic wastewater renovation:

1. Insufficient depth, and

2. Unsatisfactory hydraulic conductivity.

Soils with insufficient depth cannot provide adequate contact time for the renovation processes to be effective. Insufficient depth is a problem in both thin soils and soils in which there is a high water table.

The hydraulic conductivity of a soil may be unsatisfactory if it is either too high or too low. High hydraulic conductivity allows wastewater to pass quickly through the unsaturated zone of the soil and enter the groundwater reservoir. Some minimum contact time is necessary for the renovation processes to be effective. Thus, soils with high hydraulic conductivity may be unsuitable for use in conventionally designed filter fields.

Low hydraulic conductivity causes a long residence time for wastewater in the soil, which is desirable from a renovation standpoint. However, low hydraulic conductivity may also limit the ability of conventional filter fields to transmit water at the rate wastewater is being applied, thus causing the wastewater to eventually be emitted to the soil surface. Therefore, soils with low hydraulic conductivity may also be unsuitable for use in conventionally designed filter fields.

Design Approaches

Insufficient Soil Depth

Locate Seepage Trench Higher

The bottom of seepage trenches in conventionally designed filter fields is a minimum of 60 cm below the soil surface (U.S. Dept. of Health, Education, and Welfare, 1967). Since soil above the wastewater cannot contribute to the renovation objective, there is both a need for and the opportunity to make more soil available by locating the seepage trench higher in the soil horizon. Limitations due to insufficient soil depth can frequently be overcome by changing the design criteria so as to keep more of the natural soil below the bottom of the seepage trench.

Several people (e.g., Harper et al., 1982; Pask et al., 1985) have reported positive results with a modified standard filter field design. The modified standard design simply locates the seepage trench as high in the soil as necessary to provide the needed depth of soil below the trench. The trench bottom in a modified standard filter field can occur anywhere between the natural soil surface and a depth of 60 cm (Figure 1). Available soil at some sites may be so limited that moving the bottom of the seepage trench all the way to the soil surface will not provide the needed soil depth below the trench. In such cases the mound filter field design may be used to raise the trench bottom above the natural soil surface (Figure 2). Mounds, as described by Converse et al., (1978), have seepage trenches (or beds) installed in soil material, usually sand, hauled in from a remote site and piled onto the natural soil surface. Mounds give the system designer the opportunity to provide an adequate amount of soil at the site to meet the objectives of the filter field.

Lower The Water Table

For soils with high water tables, an optional approach is to make more soil available by lowering the water table (Uebler et al., 1984). Properly placed drain lines may be able to make enough soil available to meet the needs of the filter field by preventing the water table from

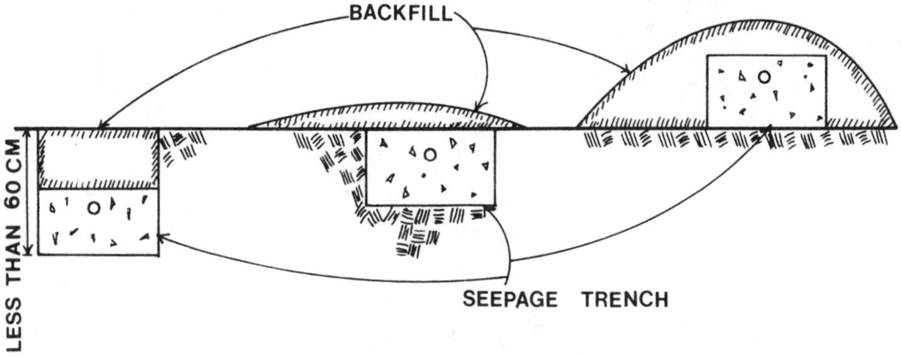

Figure 1. Modified standard filter field.

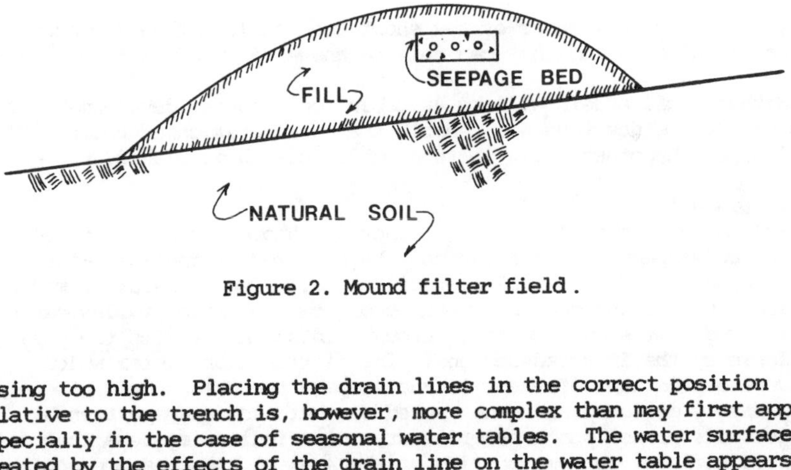

Figure 2. Mound filter field.

rising too high. Placing the drain lines in the correct position relative to the trench is, however, more complex than may first appear, especially in the case of seasonal water tables. The water surface created by the effects of the drain line on the water table appears in cross-section like the draw-down curves shown in Figure 3. The surface will be steeply sloped near the drain, and will have a reduced slope as the distance from the drain increases. The consequence of such a shape is that even though the water table can be kept away from the seepage trenches most of the time, there will be a delay in removing perched water near a trench after a heavy rainfall. The magnitude of the delay can be predicted by mathematical modelling, but the importance of the

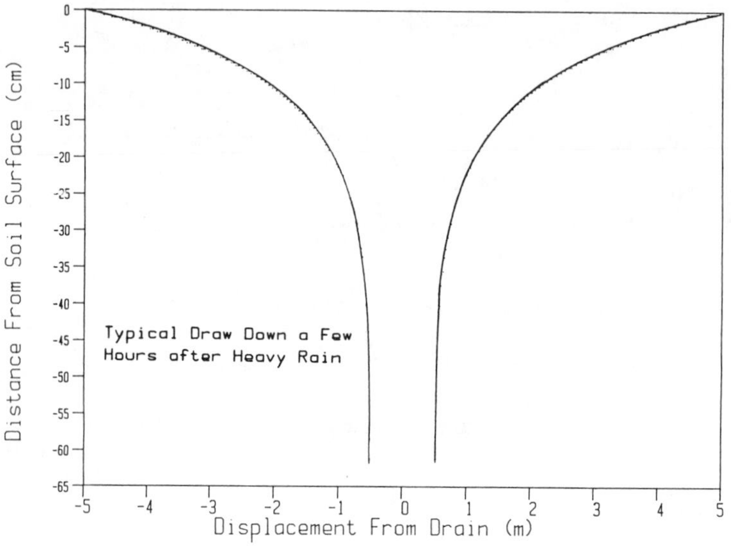

Figure 3. Draw down curves.

delay and the significance of the short-term failure that may occur is a philosophical question that can only be answered in a political context.

Mathematical models will be valuable tools in the development of this concept, since they will enable the consideration of philosophical implications hypothetically, before public health is risked.

Proper Distribution

Full advantage of the efforts described above for providing adequate soil below seepage trenches can only be realized if the wastewater is loaded onto all of the available soil. Limited soil resources made available by raising seepage trenches or lowering water tables can only be utilized if the wastewater is brought into contact with the soil made available by the improved designs. The distribution system which delivers septic tank effluent to the seepage trenches should, therefore, be carefully designed and constructed so that each seepage trench receives its proper share of effluent. Distribution systems that fail to apply effluent to some portion of the seepage trenches prevent the full utilization of the limited resources. In addition, an appropriately designed distribution system can force wastewater into all the soil above a seasonal water table (i.e., all the way to the soil surface). The ability to store water all the way to the soil surface above a seasonal water table will improve the ability of systems to accommodate stress periods, which produce seasonal water tables, without failing.

Low Hydraulic Conductivity

Proper Distribution

The limiting factor in soils with low hydraulic conductivity is that the volume of water movement per unit of soil area in a given amount of time is low. Designers of filter fields for installation in such soils face a special challenge in designing a system that will achieve the objective of conveying water to the groundwater reservoir. The slow rate of water movement through these soils should provide enough contact time that achieving adequate renovation is not of concern. To overcome the limitations presented by low hydraulic conductivity soils, a distribution system must expose the volume of wastewater to as many units of soil area as needed.

In soils with low hydraulic conductivity, it is also important that the infiltration rate of the soil in the bottom of the seepage trenches be maintained as high as possible. Several reports in the literature (e.g., Otis et al., 1978, and Otis, 1985) indicate that the formation of an infiltration-limiting crust at the seepage trench/soil interface may be managed by controlling the wastewater loading regime. The desirable loading strategy is to apply the wastewater in doses with an interval between doses sufficiently long to permit the interface to dry.

High Hydraulic Conductivity

Proper Distribution

The potential for rapid conveyance of wastewater through macropores to the groundwater in soils with a high hydraulic conductivity is not compatible with achieving the renovation objective. Application of wastewater in small doses over all the available soil will not saturate all the macropores at each dose. Unsaturated macropores permit the wastewater to flow along the surfaces of the large pores and be drawn into the intersecting micropores. Flow through the micropores is relatively slow. Slow movement of the wastewater through the soil provides time for the renovation processes to be effective. Therefore, a distribution system that will dispense, and evenly distribute, small-volume doses over all the seepage beds will help to overcome the limitations presented by soils with high hydraulic conductivities.

Review

The soil limitations that have been identified, and the suggested design approaches for overcoming each limitation, are summarized in Table 1. A review of the design approaches shows clearly that providing proper distribution is an essential part of the design approach for overcoming all problem soil conditions.

Distribution System Design And Maintenance

Mounds, discussed earlier as a means of overcoming limitations due to insufficient soil depth, can also be considered as a type of distribution system. A mound has a parallel distribution system

Table 1. Summary of soil limitations and suggested design approaches.

Problem	Solution
1. Insufficient soil depth	
a. Thin soils	1. Locate trenches higher in soil[a] 2. Provide proper distribution
b. Seasonally high water table	1. Locate trenches higher in soil[a] 2. Lower water table 3. Provide proper distribution
2. Unsatisfactory hydraulic conductivity	
a. Too low	1. Provide proper distribution
b. Too high	1. Provide proper distribution

[a]Mounds permit trenches to be located above the natural soil surface.

installed within it for applying septic tank effluent to the soil in the mound. But the mound itself serves to distribute water over a portion of the surface of the natural soil on which it is sitting. The area of the natural soil surface over which water may be distributed is proportional to the area of the base of the mound. Because of this distributing effect of mounds, they are a recommended design for overcoming limitations due to low hydraulic conductivity (Converse et al., 1978), and would probably also be effective in overcoming limitations due to high hydraulic conductivity.

The more typical distribution systems can be categorized as either serial distribution systems or parallel distribution systems. In a serial distribution system, wastewater is presented to each of the several seepage trenches in sequence. However, in a parallel distribution system, wastewater is presented simultaneously to each of the several seepage trenches at specified rates. Each trench in a parallel system may be loaded at the same rate or at different rates, if there is reason to load them nonuniformly.

Serial Distribution

In a serial distribution system a line from a pretreatment device, usually a septic tank, is connected to an initial seepage trench. All of the effluent either enters, or is given the opportunity to enter, the initial trench. Subsequent trenches are connected by overflow lines, drop boxes, etc., such that they only receive the effluent that does not soak into the soil from any previous trench. Thus, serial distribution systems do not necessarily distribute water over all available soil. Water will not reach the end of the last seepage trench until some water has failed to enter the soil from all previous trenches. Since the end of the last trench is the last resort, the entire system is not placed into service until the point of failing to meet the conveyance objective is approached.

Because a serial distribution system cannot routinely apply water to all available soil, it is probably not acceptable for filter field designs in situations where soil resources are limited.

Parallel Distribution

A distribution box with single inlet and multiple outlet lines has traditionally been used in filter fields where parallel distribution is specified. A distribution box is normally connected to a septic tank by a 10-cm pipe. Also, 10-cm pipes exit the box and connect to each seepage trench. Effluent trickles into the distribution box from the septic tank as a result of wastewater being added to the tank from the household. In theory, the flow into the box is uniformly divided among the several distribution lines going to the seepage trenches and is thus uniformly distributed among the trenches. In fact, the heads developed in a distribution box are so minor that most of the flow goes to the one or two lines whose inlets happen to be at the lowest elevations, even if elevation differences are only due to variations in the pipe wall thickness resulting from pipe manufacturing tolerances. The inability of systems relying upon the distribution box and 10-cm diameter distribution lines to provide uniform distribution has been well documented (Coulter and Bendixen, 1958; Converse, 1974).

In recent years, distribution system designs that will provide proper parallel distribution have been demonstrated (Mote, 1984). Properly functioning parallel distribution systems are characterized by small diameter, pressurized pipes (Figure 4). For systems serving a single residence, pipe diameters range between 2.5 and 3.8 cm for perforated distribution laterals, and 5 to 10 cm for manifolds and headers. Perforations in the distribution laterals are also small, normally ranging between 32 and 64 mm in diameter.

Parallel distribution system operating pressures normally are within the range of 0.3 to 1.5 m of water head. Sloping sites present an extra challenge to designers due to variations in head among the laterals resulting from elevation differences. One of two approaches commonly used to compensate for slope induced head variations is to vary the orifice spacing and/or diameter among the laterals (Cogger et al., 1982; Mote et al., 1981). The other approach is to judiciously insert lengths of small diameter pipe in lines supplying water to the distribution laterals so that the head lost to friction in the supply

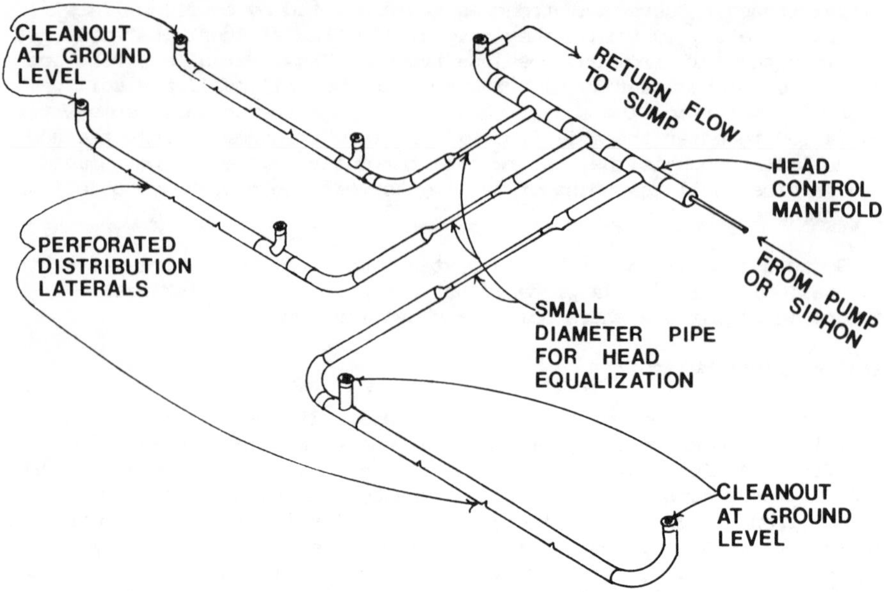

Figure 4. Typical pressurized system for parallel distribution on a sloping site.

line equals the incrementally higher head on a lateral due to its elevation being lower than the elevation of a reference lateral (Mote, 1984).

To establish pressure in distribution systems, effluent from the septic tank (or other pretreatment unit) must be accumulated and transferred in relatively large volume increments or doses. Dosing frequency, and thus dose volume, can be varied somewhat from system to system. Designers have the opportunity to use dosing frequency and volume as a means of overcoming some of the soil limitations described earlier. For example, small volume doses and a high dosing frequency may be desirable for soils with high hydraulic conductivity, whereas soils with low hydraulic conductivity will benefit more from a large volume dose and a low dosing frequency.

Doses of effluent may always be transferred to the distribution system with a pump. On sloping terrain, where the filter field is at a lower elevation than the septic tank, a dosing siphon, which uses no electrical power, may also be used to transfer effluent into the distribution system.

Parallel Distribution System Maintenance

A pressurized system that will achieve parallel distribution is not very complicated. However, it is considerably more complex than some systems traditionally used in filter fields, and will thus require more maintenance than is usually provided distribution systems. One of the conditions that results in a requirement for regular maintenance is the growth of slime on the walls of the pipe.

Growing, or living, slime has a soft, gelatinous consistency that is readily scoured by flow through orifices and appears to cause no problem. Dead slime, after sloughing off the pipe wall, develops a tougher consistency which can seal an orifice. Under steady operating conditions, slime growth and death rates are directly related to dosing frequency. A system with a high dosing frequency can be expected to require more frequent maintenance than a system with a low dosing frequency. Systems dosed once daily in Arkansas have been found to need cleaning approximately at two-year intervals. Operating systems left idle for a few weeks may experience a massive plugging when use is reinitiated, due to the death of all slime in the system during the downtime. An observed increase in the time required to transfer a dose of effluent into the filter field is diagnostic of plugged orifices.

Slime may be removed from lines by pushing a swab through each line with a stiff wire, or passing a high pressure stream of water through one open end and out the other of each line. Long radius bends installed at the time of construction to bring the end of each line to the soil surface greatly facilitate line cleaning.

Summary

There are basically two characteristics that limit the utility of soils for conventionally designed filter fields:

1. Insufficient depth, and

2. Unsatisfactory hydraulic conductivity.

Two design approaches available for overcoming depth limitations are locating seepage trenches higher in the soil, and lowering water tables in soils where depth limitations result from groundwater occurring at shallow depths. Proper distribution is also important for overcoming depth limitations, and it is the primary means available for providing properly functioning systems in soils with hydraulic conductivity limitations.

References

Cogger, Craig, Bobby L. Carlile, and Dennis Osborne. 1982. Design and Installation of Low-Pressure Pipe Waste Treatment Systems. UNC Sea Grant College Publication UNC-SG-82-03. Office of Sea Grant, NOAA, U.S. Department of Commerce, Washington, DC.

Converse, James C. 1974. Distribution of Domestic Waste Effluent in Soil Absorption Beds. Transactions of the ASAE 17(2):299-304.

Converse, J. C., B. L. Carlile, and G. W. Petersen. 1978. Mounds for Treatment and Disposal of Septic Tank Effluent. Home Sewage Treatment. ASAE Publication 5-77, ASAE, St. Joseph, MI, pp. 100-120.

Coulter, James B., and Thomas W. Bendixen. 1958. Effectiveness of the Distribution Box. Report to the Federal Housing Administration, Public Health Service, Robert A. Taft Sanitary Engineering Center, Cincinnati, OH.

Harper, M. Dean, Mary S. Hirsch, C. Roland Mote, E. Moye Rutledge, H. Don Scott, and Dee T. Mitchell. 1982. Performance of Three Modified Septic Tank Filter Fields. On-Site Sewage Treatment. ASAE Publication 1-82, ASAE, St. Joseph, MI, pp. 187-196.

Mote C. Roland. 1984. Pressurized Distribution for On-Site Domestic Wastewater-Renovation Systems. Agricultural Experiment Station Bulletin 870, University of Arkansas, Fayetteville, AR.

Mote, C. Roland, Earl Rausch, Jonathan Pote, and Richard Estes. 1981. Design of Systems for Pressure Distribution of Septic Tank Effluent in Sloping Filter Fields. Agricultural Expriment Station Report Series 257, University of Arkansas, Fayetteville, AR

Otis, Richard J. 1985. Soil Clogging: Mechanisms and Control. On-Site Wastewater Treatment. ASAE Publication 07-85, ASAE, St. Joseph, MI, pp. 238-250.

Otis, R. J., J. C. Converse, B. L. Carlile, and J. E. Witty. 1978. Effluent Distribution. Home Sewage Treatment. ASAE Publication 5-77, ASAE, St. Joseph, MI, pp. 61-85.

Pask, D. A., P. J. Casey, J. G. Vaughn, and D. Thirumurthi. 1985. Recent Research and Developments in On-Site Sewage Disposal in Nova Scotia. On-Site Wastewater Treatment. ASAE Publication 07-85, ASAE, St. Joseph, MI, pp. 155-164.

Uebler, R. L., S. J. Steinbeck, and J. D. Crowder. 1985. Septic System Failure Rate on a Leon (Hardpan) Soil and Feasibility of Drainage to Improve System Performance. On-Site Wastewater Treatment. ASAE Publication 07-85, ASAE, St. Joseph, MI, pp. 111-118.

U. S. Department of Health, Education, and Welfare. 1967. Manual of Septic-Tank Practice. Public Health Service Publication No. 526, Superintendent of Documents, U.S. Government Printing Office, Washington, DC 20201.

Soil treatment systems for small communities

B. L. Carlile, Department of Soil and Crop Science, Texas A&M University, College Station, TX 77843

On-site soil absorption systems are used for wastewater treatment in some 20 million rural and suburban households in the United States. When properly designed, installed, and maintained, such systems provide low-cost wastewater treatment superior to that of the most sophisticated secondary treatment plant. Failing septic systems, however, can pollute surface and ground waters and pose a risk to public health.

A conventional septic system consists of three parts: the septic tank, the absorption trenches or drainfield, and the surrounding soil. A typical septic tank is a concrete container which serves as a primary settling and digestion chamber. The solids are partially digested by anaerobic bacteria in the tank, while a liquid effluent flows from the tank into the absorption trenches.

The purpose of the absorption trenches is to transport the effluent to the soil where final treatment occurs. A typical system consists of several trenches which are 2-3 feet deep, 2-3 feet wide, and 50-100 feet long. They contain a 4-inch plastic distribution pipe and are lined with about 1 foot of gravel prior to being backfilled with soil.

Final treatment and disposal of the wastewater are done in the soil. Treatment occurs rapidly under aerobic conditions. Organic matter is consumed and anaerobic sewage bacteria are attenuated. Nitrogen is usually not fully treated, however. Ammonium is oxided to nitrate, which is then subject to leaching into the ground water. If the nitrate enters a reducing zone, it may be removed through denitrification. Phosphate is absorbed strongly to most soils, and is rapidly removed from wastewater. Coarse-textured soils with limited surface area will eventually become saturated with phosphate, however, and lose their treatment capacity.

Wastewater disposal is done through percolation of the treated wastewater into the ground water and evapotranspiration to the atmosphere. Evapotranspiration is often negligible during the winter months in most areas, so systems are designed based on disposal by percolation.

A number of problems can lead to the failure of septic systems. These include high water tables, shallow or impermeable soils, excessively permeable soils, overloading with wastewater, and poor design, installation, or maintenance. When the water table is too high, conditions become anaerobic beneath the absorption trenches, leading to incomplete wastewater treatment and the discharge of poorly treated wastes into the shallow ground water. Extremely high water tables can cause untreated effluent to surface, causing a direct public health risk and increasing the probability of runoff into lakes or streams. Impermeable soils and effluent overloading can also lead to surface failure, where the rate of wastewater flow into the trenches exceeds the rate of subsurface disposal. Systems in very permeable soils seldom experience surface failure, but the treatment capacity of the soil is limited by its relatively small active surface area.

COMMUNITY SOIL ABSORPTION SYSTEMS

The major factors in design of a satisfactory on-site waste disposal system for community waste flows can be summarized as follows: 1) distribution, 2) dosing, 3) sewage placement, and 4) water diversion.

Distribution cannot be over-emphasized in the design of any on-site system for large waste flows due to the need to spread sewage over a large land area. The effluent must be distributed evenly over this large area so as not to exceed the capacity of the soil to absorb the hydraulic load. Adequate distribution is extremely hard, if not impossible, to achieve in a conventionally designed gravity flow system. Some portion of the system is inherently overloaded which results in initiation of the clogging phenomena and hence the "progressive failure" observed in many such systems.

Dosing of effluent is equally important in maintaining the aerobic status of the soil system in and around the distribution trench and thus preventing the clogging or "slimming up" of soil interfaces and subsequent failure. Dosing concepts can be described as either 1) short-term dosing or 2) long-term dosing. Short-term dosing usually refers to multiple daily dosings of effluent into a single system with several hours or even days of resting and reaeration between each dose. Two to three doses per day have been shown to be optimum in systems designed and operated in North Carolina.

Long-term dosing refers to dual or multiple systems where one system receives all of the effluent for a specified time or until a problem occurs, at which time the effluent is switched to the alternative system. The assumption here is that the original system, when adequately rested, will regain its original capacity to accept and treat the sewage load and allow long-term, continuous operation by switching from one system to another.

Both dosing concepts as well as combinations and modifications of the two have been successfully utilized in several states to treat and dispose of sewage from individual homes as well as cluster developments, school systems, and mobile home parks with flows of up to 50,000 gpd (Carlile, 1979).

The design factor of <u>sewage placement</u> refers to the concept of placing the sewage in the soil zone or horizon most conducive to absorption, treatment, and reaeration. In soils with hardpans or restrictive clay horizons, this usually means at least a foot separation between the restrictive zone and the point of sewage injection. This minimum separation allows for lateral or horizontal flow of effluent away from the distribution trench before interception by the hardpan and allows for more uniform absorption through the restricting layer. This, coupled with enhanced treatment of the sewage in the soil above the restrictive horizon, greatly reduces the clogging of the soil at the "pan" interface. Generally, hardpans and restricting layers must be deeper than 36" for conventional gravity systems to function adequately on such sites.

The final design factor is that of <u>water diversion</u>. This factor becomes most important on sites located on concave or lower slope positions with soils having restrictive horizon somewhat near the surface. This combination results in the infiltration of rainwater into the more permeable surface soil and subsequent buildup or saturation of the zone just above the restrictive horizon. The water in the saturated zone then moves downslope through gravitational forces resulting in temporary perched water tables near the surface as the water accumulates at the lower slope positions. Without diversion of this water around and away from any sewage system located in the mid or lower slope position, the absorption trenches could be flooded and caused to temporarily fail during seasonable wet periods. Any system and particularly large systems placed in such positions must have interceptors or curtain drains placed immediately up-slope from the upper trenches to divert this lateral flow of rainwater from the site.

Low Pressure Pipe System

In consideration of the above design factors and the most practical and economical systems for development, one or more techniques can be utilized to achieve optimum system design for large waste flows. One system which has been utilized extensively in community systems is the Low Pressure Pipe System (LPP). This system optimizes all of the design parameters listed above and experience has shown the LPP system to offer improved efficiency in wastewater treatment as well as disposal in problem soil areas (Simon and Reneau, 1985).

The LPP system is a network of small diameter pipes placed 8 to 16 inches below the ground in narrow trenches of 6 to 12 inches width (Fig. 1). Effluent from septic tanks or other pretreatment systems is pumped uniformly throughout the LPP system in controlled doses to insure that each segment of the field is receiving equal loading. The pipes are usually 1-1/4 inch polyvinyl chloride (PVC) placed in a gravel envelope in each trench and contain small holes, usually 1/8 to 3/16 inches in diameter, spaced 3 to 5 feet apart. Under a low pressure (0.5 to 2.5 psi) supplied by a pump, the effluent flows uniformly through the holes and into each narrow trench containing the pipe. The dosing volume is controlled by dual float level controls or timers and provides for two or three pumping cycles each day which allows the soil to rest and reaerate between the doses. The effluent is distributed evenly throughout the entire system and is rapidly absorbed into the trench sidewall. Hundreds of such systems employed in large and small systems have shown that the major advantage of this system is the lack of development of "slimes" or clogging at the soil-gravel interface which is typical of a conventional gravity-fed system.

Soil and site criteria have been developed by evaluating the performance of LPP systems installed under various soil and site conditions in several states. Soil requirements generally require that at least the upper 16-18 inches be free of restrictive pans or a seasonal water table. This can be modified by bringing in a maximum of 6 inches of fill on those sites of inadequate depth. Sloping sites require extra care in design and installation to insure that the lower slope trenches do not receive excessive sewage doses during the pumping cycle or by lateral flow downslope from trenches above.

Another advantage of systems with pumps is the flexibility of multiple system combinations. For community developments where choice home sites may have only one common area suitable for waste disposal, there are operational and economic advantages to combining systems where a single disposal field will serve multiple living units. This is easily accomplished with the LPP system where each house has its own septic tank and the effluent is pumped or gravity fed to a common holding tank serving 5 to 15 household units. From here, the effluent is dosed into a split field system through use of dual, alternating pumps, insuring equal distribution and dosing over the entire field and allowing maximum resting and reaeration between doses. This type of system can be built in modules to allow the field to be expanded or added to as additional living units come on-line.

An additional advantage of the cluster system concept is the initial installation of a collector system serving the LPP fields. If at a later date, central sewage is available to the area, the collection system can be easily diverted to the central system without any disruption, construction, or major cost to the community.

Surface Irrigation Systems

Wastewater irrigation, particularly by surface spraying, has been viewed as an unacceptable system for individual home sewage disposal in most states. This was due to the danger to public health and the

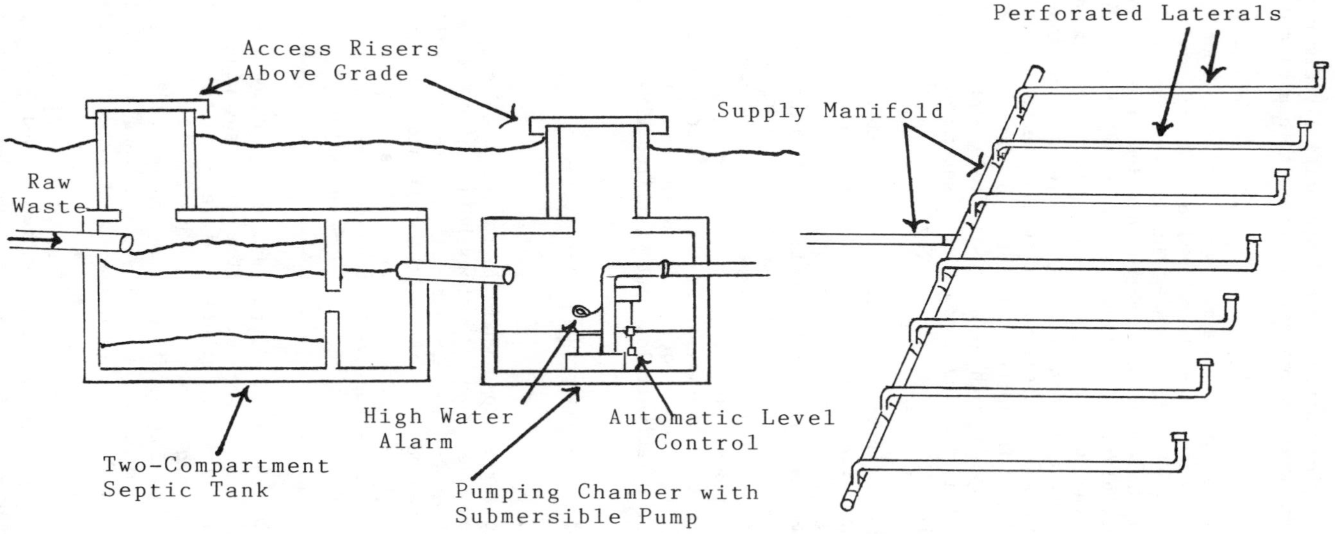

Figure 1. Basic Components of a Low Pressure Pipe System.

environment of applying poorly treated and partially disinfected sewage on the soil surface where potential contact with humans and vectors was possible.

With the improved pretreatment of effluent found in today's advanced aerobic treatment systems or improved sand filters, such as the recirculating sand filter, there is an increase in use of spray irrigation systems on soils where subsurface injection is impossible or impractical, particularly for large waste flows. Systems where the wastewater from the development is treated to a high degree in a pretreatment system, disinfected, and sprayed over a dense vegetative ground cover, offers one of the more effective means of waste treatment for problem sites. The potential for protecting public health and environmental quality is much greater with this approach than the more common practice of installing conventional septic systems and letting them "relieve" themselves into the nearby lake.

Uniform applications of treated, disinfected effluent over a vegetative surface with no runoff offers the most effective means of nutrient removal, pathogenic organism die-off, and organic degradation of the wastewater as any system available (Uebler, 1980). Sewage effluent pumps are used to recirculate sewage from various points in the treatment process, to lift sewage from tanks and distribute it uniformly over sand beds, and to pressurize and activate spray nozzles at timed intervals to uniformly apply the wastewater over a vegetative surface. All of this can be accomplished automatically, with minimum maintenance and at relatively low cost (Carlile, 1984).

Modified Gravity System

On small cluster systems where pumps are not desired to transfer and distribute sewage from the household units, improved system design, over and above the conventional septic tank-soil absorption trenches, may be needed to overcome the soil and site limitations of increased flows. At a minimum, these systems must be kept shallow, the distribution must be improved, heavy equipment must be restricted from the area of the field lines, and upslope water has to be diverted around the trenches.

The V-trench system, developed and utilized by the authors in several bay shore developments, offers some advantages in shallow placement and improved distribution where gravity-fed systems are utilized. The V-trench allows quicker distribution of effluent down the trench bottom, maximizes sidewall absorption (40% greater sidewall area), minimizes compaction of the sidewall and the destruction of soil permeability, and requires only about 60 percent of the gravel of a conventional system of equivalent size. In clayey soils, recent data from research projects where several modified systems were compared suggests that the V-trench system showed a 30 to 50 percent enhancement in sewage absorption when compared to the conventional vertical wall trench.

Another system being used as an alternative to conventional systems in cluster development is the gravelless pipe system, developed and tested initially in Texas as the SB2® system. This system utilizes 8 or 10 inch diameter rigid pipe wrapped with Drainguard® fabric to prevent soil particles from entering the pipe holes, since no gravel is used around the pipe or in the soil trenches. This pipe is usually placed in trenches 18 inches deep and 18 inches wide and backfilled with the soil material from the trench. The concepts of this gravelless system are to: (a) improve distribution of sewage in the trench, (b) improve solids retention in the pipe, (c) promote better aeration in the line, (d) eliminate the need for gravel and reduce heavy equipment traffic on the site, (e) increase capillary movement of sewage toward the soil surface, and (f) offer a system of easier installation for the septic contractor and of longer life for the homeowner. Studies conducted by the authors in both Texas and North Carolina indicate some enhanced performance of this system, over the conventional gravel system, if certain design and installation criteria are closely followed (Carlile and Osborne, 1981).

SUMMARY

From the standpoint of new improved technology, the future of community soil absorption systems has never been brighter. However, for the average community development to profit from this new technology and to advance the systems, they must be permitted and encouraged more by regulatory agencies throughout the state and nation. More and more regulatory agencies are now recognizing the limitations of all components which were conventionally used for large flow systems, including the septic tank, distribution box, and the field distribution lines. However, the overall effect of recognition of these limitations has been more restricted use of soil absorption systems. While there are some areas that could never properly function as a soil absorption site, there are many other areas that are being turned down or discouraged for on-site waste disposal needlessly. These sites could be designed for an alternative soil absorption system that would provide greater service life, low-cost development, and reduced environmental contamination when compared to other systems for waste disposal.

It is important to remember that the use of community on-site systems or any other alternative does not guarantee that water quality will be maintained. Septic systems are not the only source of pollution from community developments. Soil permeability is decreased and surface runoff dramatically increased when an area is developed. Drainage within the watershed may be changed. Population density greatly increases. All of these factors can affect water quality apart from septic systems.

Community on-site systems are more likely to fail if special precautions are not taken in their siting, design, installation, and management. Soil and site requirements need to be more restrictive than for individual systems. The designs are more sophisticated with multiple fields, pressure-dosing, and backup systems required. Construction must also be carefully done to avoid soil compaction over the large absorption area.

Finally, regular monitoring and maintenance of the system is vital. Because the entire distribution system lies underground, many problems with the septic system are hard to detect and easy to neglect. Although many conventional household systems may work for years with almost no maintenance, an unmanaged community system is not likely to be so forgiving. Pumps and electrical systems, loading rates, distribution efficiency, and condition of the field need to be checked regularly, and necessary maintenance performed promptly. This requires the funds and manpower to do the job properly.

Although a failing community septic system may pose less of a risk to water quality than a number of failing individual systems located adjacent to a water source, it nonetheless is a serious environmental and health problem. Without management and maintenance, community systems will fail, and if this happens we will lose an important wastewater treatment alternative.

REFERENCES

1. Carlile, B. L. "Sewage Effluent Pumps - The Future in Small Scale Waste Systems," Contractor Magazine, March 1, 1984.

2. Carlile, B. L. "Use of Shallow, Low Pressure Injection Systems for Large and Small Installations" in Individual Onsite Wastewater Systems, N. McClelland Ed. (Ann Arbor, MI: Ann Arbor Science Publishers, Inc. 1979).

3. Carlile, B. L. and D. J. Osborne. "Evaluation of SB2 Wastewater Disposal Systems in Montgomery County, Texas." (Published by ADS, Inc., Columbus, Ohio, May 1981).

4. Simon, J.J., and R.B. Reneau, Jr. 1985. Response to three clayey soils to a range of effluent flux densities. Agron. Abstr., American Society of Agronomy, Madison, WI, p. 31.

5. Uebler, R. L. "Demonstration and Evaluation of Alternative Treatment and Disposal Methods," Task C Report to Triangle J Council of Government. (Research Triangle Park, N.C., 1980).

Soil systems for community wastewater disposal--Treatment and absorption case histories

E. Jerry Tyler, Wisconsin Geological and Natural History Survey and Department of Soil Science, Small Scale Waste Management Project, University of Wisconsin, Madison, WI 53706
James C. Converse, Department of Agricultural Engineering, University of Wisconsin, Madison, WI 53706
Dale E. Parker, Parker and Associates, Inc., Madison, WI 53719

Subsurface soil absorption is commonly used for disposal of domestic wastewater. Wastewater from these systems usually enters the soil below the root zone. Therefore, these systems are treatment disposal systems offering no mechanism for waste utilization or storage. For maximum treatment, the wastewater must pass through the soil slowly allowing chemical and biochemical processes to occur and the transfer of gases into and out of the reaction zone.

There are three major components of a subsurface soil absorption system (Fig. 1). The wastewater from residences, and the domestic portion of the wastewater from small businesses is collected and conveyed to a location for pretreatment. The pretreatment unit is commonly a septic tank to collect solids; however, other devices such as aeration units have been employed. The soil absorption area is selected such that wastewater will infiltrate and percolate through the soil. Regardless of design, system goals are to absorb the wastewater so that it goes away, to treat potential pollutants protecting public health and the environment, and to have low maintenance and cost.

Early wastewater soil absorption systems served single family homes with very simple designs. As experience accumulated and later with research results, design, and management has been based on specific soil and site conditions. Also, new designs were developed and former ones were refined. Recently, with increased concern for public health and environmental protection from failing individual systems in small communities, and with trends in housing toward clusters, and multiple family units, larger soil absorption systems are being designed and installed.

Many of the early large soil absorption systems were installed using the criteria of small systems without regard for differences in scale. Some of these are performing satisfactorily; however, many are not performing as anticipated. The actual proportion of systems operating correctly is not known.

The purpose of this paper is to review some thoughts and experiences concerning community wastewater soil absorption disposal systems in Wisconsin. Two case histories are presented.

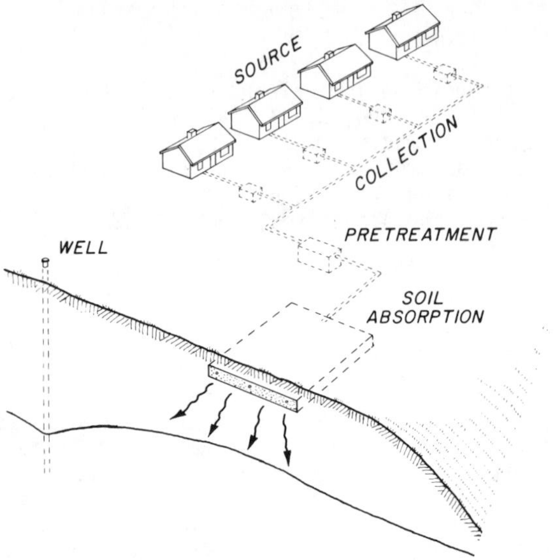

Fig. 1. Schematic of soil absorption system.

SOIL SYSTEM OPERATION

Wastewater for soil absorption usually has had minimal pretreatment. Septic tanks are used to collect solids; however, considerable amounts of pollutants go to the soil absorption area. Table 1 lists pollutant loadings of septic tank effluents from two communities and a typical household.

Table 1. Septic tank effluent character.

Parameter	Household (U.S. EPA, 1978)	Communities Kingston (Swed, 1985)	Westboro (Siegrist et al., 1983)
BOD_5, mg/L	138.0	219.0	168.0
SOC, mg/L	--	86.6	--
TSS, mg/L	49.0	72.6	85.0
TP, mg/L	13.0	6.02	8.1
TKN, mg/L	45.0	55.9	57.0
NH_3-N, mg/L	31.0	55.0	44.0
NO_3-N, mg/L	0.4	0.6	6.4
Cl, mg/L	--	148.0	--

Pretreated wastewater flows by gravity directly into the soil absorption area or into a dosing tank where it is collected and periodically pressure dosed to the soil absorption unit. For large soil absorption systems, wastewater is usually distributed uniformly over the soil infiltration surface with a pressure distribution network. The plane of infiltration is at the sand fill-aggregate interface of a mound or the soil-aggregate interface of subsurface systems.

Usually after a few months of operation a clogging layer forms at the plane of infiltration. This zone of soil clogging which restricts infiltration may be due to several factors and in most cases is probably due to a combination of factors. The most recognized factors are bacterial slimes and solids carry-over from the septic tank. Other clogging factors include decrease and reorientation of soil pores due to shearing and compressive forces during construction, fines from dirty aggregate placed in the absorption bed, and swelling of the soil. The layer is usually from one to several centimeters in thickness, black and slimy with smaller pores than the underlying soil. The clogging layer is depicted in Fig. 2 by dashed lines at the gravel soil interface.

Wastewater passing the clogging layer moves through the unsaturated soil matrix and then into the saturated zone. The effluent keeps moving to where it no longer has an influence on the infiltration of additional wastewaters.

During transport, wastewater constituents react with the soil. Some precipitate, while others are adsorbed, or become part of the microbial activity. Phosphorus is precipitated and adsorbed, nitrogen passing through the soil may first be retained by the CEC as NH_4^+ and later is nitrified passing to the groundwater as NO_3^-, and organic compounds are consumed by the microorganisms. Microorganisms utilize O_2 from the soil atmosphere while degrading the organic materials and releasing gaseous by-products. Normal operation of the soil microbial environment assumes that continued additions of nutrients, organics, and oxygen are available and that the gases and treated effluent containing such soluble substances as NO_3^- are transported away. If any of these mechanisms are retarded, chemical and biochemical processes will be different.

FLUID AND CHEMICAL MOVEMENT

Movement of soil gas and water is in response to a driving force and the ability of the soil matrix to transmit the materials (Fig. 2). The nature of the soil materials and the wastewater being applied combine to define the type and relative influence of processes that will occur under soil absorption systems. The dynamics of the reactions and flow should ultimately control system design dimensions.

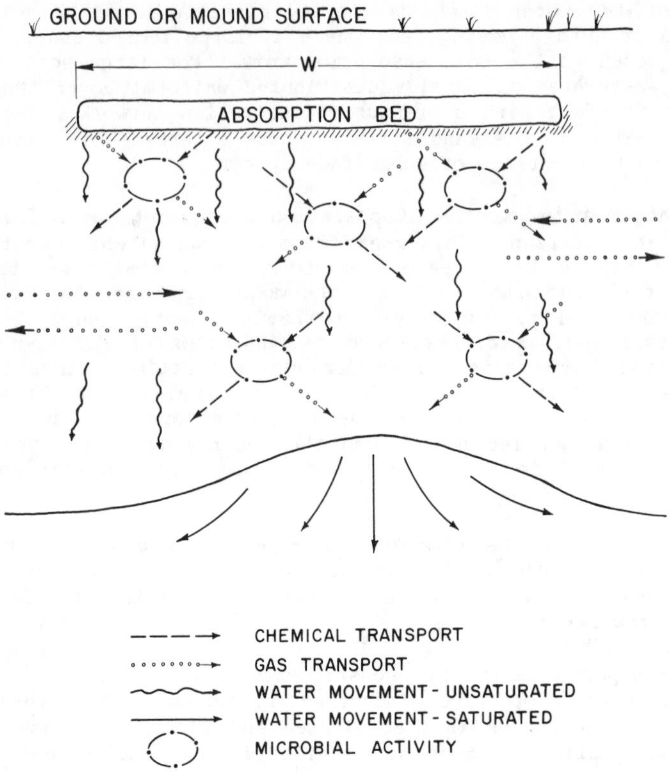

Fig. 2. Water, chemical, and gas transport and microbial activity beneath a soil absorption system.

Wastewater Movement

Wastewater soil absorption systems are installed in naturally unsaturated soils to hydraulically accept the effluent and to provide for better treatment. In most cases, the wastewater will reach the groundwater. In the absence of a clogging layer as in new systems, systems with low loading rates and high quality wastes, or systems after long resting, wastewater may enter the soil intermittently and locally at high rates inducing saturated soil conditions. In large and many small soil absorption systems, dosing and uniform distribution of wastewater is employed and often a clogging layer is present. This situation, particularly in fine-textured soils, induces a constant recharge of wastewater and the water flow system approaches steady state.

The average vertical hydraulic conductivity is influenced by the zone of slowest conductivity which is normally the clogging layer. Although the nature of the layer is not well known, the effect on wastewater flow has been defined (Bouma, 1975). Flow through a clogging layer is dependent on the wastewater head over the layer, the hydraulic conductivity of the layer, the layer thickness and the moisture potential of the soil. Unsaturated soil hydraulic conductivity of less than 0.01 m/d is not uncommon if intense clogging is present or system loading rates are low.

After passing the unsaturated soil, wastewater enters a zone of saturation at a restricting horizon or the groundwater. Under saturated flow conditions, the hydraulic conductivity is higher than in the unsaturated zone of the same material since all pores are water filled. Hydraulic gradients are controlled by the groundwater surface. Movement may be either vertical or horizontal depending on the nature of the materials and the hydraulic gradients. Wastewater continues to move away from the system to a point called the system boundary where it will no longer influence the infiltration of additional water (Tyler and Converse, 1985).

Chemical and Microbial Movement

Water moving through the soil transports microbes, waste chemicals added with the wastewater and by-products of microbial activity. Net transport depends on the water flow, the nature of the chemicals or microbes, and the adsorption and filtering capacity of the soil. Pathogenic organisms may die in the soil because of the hostile environment. Because soils have a net negative charge, anions such as NO_3^- and Cl^- are transported more easily than cations. For domestic wastewater, finer-textured, high organic matter soils have higher adsorptive and filtering capacity and therefore treatment capacity than coarse-textured soils. Long residence time due to unsaturated flow maximizes treatment. More research is needed concerning the amount and transport of volatile organic compounds.

Gas Movement

Microorganisms consume waste constituents and soil gases releasing chemicals and gaseous by-products (Fig. 2). There is a relationship between gas movement, microbial activity, and degradation of organic waste chemicals.

The transfer of O_2 to beneath the soil absorption area is by diffusion. Since gas transfer from the groundwater or through the absorption area when a clogging mat is present is unlikely, transfer to beneath the system can only be from the sides as depicted with the arrows in Fig. 2. The diffusion coefficient for the given soil moisture content beneath the system and the concentration gradient of the gas control the rate of O_2 transfer. If the rate of transfer from the edges of the system to the center is slower than the microbial consumption rate, then a volume of soil within system limits will be anaerobic.

Sikora and Corey (1976) estimated that in clay loam and finer soil, anaerobic conditions may prevail even under small soil absorption systems.

As systems get larger and the distance from the edge to the center under the absorption bed increases, the rate of diffusion from the edge to the center may not keep up with the consumption of O_2 by the microorganisms. Therefore, the center of the area would become anaerobic and the microorganisms would produce CH_4 and not nitrify the NH_4^+. This condition is different than what has been experienced under small systems and may lead to different treatment and more intensive clogging than if aerobic subcrust soil conditions existed.

If the soil beneath the system is to be aerobic, the total oxygen demand of the wastewater must be reduced, the O_2 diffusion rate increased or the distance from one side of the system to the other and the ground surface reduced. Reduction of the BOD is possible but would require additional pretreatment, and increasing the diffusion rate would occur with reduced water content of the soil or selection of sites with greater pore size. The easiest way to maintain aerobic conditions is to make systems narrow and shallow such that the diffusion of O_2 can meet the demand of the organisms for the system width. Mound systems are constructed above original grade with medium to coarse sands and are usually narrow. Therefore mounds are more likely to maintain aerobic conditions than subsurface systems.

Linear Loading Rate

Linear loading rate, the amount of wastewater that can be added per unit length of soil absorption system, has been proposed (Tyler and Converse, 1985) and methods for applying the principles to design have been suggested (Converse and Tyler, 1985). Linear loading rate is based on soil and site characteristics. During a site evaluation, the system boundary is determined. System boundary is the plane at which wastewater is totally assimilated into the environment or the extent of site evaluation. From the direction and magnitude of wastewater flow in each soil segment between the point of infiltration and the system boundary and the flow across the system boundary, the estimated maximum allowable wastewater addition is calculated. The sum includes estimation of the horizontal and vertical acceptance flow rates in the unsaturated soil and groundwater. Groundwater mounding height is also calculated to determine the operational separation distance between the soil absorption system and the groundwater. Wastewater additions, that can be transported in the most restrictive flow zone and maintain a desired separation between the system and the saturated zone, is the maximum amount of wastewater that can hydraulically be applied for a given length of system; this is system LLR. The system width is determined by dividing the LLR by the infiltration or loading rate. The length of the system is the quotient of the total wastewater volume and LLR.

Similar concepts apply to movement of soil gases. Assuming a horizontal diffusion rate and concentration gradient of O_2, the rate of supply per length of system or perimeter can be estimated. Knowing BOD of the wastewater the width can then be estimated. The amount of BOD that can be added with these assumptions is the BOD linear loading rate (BLLR). It is important to determine the characteristics of the waste and not assume all septic tank effluent has the same waste strength (Siegrist et al., 1985). BLLR and LLR may not be the same, and the most limiting should be used for design width.

Using these concepts suggests that systems should be narrow. Others have also suggested narrow systems (Siegrist et al., 1983; Swed, 1985). Just how narrow depends on the soil and site conditions and the wastewater quality. An estimate of the BLLR for the maintenance of aerobic conditions and LLR for the hydraulic disposal of the wastewater determines the width of the system. The length of soil absorption will vary depending on the total volume of wastewater.

For example, Fig. 3 depicts two equal area systems. The left system has two absorption areas of width W and length L, and the right system two trenches of width 2W and length L/2. The system of width W is narrow and the horizontal flow over the sloping restricting horizon is one half that of the wider system. The trenches on the left can be closer together than for the system with width of 2W. It will be easier to maintain aerobic conditions for the narrow system since the gas transport distance is shorter. Designing for narrow large systems should provide a soil environment similar to that expected beneath an individual household system.

TWO EXAMPLES

Two soil absorption systems which were state-of-the-art at the time of design and construction are presented as examples. Both systems have operated very well and considerable volumes of wastewater have been absorbed from each system; however, each has experienced some problems.

A conventional collection system and central septic tank followed by three subsurface soil absorption beds were put into operation in 1981 in Kingston Wisconsin. The central septic tank consists of three cells in series followed by a pump chamber. The 124,000 L/d design wastewater volume is distributed by a pressure distribution network to the 30 m x 55 m soil absorption beds (Fig. 4). The beds were installed in the subsurface horizons of the Gotham loamy fine sand, a sandy, mixed, mesic Typic Argiudoll. The soil beneath the absorption area is stratified fine and medium sands, but not as stratified as other soils of Wisconsin with water deposited parent materials. Beneath the system is 4 to 6 m of sand before reaching the groundwater. Bedrock is several meters below the groundwater surface.

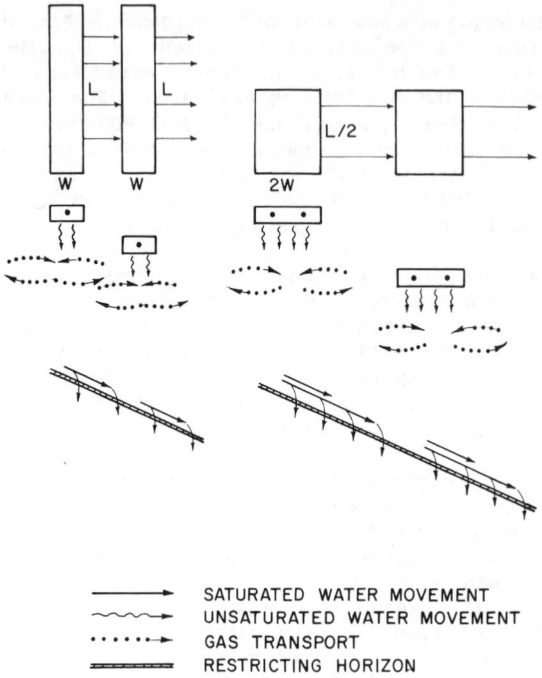

Fig. 3. Relationship of system width, length, and water and gas transport.

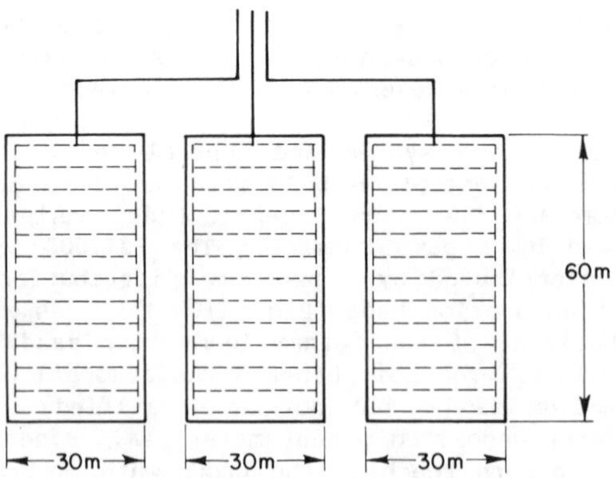

Fig. 4. Plan view of soil absorption area for the Village of Kingston.

Current loading is 58,100 L/d or about one-half the design load rate. The design called for one area to be rested on a rotating schedule; however, at the current time all areas are ponded and needed for the disposal of the wastewater. On several occasions wastewater has surfaced for short periods of time especially in wet weather. Community managers have been able to minimize the amount of surfacing by controlling applications to the separate cells.

After about one year of use, ponding was observed in the beds. Excavation to the infiltrative surface revealed the typical black soil clogging zone. Silt sized carbonate minerals were observed in thin sections prepared from the gravel soil interface. The carbonate particles probably came from the crushed, screened, and washed hard aggregate. The direct observation of the clogging layer has not offered an explanation for the rapid and intense clogging of the system. Soil clogging in combination with the underlying fine sand resulted in the low infiltration rates.

Based on the site evaluation, estimates of groundwater mounding were 15 to 45 cm above normal groundwater levels. However, no mounding was observed (M. Kopchak and J. Scarrow, 1985, Personal Communications), indicating that system operation is well within estimates made.

Monitoring soil solution and gases beneath the beds indicates that the soil environment is differnet than would be expected for a small soil absorption system (Swed, 1985). Oxygen was as low as 4%, CO_2 was elevated, and CH_4, while detected only under one bed, was as high as 2.5%. Nitrate-nitrogen was low at 30cm below the system, but increased to about 11 mg/L at 2 m. Total N decreased in this distance. Swed (1985) found that the oxygen was reduced beneath the soil absorption areas even after a three-month resting period. Another Wisconsin system had more extreme yet similar subsoil gas conditions (Siegrist et al., 1985).

Clogging developed earlier with higher intensity than expected. In addition to possible physical clogging, it is likely that the anaerobic conditions in the soil beneath the system contributed to the biological clogging. It appears that the clogging mat has reduced the hydraulic conductivity to about one-half the anticipated conductivity, and that the LLR is not the limiting condition on this sandy soil. The aeration status may be contributing to the early and intense clogging and therefore the BLLR may be the limiting factor.

A mound wastewater soil absorption system for a mobile home court was designed using the principles of LLR (Fig. 5). The system consists of six separate mounds (three groups of two) used on a daily rotating basis. The mounds are located in a three-hectare area. The site is a slightly convergent headslope with slopes of 6 to 8%. The soil is mapped as the Rockton silt loam, a fine-loamy, mixed, mesic Typic Argiudoll. The soil is platty, mottled and underlain by dolomitic bedrock.

The linear design rate is 230 L/m per d for the 61,650 L/d of wastewater. The total length of the six mounds is 268 m with an application width in the mound of 4.6 m or the edge-to-center distance of 2.3 m. Based on BOD_5, the BLLR is about 50 g/m per d (Table 2). The ultimate BOD needs to be satisfied and is not included in these figures. After two years of operation there is no evidence of ponding in the beds. The only known problem with the system has been pump failure.

The Kingston subsurface system is about six times wider than the mobile home park mounds. The LLR and BLLR differ by a factor of five. If the BOD_5 were not equal the ratio of BLLRs of the systems would have differed from the ratio of the LLRs. More research is needed to determine BLLRs and LLRs that should be used for design purposes, however, they should be kept low and systems should be narrow.

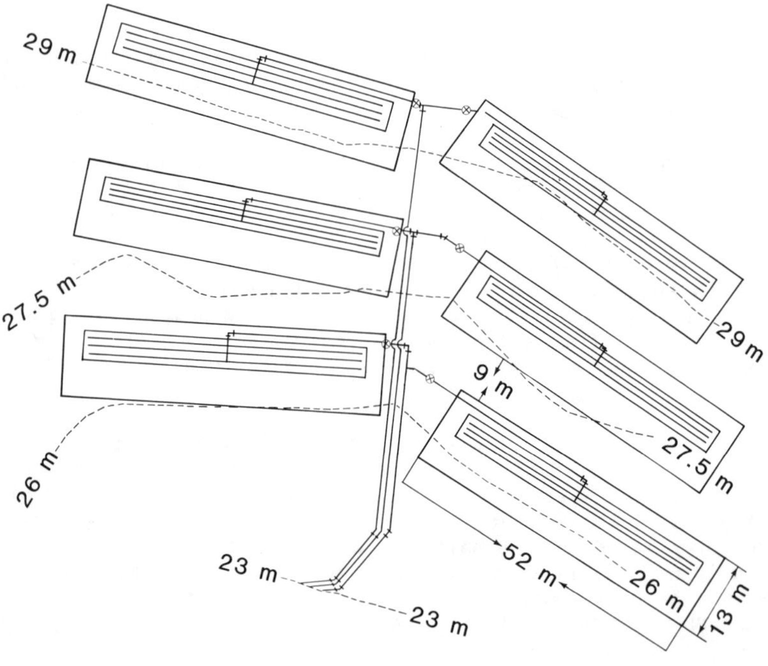

Fig. 5. Plan view for a trailor court mound wastewater disposal system.

Table 2. Design flow, system width, and design LLR and BLLR for example systems.

System	Flow	Width	LLR	BLLR*	Edge to Center
	L/d	m	L/m per d	g/m per d	m
Kingston	124,000	30	1130	247	15
Mobile homes	61,650	4.6	230	50	2.3

*Based on BOD_5 of 219 mg/L.

IMPLICATIONS TO SITE EVALUATION AND USE

Large wastewater soil absorption system design selection should be based on the maximum hydraulic acceptance of the landscape and landscape segments. If aerobic soil is desired beneath the absorption bed, then an estimate of the oxygen demand of the wastewater and rate of transfer of the oxygen toward the center of the absorption area must be made. Each process will estimate the hydraulic or oxygen demand linear loading rate that is used to estimate system width. Knowing the system width and the total wastewater to be applied, the system length can be determined.

Procedures for collecting soil and site information to be used as a basis for the design selections of the large system are the same as for the small system. However, the extent and detail needed for an adequate large system site evaluation will often be greater than for small systems and more information about the saturated zone and gas transport is necessary. Collection of information inside the system boundaries needed to estimate infiltration, saturated vertical and horizontal hydraulic conductivities, and zones of soil saturation must be based on a knowledge of soil absorption system operation and soil science.

Applying the site evaluation information to the concepts of the operation of the systems will offer more reliable system design than using the guidelines for small systems. Expecting low tolerance of calculated design values based on site evaluation information is unrealistic for most sites, but in some cases design based on the concepts with limits set by experience may be adequate. Personnel doing site evaluation and design selection for large system design must understand the principles and processes for successful soil absorption of wastewaters.

The concept of linear loading rates has had limited use and verification. More research and experience is needed to test the concepts and refine the procedures for application.

REFERENCES

Bouma, J. 1975. Unsaturated flow during soil treatment of septic tank effluent. Journal of the Environ. Engr. Div., ASCE, 101(EE6):967-983. Proc. Paper 11783.

Converse, J.C., and E.J. Tyler. 1985. The Wisconsin mound systems--siting, design and construction. 5th Northwest Wastewater Treatment Short Course On-site Wastewater Proceedings of the Treatment: Environmental Significance. Seattle, WA.

Siegrist, R.L., D.L. Hargett, and D.L. Anderson. 1983. Vadose zone aeration effects on the performance of large subsurface wastewater absorption systems. Pages 223-244 in Characterization and Monitoring of the Vadose (Unsaturated) Zone. Proceedings of NWWA/W.S. EPA Confer- ence on Characterization and Monitoring of the Vadose (Unsaturated) Zone, Dec. 8-10, 1983, Water Well Journal. 223-244.

Siegrist, R.L., D.L. Anderson, and J.C. Converse. 1985. Commercial wastewater on-site treatment and disposal. Pages 144-154 in On-site Wastewater Treatment, Proc. 4th Natl. Symp. on Individual and Small Community Sewage Systems. ASAE, St. Joseph, MI.

Sikora, L.J., and R.B. Corey. 1976. Fate of nitrogen and phosphorus in soils under septic tank disposal fields. Trans. of the ASAE, 19(5):886, 867, 869, 870, 875.9, 870, 875.

Swed, F.M. 1985. Performance of a community soil absorption field. An independent study report. Dept. of Civil and Environmental Engineering, University of Wisconsin.

Tyler, E.J., and J.C. Converse. 1984. Soil evaluation and design selection for large or cluster wastewater soil absorption systems. Pages 179-190 in On-site Wastewater Treatment, Proc. 4th Natl. Symp. on Individual and Small Community Sewage Systems. ASAE, St. Joseph, MI.

U.S. EPA. 1978. Management of small waste flows. EPA 600/2-78-173. Cincinnati, Ohio.

Development of minimum technology for hazardous waste landfills--Case history

Olaf L. Weeks, Environmental Management Department, Waste Management, Inc., Oak Brook, IL 60521
W. R. Schubert, Waste Management of North America, CID Corp., P.O. Box 1309, Calumet City, IL 60409

In recent years, much attention has been focused on the isolation of hazardous waste in landfills for the protection of human health and the environment. The most prevalent mechanism for isolation has been various landfill liner systems. Prior to the Resource Conservation and Recovery Act (RCRA) in 1976, owners and operators of waste disposal facilities had no federal regulations to specify what type of liner, if any, was used. Many of the owner/operators relied largely on in-situ or recompacted soil liners. The reliance on soil liners has been largely due to the economy of the installation and the idea that soil, being the weathered end product of geologic materials, was chemically inert and resistant to chemical attack by the waste to be isolated. Only in instances where suitable soils were not available would other liner materials be used. Since the inception of RCRA, the requirements for liners have evolved into composite systems consisting of synthetic membranes and soil materials.

In the following sections the various liner types which have been specified are presented. Also discussed is the recent U.S.EPA Minimum Technology requirements and various approaches to meeting these requirements in practice. Among these is the system presently being utilized at the Waste Management of Illinois, Inc., CID Landfill in Calumet City, Illinois.

Liner Systems in Waste Disposal Facilities

A liner system in a modern waste disposal facility should be considered as an element in an overall engineered system. To understand this element, it is beneficial to look at a water balance of a landfill (Figure 1).

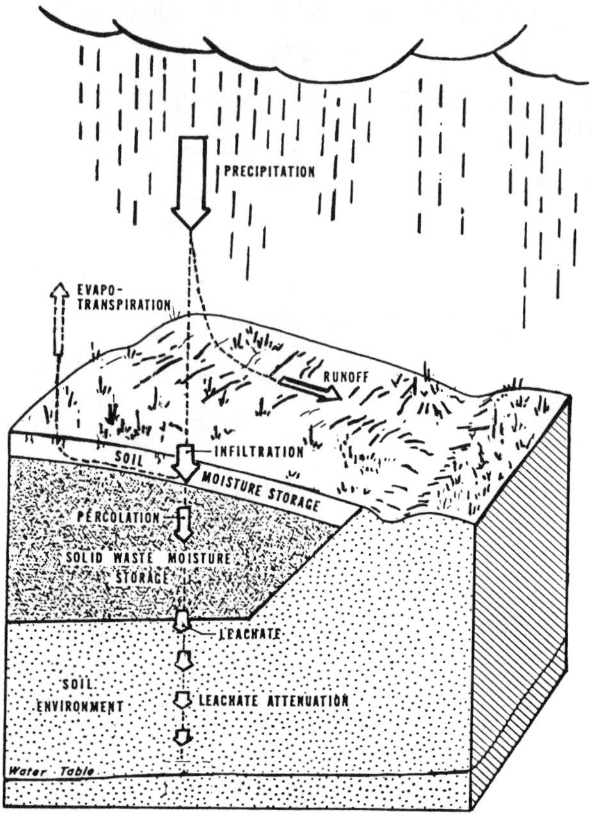

Figure 1. Landfill water balance (EPA 1983).

Precipitation falling on a landfill can either flow over land surface as surface runoff or infiltrate the landfill cover. Part of the infiltration is returned to the atmosphere through evapotranspiration, while the remainder is either stored in the cover or enters the landfill. The water which eventually reaches and percolates through the waste material is transformed into leachate by collecting soluble chemical species and particulate matter from the waste. The leachate moves vertically through the landfill, controlled by the waste permeability and internal landfill gradients, until encountering a physical or hydraulic barrier. Unless such a barrier is present, a potential exists for leachate migration into underlying soils and groundwater.

In order to prevent or reduce the amount of leachate produced and the amount which can accumulate at the base of a landfill, the features shown in Figure 2 should be incorporated into the landfill.

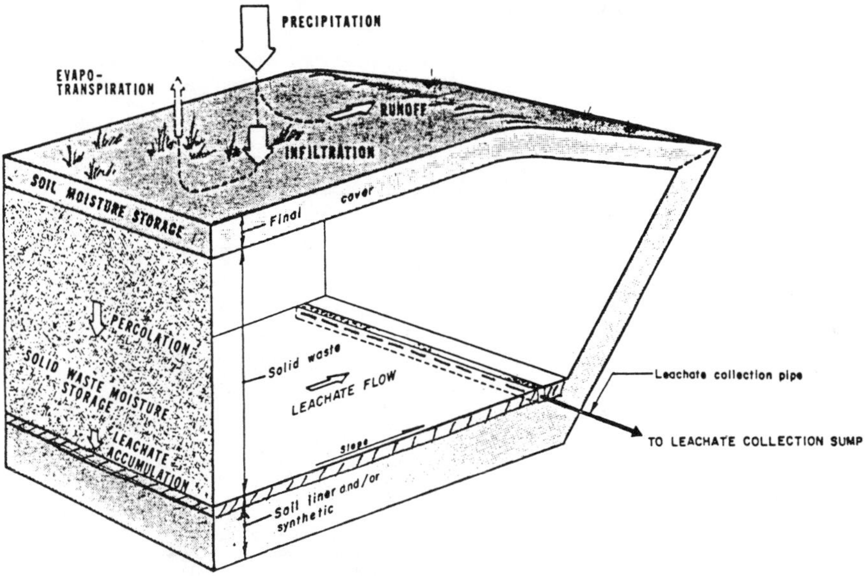

Figure 2. Engineered features for reduction and collection of leachate (EPA 1983).

These features are as follows:

-- A final cover/top liner which reduces infiltration and promotes runoff and evapotranspiration.
-- A low permeability bottom liner which will contain any leachate generated by water infiltrating through the final cover/top liner.
-- A leachate detection/collection system which will allow monitoring for and removal of leachate if and when it appears.

Both the final cover/top liner and bottom liner systems can range from a simple soil liner to a system with multilayers of soils, polymeric membranes, geotextiles, and drainage media.

Regulatory Requirements

Table 1 chronologically outlines the U.S.EPA regulatory requirements for hazardous waste landfill liners.

Table 1. Regulatory requirements for landfill liners.

Permitted Facilities				
Year	Promulgation date	Effective date	Subpart and section	Summary
1976	Resource Conservation and Recovery Act			
1982	July 26	Jan. 26, 1983	N (Landfills)	
			264.301 Design and operating requirements	Liner and leachate collection system required unless site is unique.
			264.302 Double-lined landfills	Exempt from ground water monitoring requirements if double liner is used.
			264.310 Closure and post-closure care	Provide and maintain final cover.
1984			Hazardous and Solid Waste Amendments	
	Nov. 8	May 8, 1985	Minimum technology requirements for landfills - two or more liners and leachate collection systems. - groundwater monitoring.	

Interim Status Facilities				
Year	Promulgation date	Effective date	Subpart and section	Summary
1976	Resource Conservation and Recovery Act			
1980	May 19	Nov. 19	N (Landfills)	
			265.310 Closure and post-closure	Place and maintain final cover.
			265.314 Liquid waste	Install liner and leachate collection system.
1982	July 26	Jan. 26, 1983	N (Landfills)	
			265.310 Closure and post-closure	Final cover required.
			265.314 Special requirements for liquid waste	Liner and leachate collection system required.
1984	Nov. 8	May 8, 1985	Expansion of Interim Status landfills to meet same technical requirements as permitted facilities	

The Resource Conservation and Recovery Act of 1976 is the primary federal law which regulates hazardous waste disposal facilities. The law charges USEPA with the administration of the program. The first regulations pertaining to hazardous waste landfill liners were promulgated on May 19, 1980. The liner requirements are found in the 40 Code of Federal Regulation (CFR) Part 264 (permitted facilities) and Part 265 (interim status facilities). Changes and additions to the regulations were made on July 26, 1982. On November 8, 1984 the Hazardous and Solid Waste Amendments were enacted which made further changes to the liner requirements of 40 CFR 264 and 265.

A brief description of these regulations and the liner designs used to meet the regulations appear below.

May 19, 1980 (Interim Status Facilities.) -- As part of closure and post closure requirements, a final cover was to be placed and maintained. In instances where liquid waste was to be disposed of in landfills, a liner and leachate collection system was to be installed. Figures 3 and 4 show a typical bottom liner and final cover design used to meet these requirements.

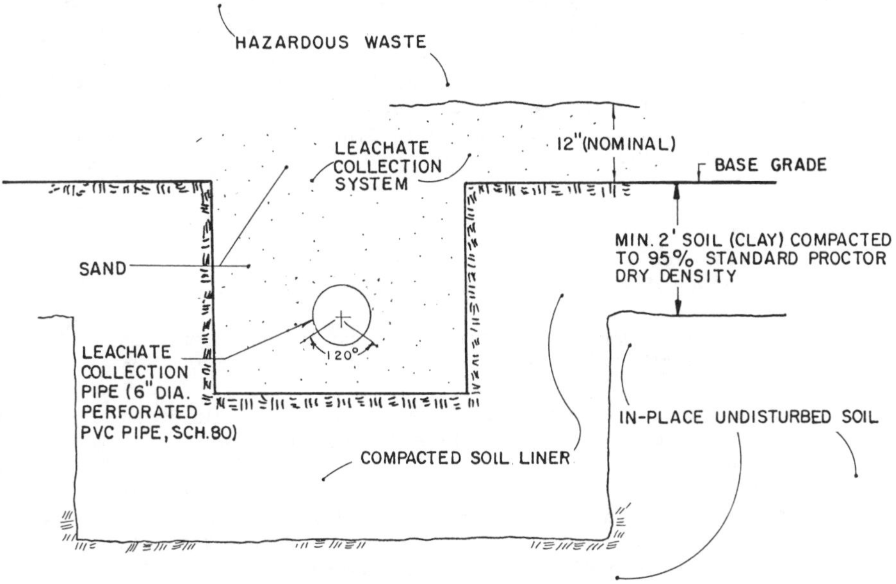

Figure 3. RCRA 1980-Bottom liner and leachate trench detail.

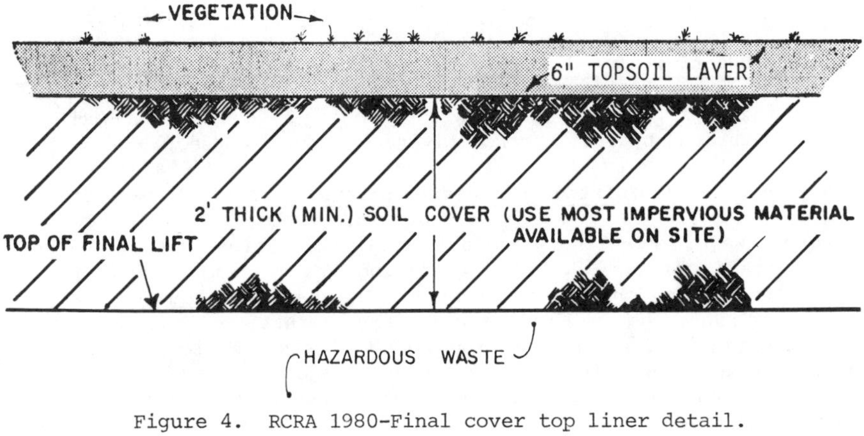

Figure 4. RCRA 1980-Final cover top liner detail.

July 26, 1982 (Permitted and Interim Status Facilities) -- These regulations required both types of facilities to have liner and leachate collection system as well as a final cover. In the preamble to these regulations, it was stated that only "de minimus" penetration of the permeant into the liner would be acceptable during the active life of the facility. "De minimus" penetration was described as that degree which occurs in synthetic membrane liners. Penetration that normally occurs in clay liners was not considered "de minimus." In this manner, USEPA effectively set best available technology standards for liners, specifically the use of synthetic membrane liners. Much of the impetus for using synthetic liners was based on research performed at a number of institutions on the compatibility of soils with liquid hazardous waste. Much of this research involved the use of pure organic solvents which resulted in dramatic increases in permeability of soil materials. The validity of this research in application to disposal sites is suspect and has been the subject of much debate.

U.S.EPA issued a guidance for liners (U.S.EPA, 1982) used in surface impoundments and landfills. The liner system would include the following:

Bottom liner

-- a single soil (clay) or synthetic material as a minimum. If waste would remain at closure, a synthetic liner would be used.

-- if an impoundment was designed to be in use longer than 30 years, a primary synthetic liner and a secondary clay liner would be used.

-- the synthetic liner would be a minimum of 30 mils thick in addition to being physically and chemically resistant to the waste.

-- soil liners would be a minimum of 2 feet thick with a saturated hydraulic conductivity of 1×10^{-7} cm/sec. The soil liner would be thick enough to contain waste within the liner during the operating life of the system.

Cap (final cover) liner system

-- The cap would be no more permeable than the bottom liner and consist of a two foot thick vegetated top cover, a 30 centimeter thick granular layer with a minimum hydraulic conductivity of 1×10^{-3} cm/sec., and a lower permeability layer consisting of a synthetic membrane liner, 20 mil minimum thickness, and a soil layer, minimum 2 feet thick, with saturated hydraulic conductivity of $1\text{x}10^{-7}$ cm/sec.

Nov. 8, 1984 (Permitted and Interim Status Facilities) -- The Hazardous and Solid Waste Amendment requires, at the base of landfills, the use of two or more liners and a leachate collection system. USEPA issued a draft guidance document outlining the minimum technology requirements of the double liner system. The document required that a double liner system be used with a secondary leachate collection system between the two liners. It also required that a synthetic membrane be used for the uppermost liner and that clay or a composite clay/synthetic membrane be used for the bottom liner. In subsequent guidance,USEPA largely discounted the suitability of clay alone as a bottom liner and

Figures 5 and 6 show details of liner and cap (final cover) systems designed to meet the above requirements.

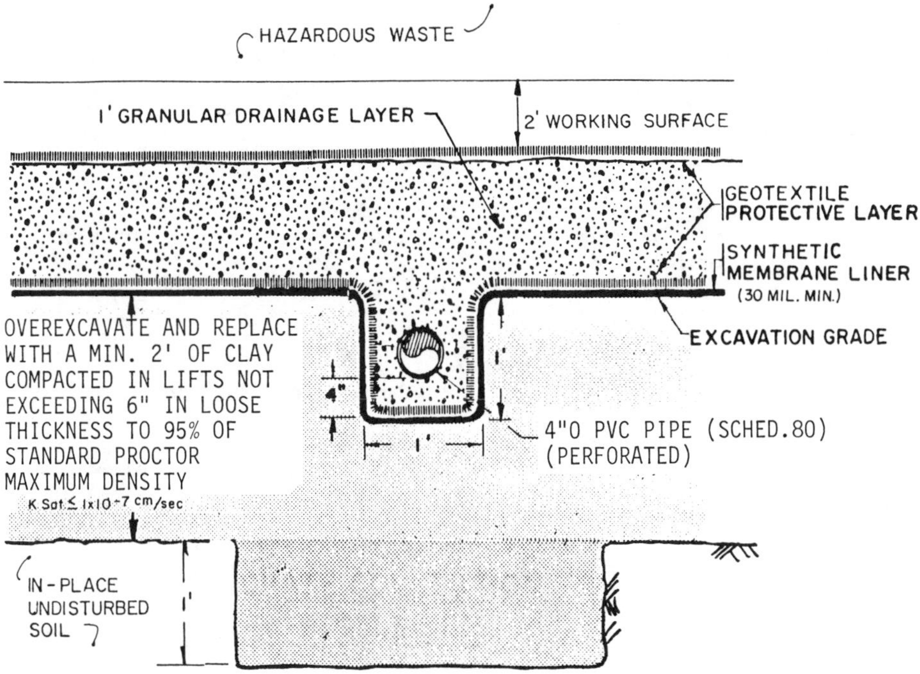

Figure 5. RCRA 1982-Bottom liner and leachate collection trench detail.

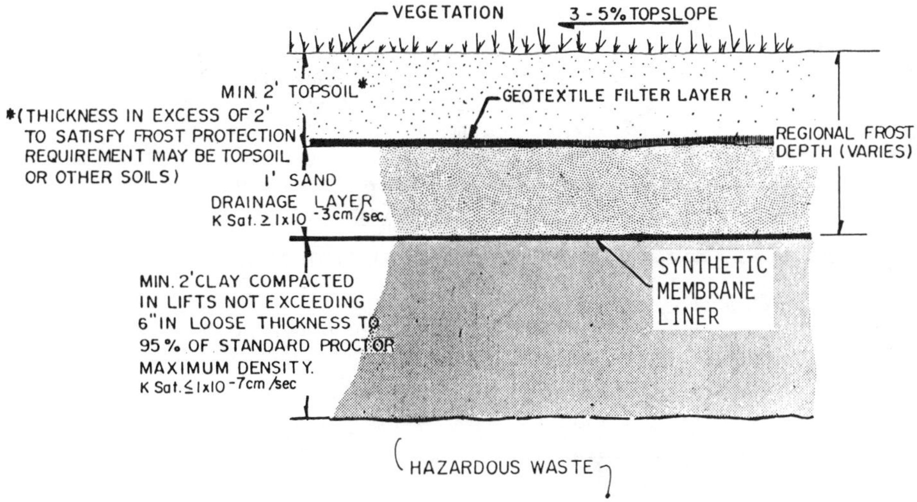

Figure 6. RCRA 1982-Final cover top liner detail.

provided breakthrough time requirements be used as bottom liner performance criteria. In evaluating breakthrough in clay liners, very conservative assumptions were required to be made regarding hydraulic properties of geologic materials.

The elements of the double liner system consist of a primary leachate collection and removal system, a primary synthetic liner, a secondary leachate collection system, and a secondary composite synthetic/clay liner. The main requirements of the liner are given below.

Primary and secondary synthetic liners

-- must be a minimum of 30 mil thick.
-- must be chemically resistant as determined by EPA test method 9090.
-- design must protect liner from operating and service loading.

Composite synthetic/clay liner

-- The synthetic component of this liner must prevent liquid penetration of the liner during the period of post closure monitoring.
-- Clay liner must be a minimum of three feet thick with a saturated conductivity of $1x10^{-7}$ cm/sec or less.

Liner systems designed to meet minimum technology requirements are illustrated in figures 7 through 10.

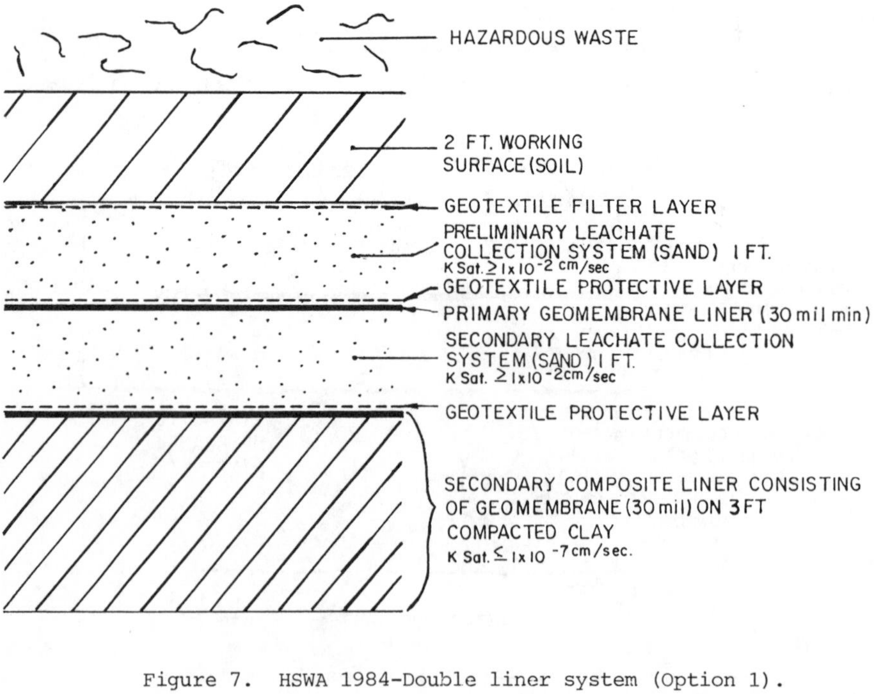

Figure 7. HSWA 1984-Double liner system (Option 1).

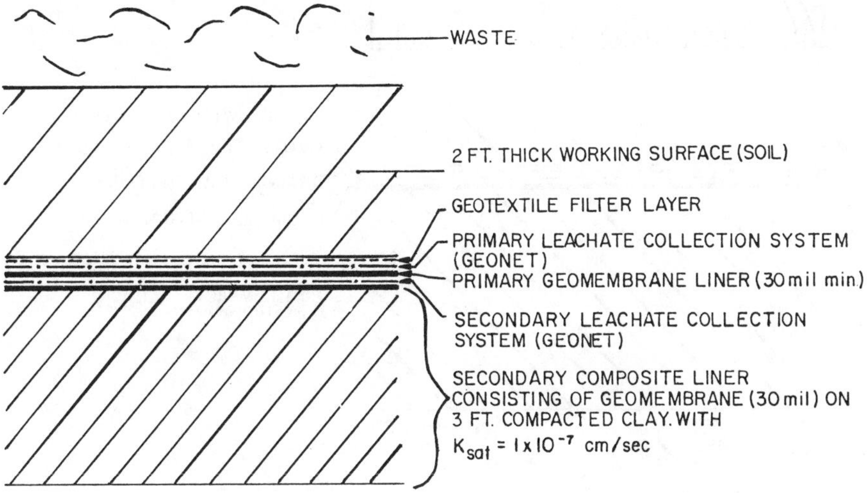

Figure 8. HSWA 1984-Double liner system (Option 2).

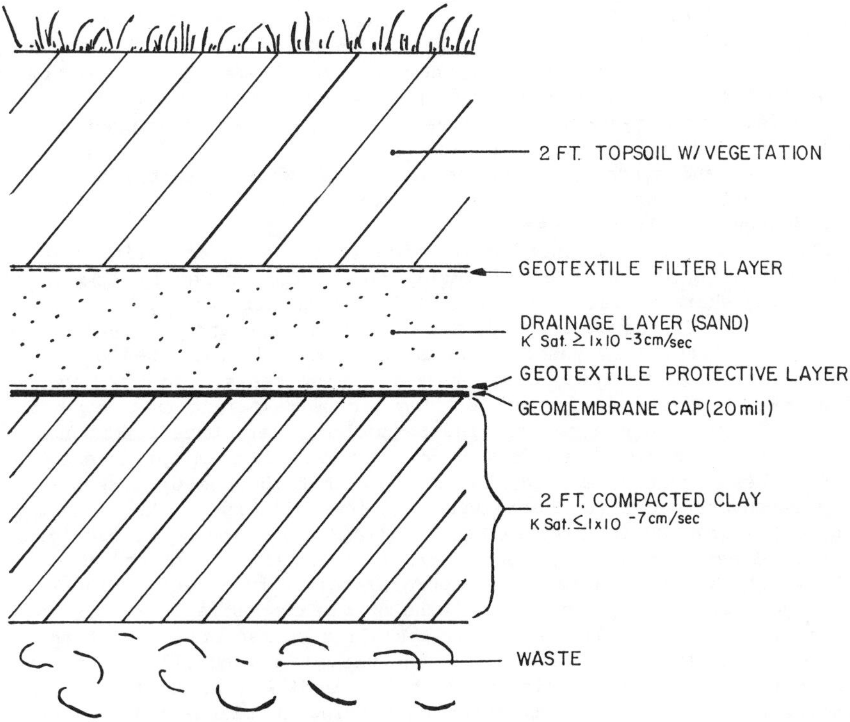

Figure 9. HSWA 1984-Final cap (Option 1).

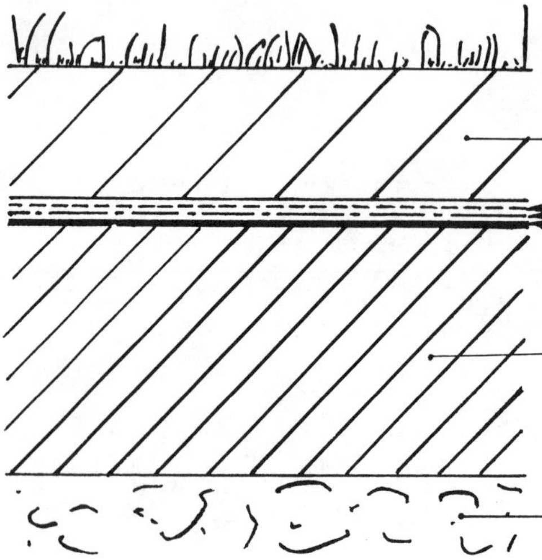

Figure 10. HSWA 1984-Final cap (Option 2).

Liners Designed to Minimum Technology Requirements

After a review of the evolution of the liner requirements, it should be obvious that the key element in any liner system is the synthetic and clay liners. The other key element is the leachate collection/detection system, but this will not be discussed here.

The clay and synthetic membrane materials, while both capable of hydraulic containment, exhibit distinct properties which are pertinent in the design of a lining system.

USEPA has contended that penetration into the liner during the active life should be severely restricted. The extremely low permeability characteristics of synthetic membrane liners appear to meet that criteria. On the other hand, many designers are uncomfortable with the susceptibility of synthetic membranes to failure, due to manufacturing defects, punctures, tears, faulty seams, etc. Clays exhibit characteristics such as swelling and self-sealing that are not present in synthetic membranes. The composite liner concept combines the qualities of minimum penetration and forgiveness to minor defects.

For these reasons, many designers feel that the use of composite liners is advantageous for both upper and lower liners in the minimum technology design. Due to material handling landfill space considerations, composite liner designs for top liners have generally limited the thickness of the clay component to approximately 2 to 5 feet. Figure 11 shows a generalized cross-section of the double composite liner system.

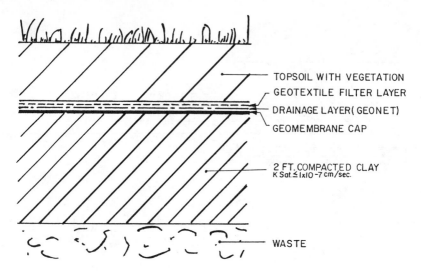

In an effort to maximize the self-sealing characteristics of the upper composite liner and to minimize the volume of landfill lost to the liner system, a design is being used at CID Landfill, in Calumet City, Illinois, using prefabricated bentonite matting in place of the clay

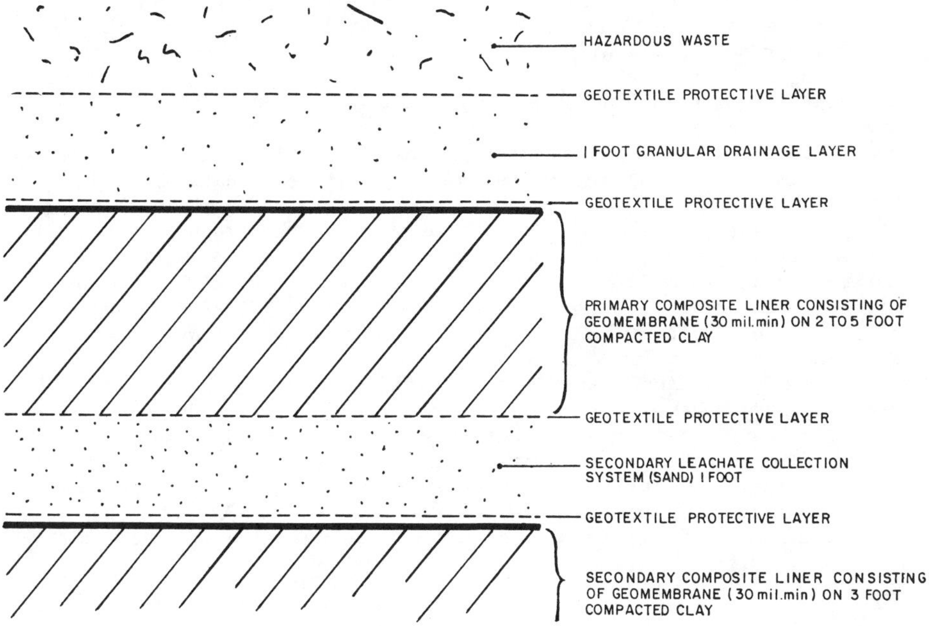

Figure 11. Double composite liner system.

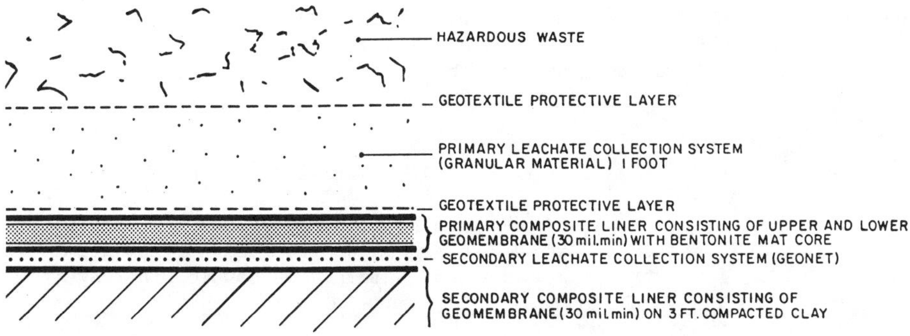

Figure 12. Double composite liner system utilizing bentonite mat.

component of the upper composite liner. The bentonite clay minerals optimize the swelling and, therefore, the self-sealing characteristics of the clay component. The thickness of the mat is about 3/8 of an inch. The bentonite matting is sandwiched between two layers of polyethylene membrane in order to prevent swelling of the matting into the drainage media of the secondary leachate collection system. A generalized cross-section is shown in Figure 12.

The bentonite matting is comprised of granular bentonite adhered to a polyester geotextile with a water soluble glue. A protective paper sheet is also glued to the back in order to provide containment of the granular bentonite during shipping and placement. Care must be taken during shipping and installation in order to keep the matting dry. If the matting becomes wet, placement problement problem will occur due to the extreme weight of the wetted matting and delamination of the bentonite, geotextile, and paper layers. More uniform in-place thicknesses of bentonite also can be achieved if the granular bentonite can remain rigid until activated in actual service. Seaming of the matting is accomplished by simple lapping of the material to avoid gapping. In order to minimize wetting of the material after placement, construction specifications required covering of the mat with synthetic membrane at the end of each working day.

The installation was subject to rigid third party quality control procedures and has been installed in accordance with the design specifications and permits. Laboratory tests performed on the composite liner indicate that the bentonite will be properly activated and perform the self-sealing function when permeated with the leachates that are being generated at the site. This installation has recently been completed. There is no long-term performance documentation available.

Conclusions

This paper reviews the evolution of liner design for hazardous waste landfills since the first RCRA regulations in 1980. We have also presented a discussion of different approaches used in meeting recent minimum technology requirements. The use of composite clay/synthetic membrane liners seems prudent in implementation of these standards. In some cases, prefabricated bentonite matting may be used as a soil component in a composite liner system.

Reference -

1) U.S.EPA, 1983, "Lining of Waste Impoundment and Disposal Facilities," SW-870, Office of Solid Waste and Emergency Response, Washington, D.C.

2) U.S.EPA, 1982, "RCRA Guidance Document, Surface Impoundments, Liner Systems, Final Cover, and Freeboard Control." Draft Document.

3) U.S.EPA, 1985, "Minimum Technology Guidance on Double Liner Systems for Landfills and Surface Impoundments--Design, Construction and Operation" Draft Document.

Land treatment of food processing wastewater--Case history

Larry W. Keith and Walter D. Lehman, Anheuser-Busch Companies, Inc., St. Louis, MO 63118

Two significant factors have been driving forces for Anheuser-Busch and other food processing companies to consider land application as a viable wastewater treatment alternative. First was the passage of PL92-500 in 1972 with emphasis being placed on nutrient recycling as a means of eliminating pollutant discharge. It was the intent of Congress in the passage of that law to encourage the development of wastewater management policies that would be consistent with the principal of returning nutrients to their place of origin.

The second driving force was pure economics. With the energy crisis of the mid '70's, energy costs have nearly tripled in the past 10 years. Comparing an actual Anheuser-Busch installed activated sludge wastewater treatment system for 20,455 kg BOD/d (45,000 lb/d) with a similarly sized land application system shows an energy utilization of 2,088 kW (2800 hp) and 298 kW (400 hp), respectively, for the two systems. At an energy cost of 5¢/kWh the mechanical treatment system's annual energy cost will be $785,000 greater than land application ($915,000 vs. $130,000). When total costs are considered, the economics of land application are even more significant. Based on a wastewater treatment system for 20,455 kg BOD/d (45,000 lb/d) the following annualized costs (average costs based on Anheuser-Busch experience -- case by case analysis must be made) result for mechanical treatment versus land application:

	Capital ($MM)	Debt Service (20 yr @ 10%) ($MM/yr)	O&M ($MM/yr)	Total Annualized Cost ($MM/yr)
Mechanical Treatment	45	5.3	3.2	8.5
Land Application	22	2.6	0.5	3.1

Anheuser-Busch experience has shown, where land application is a viable alternative, capital costs for land application systems are typically less than half that of a similarly sized mechanical wastewater treatment plant. When annual operation and maintenance (O&M) cost differences for the two systems are considered, the difference can be even greater than a two to one advantage for land application.

Previous Anheuser-Busch Land Application Experience.

Since 1975, Anheuser-Busch has expended over one million dollars on research and pilot scale land application projects at brewery and yeast plant locations in California, Virginia, Texas, Colorado, and Florida. Based on the experience and positive results obtained from this pilot work, Anheuser-Busch has spent or committed more than $50 million dollars on full-scale land application systems. The site loadings for the existing Anheuser-Busch land application facilities are presented in Table 1.

Table 1. Loadings for existing Anheuser-Busch land application facilities.

Facility	Hectares	Loadings cm/wk	Kg/ha per d BOD	TSS	N	System Type
Houston, TX Brewery	127	0.2	112	28	4.1	Side Rolls
Jacksonville, FL Brewery	141	2.5	153	50	2.7	Center Pivot & Solid Set Sprinkler
Bakersfield, CA Yeast Plant	162	1.1 (2.7)*	67	28	3.7	Flood & Hand Move Sprinkler
Robersonville, NC Snack Food	25	1.7	66	48	1.3	Center Pivot

*Numbers in parentheses are for total flow to land application site which includes 3,785 m^3/d of City of Bakersfield secondary effluent.

All the projects listed in Table 1 were justified on the basis of economics. Prior to installation of the projects, the process wastewaters from these facilities were treated in municipal wastewater treatment plants (WWTP). In each case, even though Anheuser-Busch was operating well within agreed upon wastewater discharge limits to the municipal system, the removal of Anheuser-Busch process wastewater from the municipal system relieved an overloaded municipal system that was consistently experiencing problems meeting its NPDES discharge limits. The removal of wastewater loading from these municipal systems not only alleviated treatment problems at the municipal plants but allowed the municipality to accept new customers to the plant with no or very little additional capital cost required for upgrading of the municipal WWTP. In addition, the Robersonville, Bakersfield, and Jacksonville land application projects were constructed in conjunction with expansion of the production facilities. Therefore, timely expansion, which would have been difficult at best with required associated expansions of the municipal WWTP, of the production facilities was accomplished.

The bottom line for all the existing Anheuser-Busch land application facilities is that everyone came out a winner. The projects were economically attractive to Anheuser-Busch and in three cases allowed timely expansion of production facilities. Each municipality benefited by having significant loading removed from its overloaded WWTP. In most cases, the loading removed from the municipal WWTP was significant enough that additional loading could be accepted into the WWTP without expending capital for upgrading of the facilities.

Table 2 lists the site loadings for Anheuser-Busch land application facilities that are presently approved and either in the design or construction phase.

Table 2. Loadings for future Anheuser-Busch land application facilities.

Facility (Yr. in Service)	Hectares	Loadings cm/wk	Loadings Kg/ha per d BOD	TSS	N	System Type
Jacksonville, FL* Brewery ('86)	443	1.5	98	35	1.8	Center Pivot & Solid Set Sprinkler
Robersonville, NC* Snack Food ('86)	43	1.1	49	40	2.0	Center Pivot & Solid Set Sprinkler
Fayetteville, TN Snack Food ('87)	47	0.9	50	34	1.0	Linear
Fort Collins, CO Brewery ('88)	325	1.6	50	23	1.0	Center Pivot

*Expansion of existing systems - numbers include existing operations.

Summary.

Anheuser-Busch's land application experience has been very diverse. Site soil conditions have ranged from tight clays in Houston with permeabilities less than 0.15 cm/hr to sands in Jacksonville with permeabilities in excess of 50 cm/hr. Table 3 lists the range of effluent BOD concentrations that Anheuser-Busch has successfully utilized for land application.

Table 3. Range of bod concentration.

Concentrated High Strength Streams	20 kg/m^3 BOD_5
Selected Process By-Product Streams	1-20 kg/m^3 BOD_5
Entire Production Facility Effluents	1-5 kg/m^3 BOD_5
Cooling Waters	0.05 kg/m^3 BOD_5
Treated Municipal Effluent	0.05 kg/m^3 BOD_5

Table 4 lists the range of irrigation equipment with which Anheuser-Busch has experience.

Table 4. Range of irrigation equipment utilized.

Center Pivots
Side-Roll Sprinklers
Hand-Move Sprinklers
Solid-Set Sprinklers
Traveling Gun

The primary objectives of Anheuser-Busch land application projects are the following:

1) Allow timely production facility expansion.
2) Alleviate overloading problems of existing municipal wastewater treatment systems.
3) Reduce wastewater treatment costs.
4) Recycle nutrients and water from the production process.
5) Provide an aesthetically appealing wastewater treatment system with the elimination of discharge of effluents to surface waters.

Anheuser-Busch had been successful in achieving these objectives with its land application projects through implementation of knowledge developed over the past eleven years, coupled with the utilization of intensive agricultural operations in conjunction with the land treatment process. Anheuser-Busch is proud of the recognition it has received for its land application efforts to date. It has received awards from local civic groups as well as state regulatory agencies. In 1980, the Jacksonville project received "The Presidents Award for Energy Efficiency."

CASE STUDY - JACKSONVILLE EXPANSION DESIGN

Background.

The Anheuser-Busch Jacksonville brewery has been utilizing land application to treat half of its process effluent since 1979. The system is located just north of the brewery on a site having soils that consist primarily of fine sand. The site soils have SCS permeabilities ranging from 15-150 cm/hr (6-60 in/hr). The system includes 121 ha (300 ac) of center pivot irrigators and 20 ha (50 ac) of solid set sprinklers used to apply an average of 4921 m^3/d (1.3 mgd) or 2.54 cm/wk (1 in/wk). Along with the applied effluent (liquid nutrient or LN), the site also receives an average of 137 cm/yr (54 in/yr) of rainfall. The site was originally designed as a zero discharge facility. However, it was soon realized that a surface water management plan was needed. The system's permit was revised to include discharge of stormwater runoff, yet remaining a zero discharger of the applied LN. Because the site was initially designed as a zero discharge facility, the retrofitted surface water management system necessitated on-pivot storage of rainfall runoff at many locations.

Following the initial design changes, the Jacksonville system has performed exceptionally well. This performance is exhibited by the actual discharge characteristics compared to the permitted limits. A comparison of the 1983 data to permitted levels shows the stormwater discharges to be well within the permitted limits:

Parameter	Permitted Annual Average*	1983 Annual Average
BOD	20 mg/l	7.5 mg/l
TKN	8 mg/l	2.05 mg/l
TSS	20 mg/l	14.6 mg/l
pH	6-8.5	6.4
DO	5 mg/l	8.9 mg/l

*Average of all 5 outfalls.

An evaluation in 1983 of the groundwater quality as monitored in the site's compliance well also showed that the site was performing well. The BOD of the compliance well (see Figure 1) had only exceeded 6 mg/l twice (9 and 11 mg/l) and the NO_3 exceeded 1.0 mg/l only once and was typically 0.1 mg/l. These peaks also occurred in the background well indicating a natural condition and not one due to the land application operation.

The performance of the existing system since start-up and the cost to send the remaining brewery process effluent to the City's WWTP ($1.57 MM in 1983) provided justification to fully evaluate expansion of the land application system.

Pilot Study.

A two phase pilot study was initiated in 1984 to evaluate the site's capability to handle higher hydraulic, organic, and nitrogen loadings. A 4.6 ha (11.3 ac) portion of pivot 5 (see Figure 1) was isolated to apply 4.5 cm/wk (1.75 in/wk). The corresponding BOD

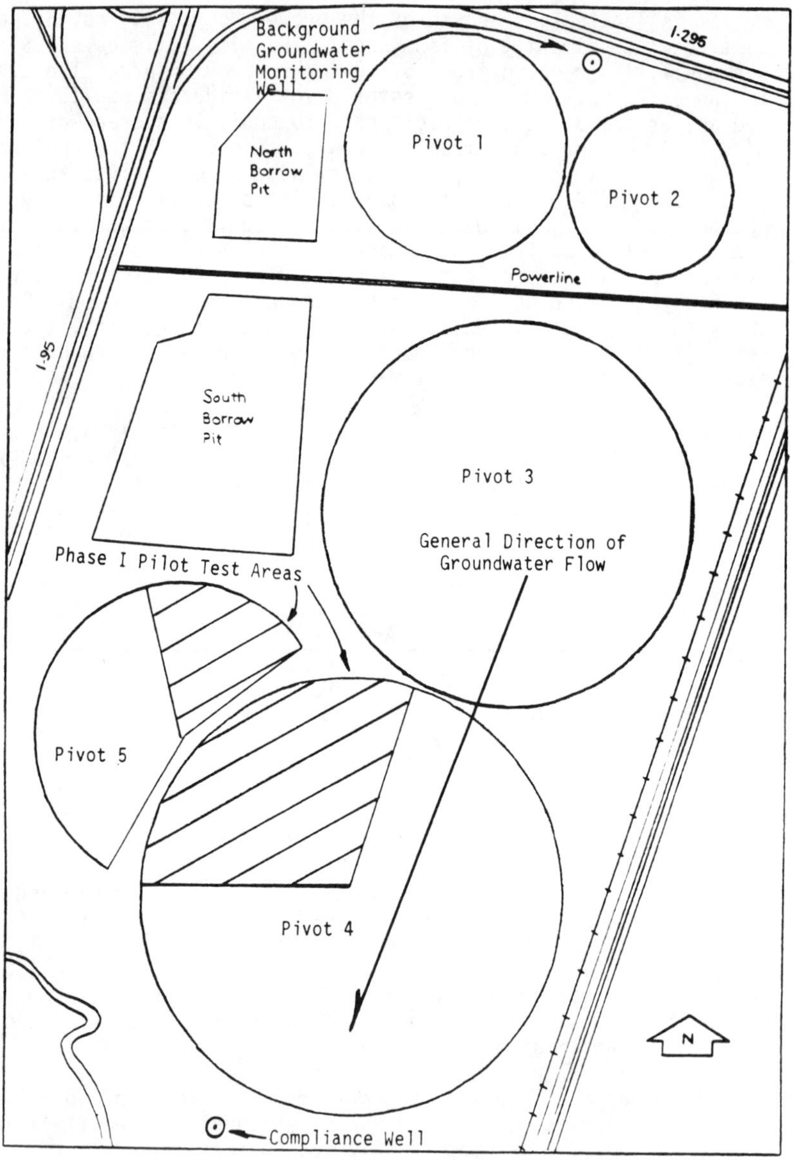

Fig. 1. Existing Jacksonville land application site.

and TKN loadings that accompanied this higher hydraulic loading were 100 kg/ha/d (90 lb/ac/day) and 728 kg/ha/yr (650 lb/ac/yr), respectively. A 13.7 ha (33.9 ac) portion of pivot 4 was isolated to apply 840 kg TKN/ha/yr (750 lb/ac/yr). The corresponding hydraulic and BOD loadings were 1.9 cm/wk (0.75 in/wk) and 75 kg/ha/d (67 lb/ac/day), respectively. The underdrain water from these two areas was monitored throughout Phase I. By May, 1984 the annual spring NO_3 peak had passed with the peak from both of these test areas being less than 4.0 mg/l. Phase II of the pilot study was then initiated. Phase II consisted of utilizing the entire site at loadings of 2.54 cm/wk (1.0 in/wk), 100 kg BOD/ha/d (90 lb/ac/day), and 850 Kg TKN/ha/d (750 lb/ac/yr).

Due to the success of phase I, a site search and evaluation was initiated to locate a tract of land with suitable characteristics and acreage for expansion of the land application system. An irrigated area requirement of 750 acres for the expansion was used based on projected brewery discharge loadings and site loadings developed following phase I. The unit design site loadings are 1.5 cm/wk (0.6 in/wk), 98 kg BOD/ha/d (90 lb/ac/d), and 670 kg TKN/ha/yr (600 lb TKN/ac/yr). These loadings, which are lower than the design loadings on the existing 121-ha site, were selected in order to provide the land application system with greater flexibility than was available in the past. This flexibility was felt to be necessary because the brewery would be 100% dependent on land application once the expansion was completed.

Available tracts of land within a ten-mile radius of the brewery were looked at following screening with the aid of SCS soils maps and USGS topography maps. After 15 months of searching, evaluating, and negotiating, four contiguous tracts of land totaling 484 ha (1,200 ac) were placed under option. This property was located approximately 16.8 km (10.5 mi) northwest of the brewery.

Based on the results of the pilot study and the characteristics of the optioned property, a preapplication package was prepared and presented to the Florida Department of Environmental Regulation (DER) and the Jacksonville Bio-Environmental Services Division (BESD). Because of the performance record of the existing site and the fact that Anheuser-Busch was requesting a permit with lower unit loadings than those approved for the existing site, Anheuser-Busch received preliminary approval. The characteristics of the brewery effluent for the full-scale project are:

BOD	-	4,100 mg/l
COD	-	6,200 mg/l
TSS	-	1,100 mg/l
TKN	-	115 mg/l
pH	-	3.7

Effluent with these characteristics has been applied to the existing site since May 1984 when Phase II of the pilot study began.

Preliminary Site Selection/Evaluation.

The services of a surveyor and a hydrogeologist were employed to perform a preliminary site evaluation on the optioned property as part of the project Conceptual Design Report (CDR). A 0.3m (1 ft) contour topography map, a groundwater contour map, and site soil characteristics were developed. The CDR was prepared to address such items as existing site characteristics, design, and performance, along with the new site surface hydrology, loadings, layout, drainage, and operation. The CDR was submitted as part of the permit package to DER/BESD.

Permitting.

The permitting process became more involved than was first indicated by the preapplication approval. In October, 1984 the "Warren S. Henderson Wetlands Protection Act of 1984" went into effect. The vegetative indicator portion of Florida Administrative Code 17-4.02(17), defining the waters of the State, was extensively increased and the procedures for determining these jurisdictional lands were substantially modified. The Jacksonville land application expansion became the first major project to be regulated by this new law. Instead of having to obtain one permit through DER as was the case with the existing site, Anheuser-Busch was now faced with obtaining permits from the DER Industrial Waste Section, DER Dredge and Fill, the Corps of Engineers (COE), Florida DNR, and the St. Johns River Water Management District (SJRWMD). Following four months of evaluation by the various agencies, it was finally determined that the original site layout as presented to DER was not acceptable. That layout included an extensive surface water management system that utilized the low lying areas of the site for temporary retention of stormwater runoff. These areas were determined to be jurisdictional by the various agencies. In all, 141 ha (350 ac) of the 484 ha (1200 ac) site were classified as jurisdictional. Fortunately, an additional 121 ha (300 ac) adjacent to the optioned property was found to be available. This property was optioned and the CDR was revised, submitted, and approved by all the agencies. The approved layout is shown in Figure 2. Because the low-lying areas of the site could not be used for stormwater runoff collection and retention, a system of pumping stations had to be included in order to utilize ponds located at higher elevations. An extensive surface water management plan evaluating the stormwater drainage plan and detailing its design requirements had to be prepared and approved by the SJRWMD.

The CDR also had provisions for a 16.8 km (10.5 mi) pipeline to transport the process effluent to the site. Easements and permits from Florida Department of Transportation (D.O.T.), DNR, DER, BESD,

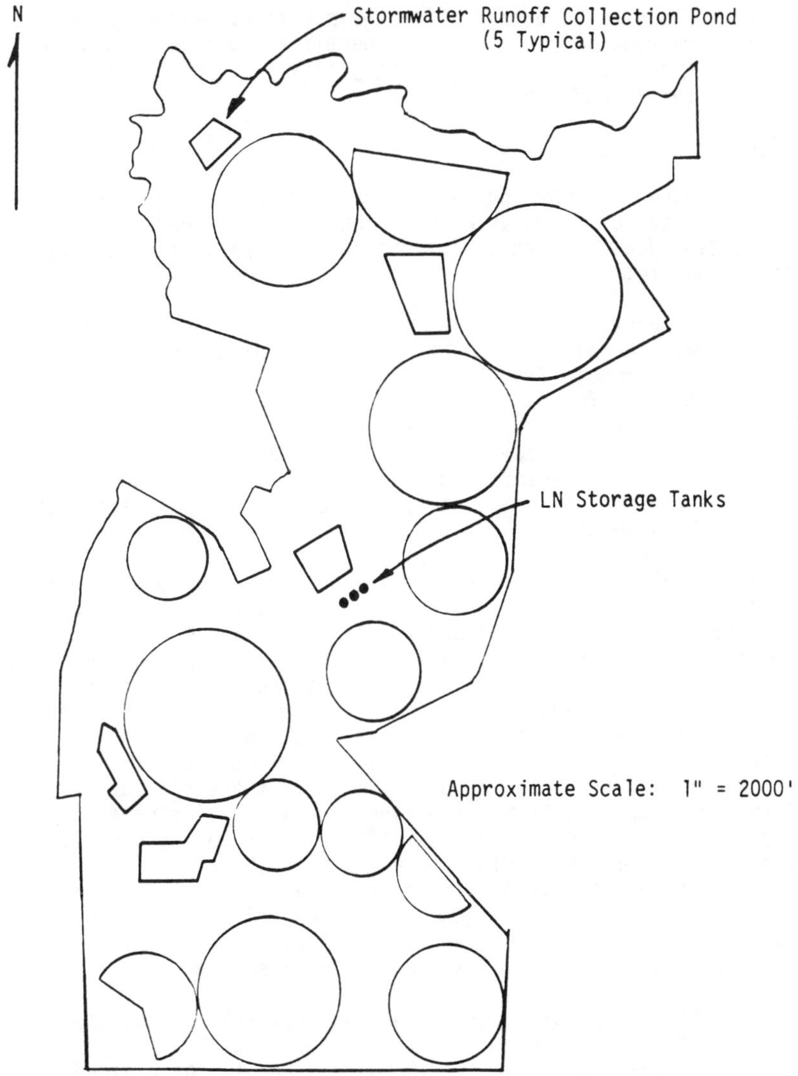

Fig. 2. Jacksonville land application expansion site.

and the COE had to be obtained for the pipeline. All of these permits were routine except one DNR permit for a creek crossing. This involved preparing and submitting an extensive and time consuming survey and appraisal.

Design.

Having prepared the CDR and having obtained regulatory approval, Anheuser- Busch employed an engineering firm to prepare construction drawings and specifications. As part of this work the engineer had to:

1) prepare a site grading plan that:
   a) provided acceptable surface drainage of stormwater runoff from the pivot areas, and
   b) conformed to the surface water management plan as approved by the SJRWMD
2) finalize all pump sizes and pipe lengths and sizes
3) provide Anheuser-Busch with all required design details necessary to complete specifications for purchasing the equipment.

An evaluation by Anheuser-Busch of the preliminary hydrogeological investigation performed as part of the preliminary site evaluation determined that an extensive underdrain system would be necessary for the site to operate properly. The services of Dr. Wayne Skaggs (North Carolina State University) were retained to design this underdrain system. Auger borings taken on a 61m by 91m (200 ft x 300 ft) grid revealed that a very fine sand, consistent throughout the entire site, existed over a sandy clay layer. The K values (permeability or hydraulic conductivity) for these two soils were determined to be 6 cm/hr for the fine sand and 0.3 cm/hr for the sandy clay. A contour map indicating the depth to the sandy clay was prepared for each pivot. The average depth of the fine sand over the entire site was determined to be approximately 76 cm (30 in). An extensive process then began to optimize the underdrain design based on performance versus installation costs. The criteria established by Anheuser-Busch included:

1) Three days of storage would be available through the on-site liquid nutrient (LN) storage tanks.
2) An average of 0.6 to 0.9 meters (2-3 ft) of drained soil must be obtained within 72 hours of the design rain event.
3) One-third of the underdrains were to be designed for the 5-year event and two-thirds of the underdrains were to be designed for the 1-year event.

The underdrains were laid out to fit the surface slopes, the underdrain outlets, and areas having a constant depth to the sandy clay layer. The spacings were based on the depth to the sandy clay layer.

The one final criteria that impacted the underdrain design was the potential for iron ochre formation. Dr. Harry Ford (former Professor, University of Florida) was consulted to determine the iron ochre potential of the site and provide possible design parameters that would allow the underdrains to better handle the ochre formation. The site was found to have severe ochre formation potential. To help prevent drain pipe clogging, Dr. Ford recommended the use of round holes rather than slots to hinder the ochre bridging the openings.

The final underdrain design included over 120 km (75 mi) of drain pipe. A sock and a special filter sand were specified for the perforated portion of this drain pipe. The special filter sand was required because of the extremely uniform fine sand found consistently throughout the site. The sand particle size was 100-200 microns as shown in Figure 3. Because of its uniformity and size, the sand is quite fluid. Lab tests performed by Dr. Skaggs found the sand to "seal" the drain pipe with and without the sock. Flow rates of about 25 cm^3/min were found to develop in both cases. By using filter sand that met the SCS and Spaulding gradation criteria, flows of over 100 cm^3/min could be maintained.

Other design work that was performed included extensive sanitary sewer segregation at the brewery and the pipeline to the new site.

CONSTRUCTION

Construction for the project began in July 1984. Center pivots, surface drainage ditches, and runoff collection ponds located within wooded areas were surveyed and flagged. The surveying was performed to 1) maintain as much of the natural vegetation as possible for conservation, aesthetics, and buffer, and 2) prevent construction activities from occurring in the jurisdictional lands to meet permit requirements for preserving the site's wetlands. Timber removal, clearing and grubbing, and site grading have followed in that order. The system is scheduled for start-up in February 1986. Meeting the permitting constraints of the SJRWMD and excessive rainfall have created construction delays. From August 28 through October 30 the site has received almost 76 cm (30 in) of precipitation. A review of the historical precipitation data for Jacksonville showed that the combined September and October rainfall was the fourth largest on record.

Two unique systems are being utilized for the first time by Anheuser-Busch on the Jacksonville land application expansion. First, a PVC underslung irrigation header will replace the standard galvanized header on the center pivots. Organic acids formed by bacterial action in the LN will result in an effluent pH of 3-4 going to the irrigation system. It is anticipated the PVC underslung system will prolong the life of the center pivots by 3-4 times. The other first for Anheuser-Busch is the use of an

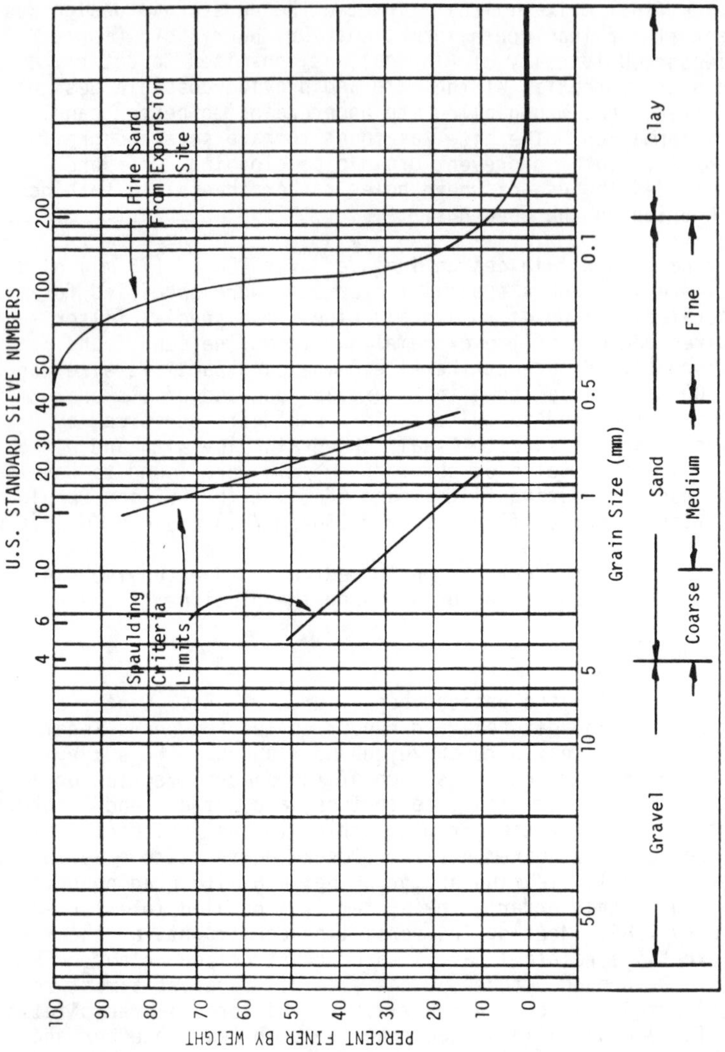

Fig. 3. Gradation curves.

underdrain pipe containing a biocide to retard iron ochre formation. It is anticipated that the biocide will prevent iron ochre formation for a sufficient period of time to allow leaching of the soluble iron from the soil profile above the drain pipe. Following start-up the performance of both of these items will be evaluated.

CONCLUSIONS

Land application treatment of food processing wastewater has been tested and used by Anheuser-Busch since 1975. A variety of streams have been applied to sites having diverse soils and climatic characteristics. The conclusions resulting from testing, design, and operation of these projects are:

1) Full-scale land application of high strength effluent is viable.

2) Land application treatment is complex and requires detailed analysis, design, and operations management.

3) Intensive agricultural activities at a land application site result in the operational management necessary for successful land application treatment.

4) The effluent provides nutrients and water required in an agricultural operation.

5) Land application, when properly designed and operated, does meet the primary objective of treatment through recycling nutrients and water to the land in an environmentally acceptable, energy-efficient, and cost effective manner.

6) Land application is a viable treatment alternative for numerous Anheuser-Busch effluents around the country and will continue to be evaluated as a treatment alternative for all future Anheuser-Busch production sites.

Classification of hazardous wastes disposed on land

Walter E. Grube, Jr., Hazardous Waste Engineering Research Laboratory, U.S. Environmental Protection Agency, Cincinnati, OH 45268

Waste by-products result from nearly every commercial activity. Environmental legislation mandates that wastes which are deemed to be hazardous must be disposed of in an environmentally acceptable manner. Regulations published (3) define hazardous waste as any solid waste which because of its quantity, concentration or characteristics, "may cause ... an increase in serious irreversible or incapacitating reversible illness" or "may pose a substantial ... hazard to human health or the environment," when improperly managed. A solid waste was defined as "any garbage, refuse, sludge ... and other discarded material, including solid, liquid, semisolid, or contained gaseous material."

Hazardous wastes have been generated by several activities:

- Manufacturing processes that produce an unwanted or unusable by-product
- Finished products that are unusable, for a variety of reasons
- Spills of materials during shipment
- Used residuals or product remaining after use.

A fundamental policy for hazardous waste management in the United States is to preserve and protect our land and groundwater resources. The hazardous waste control program under the Resource Conservation and Recovery Act (RCRA) and the Superfund program under the Comprehensive Environmental Response, Compensation, and Liability Act (CERCLA) are the two principal programs that implement this policy.

RCRA CONTROL OF LAND DISPOSAL

The RCRA hazardous waste program controls hazardous waste generators and transporters and the operators of hazardous waste treatment, storage, and disposal facilities (TSDF). The scope of this control program is impressive. Over 240 million metric tons of hazardous waste are generated each year in the United States by about 50,000 generators. There are about 14,000 hazardous waste transporters, and about 5,000 treatment, storage, or disposal facilities, most located at waste generator sites.

Generators of hazardous wastes are confronted with listings of hazardous waste materials promulgated by federal regulations. These lists are comprised of particular hazardous waste materials, with assigned EPA hazardous waste numbers, and are arranged in the following way:

Regulatory Part Number	Description	EPA Hazardous Waste Number
	Process Wastes:	
261.31	Hazardous waste from nonspecific sources	F _ _ _
261.32	Hazardous waste from specific sources	K _ _ _
261.33	Commercial chemical products	
261.33e	Acute hazardous wastes(H)	P _ _ _
261.33f	Toxic wastes(T)	U _ _ _

An example of a listing is:

"K001 .... Bottom sediment sludge from the treatment of wastewaters from wood preserving processes that use creosote and/or pentachlorophenol."

Under RCRA, EPA was charged with identifying wastes which pose a hazard to human health and the environment when improperly managed. In fulfilling this mandate, EPA identified a number of characteristics which, if exhibited by a waste, would indicate that the waste is hazardous. These characteristics include: Ignitability, Corrosivity, Reactivity and Extraction Procedure (EP) Toxicity. In developing the toxicity characteristic, the Agency's concern was that potentially hazardous industrial waste might, unless subjected to regulatory control, be sent to a sanitary landfill with resulting high level of potential groundwater contamination. Although less industrial solid wastes are being disposed of in this manner, as compared to a few years ago, the Agency believes that the codisposal scenario still represents a reasonable mismanagement possibility. Thus, proper disposal of wastes characterized as toxic is an especially important consideration where the disposal is onto the land or into earthen containment structures.

It should be noted that legal notice of EPA's position on waste classification, and other proposed or final regulation is published in the Federal Register and the Code of Federal Regulations (CFR). The Federal Register is published daily, weekdays, and provides a system for making regulations and legal notices issued by all federal agencies available to the public. The CFR, published annually, is a compilation of the general and permanent rules and regulations that have been previously released in the Federal Register. The Federal Register is usually subscribed to by all public libraries. Technical persons needing to be aware of EPA's regulatory status on any aspect of hazardous waste must regularly examine the CFR and Federal Register to be assured of the most current information. Hazardous waste regulatory actions appear within Part 260 of Title 40, CFR; numerous sections and subsections also deal with a wide array of regulatory actions that may not be of direct relevance to land disposal.

In addition, individual state regulations, which are in some instances more conservative than those of the USEPA, need to be well known by both those involved in hazardous waste management businesses, and the soil scientists who are advising such businesses.

The RCRA land disposal regulations are based on several principles. They attempt to control and manage liquid hazardous wastes or hazardous leachates from solid wastes within a land disposal environment. This is accomplished with liner systems, leachate collection systems, capping systems over the material at site closure, and through run-on and run-off controls to prevent surface water contamination. Regulations ban the disposal of liquid waste or waste-containing free liquids.

EPA regulations require that a permit be obtained to operate a facility which treats, stores, or disposes of hazardous wastes. The operating permit consists of Parts A and B. Part A includes general facility information and the permit application. Part B has no form provided. A specific form has not been developed because the information required is not readily adaptable to a standardized format. The requirements of a Part B application are site specific and are detailed in Section 270.14 of 40 CFR. These detailed data requirements are addressed to new landfills, surface impoundments, and land treatment facilities. Siting, including geographic and geological conditions, and soil structures, such as foundations, liners, dikes, and cover systems, are factors which undoubtedly require technical input from the soil science community.

Major amendments to RCRA were enacted in November 1984. These amendments in several ways reinforce the fundamental policy for hazardous waste management which the U.S. established almost 10 years ago. Probably the most important provisions of the Amendments are those that ban or restrict disposal of hazardous wastes on the land. Today, in the United States, most hazardous wastes are managed by depositing them into or on the surface of the land. For example, 133 million metric tons per year of hazardous waste are treated, stored, or disposed of in surface impoundments. These ponds or lagoons contain liquid hazardous wastes which pose significant potential for groundwater contamination. Over 5 million metric tons of waste per year are disposed of in landfills, land treatment facilities, waste piles, and similar means.

Through the land disposal restrictions included within the 1984 Amendments to RCRA, the Congress has set up a presumption against land disposal of hazardous wastes. Under this program, EPA must show that land disposal of hazardous waste is safe. Otherwise, a ban on land disposal goes into effect on a phased schedule for specified waste categories over a 2-5 year period. Long-term storage of banned waste (for more than 1 year) is also restricted by the 1984 Amendments. It is clear that Congress wants hazardous waste pretreatment prior to land disposal to become the preferred management option in this country.

According to provisions of the 1984 Amendments, effective 24 months after enactment, further land disposal of solvents is prohibited unless the EPA determines that such prohibition is not required in order to protect human health and the environment. The statute specifically addresses the list of spent solvents in 40 CFR 261.31 -- EPA Hazardous Waste Nos. F001, F002, F003, F004, and F005. Additional solvents listed in other categories are also included. However, wastes that meet treatment standards established by the EPA are not subject to land disposal prohibitions. The 1984 Amendments to RCRA requires EPA to set "... levels or methods of treatment, if any, which substantially diminish the toxicity of the waste or substantially reduce the liklihood of migration of hazardous constituents from the waste so that short-term and long-term threats to human health and the environment are minimized." The EPA recently published the first Proposed Rule (5) specifying standards to which solvent-containing hazardous wastes must be treated in order to qualify for land disposal. This Proposed Rule publication contains a lengthy description of the background and rationale for the concentration limit below which treatment processes must reduce a hazardous constituent, primarily solvents in this case. This Proposed Rule will be open for public comment, after which the technical merit of any comments received by the Agency will be considered, and a Final Rule will substantially be promulgated.

The legislative history of the 1984 Amendments to RCRA indicates that a waste may be restricted from land disposal not only on the basis of hazards posed by its inherent toxicity, but also because of its ability to degrade clay and synthetic liners, and to mobilize relatively nonmobile hazardous constituents, when co-disposed with other hazardous waste. Numerous recent studies sponsored by EPA's Office of Research and Development have shown that concentrated solvents with properties similar to those in F001 and F005 can greatly increase the permeability of clay soils, when these soils are compacted as soil liners. Consequently, if any of these undiluted solvents are placed in soil-lined land disposal facilities, the permeability of the soil liner may be greatly increased. Recent research data have shown that the effect of aqueous solutions containing these solvents on the permeability of soil liners is related to the concentration of the solvents. Although each solvent or waste mixture represents a unique disposal situation, it appears that a solvent concentration of at least several percent needs to be present before the mixture is sufficiently aggressive, in a chemical sense, to degrade the clay liner. However, few solvents in F001 and F005 are soluble in water at concentrations greater than a few percent. These solvents separate and form a concentrated floating or sinking phase, which may be expected to cause clay liner degradation as shown in our research. Overall, however, there is little really known about the long-term effects of dilute solvents on the permeability of soil liners.

Other recent studies supported by EPA (1,2,6) showed that solvents may increase the mobility of other compounds. Wastes containing aromatic solvents and chlorinated solvents, as well as surfactants, generally increase the mobility of compounds adsorbed onto a synthetic waste. Further research is obviously needed to confirm the actual physical-chemical mechanisms involved in solvent enhancement of waste constituent mobilization.

Other waste categories, such as used oils from industry and domestic consumption, and high volume wastes such as residues from mining, mineral extraction and processing, and the utility industry are being examined for suitability for continued land disposal (4). These wastes are included in reports to Congress which the EPA has scheduled in the very near future.

SUPERFUND

In response to numerous uncontrolled hazardous waste site problems, such as the Love Canal in New York, Congress in 1980, enacted the Comprehensive Environmental Response, Compensation, and Liability Act (CERCLA). This is also known as the Superfund program. The scope of this program in the United States, and the progress made over its first five-year period are impressive. Over 14,000 sites have had a preliminary assessment regarding potential releases to the environment. Almost 5,000 site inspections have been completed. Over 1,100 sites have been evaluated for placement on the National Priorities List (NPL). The National Contingency Plan (NCP) is a regulation that provides a framework for implementing the response powers and responsibilities established under CERCLA. The NCP prescribes five steps in the remedial response process:

- Site discovery or notification
- Preliminary assessment and site inspection
- Establishment of priorities for remedial action
- Remedial investigation/feasibility study
- Remedial action, design, and construction

Technical options for remedial action implemented under CERCLA can be categorized as either waste control or environment control. Waste control refers to the removal of hazardous material from a site, followed by some treatment that reduces the potential harm of hazardous compounds and subsequent disposal of the waste or treatment residue in an appropriate facility. The treatment can involve destruction of toxic components of the excavated material through chemical, physical, or biological processes, or immobilization of the hazardous components. Environment control options include those techniques that contain or isolate hazardous material, divert water movement away from a site, or treat contaminated water sources. The technologies for containment are adapted from structural or civil engineering procedures and consist of caps, barrier walls, and drainage systems. A major concern associated with either type of remedial action is the limited experience with use of these techniques for this purpose. Sites using either waste or environment control have not been in existence long enough to provide sufficient data about the long-term integrity of the methods.

The USEPA's experience in cleanup of contaminated soils is being broadened by associations with several Department of Defense services. We have a number of joint research studies with these services to try to expedite the cleanup of soils polluted by waste residues from munitions synthesis and waste solvents and fuels from routine maintenance operations. These studies are providing valuable operational experience for remedial technologies which have not yet been widely proven through the experience of the private sector, but have shown promise from bench-scale research.

Some of the remedial response technologies are of direct interest to soil scientists, both from the standpoint of direct technical involvement, and the advancement of a better understanding of soil chemical and physical processes. Soil/waste interactions involving solvent chemicals, additives, or amendments useful in improving the soil's mechanical structural properties, and effect of construction techniques on soil/pollutant interactions, are still young areas of soil research. Processes that enhance miscibility or immiscibility of liquid pairs and occur within a soil mass are important in understanding solvent concentration processes in the subsurface. It might be interesting to evaluate the utility of octanol-water partition coefficients in subsoils that do not contain the high amounts of organic matter present in most soil surface horizons, where carbonaceous residues have a strong influence on pollutant mobility. Soil-catalyzed decomposition of pollutant molecules is being examined by a few scientists familiar with catalytic reactions of clay minerals. Fixation onto and desorption from soil mineral surfaces remains an area of data paucity. The enhancing effect of surface active compounds in aiding soil flushing techniques and other in-situ treatments to detoxify a contaminated soil site has been demonstrated only in laboratory column studies at this time. All soil cleaning technologies need to be proven in full-scale field trials and demonstrations.

The public needs to be assured of the effectiveness of remedial technologies proposed or actually applied to soil contaminated by past hazardous waste disposal practices.

CONCLUSIONS

In conclusion, it is clear that soil scientists, both practicing advisors in the field, and research scientists in laboratories, need to recognize the many categories or classes of hazardous wastes. From a regulatory or legal standpoint, some types of waste are of much greater agency concern than others. Decontamination processes are much more effective in removing some wastes than others. The intereactions of some wastes with soils are unknown.

Fundamental soil chemistry, physics, and engineering studies of soil/waste interactions have been tremendously fruitful to the support of EPA's regulatory mission. Further work is needed, however, to document interactions where more complex waste mixtures are in soil. Soil decontamination processes that can be economically scaled up for field use with a minimum of change from theoretical or engineering designs are needed for all classifications of hazardous wastes. Although refractory organic compounds are of great current interest, the questions of how we can most safely allow metals to remain in soils is a continuing undercurrent of concern.

Where soil scientists are involved with private companies or as regulatory agency advisors, they need to be knowledgeable of the most current agency laws that apply to contaminated soil or land disposal sites. State-specific waste classifications and land disposal regulations are sometimes more conservative than those of the Federal EPA. Both those in the business of hazardous waste management and the soil scientists advising such businesses must know the local as well as the national regulations. Interdisciplinary technical interactions and societal impacts of research and field investigation data must be anticipated, and recognized as a necessary part of data reports and interpretations.

As is true with any classification system, the categories applied in EPA's published regulations provide science, industry, and citizens with a uniform system of definitions. These categories make it possible to direct resources to solution of the most important problems in protecting human health and the environment.

DISCLAIMER

Although the research described in this article has been funded wholly or in part by the U.S. Environmental Protection Agency, it has not been subjected to Agency review and therefore does not necessarily reflect the views of the Agency and no official endorsement should be inferred.

REFERENCES

1. Griffin, R. A., and E. S. K. Chian. 1980. Attenuation of Water-Soluble Polychlorinated Biphenyls by Earth Materials. EPA-600/2-80-027, SHWRD, MERL, Cincinnati, Ohio; available from NTIS as PB 80-219652.

2. Griffin, R. A., and F. J. Chou. 1982. Attentuation of Polybrominated Biphenyls and Hexachlorobenzene by Earth Materials. Available from NTIS as PB 82-107558.

3. USEPA. 1985. Criteria for Identifying the Characteristics of Hazardous Waste and for Listing Hazardous Waste. 40 CFR Section 261, Subparts B, C, and D; p.362-376 of 7/1/85 edition.

4. USEPA. 1985a. Wastes from the Extraction and Beneficiation of Metallic Ores, Phosphate Rock, Asbestos, Overburden from Uranium Mining, and Oil Shale. EPA/530-SW-85033, available from Supt. of Documents, GPO, Washington, D.C. 20402.

5. USEPA. 1986. Hazardous Waste Management System: Land Disposal Restrictions. Federal Register (FR) Vol. 51, No. 9, p. 1602-1866, Jan. 14, 1986.

6. Griffin, R. A., and W. H. Fuller. 1986. Transport of Cadmium by Organic Solvents Through Soil. Soil Sci. Soc. Am. J. 50:24-28.

Constraints to land application of sewage sludge

Lee E. Sommers and Ken A. Barbarick, Department of Agronomy, Colorado State University, Fort Collins, CO 80521

Land application of sewage sludge has been a viable option for sludge management in many municipalities. However, as our knowledge of sludge composition improves, not only the short- and long-term environmental consequences but also the human health aspects of sludge application to soils, particularly agricultural cropland, have been increasingly questioned. The recurring problems with hazardous waste disposal facilities has also sensitized the public to the potential for soil and water pollution problems during sludge utilization on cropland. The constraints associated with sludge utilization are directly related to sludge composition and treatment methods. The objectives of this chapter will be to summarize data on sludge composition and to describe the principal constraints associated with land application of sewage sludge. Our discussion will emphasize the technical aspects of constraints involved with sludge use on cropland including 1) pathogens; 2) synthetic organic compounds; 3) nitrogen and; 4) metals and other trace elements. However, it should be emphasized that non-technical issues can have a marked impact on the success of a land application program. For example, the authors are aware of cities who had viable sludge use programs until they attempted to purchase land in an adjacent township. The attempted purchase elicited a negative response from many citizens and the ensuing public hearings and other legal proceedings had a negative impact on the otherwise environmentally sound and well-managed sludge recycling program. In essence, social-political-institutional constraints many in some cases be far more important in determining the viability of a land application program than the technical constraints.

Research has been summarized at several times during the past 15 years in a series of conference proceedings. The initial concerns related to land application were summarized in a 1973 conference (NASULGC, 1973) and then updated in 1983 (Page et al., 1983). Other useful sources of general information are extension bulletins and guidelines developed by the appropriate regulatory agency in many states. A comprehensive development of land treatment systems is described in design manuals for wastewater (U.S. EPA, 1981), sewage sludge (U.S. EPA, 1983) and industrial wastes (Overcash & Pal, 1979). Even though all regulations are currently being reevaluated, most sludge utilization systems are impacted by the "Criteria" document adopted by U.S. EPA in 1979 (U.S. EPA, 1979) which addressed pathogens, PCB's and cadmium in wastes applied to land used for growing crops entering the human food chain. The results of a recent conference indicate that U.S. and European scientists have reached similar conclusions concerning the impacts of sludge components on the environment (Dean & Suess, 1985).

Pathogens

One of the major objectives of sewage treatment processes is the destruction of pathogens present in domestic sewage. Feacham et al. (1980) have summarized the survival of pathogens during sewage treatment processes. Even though >90% of many pathogens contained in sewage are destroyed during treatment, specific sludge stabilization processes are needed to further reduce the pathogen content of sludges applied to land. The pathogens of major concern in sludges are bacteria, protozoa, helminths, and viruses. A summary of the common organisms and associated diseases would include (Gerba, 1983):

1. **Bacteria**
   Salmonella - typhoid, paratyphoid, salmonellosis
   Shigella - bacillary dysentery
   Escherichia coli - gastroenteritis
   Yersinia enterocolitica - gastroenteritis
   Campylobacter jejuni - gastroenteritis
   Vibrio cholerae - cholera
   Leptospira - Weil's disease

2. **Protozoa**
   Entamoeba histolytica - Amoebic dysentery
   Giardia lamblia - diarrhea
   Balantidium coli - mild diarrhea

3. **Helminths**
   Ascaris lumbricoides - ascariasis
   Ancyclostoma duodenale - anemia
   Necator americanus - anemia
   Taenia saginata - taeniasis

4. **Viruses**
   Poliovirus, echovirus, coxsackievirus, hepatitis, norwalk virus, calicivirus, astrovirus, reovirus, rotavirus, adenovirus

Common concerns relative to pathogens added to cropland by sludge application include contamination of surface water due to runoff, contamination of ground water due to leaching, adherence to crops and direct ingestion by grazing livestock. All of these potential problems are directly impacted by the survival of pathogens in soil and on crops. Survival of pathogens in relation to land application has been reviewed by several authors (Elliott & Ellis, 1977; Burge & Marsh, 1978; Bitton et al., 1980; Gerba, 1983).

Reddy et al. (1981) reviewed numerous studies of pathogen survival. Table 1 summarizes the half-lives of pathogens based on the assumption that the rate of microbial die-off in soils was a first-order process. As shown by these data, the large populations of many pathogens will not persist in soils for lengthy periods of time. Some pathogenic protozoa and helminths, however, can survive in soils for months to years (Feachem et al., 1980; Burge and Marsh, 1978).

The survival of pathogens in soils is related to the biological, chemical and physical environment present. Many pathogens have been added to soils through natural wildlife excrement for many centuries. Thus, if microorganisms such as *Salmonella* were competitive with the normal soil microflora, significant populations of pathogens would be found in essentially all soils without the introduction of sludge or other human-derived wastes. As summarized by Gerba (1983) and Reddy et al. (1981), temperature, moisture, pH, nutrient availability, texture and method of sludge application influence the survival of pathogens in soil. The die-off rate of many organisms is doubled for each 10 °C increase in temperature and is accelerated by desiccation and exposure to UV radiation. A near neutral pH tends to promote growth and thus survival of pathogenic microbes other than viruses. Since exposure to elevated temperature conditions and accelerated drying would be encountered at the soil surface, it is anticipated that pathogen destruction would be maximized after surface application of sludge. Temperature and drying are also likely responsible for the enhanced destruction of most pathogens adhering to plant surfaces. The longest survival times would be encountered following incorporation of sludge into a soil maintained under cool and water saturated conditions.

The U.S. EPA (1979) established sludge treatment methods that would reduce the pathogen content of sludges prior to their application to soils used for growing food chain crops. Waiting periods were also imposed for the elapsed time needed between sludge application and growth of specific types of crops and grazing by livestock. After reviewing the health effects associated with sludge application and pathogen survival, Kowal (1983) proposed the summary shown in Table 2 for the maximum survival times of various pathogens. Based on this summary, it was

Table 1. Survival of selected pathogens in soil.

Organism	Half-life, d	No. observed
Escherichia coli	0.75	26
Fecal coliforms	0.45	46
Fecal streptococci	1.87	34
Salmonella sp.	0.52	16
Shigella sp.	1.02	3
Staphylococcus sp.	4.3	2
Viruses	0.48	11

Source: Reddy et al., 1981

Table 2. Summary of pathogen survival in soil and on plants.

	Soil		Plants	
Pathogen	Absolute maximum	Common maximum	Absolute maximum	Common maximum
	survival time			
Bacteria	1 yr	2 mo	6 mo	1 mo
Viruses	6 mo	3 mo	2 mo	1 mo
Protozoa	10 d	2 d	5 d	2 d
Helminths	7 yr	2 yr	5 mo	1 mo

Source: Kowal, 1983

recommended for plants consumed by humans that aerial crops not contacting the soil (e.g., wheat) could be harvested 1 mo after sludge application while 6 mo should elapse before harvesting root or low-growing crops. The potential for contamination of ground water with viruses (Zenz et al., 1976), coliforms (Higgins, 1984) and protozoa or helminths (Kowal, 1983) appears to be minimal following the application of sludges. Recent studies on surface application of sludge on pastures have shown that, although secondary pathogens (*Enterobacteriaceae*) were present, the health hazards to humans and animals were similar in untreated and sludge-treated pastures (Wallis et al., 1984). The greatest threat for ground water contamination appears to be in rapid infiltration systems used for the processing of wastewater (Kowal, 1983; Moore et al., 1981) and in soils where a significant fraction of the flow is through macropores allowing movement of bacteria (Smith et al., 1985) and potentially other microbes.

The data summarized above indicate that sludge treatment methods and required waiting periods after sludge application as described in the 'Criteria' will minimize problems related to pathogens contained in sludges applied to cropland. However, there appears to remain a need for further epidemiological information from sludge application sites. The potential for movement of pathogens into water supplies and for contamination of human food is nearly always a major concern when discussing sludge use on cropland with the general public.

Organics

The majority of sewage sludges originate from municipalities whose treatment systems receive wastewaters from industries. In addition to indigenous organics found in sewage (e.g., carbohydrates, lipids, proteins), a variety of organics used in industrial processes may be present. Sewage can contain paints, petroleum products, oils, waxes, solvents, plasticizers, pigments and so forth. These organics can be separated into several categories including pesticides, polychlorinated biphenyls (PDB's), halogenated aliphatics, ethers, phthalate esters, monocyclic aromatics, phenols, polycyclic aromatic hydrocarbons (PAH's), and nitrosamines. Organics have been analyzed in sludges obtained from numerous cities (U.S. EPA, 1982; U.S. EPA, 1983). In developing indices for the environmental impact of various organics, U.S. EPA (1985) summarized data on the typical and worst-case concentrations for selected organics found in sewage sludges (Table 3).

The data in Table 3 show that the concentration of most organics in sludges is less than 10 mg/kg although elevated levels are possible where an industry is discarding a specific compound into the sewage treatment system. After application to cropland, the fate of an organic compound is controlled by several chemical, physical, and biological processes. Common pathways include volatilization, photodecomposition, microbial decomposition, adsorption, leaching into ground water, runoff into surface water and plant uptake (Overcash, 1983; Overcash & Pal, 1979).

Microbial decomposition influences the concentration of organics not only after land application of sludges but also during sludge digestion and handling. Microbial decomposition has been evaluated for many of the compounds listed in Table 3. Using static-cultures inoculated with settled sewage water, Tabak et al. (1981) determined the degradation of 96 priority pollutants. They found that PCB's and chlorinated hydrocarbon pesticides were resistant to decomposition in a 7 d period. Considerable variability

Table 3. Typical and worst-case levels for organics in sludge (data presented on a sludge dry weight basis).

Compound	Typical	Worst
	------ mg kg^{-1}	-----
Aldrin/Dieldrin	0.07	0.81
Benzene	0.33	6.58
Benzidine	-	12.7
Benzo(a)anthracene	0.68	4.8
Benzo(a)pyrene	0.14	1.94
Bis(2-ethylhexyl)phthalate	94.3	459
Carbon tetrachloride	0.05	8.01
Chlordane	3.2	12
Chloroform	0.05	1.18
DDT/DDE/DDD	0.28	0.93
3,3'-Dichlorabenzidine	1.64	2.29
Dichloromethane	1.6	19
2,4-Dichlorophenoxyacetic acid	4.64	7.16
Dimethyl nitrosamine	-	2.55
Endrin	0.14	0.17
Heptachlor	0.07	0.09
Hexachlorabenzene	0.38	2.18
Hexachlorobutadiene	0.3	8
Lindane	0.11	0.22
Methylenebis (2-chloro aniline)	18	86
Malathion	0.05	0.63
PCB's	0.99	2.9
Pentachlorophenol	0.09	30.4
Phenanthrene	3.71	20.7
Phenol	4.88	82.1
Tetrachloroethylene	0.18	13.7
Toxaphene	7.9	10.8
Trichloroethylene	0.46	17.8
2,4,6-Trichlorophenol	2.3	4.6
Tricresyl phosphate	6.85	1,650
Vinyl chloride	0.43	312

Source: U.S. EPA (1985)

was observed in the resistance to decomposition of compounds having similar chemical structures, e.g., carbon tetrachloride was rapidly degraded while trichlorofluoromethane was not decomposed. Many of the aromatic compounds (benzene, phenol), PAH's and phthalates found in sludges will be rapidly decomposed in soils (see Gibson, 1983). Microbial degradation of chlorinated aromatics will also occur under anaerobic conditions resulting in the release of primarily methane and carbon dioxide (Suflita et al., 1982). Recent results indicate that plasmids are important in determining the microbial degradation of halogenated organics and that specific bacterial strains offer promise to accelerate degradation of organics in the environment (Ghosal et al., 1985). However, Brunner et al. (1985) found that enhanced degradation of PCB was obtained only when soils were amended with Acinetobacter sp. plus a substrate analog (biphenyl). Obviously, it is difficult to generalize about the rate and extent of decomposition for all organics found in sludge.

Movement of organics from the point of sludge application is possible through volatilization, leaching, and runoff. Using soil columns in the laboratory, volatilization was shown to be a major pathway for loss of chloroform, dichlorobromomethane, 1,2-dichloromethane, tetrachloroethane, 1,1,2-trichloroethane, trichloroethane, chlorobenzene and toluene (Wilson et al., 1981). For those compounds not degraded, a simple model based on water solubility and sorption by soil organic matter was developed to predict the leaching of small molecular weight organics. Trace levels (<1 ug/l) of organics have been found in groundwater beneath a site used for rapid infiltration treatment of wastewater (Tomson et al., 1981). It appears that many organics are sorbed by the organic-enriched solids found in sludges and that sludge additions to soils enhance the sorption of PCB's (Fairbanks & O'Connor, 1984) and phenoxyacetates (O'Connor et al., 1981). Even though the potential for movement of sludge-derived organics is not likely a major concern, the presence of many commonly used pesticides in groundwater beneath agricultural areas does indicate that additional studies are needed to evaluate the fate of organics in sludge-amended soils.

One significant problem arising from organics in sludge is the transfer to animals and humans. In an analysis of the potential problems due to sludge-borne PCB's, Fries (1982) evaluated direct plant contamination, indirect plant contamination and soil ingestion by animals. This study indicated that the greatest potential for PCB contamination occurred when lactating dairy cows are grazed immediately after surface applying sludge to a pasture. To maintain milk fat below the 1.5 mg L^{-1} PCB tolerance of FDA, it was shown that the sludge must contain <1 mg PCB kg^{-1}. Since the existing U.S. EPA (1979) regulations specify a 30 d waiting period before grazing dairy cattle, the sludge could actually contain 3 mg PCB kg^{-1} and not result in milk contamination. Based on the PCB concentrations shown in Table 3, significant contamination of soils, plants, and animal products is not anticipated if current regulations are used. Naylor and Loehr (1982a, 1982b) also conclude that the organics found in sludges do not pose a serious threat to plant, animal, or human health.

The presence of mutagens in sludges has been evaluated in recent studies. The sludge constituents responsible for the mutagenic reaction can be either organic or inorganic compounds. Using several assays to demonstrate mutagenic activity, Hopke et al. (1982) showed that sludge from Chicago contains components mutagenic to several plant species. Subsequent research indicated that mutagenic activity was more pronounced in sludges from industrialized municipalities (Hopke et al., 1984). However, analysis of dichloromethane extracts from 34 sludges with the Salmonella/mammalian microsome (Ames) assay indicated that mutagenicity was not related to degree of industrialization or type of wastewater treatment (Babish et al., 1983). The Ames assay has also been used to show that sludge components responsible for mutagenicity are degraded within 2 to 3 weeks after sludge incorporation into soil (Angle and Baudler, 1984). When evaluating data on mutagenicity of complex materials such as sludge it must be realized that a variety of compounds may be eliciting the positive response observed and that analysis of the sludge itself represents a worst case situation, being analogous to direct consumption of sludge. In addition, the commonly used Ames test indicates that essential biochemical compounds (e.g., glutathione and cysteine) are mutagenic at physiological concentrations (Glatt et al., 1983).

The U.S. EPA (1985) is in the process of evaluating all major sludge components and their impact on the environment. Indices are being developed for two concentrations (see Table 3) of sludge components and sludge application rates of 5, 50, and 500 mt/ha. The indices being considered are 1) soil concentration increase; 2) soil biota toxicity; 3) soil biota predator toxicity; 4) phytotoxicity; 5) plant concentration increment cause by intake; 6) plant concentration increment permitted by phytotoxicity; 7) animal toxicity resulting from plant consumption; 8) animal toxicity resulting from sludge ingestion; 9) human toxicity/cancer risk resulting from plant consumption; 10) human toxicity/cancer risk resulting from consumption of animals feeding on plants; 11) human toxicity/cancer risk resulting from consumption of animal products derived from animals ingesting soil; 12) human toxicity/cancer risk resulting from soil ingestion and; 13) aggregate human toxicity/cancer risk. Evaluation of worst case situation indicates that organics do not pose a threat to the soil biota, plants and animals but that the major potential problems arise from human toxicity resulting from plant or animal products and from direct soil ingestion. Indices are typically based on a linear partition coefficient to calculate the transfer of organics between components of the soil-plant-animal system. Even though it is difficult to generalize, the chlorinated hydrocarbon pesticides and other chlorinated organics tend to have the greatest hazard indices. This type of analysis is useful to identify organics needing more critical study in sludge-amended soils. Unfortunately, availability of data on sludge organics is limited and undue restrictions on sludge use may result if the above hazard analyses is not consistent with current sludge composition.

Metals

The trace metal content of sewage sludges limits application of sludges to cropland from both the annual and long-term standpoint. This report will concentrate on information published since the excellent review of Logan & Chaney (1983) who summarized the impact of sludge-borne metals when applied to a soil plant system. The trace element content of sludges is illustrated in Table 4.

The elements of primary concern to plant, animal, and human health include As, B, Cd, Cu, Hg, Mo, Ni, Pb, Se, and Zn. Of these, Cd poses the greatest long-term threat to human health; consequently, recommendations on limitations of sludge application to land typically involve annual and cumulative loadings of Cd (U.S. EPA, 1979, 1983). A CAST report (1976) implicated Cd, Cu, Ni, Zn, and Mo as the elements that plants could accumulate and then create problems throughout the food chain. Copper, Ni, and Zn can cause phytotoxicity at elevated levels in soils. Boron, Mo, and As, are of concern for both plant and animal health and Se is of concern for animal health. Mercury and Pb have not posed a problem for land application of sewage sludge since

Table 4. Trace element levels in dry sludges.

Element	Median from Chaney (1983)	U.S. EPA (1985) Typical	U.S. EPA (1985) Worst
	mg kg^{-1}		
As	10	4.6	20.8
Be	-	0.31	1.17
Cd	10	8.2	88
Cr	500	230	1499
Co	30	11.6	40
Cu	800	410	1427
F	260	86	739
Fe	17,000	28,000	79,000
Hg	6	1.5	5.8
Mn	260	-	-
Mo	4	9.8	40
Ni	80	45	663
Pb	500	248	1071
Se	5	1.1	4.8
Sn	14	-	-
Zn	1700	678	4580

these elements form relatively insoluble minerals in the soil, are non-mobile in fibrous root systems and phytotoxicity occurs at plant tissue concentrations that are not injurious to animals. The above elements have been categorized based on their impact on plants and animal health by Chaney (1983). As summarized in Table 5, the "Soil-Plant Barrier" concept indicates that environmental or phytological factors prevent certain elements from concentrating at harmful levels in above-ground plant growth. The relative sensitivity of crops to metal toxicity is summarized in Table 6.

The plant available forms of Cd, Cu, Ni, and Zn are generally cationic solution or exchangeable species while plant available forms of Mo, B, As, and Se, tend to be anions. Emmerich et al. (1982b) showed that 50 to 60% of the Cd, and 60 to 70% of the Ni and Zn in solution were in the free ionic form (Cd^{2+}, Ni^{2+}, Zn^{2+}), while the remainder of the solution forms existed as organic or inorganic complexes or ion pairs in sludge-amended soil columns that had been leached with 5m of water over 25 mo.

Shifts in the forms of the elements have also been observed. Emmerich et al. (1982a) showed, using a sequential extraction procedure, that an applied sludge originally contained 36% of Cd, Cu, Ni, and Zn in the residual form but at the end of the 25 mo column study, greater than 65% of all the Cd, Ni, and Zn were in this stable, phase. Most of the Cu remained in the organically bound phase. Composting can change the type of organic bonding associated with the metals. Simeoni et al. (1984) showed that composting an anaerobically digested sludge shifted Cu from -humus to humic acid. This change appeared to increase the availabity of Cu to lettuce (Latuca sativa cv. Waldmann's Green) and oats (Avena sativa cv. Russel).

The plant availability of most trace elements is strongly influenced by soil pH. Generally, the concentration of cationic species (Cd^{2+}, Cu^{2+}, Ni^{2+}, and Zn^{2+}) tends to decrease in availability as soil pH increases since insoluble phases of these elements are more stable at higher pH. In fact, a one unit pH increase would theoretically decrease the solution level of Cd, Cu, Ni, and Zn by 100-fold (Lindsay, 1979). Sommers and Lindsay (1979) also showed that pH will influence the degree of chelation of metals by altering the relative concentrations present. Emmerich et al. (1982a) showed, however, that the total concentration and activities of metals in the soil solution of sludge-amended soils were relatively independent of pH. They suggested that this apparent anomaly was caused by lack of equilibrium conditions, not knowing the controlling solid phase for each metal, errors in the thermodynamic data read into the GEOCHEM computer program, and/or the controlling

Table 5. Maximum tolerable levels of dietary minerals for domestic livestock in comparison with levels in forages (Logan and Chaney, 1983).

Element	"Soil Plant Barrier"	Level in Plant Foliage*		Maximum Levels Chromically Tolerated			
		Normal	Phytotoxic	Cattle	Sheep	Swine	Chicken
		-- mg kg^{-1} dry foliage --		------------- mg kg^{-1} dry diet --------			
As, inorganic	Yes	0.01-1	3-10	50	50	50	50
B	Yes	.7-75	75	150	(150)	(150)	(150)
Cd	Fails	0.1-1	5-700	0.5	0.5	0.5	0.5
Cr^{3+}, oxides	Yes	0.1-1	20	(3000)	(3000)	(3000)	3000
Co	Fail?	0.01-0.3	25-100	10	10	10	10
Cu	Yes	3-20	25-40	100	25	250	300
F	Yes?	1-5	--	40	60	150	200
Fe	Yes	30-300	--	1000	500	3000	1000
Mn	?	15-150	400-2000	1000	1000	400	2000
Mo	Fails	0.1-3.0	100	10	10	20	100
Ni	Yes	0.1-5	50-100	50	(50)	(100)	(300)
Pb	Yes	2-5	--	30	30	30	30
Se	Fails	0.1-2	100	(2)	(2)	2	2
V	Yes?	0.1-1	10	50	50	(10)	10
Zn	Yes	15-150	500-1500	500	300	1000	1000

* Based on literature summarized in Chaney (1983).
Based on NRC (1980). Continuous long-term feeding of minerals at the maximum tolerable levels may cause adverse effects. Levels in parentheses were derived by interspecific extrapolation by NRC.
Maximum levels tolerated based on human food residue considerations.

Table 6. Relative sensitivity of crops to sludge-applied heavy metals (Chaney and Hundemann, unpublished as cited by Logan and Chaney, 1983).

Very Sensitive	Sensitive	Very Tolerant	Tolerant
chard	mustard	cauliflower	corn
lettuce	kale	cucumber	sudangrass
redbeet	spinach	zucchini squash	smooth bromegrass
carrot	broccoli		'Merlin' red fescue
turnip	radish	flatpea	
peanut	tomato		
	marigold	oat	
ladino clover		orchardgrass	
alsike clover	zigzag, Red, Kura and crimson clover	Japanese bromegrass	
crownvetch	alfalfa	switchgrass	
'Arc' alfalfa	Korean lespedeza	red top	
white sweetclover	sericea lespedeza	buffelgrass	
yellow sweetclover	blue lupin	tall fescue	
	birdsfoot trefoil	red fescue	
weeping lovegrass	hairy vetch	Kentucky bluegrass	
Lehman lovegrass	soybean		
deertongue	snapbean		
	Timothy		
	colonial bentgrass		
	perennial ryegrass		
	creeping bentgrass		

* Sassafrass sandy loam amended with a highly stabilized and leached digested sludge containing 5300 mg Zn, 2400 mg Cu, 320 mg Ni, 390 mg Mn, and 23 mg Cd/kg dry sludge. Maximum cumulative recommended amounts of Zn and Cu are applied at 5% sludge
Injured at 10% of a high metal sludge at pH 6.5 and at pH 5.5.
Injured at 10% of a high metal sludge at pH 5.5, but not at pH 6.5.
Injured at 25% high metal sludge at pH 5.5, but not at pH 6.5, and not at 10% sludge at pH 5.5 or 6.5.
Not injured even at 25% sludge, pH 5.5.

solid phase not existing in its standard state. Also, complexation of metals by insoluble organics (i.e., humics) could alter the effect of pH on metal concentrations in solution. Copper, for example, is very strongly complexed and thus solubility is reduced by soil organic matter interactions.

Soil pH is one factor that the U.S. EPA (1983) and many state agencies recommends controlling in land disposal sites. Acidic soils (pH < 6.5) must be limed to pH 6.5 to reduce metal availability. Generally, areas that receive more than 50 cm of precipitation may have to be limed in order to permit land application of sludges. Acidity could also develop in normally neutral to basic soils due to the long-term input of NH_4-type fertilizers. Mahler et al. (1985) has shown that continued addition of ammonium fertilizers to soils in the Palouse region of Washington and Idaho have lowered soil pH from near neutral to less than 6.0 since the 1940's. For the elements that exist primarily as anionic species in sludge (Mo, As, and Se), their availability theoretically increases as soil pH increases. Overliming, therefore, could increase the availability of these elements and their potential hazard; however, liming to pH 6.5 is recommended for growing many crops and should not produce a severe risk in increased plant availability of these elements.

Also of concern is the influence of sludge on animals consuming crops grown on sludge-amended soils. Beyer et al. (1982) showed that earthworms (Lumbricidae, spp.) from sludge-amended sites contained 12 times more Cd, 2.4 times more Cu, 2.0 times more Zn and 1.2 times more Pb than earthworms found in control plots. Pietz et al. (1984) found similar results for earthworms extracted from sludge-amended soils at a strip mine reclamation site. Anderson et al. (1982) measured higher Cd levels in the kidneys and livers of meadow voles (Microtus pennsulvanicus) found in sludge treated areas. Baxter et al. (1982), on the other hand, showed that feeding sludge directly to cattle (Bos tarus) caused no apparent ill effects even though significant Cd accumulations were found in the liver and kidneys and the Cd was not released from these tissues after the sludge was removed from the diet. Studies by Dowdy et al. (1983a, 1983b) and Bray et al. (1985) indicated that high Cd corn silage grown on sludge-amended soils and fed to dairy goats (Capra hircus) and lambs (Ovis aries) for three years did not affect the performance or trace metal content of the tissues of these animals. Apparently, forage or grain crops grown on sludge-amended soils that do not have an excessive accumulation of trace metals can be fed safely to animals; however, burrowing organisms found in the treated soils may accumulate significant levels of trace metals.

The annual sludge application rate can be limited by an annual Cd loading. The annual Cd limitations developed by U.S. EPA (1979) are 0.5 kg $ha^{-1}y^{-1}$ for tobacco, root crops, and leafy vegetables and 2 kg Cd $ha^{-1}y^{-1}$ for other food chain crops (e.g., grains). The latter limit may be reduced to 0.5 kg $ha^{-1}y^{-1}$. In addition, plant uptake is minimized by recurring that soil pH be $\geq$ 6.5. If only animal feed is grown, then no annual Cd limit is imposed but a detailed plan and deed restrictions are imposed. An example calculation for metal loading limits is provided in the appendix.

The Total amount of sludge that can be applied over a period of years is controlled by limits on cumulative amounts of Pb, Zn, Cu, Ni, and Cd. U.S. EPA (1979) established regulations for cumulative Cd additions while only recommendations are available for Pb, Zn, Cu, and Ni (U.S. EPA, 1983). The cumulative limits shown were designed to allow growth of any crop after cessation of sludge application provided that soil pH is 6.5 or above. Several states use the metal limits in Table 7.

Some inconsistencies are encountered in using cation exchange capacity (CEC) for limiting sludge metal loading rates. On calcareous soils (pH > 7), accurate CEC measurements are difficult plus metal availability to plants is rather independent of soil properties due to pH buffering of $CaCO_3$. For example, it is well known that Fe deficiency in crops can occur in calcareous soils containing more than adequate amounts of total Fe and that soil pH alone controls the availability of Fe. A similar problem exists for soils dominated by allophanes (amorphous volcanic minerals), which typically have low CEC's but very high pH buffering capacities. For most temperate soils in the United States, however, CEC does represent a gross measure of the total

Table 7. Suggested cumulative limits for metals applied to cropland.

Metal	Soil cation Exchange Capacity cmol(+) kg^{-1}		
	< 5	5-15	> 15
	----------- kg ha^{-1}	-----------	
Cd	5	10	20
Cu	140	280	560
Ni	140	280	560
Zn	280	560	1120
Pb	560	1120	2240

Soil pH must be maintained at $\geq$ 6.5.

metal retention capacity of a soil (i.e., an indication of organic matter, Fe and Al oxide and clay content) which may or may not be related to metal availability for plants. Additional research is needed to develop more appropriate means of controlling plant uptake of metals in sludge-amended soils.

The discussion above applies to privately owned land and the regulations that allow for the growth of any crop in the future. Another approach is to use sites dedicated to land disposal of sludge (U.S. EPA, 1983). The primary purpose of these sites is long-term sludge disposal so that rates higher than those needed for crop production are utilized. These sites are carefully controlled and monitored so as to avoid any environmental contamination (e.g., groundwater with NO_3^-). The dedicated sites allow use of smaller land area and may present the only option for disposal of sludges if the potential contaminent concentrations are above recommended limits. They also require a long-term land commitment, extensive control and monitoring devices and possible preclusion of future use of the land for agricultural production. Consequently, one of the greatest concerns for dedicated sites is the Cd loading.

One limitation to assessing the plant availability of trace elements in sludge-amended soils is soil testing. Numerous studies have shown that DTPA (diethylenetriaminepento-acetic acid) can extract trace metals at levels that correlate well with metal concentrations in plants (Street et al., 1977; Korcak & Fanning, 1978; Latterell et al., 1978; Haq et al., 1980; Singh & Narwal, 1984; Kuo et al., 1985; Barbarick & Workman, 1986). Baxter et al. (1983) found poor correlations between plant concentrations and DTPA soil extracts but this work involved the extraction of soils with relatively low concentrations or additions of the trace metals (Barbarick & Workman, 1986). The NH_4HCO_3-DTPA procedure (Soltanpour et al., 1979) also extracted Cd, Cu, Ni, and Zn in sludge-amended soils at levels that were significantly correlated (r^2 generally > 0.80) with concentrations in swiss chard (*Beta vulgaris*) (Barbarick & Workman, 1986). The NH_4HCO_3-DTPA extraction can also be analyzed for P, K, and other elements and many elements extracted by this solution can be determined simultaneously on an Inductively Coupled Plasma-Optical Emission Spectrometer. Barbarick & Workman (1986) suggest that the NH_4HCO_3-DTPA procedure be adapted as a standard test for routine analysis of sludge-amended soils.

Nitrogen

To reduce problems of NO_3^- leaching into ground water, it is commonly recommended that sludge application rates provide adequate N for the growth of the desired crop. The plant available N in sludge is assessed by considering all of the NO_3-N as available, accounting for a certain loss of NH_4-N by volatilization and estimating a mineralization rate for the organic N. A net removal of N from the sludge application site will result from crop uptake and subsequent harvest.

Proper estimation of plant available N and the crop needs will reduce the risk of leaching losses of NO_3-N. In citing Duncomb et al. (1982), Linden et al. (1983) state that to prevent NO_3-N leaching below the root zone, the total N applied by waste water disposal should not exceed two times the potential crop removal of N. A more common approach is to determine the N fertilizer recommendations from routine soil tests or crop yield level considerations.

Based on prior work, the U.S. EPA (1983) estimated that NH_3 volatilization from surface-applied liquid sludge is 50% of the NH_4-N present and 0% for incorporated liquid sludge. The actual loss depends on soil, sludge, and climatic factors. Beauchamp et al. (1978) indicated that the "half-life" of NH_4-N in sludge was from 3.6 to 5.0 days. Donovan & Logan (1983) found that cumulative volatilization increased linearly with time and 32% of the NH_3-N was lost in the first 24 h before liquid sludge was incorporated. Barbarick et al. (1986) found that total plant recovery of N in liquid sludge was 42% for immediately incorporated sludge as opposed to a significantly lower recovery of 34% if incorporation was delayed for at least 5 days. This study also indicated that not all of the NH_4-N is conserved with immediate incorporation. By delaying incorporation, more N is lost by volatilization resulting in a potentially larger application rate of sludge as long as a metal does not limit the application.

The mineralization rate of organic N in sewage sludge is a function of the wastewater treatment process. The U.S. EPA (1983) uses values adapted from Sommers et al. (1981) that estimate the first year mineralization rate at 40% for unstabilized primary and waste activated, 30% for aerobically digested, 20% for anaerobically digested and 10% for composted sludge. Parker & Sommers (1983) developed a regression equation showing that N mineralization rate over a 16-week period was related to the organic N content of the sludge. They also concluded that N-release predictions could be based either on the type of sludge or the organic N content of the material. King (1984) also developed regression models to predict plant available N and used these models with results of other researchers. The models did not work well for a range of soils and wastes suggesting that equations developed for a given set of soils and waste materials cannot be applied to more general conditions. Harding et al. (1985) found that in sites where sludge amended soils did not receive sludge for five years the residual mineral N (e.g., soil NO_3^- and NH_4^+ measured before planting) and an estimate of the amount of N that would be mineralized during the growing season could accurately predict N uptake by a crop (r^2 = 0.92). The U.S. EPA (1983) suggests an estimated N mineralization rate of 3% of the remaining organic N per year for all types of sludge at four or more years after their initial application. This N release rate would be comparable to the rate found for native soil humus.

Other Constraints

Typically, amounts of trace elements, N or P in sewage sludge sludge will dictate the land application rate. Occasionally, other factors are equally important in determining sludge rates. Many if not most of these constraints are aesthetic or sociological in nature. This section, however, will present some of the physical and chemical limitations other than pathogens, organics, metals, and N.

Large application rates of sludge could add a considerable quantity of soluble salts to the soil. Simeoni et al. (1984) found that saturated pastes of the Fort Collins, Colorado sludge had an electrical conductivity (EC) of 7.5 dS m^{-1}. When a rate equivalent to 240 dry Mg ha^{-1} was added to the acidic Redfeather soil (Lithic Cryoboralf), the EC of saturated soil pastes was elevated from 0.6 (for the untreated control) to 6.0 dS m^{-1}. This increase in salinity level could influence yields of salt sensitive plants. Higgins (1985) also showed that 44.8 Mg ha^{-1} of sludge applied for 3 years to Sassafras sandy loam soils (Typic Hapludults) resulted in soluble salts and NO_3^- movement into ground water.

Problems can develop from planting a crop too soon after sludge application. Rapid decay of sludge organic matter can decrease soil O_2 and the resulting anaerobic soil conditions could restrict seed germination. High volatilization rates of NH_3 that occur after seed germination could produce NH_3 toxicity to seedlings. Toxicity from NH_3 was observed on dryland winter wheat treated with liquid digested sludge in a field and a greenhouse study (Utschig, 1985). Utschig et al. (1985) also found that as the sludge dries, crusting can result and this could restrict seedling emergence. Application of liquid sludge could seal the soil surface and greatly reduce infiltration rates (Linden et al., 1983). Runoff problems can result from soil sealing and steep slopes. The U.S. EPA (1983) states that surface application of liquid sludge is limited to less than 6% slopes while sludge injection can be used up to 12% slopes. The liquid loading limits of a soil must be carefully determined to avoid runoff problems. Another consideration for land disposal is scheduling the land application. Sometimes, weather will preclude the application, especially in the case of liquid sludge. Also, the application has to be timed so as not to interfere with planting, forage cuttings or harvest in general. In all land spreading operations, the logistics of scheduling applications must be carefully considered.

REFERENCES

Anderson, T. J., G. W. Barrett, C. S. Clark, U. J. Elia, and U. A. Majeti. 1982. Metal concentrations in tissues of meadow voles from sewage sludge-treated fields. J. Environ. Qual. 11:272-277.

Angle, J. S., and D. M. Baudler. 1984. Persistence and degradation of mutagens in sludge-amended soil. J. Environ. Qual. 13:143-146.

Babish, J. G., B. E. Johnson, and D. J. Lisk. 1983. Mutagenicity of municipal sewage sludges of American cities. Environ. Sci. Technol. 17:272-277.

Barbarick, K. A., and S. M. Workman. 1986. NH_4HCO_3-DTPA and DTPA extractions of sludge-amended soils. Unpublished data.

Barbarick, K. A., J. M. Utschig, D. G. Westfall, and R. H. Follett. 1986. Liquid sludge incorporation and nitrogen availability to dryland winter wheat. Unpublished data.

Baxter, J. C., B. Barry, D. E. Johnson, and E. W. Kienholz. 1982. Heavy metal retention in cattle tissue from ingestion of sewage sludge. J. Environ. Qual. 11:616-620.

Baxter, J. C., M. Aguilar, and K. Brown. 1983. Heavy metals and persistent organics at a sewage sludge disposal site. J. Environ. Qual. 12:311-316.

Beauchamp, E. G., G. E. Kidd, and G. Thurtell. 1978. Ammonia volatilization from sewage sludge applied in the field. J. Environ. Qual. 7:141-146.

Beyer, W. N., R. L. Chaney, and B. M. Mulhern. 1982. Heavy metal concentrations in earthworms from soil amended with sewage sludge. J. Environ. Qual. 11:381-385.

Bitton, G., B. L. Damron, G. T. Edds, and J. M. Davidson. (ed.) 1980. Sludge-Health Effects of Land Application. Ann Arbor Science Publishers, Ann Arbor, MI.

Boyce, J., K. A. Barbarick, D. G. Westfall, and R. H. Follett. 1985. Application of sewage sludge to dryland winter wheat. Quarterly Report May to August 1985 to Cities of Littleton and Englewood. Colorado State University, Fort Collins.

Bray, B. J., R. H. Dowdy, R. D. Goodrich, and D. E. Pamp. 1985. Trace metal accumulations in tissues of goats fed silage produced on sewage sludge-amended soils. J. Environ. Qual. 14:114-118.

Brunner, W., F. H. Sutherland, and D. D. Focht. 1985. Enhanced biodegradation of polychlorinated biphenyls in soil by analog enrichment and bacterial inoculation. J. Environ. Qual. 14:324-328.

Burge, W. D., and P. B. Marsh. 1978. Infectious disease hazards of landspreading sewage wastes. J. Environ. Qual. 7:1-9.

CAST. 1976. Application of sewage sludge to cropland: Appraisal of potential hazards of the heavy metals to plants and animals. Council for Agric. Sci. Tech. No. 64. Ames, IA.

Chaney, R. L. 1983. Potential effects of waste constituents on the food chain. pp. 152-240. In J. F. Parr et al. (eds.) Land Treatment of Hazardous Wastes. Noyes Data Corp., Park Ridge, NJ.

Dean, R. B. and M. J. Suess (ed.) 1985. The risk to health of chemicals in sewage sludge applied to land. Waste Management Research 3:251-278.

Donovan, W. C. and T. J. Logan. 1983. Factors affecting ammonia volatilization from sewage sludge applied to soil in a laboratory study. J. Environ. Qual. 12:584-590.

Dowdy, R. H., B. J. Bray, R. D. Goodrich, G. C. Marten, D. E. Pamp, and W. E. Larson. 1983a. Performance of goats and lambs fed corn silage produced on sludge-amended soil. J. Environ. Qual. 12:467-472.

Dowdy, R. H., B. J. Bray, and R. D. Goodrich. 1983b. Trace metal and mineral composition of milk and blood from goats fed silage produced on sludge-amended soil. J. Environ. Qual. 12:473-478.

Duncomb, D. R., W. E. Larson, C. E. Clapp, R. H. Dowdy, D. R. Linden, and W. K. Johnson. 1982. Effect of liquid wastewater sludge application on crop yield and water quality. J. Water Poll. Control Fed. 54:1185-1193.

Elliott, L. F. and J. R. Ellis. 1977. Bacterial and viral pathogens associated with land application of organic waste. J. Environ. Qual. 6:245-251.

Emmerich, W. E., L. J. Lund, A. L. Page, and A. C. Chang. 1982a. Predicted solution phase forms of heavy metals in sewage sludge-treated soils. J. Environ. Qual. 11:182-186.

Emmerich, W. E., L. J. Lund, A. L. Page, and A.C. Chang. 1982b. Solid phase forms of heavy metals in sewage sludge-treated soils. J. Environ. Qual. 11:178-181.

Fairbanks, B. C., and G. A. O'Connor. 1984. Effects of sewage sludge on the adsorption of polychlorinated biphenyls by three New Mexico soils. J. Environ. Qual. 13:297-300.

Feachem, R. G., D. J. Bradley, H. Carelick, and D. D. Mara. 1980. Appropriate technology for water treatment and sanitation: Health aspects of excreta and sileage management - A State-of-the-art review. World Bank, Washington, D.C.

Fries, G. F. 1982. Potential polychlorinated biphenyl residues in animal products from application of contaminated sewage sludge to land. J. Environ. Qual. 11:14-20.

Gerba, C. P. 1983. Pathogens. p. 147-187. In A. L. Page (ed.) Utilization of municipal wastewater and sludge on land. Univ. of California, Riverside.

Ghosal, D., I-S. You, D. K. Chatterje, and A. M. Chakrabarty. 1985. Microbial degradation of halogenated compounds. Science 228:135-142.

Gibson, D. T. (ed.) 1984. Microbial degradation of organic compounds. Marcel Decker, Inc. New York.

Glatt, H., M. Protic-Sabljic, and F. Oesch. 1983. Mutagenicity of glutathione and cysteine in the Ames test. Science 220:961-963.

Haq, A. U., T. E. Bates, and Y. K. Soon. 1980. Comparison of extractants for plant-available zinc, cadium, nickel, and copper in contaminated soils. Soil Sci. Soc. Am. J. 44:772-777.

Harding, S. A., C. E. Clapp, and W. E. Larson. 1985. Nitrogen availability and uptake from field soils five years after addition of sewage sludge. J. Environ. Qual. 14:95-100.

Higgins, A. J. 1984. Land application of sewage sludge with regard to cropping system and potential pollution. J. Environ. Qual. 13:441-448.

Hopke, P. K., M. J. Plewa, J. B. Johnston, D. Weaver, S. G. Wood, R. A. Larson, and T. Hinesly. 1982. Multi-technique screening of Chicago's municipal sewage sludge from mutagenic activity. Environ. Sci. Technol. 16:140-147.

Hopke, P. K., M. J. Plewa, P. L. Stapleton, and D. L. Weaver. 1984. Comparison of the mutagenicity of sewage sludges. Environ. Sci. Technol. 18:909-916.

King, L. D. 1984. Availability of nitrogen in municipal, industrial, and animal wastes. J. Environ. Qual. 13:609-612.

Korcak, R. F., and D. S. Fanning. 1978. Extractability of cadmium, copper, nickel, and zinc by double acid versus DTPA and plant content at excessive soil levels. J. Environ. Qual. 7:435-440.

Kowal, N. E. 1983. An overview of public health affects. p. 329-394. In A. L. Page (ed.) Utilization of municipal wastewater and sludge on land. Univ. of California, Riverside.

Kuo, S., E. J. Jellum, and A. S. Baker. 1985. Effects of soil type, liming, and sludge application on zinc and cadmium availability to Swiss chard. Soil Sci. 139:122-130.

Latterell, J. J., R. H. Dowdy, and W. E. Larson. 1978. Correlation of extractable metals and metal uptake of snapdragons grown on soil amended with sewage sludge. J. Environ. Qual. 7:435-440.

Linden, D. R., C. E. Clapp, and R. H. Dowdy. 1983. Hydrologic and nutrient management aspects of municipal wastewater and sludge utilization on land. pp. 79-101. In A. L. Page (ed.) Utilization of Municipal Wastewater and Sludge on Land. Univ. California, Riverside.

Lindsay, W. L. 1979. Chemical Equilibria in Soil. John Wiley and Sons, Inc. New York.

Logan, T. L., and R. L. Chaney. 1983. Utilization of municipal wastewater and sludge on land-metals. pp. 235-323. In A. L. Page (ed.) Utilizaton of Municipal Wastewater and Sludge on Land. Univ. California, Riverside.

Mahler, R. L., A. R. Halvorson, and F. E. Koehler. 1985. Long-term acidification of farmland in northern Idaho and eastern Washington. Commun. Soil Sci. Plant Anal. 16:83-95.

Moore, B. E., B. P. Sagik, and C. A. Sorber. 1981. Viral transport to groundwater at a wastewater land application site. J. Water Poll. Control Fed. 53:1492-1502.

NASULGC. 1973. Recycling municipal sludges and effluents on land. Natl. Assoc. State Univ. Land Grant Colleges, Washington, D.C.

National Research Council. 1980. Mineral tolerance of domestic animals. National Academy of Sciences, Washington D.C.

Naylor, L. M., and R. C. Loehr. 1982a. Priority pollutants in municipal sewage sludge. Biocycle 23(4):18-22.

Naylor, L. M., and R. C. Loehr. 1982b. Priority pollutants in municipal sewage sludge. Part II. Biocycle 23(6):37-41.

O'Connor, G. A., B. C. Fairbanks, and E. A. Doyle. 1981. Effects of sewage sludge on phenoxy herbicide adsorption and degradation in soils. J. Environ. Qual. 10:510-515.

Overcash, M. R. 1983. Land treatment of muncipal effluent and sludge: Specific organic compounds. p. 199-231. In A. L. Page (ed.) Utilization of municipal wastewater in sludge on land. Univ. of California, Riverside.

Overcash, M. R., and D. Pal. 1979. Design of Land Treatment Systems for Industrial Waste - Theory and Practice. Ann Arbor Science Publishers, Ann Arbor, MI.

Page, A. L., P. L. Gleason, III, J. E. Smith, Jr., I. K. Iskandar, and L. E. Sommers (ed.) 1983. Utilization of municipal wastewater and sludge on land. Univ. of California, Riverside

Parker, C. F., and L. E. Sommers. 1983. Mineralization of nitrogen in sewage sludges. J. Environ. Qual. 12:150-156.

Pietz, R. I., J. R. Peterson, J. E. Prater, and D. R. Zenz. 1984. Metal concentrations in earthworms from sewage sludge-amended soils at a strip mine reclamation site. J. Environ. Qual. 13:651-654.

Reddy, K. R., R. Khaleel, and M. R. Overcash. 1981. Behavior and transport of microbial pathogens and indicator organisms in soils treated with organic waste. J. Environ. Qual. 10:255-266.

Simeoni, L. A., K. A. Barbarick, and B. R. Sabey. 1984. Effect of small-scale composting of sewage sludge on heavy metal availability to plants. J. Environ. Qual. 13:264-268.

Singh, B. R., and R. P. Narwal. 1984. Plant availability of heavy metals in sludge-treated soil: II. Metal extractability compared with plant metal uptake. J. Environ. Qual. 13:344-349.

Smith, M. S., G. W. Thomas, R. E. White, and D. Ritonga. 1985. Transport of Escherichia coli through intact and disturbed soil columns. J. Environ. Qual. 14:87-91.

Soltanpour, P. N., S. M. Workman, and P. A. Schwab. 1979. Use of inductively-coupled plasma spectrometry for the simultaneous determination of macro- and micronutrients in NH_4HCO_3-DTPA extracts of soils. Soil Sci. Soc. Am. J. 43:75-78.

Sommers, L. E., and W. L. Lindsay. 1979. Effect of pH and redox on predicted heavy metal-chelate equilibria in soils. Soil Sci. Soc. Am. J. 43:39-47.

Sommers, L. E., C. F. Parker, and G. J. Meyers. 1981. Volatilization, Plant Uptake and Mineralization of Nitrogen in Soils Treated with Sewage Sludge. Tech. Rpt. 133, Purdue Univ. Water Resources Research Center, West Lafayette, Indiana.

Street, J. J., W. L. Lindsay, and B. R. Sabey. 1977. Solubility and plant uptake of cadmium in soils amended with cadmium and sewage sludge. J. Environ. Qual. 6:72-77.

Suflita, J. M., A. Horowitz, D. R. Shelton, and J. M. Tiedje. 1982. Dehalogination: A novel pathway for the anaerobic biodegradation of haloaromatic compounds. Science 218:1115-1117.

Tabak, H. H., S. A. Quave, C. I. Mashni, and E. F. Barth. 1981. Biodegradability studies with organic priority pollutant compounds. J. Water Poll. Control Fed. 53:1503-1518.

Tomson, M. B., J. Dauchy, S. Hutchins, C. Curran, C. J. Cook, and C. H. Ward. 1981. Groundwater contamination by trace level organics from a rapid soil infiltration site. Water Res. 15:1109-1116.

U.S. Environmental Protection Agency. 1979. Criteria for classification of solid waste disposal facilities and practices. Federal Register 44:53438-53464.

U.S. Environmental Protection Agency. 1981. Process design manual for land treatment of municipal wastewater. EPA 625/1-81-013.

U.S. Environmental Protection Agency. 1982. Fate of priority pollutants in publicly owned treatment works. EPA 440/1-82-303. U. S. Government Printing Office Washington, D.C.

U. S. Environmental Protection Agency. 1983. Process design manual for land application of municipal sludge. EPA-625/1-83-016. U. S. Government Printing Office, Washington, D.C.

U. S. Environmental Protection Agency. 1985. Summary of environmental profiles and hazard indices for constituents of municipal sludge: Methods and Results. U. S. Environmental Protection Agency, Office of Water Regulations and Standards, Washington, D.C.

Utschig, J. M. 1985. Sewage sludge versus nitrogen fertilizer application on dryland winter wheat. M. S. Thesis. Colorado State Univ., Fort Collins.

Utschig, J. M., K. A. Barbarick, D. G. Westfall, R. H. Follett, and T. M. Bride. 1985. Application of liquid anaerobically digested sewage sludge to dryland wheat. Technical Report 85-6. Colorado State University Agricultural Experiment Station, Fort Collins.

Wallis, P. M., D. L. Lehmann, D. A. MacMillan, and J. M. Buchanan-Mappin. 1984. Sludge application to land compared with a pasture and hayfield: Reduction of biological health hazard over time. J. Environ. Qual. 13:645-650.

Wilson, J. T., C. G. Enfield, W. J. Dunlap, R. L. Cosby, D. A. Foster, and L. B. Baskin. 1981. Transport and fate of selected organic pollutants in a sandy soil. J. Environ. Qual. 10:501-506.

Zenz, D. R., J. R. Peterson, D. L. Brooman, and C. Lue-Hing. 1976. Environmental impacts of land application of sludge. J. Water Poll. Control Fed. 48:2332-2342.

Design and management of successful land application systems

Robert K. Bastian, Office of Municipal Pollution Control, U.S. Environmental Protection Agency, Washington, DC 22046
James A. Ryan, Water Environmental Research Laboratory, U.S. Environmental Protection Agency, Cincinnati, OH 45268

As a sludge disposal technique, land application of sewage sludge sludge has been practiced for many years in this country and overseas. Only in recent years, however, have the necessary research and monitoring studies been undertaken to develop sound design guidelines for recycling sludge on the land. Appropriate management practices have been developed to allow many land application systems to be properly designed and operated from a public health and environmental impacts standpoint, while protecting the long-term productivity of the sites to which the sludge is applied. A number of projects using these practices have been closely monitored and carefully evaluated for extended periods of time. The results of these and other studies have lead to the publication of numerous scientific reports and state-of-the-art symposium proceedings (Page et al., 1983; Sopper et al., 1982; Cole et al., 1986). A number of universities, State and Federal agencies have issued detailed guidance and/or requirements covering many land application practices (MDNR, 1985; PSU, 1985; Hornick et al., 1984; Simpson et al., 1984; EPA, 1983, 1979, 1978, 1977; IEPA, 1983; WDOE, 1982; OSU, 1982; EPA/FDA/USDA, 1981; Sommers et al., 1980). Design guidance and recommended management practices have been developed that will allow most sludges to be land applied in one manner or another if proper controls are implemented (Reed and Crites, 1984; EPA, 1983; Parr et al., 1983; Overcash and Pal, 1979).

Land Application and Related Practices for Sludge Management

A wide range of land application practices for recycling sewage sludge have been investigated and employed to date, including application to cropland, rangeland and forest lands, parks, golf courses, and a variety of disturbed and marginally productive areas. In addition, a number of systems have been developed to dispose of sewage sludge using high rate land application practices which include no efforts to beneficially recycle the nutrients and organic matter contained in the sludge. Land application projects are underway in many large metropolitan areas, including Washington, D.C., Philadelphia, Baltimore, Chicago, Milwaukee, Denver, San Diego, Sacramento and Seattle, as well as in thousands of smaller cities and towns across the country, especially in the Midwest. Many of these systems are prime examples of the basic land treatment, recycle/reuse concepts that have been strongly encouraged by Congress and EPA. The research and demonstration experience and guidance available covering such land application practices (Hornick et al., 1984; Page et al., 1983; EPA, 1983, 1978, 1977; Sopper et al., 1982; Knezek and Miller, 1978; Cole et al., 1986) and a growing number of successful projects across the country are a clear indication of the potential value and benefits that can be achieved through land application of sludge.

Land appliation practices have been implemented by many rural communities with adequate available land, and agreeable land owners and neighbors. In some cases "dedicated" or publicly owned and controlled sites have been used, but a more common practice has involved application to privately owned and managed farmland, strip mined sites, etc. The results of one survey suggests that, nationwide, well over 30 percent of the smaller communities have applied their sludge to the land for over 40 years.

For years many communities simply stockpiled dried sludge and allowed the public to haul it away for their own use. Now there is a growing interest in the potential for marketing sludge to farmers and others as an organic fertilizer and soil amendment. A number of communities have been heat drying, composting, or otherwise processing their sludge prior to marketing or giving it away as a soil amendment for many years. Others have established programs for applying liquid sludge to cropland at carefully predetermined rates, and in at least some cases farmers pay for the sludge or sludge application service. For years now, in West Herfordshire, England, the liquid digested sludge from a 35 MGD activated sludge secondary wastewater treatment plant has been hauled to consenting farmers as "HYDIG," and there are more requests for this organic material than can be satisfied. At least part of the sludge produced in Salem, Oregon, and in Madison and Milwaukee, Wisconsin, is managed in a similar manner, where "BIOGRO," "METROGRO," and "AGRI-LIFE" are delivered to farmers who request the material. Seattle, Washington, which currently applies much of its sludge to city owned forest land, expects to be applying "SILVIGRO" to privately owned commercial timberland in the near future. Many other cities are either considering or currently attempting to dispose of either dewatered or liquid sludge through the use of similar land application programs.

The selling or give-away of bagged or bulk, processed (composted, heat dried, etc.) sludge-containing products for use as soil conditioners, organic fertilizers, or potting media has been practiced for many years in several major urban areas (e.g., Los Angeles County, California - Kellogg Supply Co.'s "Nitrohumus" and other products; Milwaukee, Wisconsin "Milorganite"; Houston, Texas - "Hou-actinite") and have stared more recently in other areas (e.g., Philadelphia, Pennsylvania - "Earthlife"; Washington, D.C. metropolitan area - "Compro"; Oakland, California "CompGro"). Also, various types of "give-away" programs have been operated by numerous small and large communities as a common sludge management practice for years. Efforts by private companies, either using their own company owned facilities or city owned facilities, to produce and market sludge related products are also becoming a more common situation; examples of several very recent projects include those in Portland, Oregon, and Wilmington, Delaware. In Missoula, Montana, a private firm composts sludge mixed with other organic and nutrient-rich materials to produce a final product which it sells as a high nutrient content organic fertilizer /soil conditioner ("Eko-compost"); this firm is currently developing similar but larger projects in Orange County, California, and Orlando, Florida. Projects in Largo, Florida, and Winston-Salem, North Carolina have also marketed fortified sludge-derived products ("LarGrow" and "Organoform SS") through fertilizer companies. Various composted products containing sludge have also been marketed as a substitute for topsoil and peat for certain horticultural uses. A recent Biocycle survey found 79 full-time operating facilities composting sewage sludge, 8 pilot projects, and 62 additional facilities either under construction or in the planning, bid, or design stage (Goldstein, 1985).

Benefits from Land Application

The potential benefits from recycling the organic matter and nutrient resources in sewage sludge through various land application practices have been well demonstrated and have helped lead to an increased use of these practices in many parts of the country. Not only can land application help the municipalities by serving as a cost-effective sludge "disposal" technique, it can also serve the farmer or other land owners by improving soil characteristics, reducing fertilizer costs, and increasing productivity. While not a high grade fertilizer, the $30-60/dry ton worth of organic nitrogen and phosphorus in typical sewage sludges alone can make land application worthwhile as a partial replacement for commercial inorganic fertilizers in certain cases. Sludge compost has been effectively used as a substitute for topsoil and peat for certain horticultural applications, showing a yield improvement value at $35-50/dry ton over normal potting media (Hornick et al., 1984). The potential value of sludge related materials (liquid sludge, dewatered sludge cake, compost, etc.) in the production of field crops, vegetables, nursery crops and ornamentals, forages, and sod production and maintenance has been clearly demonstrated in many locations across the country. The potential for using sludge to dramatically increase productivity in commercial forests (especially areas of naturally low production due to limited nutrients) has also been demonstrated in several areas.

Generally, the greatest cost savings and value may be realized as a result of using sludge on less productive soils. However, the increase in operational costs associated with sludge use in agriculture can be more than offset, especially for crop rotations requiring relatively large amounts of nitrogen, if site conditions do not require additional contouring or other expensive measures for runoff control or other substantial investments as a result of sludge use. A study of the utilization of sewage sludge and effluent on selected agricultural crops in one area of Oregon found that the return per acre from sludge use when compared to traditional fertilizer sources ranged from a loss of $6 to an increase of $15 per acre, depending on the crop rotation involved, previous soil management practices, soil type, and level of sludge application (Schotzko et al., 1977). These calculated economic results from sludge use were limited to the fertilizer savings after subtracting costs for new production practices when the sludge was available to the farmer with no cost for application.

Municipal sludges are also being used as an effective top soil substitute, soil conditioner and organic fertilizer in forestry and in reclamation, revegetation, and stabilization of strip mined lands, mine spoils and tailings piles, construction sites, quarries, dredge spoils, borrow pits and other drastically disturbed areas (Sopper et al., 1982; Cole et al., 1986). Although limited to areas where transport to such sites is cost-effective and locally acceptable, reclamation practices offer an opportunity to solve several environmental problems at one time. Either experimental or full-scale projects using liquid, dewatered or composted sludges for reclamation and/or forestry uses have been undertaken in such areas as Illinois, Ohio, Michigan, Wisconsin, Pennsylvania, Maryland, Virginia, West Virginia, South Carolina, Alabama, Florida, Washington, Colorado, Montana, and Washington.

Major Problems Facing Acceptance of Land Application Systems

The characteristics of and constituents present in municipal sewage sludges can vary widely, depending upon such factors as the type and amount of industrial discharges to the treatment plant and the sludge treatment processes used, making it difficult to make generalizations about the physical, chemical, and biological properties of sewage sludge. In reality almost anything can be found in sewage sludge. Some of the physical, chemical, and biological characteristics and constituents of municipal sludge that are of potential importance or concern regarding sludge management are listed in Table 1. The variability in content of trace elements, toxic organics and pathogens reported in sludge has been a major drawback in gaining greater acceptance of the use of these materials as a source nutrients and as organic soil amendments.

While the potential benefits from recycling the organic matter and nutrient resources in sewage sludge through various land application practices have been well demonstrated and have led to an increased use of these practices in many parts of the country, there are a number of questions that are frequently raised when projects involving land application of sewage sludge are proposed. The issues involved generally center on concerns that pathogens and/or toxic chemicals, which may be present in the sludge at varying levels, may contaminate the soil, nearby surface or ground waters, or plants and animals grown on sites which receive sludge. Differential responses of various pathogens to specific stabilization processes, toxic organic compounds to breakdown by soil microbes, and plants to uptake of trace elements applied with sludge clearly add to the concerns over the potential for land application projects to harm public health or the environment.

Clearly, public doubts and officials' concerns about adding potentially toxic substances and pathogens found in the sludge to productive farmland, watersheds, or other land application sites must be closely examined in terms of protecting human health, crop quality, surface and ground water quality and other potential environmental impacts, and future land productivity or uses. Concerns are also often expressed about potential odors, increased truck traffic and noise, and other aesthetics problems, as well as the potential impact on neighboring property values. Also, there is often a general

Table 1. Sludge Characteristics and Constitutents of Importance or Potential Concern

Physical Characteristics

Volume	Dissolved solids
Density	Volatile solids
Particle size	Total solids

Biological Constituents

Bacteria	Fungi
Salmonella spp.	*Aspergillus* spp.
Escherichia coli	
Pseudomonas spp.	

Virus	Parasites & Protozoa
Enteroviruses	*E. histolytica*
Polioviruses	*Ascaris* spp.
Coxsackie viruses	*Toxacara* spp.
Echoviruses	
Hepatitus viruses	
Adenoviruses	
Reoviruses	

Chemical Constituents

Organic Carbon	DDT	Benzene
Total Nitrogen	DDD	Chlorobenzene
Ammonia-Nitrogen	DDE	1,4-dichlorobenzene
Nitrate-Nitrogen	Dieldrin	1,2-dichlorobenzene
Total Phosphorus	Aldrin	1,3-dichlorobenzene
Total Potassium	Chlordane	1,2,4-dichlorobenzene
Total Sulfur	Heptachlor	1,2,3-trichlorobenzene
Sodium	Lindane	1,3,5-trichlorobenzene
Calcium	Toxophene	Hexachlorobenzene
	Endrin	Benzidine
Aluminum		Phenol
Arsenic	Bis(2-ethylhexyl)phthalate	Pentachlorophenol
Beryllium	Di-n-butyl phthalate	2,4-dichlorophenol
Boron	Chloroethane (ethyl chloride)	4-chloro-m-cresol (4-chloro-3-methylphenol)
Cadmium	Methyl chloride	
Chromium	Methyl bromide	Toluene
Copper	Vinl chloride	Napthalene
Cyanide	1,1-dichloroethylene	2-chloronapthalene
Flourine	1,2-dichloropropane	Achenaphthylene
Gold	1,3-dichloropropane	Anthracene
Iron	Pentachloroethane	Benzo(a)anthracene (1,2-benzanthracene)
Lead	Hexachloroethane	
Magnesium	Carbon tetrachloride	Ideno (1,2,3-cd) phrene
Manganese	Dichlorodifluoromethane	Benzo(a)fluoranthene
Mercury	Dichlorobromomethane	PCBs (Polychlorinated biphenyls)
Molybdenum	Trichlorofluoromethane	
Nickel	Tetrachloroethylene	
Palladium	Trichloroethylene	
Platinum	Hexachlorobutadiene	
Selenium	Chloroform	
Silver	Bromoform	
Tin	1,1,2,2-tetrachloroethane	
Vanadium	1,1,1-trichloroethane	
Zinc	1,1,2-trichloroethane	

psychological opposition to the basic idea of recycling human wastes on land since many people in this country have developed an "out-of-sight/out-of-mind" attitude toward waste management problems which generally favors the more highly engineered approaches to sludge disposal.

Public acceptance by land owners and their neighbors is a key to successful projects involving land application of municipal sludge. Odors from poorly managed sludge management systems and perceived odors from anything that has to do with sewage and sludge may well be the biggest problem facing the successful establishment and operation of land application projects. General reluctance of rural areas to receive urban wastes (at least until adequate economic or other incentives are offerred) can also be a significant factor. "What do I get out of it" and "Not-in-my-backyard" (NIMBY) attitudes often become very apparent when land application projects are proposed.

Such factors can in certain circumstances lead to the passage of what appear to be overly restrictive regulations, special land use ordinances or zoning restrictions, and extensive monitoring and record keeping requirements. Concerns that additional restrictions may be placed on land application practices once they are commenced, excessive nuisance claims may be filed, contractor performance may lapse, or other factors may lead to increased operating costs and a loss of public and regulatory acceptance have often prevented serious consideration of land application practices in areas where they were clearly appropriate.

Of course in addition to the problems faced in gaining and maintaining the acceptance of both the public and regulatory officials, land application systems are faced with the usual array of equipment and operating problems associated with hauling and spreading large volumes of liquid or dewatered organic materials -- not unlike those faced by livestock producers or food processors involved in land application of their wastes.

Such concerns as those described above led to the development and issuance of detailed guidance and eventually regulations by EPA (40 CFR Part 257) and many states. In many cases, these guidelines and regulations either recommend or establish sludge contaminant concentration or loading limits for certain sludge contaminants, certain management practices when using sludge in agriculture to assure long-term soil productivity, and require specific levels of treatment to assure adequate pathogen reduction in sludges that are land applied. It is likely that such guidelines and regulations will be modified and expanded in future years as more data becomes available about the various concerns associated with land application of sewage sludge. However, through recent efforts to improve source control and industrial pretreatment, many municipal wastewater treatment plants have been able to greatly improve the quality of their sewage sludges and as a result its acceptability for utilization through land application. As a result, for the most part the highly contaminated sludges often cited in the literature and represented by the high end of the range for many of the chemical contaminants shown in Table 2 are now a thing of the past.

Table 2. Range and median concentration of trace elements in dry digested sewage sludges (Chaney, 1983).

Element	Reported Range Minimum	Reported Range Maximum	Median
	------------- mg/kg dry sludge -------------		
As	1.1	230	10
Cd	1	3,410	10
Co	11.3	2,490	30
Cu	84	17,000	800
Cr	10	99,000	500
F	80	33,500	260
Fe	1,000	154,000	17,000
Hg	0.6	56	6
Mn	32	9,870	260
Mo	0.1	214	4
Ni	2	5,300	80
Pb	13	26,000	500
Sn	2.6	329	14
Se	1.7	17.2	5
Zn	101	49,000	1,700

Examples of Design and Management Guidance

In 1983, EPA issued a Process Design Manual for Land A Application of Municipal Sludge (EPA 625/1-83-016; July 1983). In a manner not unlike several recent design-oriented texts on land application systems (Reed and Crites, 1984; Parr et al., 1983; Overcash and Pal, 1979), in general, this manual attempts to provide rational procedures for the design of various municipal sludge land application systems, including agriculture, forest land, and land reclamation uses, plus dedicated high-rate disposal systems (other than landfill). The manual offers general guidance and basic information for use in planning, design and operating land application projects that may be of considerable use when developing new projects.

As in the texts cited above and many of the university and State agency issued guidance materials (e.g., MDNR, 1985; Simpson et al., 1984; IEPA, 1983; WDOE, 1982; OSU, 1982; Sommers et al., 1980), a number of basic project design and management principles are provided which are based on the results of closely monitored and carefully evaluated long-term projects. In general, these design documents recommend designing agricultural utilization and forest land application projects in a manner which uses sludge as an N or P fertilizer supplement or partial replacement, while using Federal or State regulatory limits for contaminants as the long-term sludge loading constraint. For reclamation projects more consideration is usually given to the soil amendment/

enhancement values of sludge, and considerably higher onetime application rates are generally recommended. For dedicated high-rate disposal systems, design considerations are more like those for landfills since beneficial use of the sludge nutrients and organic matter is not a factor.

Application site evaluation considerations such as topography, slope, soil properties (e.g. type, texture, permeability, CEC, depth), susceptibility to flooding, proximity to surface water, roads and dwellings, type and depth of groundwater, crop selection and management, etc. are covered in considerable detail. Detailed procedures for calculating sludge application rates based on crop nutrient requirements and contaminant loading considerations are provided. General information and guidance on the types of sludge application equipment available, their strengths and weaknesses are provided. Recommendations on the types of monitoring and recordkeeping procedures that are appropriate for land application systems are included, as are examples of contract agreements between landowners and sludge applicators and procedures for gaining better public acceptance. Tables 3 thru 5 provide a summary of some of the more generally agreed upon key design and operation characteristics and recommendations for land application of sludge systems.

Based on our observations of numerous land application projects that have been successfully designed and implemented to date, there are a number of common features that come to mind. Sludge usually must be well stabilized before application to the land in order to reduce the numbers of pathogens and amount of odor-causing volatile organic matter present in raw sludge. A number of processes have been successfully used for this purpose, including composting, heat treatment, digestion, long-term storage, and chemical stabilization. However, efforts undertaken to date to develop processes that effectively remove chemical contaminants from sludge have generally proven to be very expensive, leaving source control and pretreatment as the most effective means of controlling the levels of most chemical contaminants in sewage sludge.

Stabilized liquid sludge can be sprayed directly onto the soil surface, incorporated into the upper layer by plowing or discing, or injected beneath the soil surface with specially designed injection systems. The equipment to apply sludge by such methods is now generally available even to the smallest communities. However, when considerable distances are involved between the treatment plant and application sites, sludges are often dewatered in drying beds or with the aid of vacuum filters, centrifuges, presses or other devices to concentrate the sludge solids and lower sludge transportation costs. Specialized equipment and management procedures have even been developed to make forest land application of sludge more efficient and easier to control.

Availability of adequate application sites may well be the most important issue facing the design and implementation of every land application project. Many communities rely on interested farmers or other land owners as a continuing source of land application sites, while others have purchased land in order to be assured that application sites would be available when needed. In some areas, the treatment plant operators or their contractors must take special steps to seek

Table 3. Key design and operation characterists of land application systems.

	Agricultural Utilization	Forestry Utilization	Land Reclamation
Application techniques	Surface (with or without incorporation) Subsurface injection	Surface (generally without incorporation	Surface (with or without incorporation)
Process objectives	Recycle nutrients and organic matter Enhance crop production Improve soil properties Disposal of sludge	Recycle nutrients and organic matter Enhance forest production Disposal of sludge	Establish vegetative cover Recycle nutrients and organic matter Improve soil properties Disposal of sludge
Application rates			
Range (DT/ac)	1 - 30	4 - 100	3 - 200
Typical (DT/ac)	5	20	20 - 40
Frequency	One time, infrequently, or yearly	One time or 3-5 yr intervals	Usually one-time unless dedicated site
Available technical literature and experience	Extensive literature, hundreds of large & small full-scale projects	Limited literature, a few demo and full-scale projects	Limited literature, a few demo and full-scale projects

Table 4. Recommended Slopes for Sludge Application Sites.

Slope (%)	Comment
0-3	Ideal
3-6	Acceptable for surface application or injection
6-12	Inject liquid sludge
12-15	Immediate incorporation of all sludges. Effective runoff control
>15	Only suitable with good permeability, short slope length, and where it is a minor part of the total application area

Table 5. Set-Back Distances for Sludge Application.

	Distance from feature to sludge application site				
	50-300 ft		300-1500 ft		1500 ft
Feature	Injec.	Surf.	Injec.	Surf.	Injec. & Surf.
Residential developmemt	No	No	Yes	No	Yes
Inhabited dweling	Yes	No	Yes	Yes	Yes
Ponds and lakes	Yes	No	Yes	Yes	Yes
Springs	No	No	Yes	Yes	Yes
Ten-year high water mark of streams, rivers and creeks	Yes	No	Yes	Yes	Yes
Water supply wells	No	No	Yes	Yes	Yes
Public road right-of-way	Yes	No	Yes	Yes	Yes

out cooperative landowners, while in other areas an ample supply of potential land application sites have been made available simply by answering inquiries and requests that have come to the treatment plant in response to local advertisement of their planned or ongoing land application program. Written agreements or contracts between the sludge applier (the city or its contractor) and the landowners are often used to help avoid uncertainties about future responsibilities, liability, and recordkeeping. Public education programs and the use of local advisory panels, demonstration projects, etc. aimed at providing factual information to help conteract rumors and vague apprehensions can result in the identification of interested landowners as well as lead to improvements in public acceptance. Special demonstration projects, participation in local fairs, university research programs and Farm Science Reviews, and hosting or cooperation with farm tours set up through local soil and water conservation districts, Grange, Farm Bureau, or other farm organizations can also go a long way towards gaining greater farmer and landowner interest in sludge use. In some cases communities or private companies add nonsludge sources of nutrients in order to enrich the nutrient content of their sludge and to attract greater interest in the use of their high analysis, fortified organic fertilizer "products."

If efforts to prevent impacts (e.g., controlling sludge quality and loading rates, careful siting of projects, truck routing, odor and noise control measures, spill prevention, etc.) fail, various means of contingency planning as well as mitigation measures can also be employed to help deal with problems that may actually occur as a result of land application project operations. In some cases compensation of unavoidable impacts may be appropriate (e.g., payments for correcting any damage to local highways, culverts, bridges, etc); in other cases offsetting benefits may be feasible (e.g., creating public parks and recreation areas, generating local jobs and business opportunities, paying local taxes, etc.). Some other contingency management measures that have been effective include the posting of performance bonds, use of liability insurance, funding of independent project monitoring, and setting up "hotlines" to facilitate access to project information.

Examples of "Successful" Land Application Projects

Across the country there are clearly many examples of successful sludge projects involving land application of sewage sludge. Detailed accounts of such projects are found in various technical journals, research reports, pollution control magazines, EPA design manuals and guidance documents. The following accounts offer a few insights into a number of land application projects which are generally considered to be highly "successful" which may help to reinforce the notion that even successful projects have their share of implementation problems.

AGRICULTURAL USE:

1. Building upon their extensive experience with research and development work in cooperation with Florence, Alabama, (Giodarno & Mays, 1981; Matthews et al., 1981), the Tennessee Valley Authority (TVA) has been working with a number of communities to establish demonstration projects in each of the TVA states in an effort to help inform other communities of the potential benefits to be gained from land application of sludge to farmland. In spite of their numerous successes in this effort, they have encountered difficulty in identifying acceptable land application sites for some communities, in some cases apparently due to a lack of interest or initiative by key local officials and POTW

personnel. Further, considerable differences are apparent in the operating expenses of what appear to be "similar" projects--a situation certainly not limited to just these TVA projects!

2. In an effort to help evaluate and demonstrate safe systems for managing sewage sludge application to farmland while investigating sludge-related health risks to rural residents and their livestock, EPA's Water Environmental Research and Health Effects Research Laboratories in Cincinnati, Ohio, helped fund a major project in Ohio involving the Ohio Farm Bureau, the Ohio State University, several different POTWs, and a large number of individual farm families. The study demonstrated that large as well as small municipalities can work cooperatively with large numbers of farmers in a mutually beneficial program and found the health risks involved with the low application rates and sludge management practices involved in this study to be insignificant (Brown et al, 1985). However, although there was already a general awareness of land application practices and many examples of existing land application projects by other communities in Ohio for this project to build on, poor project communications with at least one POTW involved with this study eventually lead to an injunction against all land application of that POTW's sludge by a local county board of health.

3. Both Salem, Oregon, and Madison, Wisconsin, implement well-publicized programs (designed by the same engineering company) involving land application of liquid digested sewage sludge to privately owned farmland (EPA, 1982, 1983). Overall, sludge has been accepted as a safe, effective fertilizer supplement and has been widely welcomed by farmers in both areas. These two cities have managed to develop and maintain strong support and acceptance of their programs by most regulators, farmers and local residents even when faced with issues that would be expected to terminate such projects - at least in part as a result of their early public education efforts and intensive monitoring/recordkeeping programs.

For example, soon after the publication of EPA's land disposal regulations (40 CFR Part 257) in 1979, concerns over potential product liability were raised by a major food processor which has a processing plant in Salem. The company actively discouraged the use of sewage sludge by the growers it purchased vegetables from and encouraged FDA to establish action levels for contaminants in vegetable crops or processed products. While such concerns lead to the issuance of a joint EPA/FDA/USDA (1981) policy statement and guidance document on the use of sludge in the production of fruits and vegetables, they continue to face land application projects in many areas. Today, Salem's sludge is applied primarily to fields used to produce grains, grasses, pasture, and silage corn. In addition, sludge is also applied to fields used for the production of seed crops and Christmas trees, and to sites used as commercial nurseries and filbert orchards. No sludge is applied to fruit and vegetable crops which will be processed by local processing plants; the Oregon DEQ requires an 18-month waiting period after sludge application before planting of fruits and vegetables which may be eaten raw.

In the Madison's case, the city's discovery that sludge they had generated many years ago (which was stored in an old lagoon they wanted to empty) contained greater than 50 ppm of PCBs created considerable controversy and concern over how to treat-or-dispose of this particular lagoon-full of sludge. When they notified the farmers involved in their land application program of this situation, they were surprised to find that the farmers were willing to continue working with the city and to accept their low-PCB content sludges while the city worked with the University of Wisconsin, State and EPA Region to develop an acceptable plan for handling the old, lagooned high-PCB content sludge.

RECLAMATION:

1. Both Chicago and Philadelphia have established and implemented major projects involving the use of sewage sludge in land reclamation activities. In Chicago's case, thousands of acres of stripmined land were purchased in a rural mining area (Fulton County) some 200 miles away from the city and developed into a major site for using the sludge to help reclaim the land and to produce crops for animal feed only (in this case applying the sludge at high rates on a site dedicated for that purpose). After several years of building lagoons and storing sludge, local concerns focused on when would they start applying sludge. Once they began applying the sludge, local concerns focused on odors, site management, the possibility of surface and groundwater contamination, etc. After some ten years of detailed monitoring and research efforts (Hinesly, 1984a, 1984b) and extensive legal battles, as well as the development of improved on-site management practices, active involvement by the local health department and citizens' advisory groups, the project had become well established and clearly locally acceptable. The site manager was even elected to the local school board. In fact, recently local concerns have become focused on the loss of jobs associated with phasing the project out of operation as a result of Chicago obtaining the Illinois EPA's permission to convert the project to a backup system to their new, much less expensive sludge management project in Chicago which involves the use of dewatered sludge to cap a large landfill located near their sewage treatment facilities.

Following-up on the results of a five year demonstration program in Pennsylvania (Sopper & Seaker, 1984; Sopper et al., 1982), involving the Appalachian Regional Commission, the Pennsylvania DNR, Penn State University and funded by EPA's Water Environmental Research Laboratory, the City of Philadelphia embarked upon a land reclamation program involving the use of composted sludge and sludge cake applied at relatively low application rates to privately owned strip mined areas in western Pennsylvania. For several years this program operated quite successfully, involving the reclamation of as much as 1,000 acres per year. Due to other long existing issues which have put many of the residents of rural western Pennsylvania at odds with the City of Philadelphia politically, the Pennsylvania DNR had encouraged the city to use a contractor rather than city employees to implement this program, and for the city to keep a low profile. This arrangement worked quite well until the city's contractor was brought up on and convicted of charges of graft and corruption associated with attempting to buy off other potential contractors from bidding on a related project they were attempting to establish with the City of Baltimore. Apparently Philadelphia also found problems with some of the billings made by their contractor as well. As a result of such contractor-related image problems, Philadelphia's program suffered a major loss of confidence by many people in western Pennsylvania and their mine land reclamation program came to a quick halt. Efforts are currently underway to revive the program, involving a new contractor, and this time having an on-site city employee present to help oversee the contractor's work and to assure the local citizens that the city will do whatever is necessary to address any problems that might occur as a result of their land reclamation program.

FOREST LAND USE:

1. In conjunction with the University of Washington, the City of Seattle has developed a sizable land application program involving forest land uses of sludge. Initial work in this area resulted in many problems associated with attempting to establish appropriate application rates as well as how to control competing vegetation growth, deer browsing, and discourage small mammals from gnawing the bark off the young trees at sites where sludge was used to help revegetate clear-cut areas. Later efforts to apply sludge in established stands of trees required the development of more efficient application equipment that was also capable of negotiating rough terrain. The issue of whether the city should purchase their own forest land for sludge application purposes (and go into competition with other forest land owners) or work toward gaining the agreement of private commercial timber companies to have sludge applied to their forest land also had to be faced. Questions concerning public access to forested sites receiving sludge, long-term land use of sites, and possible impacts of wildlife continue to be addressed. Along with work undertaken in Michigan, South Carolina, and elsewhere, considerable progress has been made by Seattle in developing systems for applying sewage sludge to forested lands (Cole et al., 1986).

HIGH RATE DEDICATED LAND DISPOSAL:

1. Sacramento, California, and a number of other locations (mainly in the West, have been involved with dedicated land disposal practices (EPA, 1983, 1979). Unlike certain agricultural, forest land, and reclamation use projects where high rates of sludge have been applied to land purchased or otherwise controlled by the municipality, most projects involving high rate land disposal of sewage sludge do not include crop production. Such systems typically involve multiple applications of liquid sludge each year, often using injection equipment, with annual loadings reaching as much as 100 dry tons per acre (224 dt/ha) or more. While these systems can often provide a cost-effective sludge disposal alternative, their high annual application rates require careful attention to such areas as the control of odors, surface runoff, public access, and especially the potential for contamination of groundwater

aquifers. In some cases, special considerations have been necessary when siting high rate dedicated land disposal projects to be able to avoid the need for special liners beneath the sites. In the case of Sacramento project, specially designed fans were installed to help control potential odor problems from developing during periods of air inversions.

Conclusions

In conclusion, land application of municipal sludge is currently practiced throughout the country. When proper steps are taken in designing and managing these systems, very few sites are totally unacceptable for land application of municipal sludge. However, a number of site-specific factors can greatly impact how effectively a site can be used for sludge application, including soil properties, slope, type and depth to groundwater, susceptibiity to flooding, crop selection and management, proximity to surface water, buildings, etc. Many of the common concerns faced when developing land application projects can be minimized by taking the following precautions:

1. use appropriate application site selection criteria
2. use sludges that are well stabilized and that have relatively low concentrations of contaminants
3. obtain appropriate permits and permission to proceed
4. use appropriate sludge application techniques and equipment
5. apply sludges at rates which meet crop nitrogen or phosphorus requirements
6. adhere to established State/Federal regulatory requirements and guidelines to limit contaminant loadings
7. maintain appropriate soil pH and soil erosion control measures
8. provide for adequate monitoring of sludge quality and soil; and where appropriate, vegetation, surface and groundwater quality
9. maintain accurate records
10. be sensitive to the concerns of application site land owners and neighbors; avoid creating unnecessary nuisance conditions - extensive noise, odors, rutting of fields, nighttime operations, etc.
11. provide information when requested
12. seriously consider contingency planning and possible mitigation measures for uncontrollable problems.

Finally, remember that even the best planned and most successfully implemented projects do continue to face problems that require flexibility in operation and creative solutions.

REFERENCES

Brown, R. et al. 1985. Demonstration of Acceptable Systems for Land Disposal of Sewage Sludge. EPA 600/52-85-062.

Chaney, R.L. 1983. Potential Effects of Waste Consituents on the Food Chain. p. 50-76. IN J.F. Parr et al (eds.) Land Treatment of Hazardous Wastes. Noyes Data Corp., Park Ridge, NJ.

Cole, D.D. et al. (ed.) 1986. Forest Land Application Symposium. Proceedings of an International Symposium on Forest Utilization of Municipal and Industrial Wastewater and Sludge held June 25-28, 1985. Seattle, WA. College of Forest Resources, The University of Washington, Seattle, WA. (In press.)

Giodarno, P.M., and D.A. Mays. 1981. Plant Nutrients from Municipal Sewage Sludge. Ind. Eng. Chem. Prod. Res. Dev. 20(2):212-216.

Goldstein, N. 1985. 1985 Biocycle Survey, Sewage Sludge Composting Facilities on the Rise. Biocycle 26(8):19-24.

Hinesly, T.D. et al. 1984a. Effects of Using Sewage Sludge on Agricultural and Disturbed Lands. EPA 600/S2-83-113.

Hinesly, T.D. et al. 1984b. Longterm Use of Sewage Sludge on Agricultural and Disturbed Lands. EPA 600/S2-84-128.

Hornick, S.B. et al. August 1984. Utilization of Sewage Sludge Compost as a Soil Conditioner and Fertilizer for Plant Growth. USDA/ARS, Ag. Info. Bull. 464.

IEPA (Illinois EPA). October 1983. Design Criteria for Sludge Application on Land. IEPA.

Knezek, B.D., and R.H. Miller (ed.) March 1978. Application of Sludges and Wastewaters on Agricultural Land: A Planning and Educational Guide. EPA Reprinted update of NC Regional Res. Publ. 235 (Oct. 1976). Ohio Agric. Res. and Devel. Center Res. Bull. 1090. Wooster, OH.

Matthews, M.R., F.A. Miller, III, and G.J. Hyfantis, Jr. 1981. Florence Demonstration of Fertilizer from Sludge. Ind. Engr. Chem. Prod. Res. Dev. 20(4):567-574.

MDNR (Missouri Dept. of Natural Resources). January 1985. Agricultural Use of Municipal Wastewater Sludge - A Planning Guide. Missouri DNR, Jefferson City, MO.

Overcash, M.R., and D. Pal. 1979. Design of Land Treatment Systems for Industrial Wastes. Ann Arbor Science, Ann Arbor, MI.

OSU (Ohio State Univ.). 1982. Ohio Guide for Land Application of Sewage Sludge. Ext. Bull. 598 (Agdex 530), Res. Bul. 1079. Coop. Ext. Serv., The Ohio State Univ., Columbus, OH.

Page, A.L. et al. (ed.) 1983. Proceedings of the 1983 Workshop on Utilization of Wastewater and Sludge on Land. Univ. of California-Riverside, CA.

Parr, J.F. et al. 1983. Land Treatment of Hazardous Wastes. Noyes Data Corp., Park Ridge, NJ.

PSU (Pennsylvania State University). March 1985. Criteria and Recommendations for Land Application of Sludges in the Northeast. Penn. State Univ. Ag. Exp. Station Bull. 851. The Penn. State Univ., University Park, PA.

Reed, S.C., and R.W. Crites. 1984. Handbook of Land Treatment Systems for Industrial and Municipal Wastes. Noyes Publ., Park Ridge, NJ.

Schotzko, R.T. et al. 1977. Projecting Farm Income Effects of Sewage Sludge Utilization in the Tualatin Basin of Oregon. Special Report 498. Agric. Exp. Station. Oregon State Univ., Corvallis, OR.

Simpson, T.W., S.N. Nagle, and G.D. McCart. June 1984. Land Application of Sewage Sludge for Agricultural Purposes. Virginia Polytechnic Inst. and State Univ., Blacksburg, VA.

Sommers, L.E., D.W. Nelson, and C.D. Spies. September 1980. Use of Sewage Sludge in Crop Production. (Fertility AY-240) Coop. Ext. Serv., Purdue Univ., West Lafayette, IN.

Sopper, W.E., and E. M. Seaker. 1984. Strip Mine Reclamation with Municipal Sludge. EPA 600/S2-84-035.

Sopper, W.E., E.M. Seaker, and R.K. Bastian (ed.) 1982. Land Reclamation and Biomass Production with Municipal Wastewater and Sludge. Proceedings of a Symposium held Sept. 16-18, 1980. Pittsburgh, PA. The Pennsylvania State Univ. Press, University Park, PA.

U.S. EPA. October 1983. Process Design Manual for Land Application of Municipal Sludge. EPA 625/1-83-016. CERI, Cincinnati, OH.

U.S. EPA. 1982. Sludge Recycling for Agricultural Use. EPA 430/9-82-008. Office of Water Program Operations, Washington, D.C.

U.S. EPA/FDA/USDA. January 1981. Land Application of Municipal Sewage Sludge for the Production of Fruits and Vegetables; A Statement of Federal Policy and Guidance.

U.S. EPA. September 1979. (40 CFR Part 257). Criteria for Classification of Solid Waste Disposal Facilities and Practices. Federal Register 44(179):53439-53468.

U.S. EPA. September 1979. Process Design Manual for Sludge Treatment and Disposal. EPA 625/1-79-011. CERI, Cincinnati, OH.

U.S. EPA. October 1978. Sludge Treatment and Disposal, Vol 2. Sludge Disposal. EPA 625/4-78-012. CERI, Cincinnati, OH.

U.S. EPA. October 1977. Municipal Sludge Management: Environmental Factors. EPA 430/9-77-004 (MCD-28). Office of Water Program Operations. Washington, D.C.

WDOE (Washington Dept. of Ecology). October 1982. Municipal and Domestic Sludge Utilization Guidelines. WDOE Rep. 82-11.

Institutional constraints on residuals use

D. L. Forster, T. J. Logan, G. M. Pierzynski, and D. D. Southgate,
Department of Agricultural Economics, The Ohio State University,
Columbus, OH 43210

Landspreading of municipal sewage sludge has been hampered by institutional constraints during the past 15 years. Among these constraints have been unfavorable public attitudes toward landspreading, a variety of government regulations applying to those activities, as well as uncertainty about the future regulatory climate. The importance of institutional constraints is underscored by many technically sound landspreading programs being delayed or even scuttled because of opposition from a group of concerned citizens, local government authorities (e.g., zoning boards, county commissioners, health boards), or state environmental protection agencies.

The technology adoption process, as depicted in Figure 1, is affected by available technologies, institutional constraints, and programs to overcome these constraints. The intent of this paper is to focus primarily on the constraints and programs to overcome these constraints. Specifically, our purposes are to provide (a) a classification of these institutional constraints, (b) evidence of their importance in hampering landspreading, and (c) suggestions for overcoming these constraints. The classification of institutional constraints comes primarily from an earlier review of the literature (Forster and Southgate, 1984). Evidence of the importance of these constraints is a product of two relatively recent surveys. The first survey elicited information from 110 local government officials responsible for wastewater treatment in communities across the country; the second survey was directed at those state government officials responsible for sludge management in 30 states. A list of methods for overcoming these constraints was obtained from the literature, and the two recent surveys provided information on the frequency of their use.

Definition of Institution

The economics literature defines an institution as collective action in control of individual action (Commons, 1934). Institutions include family, community, economic sphere (profession, business or union), religion, other social groupings and the behavioral norms of these groupings (Gordon, 1980). Formal rules and regulations are commonly thought of as institutional constraints, and indeed they are because they directly influence behavior. But an equally important set of constraints are attitudes. While harder to define than formal rules and regulations, attitudes affect group behavior. Many attitudes are held in common by individuals in a group and influence group action as surely as do more formal regulations. A community's use of sludge landspreading is affected by federal, state, or local regulations governing its use. But probably more important are the attitudes of those in the community (e.g. land owners, rural

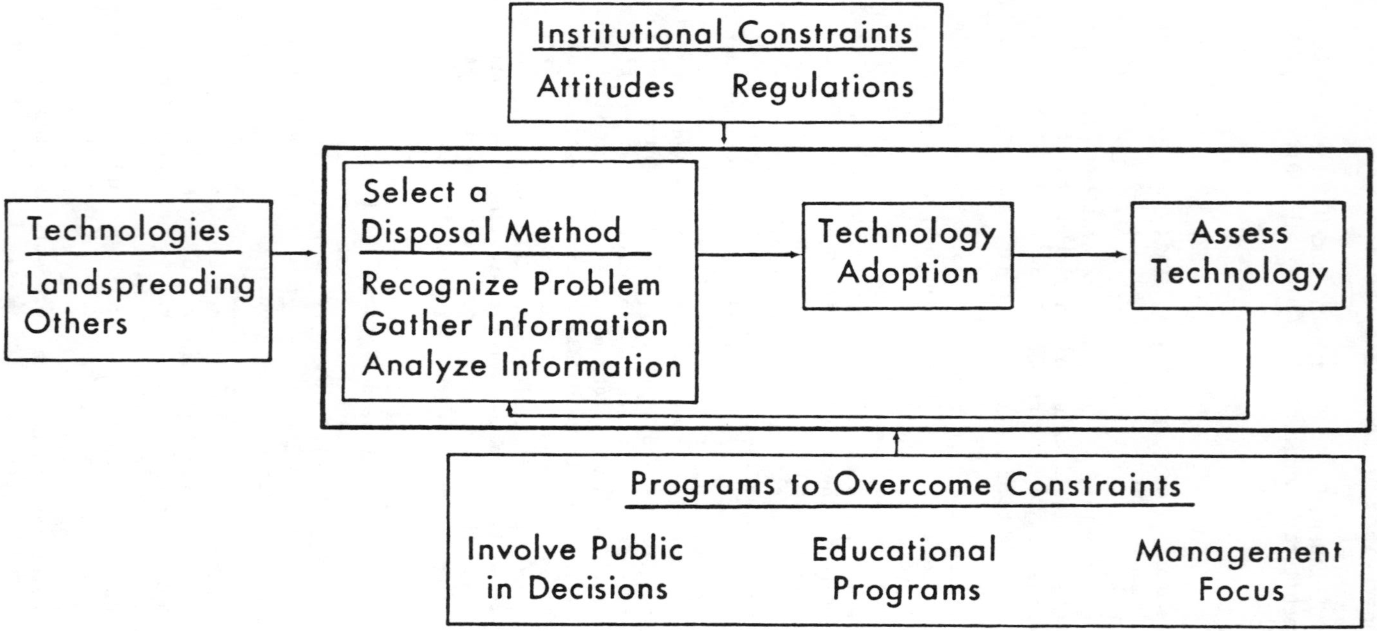

Figure 1. Disposal Technology Adoption Process

residents, community officials, and environmentalists) toward landspreading.

Institutions, both attitudes and formal regulations, tend to be static, inherited from the past, creatures of habit, and dictatorial. They inhibit change. However, technological innovation causes existing attitudes to change thereby eventually changing institutions. As an example, consider the impacts of the birth control technology on changing sexual mores, family planning, and expectations of women regarding careers outside the home. In addition, one could safely argue that the impacts extend to the formal institutions of marriage and regulations regarding equal opportunity employment. New institutions evolve as society reevaluates its rules and norms in light of new technology.

Institutions constraining the use of residuals are continuously changing as new technology develops. Attitudes and regulations regarding sludge landspreading are a product of landspreading technology. More specifically, this technology includes techniques of application, analysis, and site monitoring, as well as knowledge about optimum application rates and contractual arrangements with landowners. These constraints are also affected by other sludge disposal technologies (e.g., incineration, landfilling, and ocean disposal).

Constraining Attitudes

Concern about health risks was found in our survey of 30 state public officials to be the most constraining attitude affecting sludge landspreading. Nearly 60 percent of these officials cited it as the concern that led to most local opposition to landspreading. There is concern that cadmium, lead, and some organic compounds may enter the food chain, be directly ingested, or contaminate water supplies. The risk of pathogenic organisms and parasitic ova being distributed by landspreading is also discomforting.

The available literature suggests little hazard confronts sewage plant workers. No demonstated hazard has been documented for communities adjacent to well-managed landspreading sites (Burge and Marsh, 1978; Euga and Crites, 1980). Yet the public intuitively suspects that land application systems are unsafe. Indeed, the easiest way for opponents to thwart a landspreading program is to question whether it is safe for the community's health. Minuscule concentrations of metals, organic compounds, pathogens, and parasites may be harmless in reality, but to the uninformed rural resident they are frightening.

Concern about nuisances was mentioned as the most constraining attitude by about 20 percent of the state public officials interviewed. Odor is the primary concern because sludge is considered to be foul-smelling and may affect the surrounding community. Other potential nuisances are road damage from trucks and equipment, sludge spills and the perception that the neighborhood is a dumping ground.

Concern about the environment was sited by about 10 percent of the surveyed state public officials as being the most constraining attitude. There is fear that concentrations of zinc, copper, nickel, and industrial organic compounds may be present in sludge and that these could accumulate in the soil and become toxic to crops. Also, landspreading near streams carries the possibility of runoff of nutrients, metals, and organics to surface water supplies. Of course, such problems can be avoided with most sludges by using moderate application rates and proper placement.

Concern about accepting sludge from a neighboring political entity was mentioned by several state officials as being a constraining attitude. Causes for this concern might be uncertainty about sludge content, fear that the community will be labeled a dumping ground, or provincialism.

Concern about land values were mentioned by only a few state officials in our survey, but it is our opinion that this concern is relatively important. While a rural resident may not be personally concerned about health, nuisances, environment, or sludge from a neighboring community, he (she) may feel that sludge landspreading would threaten land values. The fear is that residential and commercial values will diminish with landspreading. In fact, for many the concern over land values may be the most constraining attitude, but it is disguised as a concern about health, nuisances, or the environment.

Farmers' attitudes may be constraining, although this was not mentioned by the state officials we interviewed. Generally, farmers tend to hold to an "agrarian ideology", the proposition that farming is the foundation of society and that rural life is better or more natural than urban life (Napier and Mast, 1981). In this light, farmers view municipal waste as an urban problem that ought to be solved by those creating the waste and not by pushing it off on the rural community.

Historical precedents also shape attitudes (Gordon, 1980). From the 1840's to the early 1900's, fear of disease and epidemics prompted communities to adopt waste treatment technology. Land treatment was the initial technology adopted and was responsible for sharply reducing death rates (Jewell and Seabrook, 1979). It was widely accepted. New treatment technologies such as sedimentation, chemical precipitation, and screening were developed in the late 1800's, and these technologies changed the philosophy of wastewater treatment. Partial treatment of effluent rather than the complete treatment of wastewater offered by land application became the norm. By the 1920's, land treatment had been replaced by other methods which emphasized partial treatment of effluents and sludge disposal by landfilling and, later, incineration. Given that attitudes tend to be static and inherited from the past, it should be no surprise that it would take time for sludge landspreading to be favored by sanitary engineers, public officials, and the public at large.

Cultural influences also shape attitudes, and these influences prevail over nearly any community (Musselman et al., 1980; Olson and Bruvold, 1982). Women are less accepting of land treatment than are men. Because of women's traditional responsibility for family health and cleanliness, it follows that they are more reluctant than men to accept a practice with possible health risks. Age is another factor; most new technologies are accepted less readily by the old, and landspreading is no exception. Formal education is positively correlated to acceptance. Landspreading is more likely to be disfavored by persons with less than 12 years of formal education. Exposure to new ideas is a product of education, and this generally leads to greater acceptance of new technologies.

Municipal officials' biases are generally against landspreading (Jewell and Seabrook, 1979). As a result of the historic evolution of waste treatment, municipal officials and the technical community are unfamiliar with land treatment. For the past half century, wastewater engineering texts have ignored land application or only briefly mentioned it. Waste is considered a disposal problem not a resource recovery problem.

Regulatory Constraints

Federal legislation and regulation have encouraged landspreading and generally have refrained from placing constraints on it. The Federal Water Pollution Control Act Amendments of 1972 encouraged the recycling of potential wastewater pollutants through agricultural, silvicultural, or aquacultural production. It also contained provisions that require consideration of "appropriate alternative waste management techniques," such as landspreading, in order to receive federal grants. The 1977 Clean Water Act further encouraged the use of "innovative and alternative" technologies such as landspreading. Only 10 percent of the state officials and practically none of the 110 community officials interviewed considered federal regulations to be the most important regulatory constraints facing communities in their state.

Federal regulation of landspreading has been debated for years, and the U.S. Environmental Protection Agency promises to develop regulations to govern the distribution and marketing of sludge. Generally, these proposed regulations are intended to encourage responsible, well-managed systems. Community officials interviewed overwhelmingly denounced federal regulation of landspreading. While relatively few expressed the fear that federal regulations would put their communities' programs out of business, the majority doubted that the regulations could be flexible enough to take special local conditions into account.

State legislation and regulation are the most important formal rules facing landspreading according to both state and community officials interviewed. In nearly every state contacted, some form of license or permit is required. In two-thirds of the states contacted, sludge analysis, application site monitoring, and application rate restrictions are required. Of course, the nature of sludge analysis and site monitoring varies widely among states and probably within each state. State regulations were imposed on just 40 percent of the surveyed communities in 1972, but more than 80 percent of the communities now must obtain permits and conform to landspreading guidelines established by state agencies. Several state officials interviewed admitted that while all communities in their state are subject to landspreading regulations, enforcement of regulations in small communities is lax.

Local constraints may include zoning or other land use controls, health codes, nuisance laws, and outright prohibition of landspreading. Our survey of 110 community officials pointed out major changes in these local constraints. Local regulations, such as limitations on sites or application rates, permit requirements, and so forth, were practically non-existent in 1972. However, over 60 percent of the surveyed community officials reported that their own or neighboring political jurisdiction had adopted local regulations by 1982. Few local prohibitions on landspreading were found. Even those communities with reservations about landspreading seemed willing to accept a well-managed program.

To summarize the nature of regulatory constraints, it appears that state government agencies provide the most important set of regulatory constraints with local political jurisdictions also being an important regulatory actor. The federal government has imposed few regulatory constraints, although federal guidelines have influenced state regulations. Most regulations have been established within the past 15 years, and the vast majority have had the intention of assuring safe, well-managed landspreading programs rather than disallowing them.

Overcoming Institutional Constraints

Opposition to landspreading, whether in the form of the public's attitudes or formal regulations, can be changed. The literature suggests that conducting educational programs and involving the public in decision making are the most important tools in winning public favor.

An educational program can remove much of the local opposition (Musselman et al., 1980; Deese et al., 1980; Dotson, 1982; Ellis and Disinger, 1981). These sources suggest that complete and unbiased information should be made available through a variety of forums, such as seminars and workshops, lectures to community groups, tours of treatment plants and application sites, newspaper articles and television and radio interviews. Written material, varying in scope from brief brochures to university publications, should be made available. The U.S. EPA , state agencies, and many land grant colleges can provide such material. Also, demonstration sites should be used to show methods of application and crop response.

During the educational program, incentives need to be clarified continually. The economic justification of land application is that benefits accrue to both community taxpayers and farmers receiving the sludge. Generally, landspreading is a low cost disposal option for the community and provides low cost plant nutrients to farmers. These benefits need to be assessed early and not exaggerated. The potential costs of an improperly managed program to rural residents need to be recognized. Local roads can be damaged and severe odor nuisances can reduce property values. However, well-managed programs avoid these costs, and nearby examples of excellent programs should be emphasized.

Odor, health, and environmental issues should be addressed early and with vigor. Unless treatment plants resolve odor problems, any landspreading program is doomed. Evaluation of the program by public health officals can help assure the public that the program will not be detrimental to local health or the food chain. Soil scientists might be used to identify the soils, application rates, and sludge contents required for an environmentally safe program.

Over 30 percent of the surveyed state officials said that "nearly all" communities in their state actively sought public acceptance of landspreading programs. Another 60 percent said that "some" of the communities in their state actively sought acceptance. An educational program was identified as the most important mechanism in gaining public acceptance. Of the 110 community officials interviewed, over one-half reported the use of educational campaigns.

The second general mechanism for gaining public acceptance is to involve the public in decision making. In the early stages of a program, ideas need to be developed, attitudes and potential support of affected groups must be assessed, and rapport with influential individuals established (Donnermeyer, 1977; Stitzlein, 1980). As the program develops, the public is involved in decision making. Problems are discussed, information is made available, and alternatives are developed and analyzed with the assistance of public representatives. Avoided is a "public hearing" situation that takes the form of a "we-they", winner-take-all battle between treatment plant officials and the public.

Advisory groups can be established to facilitate public involvement in the decision-making process. Specific tasks that these groups might perform include helping to select suitable demonstration and application sites, pointing out local problems or probable constraints, and scrutinizing the technical and management aspects of the plan.

Once a landspreading program is established, there must be continued focus on gaining and maintaining public acceptance. Land application at sites removed from residential areas; sludge and soil analysis prior to application; site monitoring after application; well-kept records; low application rates; clean vehicles with responsible operators; soil incorporation of odorous sludge; and cleaning up application sites and roads are practices that can help assure favorable public attitudes. Many communities have found these day to day management activities to be the weak link in their program. One alternative would be to contract with commercial haulers to actually do the transportation and application of sludge. Of the state officials surveyed, two-thirds thought that commercial haulers' programs were better managed than community managed programs.

Conclusions

Recent experience with sludge landspreading programs indicates that institutional evolution lags technological evolution. Even when a new wastewater treatment technique, like landspreading, is effective and environmentally sound, implementation of that technique may be stalled by attitudes, formal rules, and other institutional constraints. Surveys of officials from communities with landspreading programs and officials in state regulatory agencies offer evidence of the nature of these constraints. The attitudes most affecting landspreading are concerns about health risks, nuisances and the environment. Other attitudes shaping public acceptance of landspreading are concern about accepting sludge from a neighboring political entity, concern about land values, farmers' and municipal officials' attitudes, historical precedents, and cultural influences. Regulatory constraints most important to landspreading generally are a product of state government agencies. In addition, local government regulations have become commonplace.

The pace at which attitudes and laws constraining land application change can be accelerated in a number of ways. Most important, information about the benefits and potential risks of landspreading must be disseminated. Effective educational techniques include seminars, workshops, tours of treatment plants and application sites, and media coverage. In addition, the public needs to be involved in the decision-making process. Community officials collaborating with rural residents, environmentalists, and farmers in developing a program is the preferred strategy. Avoided is unilateral decision making by community officials, which may evolve into a we-they, winner-take-all conflict.

REFERENCES

1. Burge, W.D. and P.B. Marsh, 1978. Infectious disease hazards of landspreading sewage wastes. J. Environ. Quality 7:1-9.

2. Commons, J.R., 1934. Institutional economics. Univ. of Wisc. Press, Madison.

3. Deese, P.L., et al. 1980. Institutional constraints and public participation barriers to utilization of municipal wastewater for land reclamation and biomass production. Report to President's Council on Environmental Quality, Washington.

4. Donnermeyer, J., 1977. Socio-cultural factors associated with the utilization of municipal waste on farmland for agricultural purposes. In Wastewater management in rural communities: a socio-economic perspective. C.E. Young and D.J. Epp (eds.), Inst. Res. on Land and Water Resources, Penn.State Univ.

5. Dotson, K., 1982. Public acceptance of wastewater sludge on land. U.S. EPA, Cincinnati.

6. Ellis, R.A. and J.F. Disinger, 1981. Public outcomes correlate with public participation variables. J. Water Pollution Control Fed. 53:1564-1567.

7. Euga, A. and R.W. Crites, 1980. Relative health risks of activated sludge treatment and slow rate land treatment. J. Water Pollution Control Fed. 52:2865-2874.

8. Forster, D.L. and D.D. Southgate, 1984. Social institutions influencing land application of wastewater and sludge. J. Water Pollution Control Fed. 56:399-404.

9. Gordon, W. 1980. Institutional economics, the changing system. Univ. of Texas Press, Austin.

10. Jewell, W.J. and B.L. Seabrook, 1979. A history of land application as a treatment alternative. EPA 430/9-79-012. U.S. EPA, Washington.

11. Musselman, N.M., et al., 1980. Information programs affect attitudes toward sewage sludge use in agriculture. EPA 600/2-80-103. U.S. EPA, Cincinnati.

12. Napier, T.L. and D.S. Mast, 1981. Attitudes toward land use controls within a multi-ethnic county of Ohio. J. Community Dev. 12:103-122.

13. Olson, B.H. and W. Bruvold, 1982. Influence of social factors on public acceptance of renovated wastewater. In Water reuse. E.J. Middlebrooks (Ed.), Ann Arbor Science Publishers, Ann Arbor.

14. Stitzlein, J.N., 1980. Public acceptance of land application of sewage sludge. In Utilization of wastes on land: emphasis on municipal sewage. U.S. Dept. of Agr., Washington.

Land treatment of hazardous wastes

Michael R. Overcash, Professor, Chemical Engineering Department,
North Carolina State University, Raleigh, NC 27695

The use of a terrestrial system to treat and serve as an ultimate receiver for hazardous waste is an established extension of the technology for land treatment of numerous waste types. Since this practice and the requisite design techniques have a long and established history, this paper will focus on several perspectives and concepts related to hazardous waste land treatment (HWLT). In addition, a partial review of HWLT research and a series of conclusions and recommendations from the author's perspective will be presented.

THE HAZARDOUS WASTE ASPECT OF HWLT

A major factor in HWLT is the phenomena that define and result in the public perspective of the materials being applied to the land surface. The public attitude and regulatory response is rather abrupt in defining a waste as either hazardous or not hazardous. Hazardous wastes are of high, often extreme public concern with respect to all aspects between cradle and grave, while nonhazardous wastes are managed carefully but routinely. Once the label of hazardous is put on a waste, there seems suddenly to be a step change in the regulatory requirements imposed on waste, and the waste is no longer viewed in the context of a continuum of waste characteristics, Figure 1. However, if one looks at wastes that were originally listed as hazardous but did not remain on such lists or wastes that are delisted, it is clear that no quantum differences exist between many hazardous and non hazardous wastes. Proceeding to the left or right ends of this spectrum, there are greater differences in diversity and concentration of chemicals in these wastes. As will be described later, these greater differences do not, however, reflect a linear or order of magnitude difference in the ability of land to treat such wastes.

The definition of which waste will be labeled hazardous centers on the following:

1. a leaching test and chemical concentrations of leachate,

2. certain reactive properties, and

3. a generic list of industrial wastes.

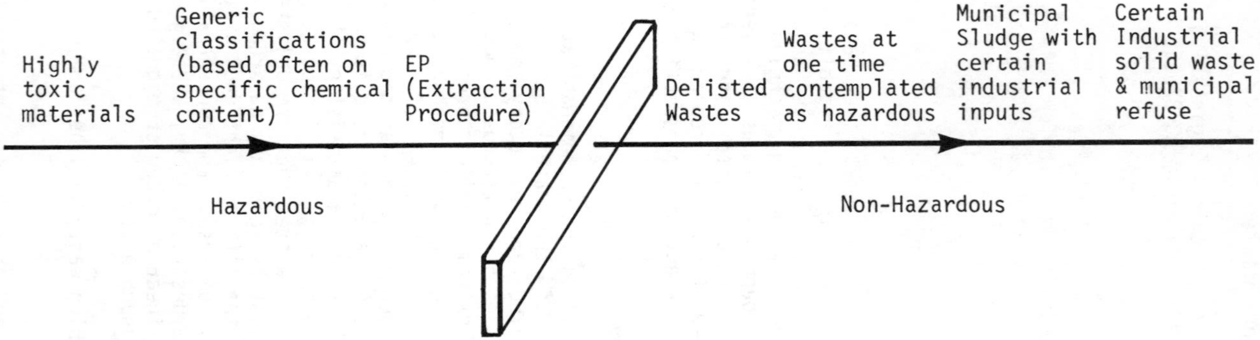

Fig. 1. Schematic of quantum differentiation associated with a solid waste declared hazardous within the actual continuum of chemical properties of such wastes (Overcash and Miller, 1981).

The first criterion is called an extraction procedure, which has undergone several revisions but remains primarily a prescribed liquid extraction of a waste using a "landfill environment extractant fluid." This is often an acid. Thus, the definition of hazardous waste centers on landfill leachability, as well as certain reactive properties such as ignitability and reactivity. These are quite focused on the behavior in a landfill environment. The generic list is derived from considerations of various wastes for which the other tests might not label hazardous, but which might still be of landfill-related concern.

There has been significant debate on the relevance of the above criteria for landfills. The relevance for land treatment, incineration, or other technologies is even more in question. One of the consequences of this debate is that the definition of hazardous waste per se reflects a separation between scientific and mechanistic basis for regulation and the path chosen by the regulatory community. These simplified definitions, surrounded by major scientific debate on relevance, is similar to the use of 5-day biochemical oxygen demand (BOD_5) in the wastewater field, and after many years the abandonment of BOD_5 as a useful parameter.

PROCEDURAL ASPECTS OF HWLT

Once a waste is labeled hazardous, there begins a complex environmental structure associated with the management of that waste. A hazardous waste must first enter a manifest system meant to provide a written record from generation through all intermediate steps to ultimate disposal. These intermediate steps are broadly as follows:

1. transportation,
2. treatment,
3. storage, and
4. ultimate disposition, typically in a landfill.

HWLT was for several years categorized on the basis of the principal characteristic of treatment (decomposition, metals immobilization,etc.). This reflects the use and capabilities of land treatment for all other wastes, including manure, refuse, municipal sludge, effluents, and industrial wastes. However, recently the word land in land treatment allowed a move to switch HWLT to the ultimate disposal or land disposal category. Thus, the following equation has evolved:

HWLT = landfill = land disposal.

Any issue that is raised from the national concern for abandoned dumps and present landfills is, thus, applied to land treatment. This includes allowable wastes, impact on groundwater, closure, public perception, etc., without regard to the obvious conceptual and technical differences between land disposal and land treatment.

The Resource Conservation and Recovery Act (RCRA) has always been a complex and evolving set of regulations. There is at present a list of wastes which have been declared hazardous on the basis of the previous criteria. Of these wastes, there are approximately 150 industrial plants involved with land treatment, Figure 2. These are mostly petroleum refinery landfarms. This is a relatively small number compared with all forms of land treatment sites. However, certain appendices in RCRA allow the expansion of the classification process to include many other sludges, including municipal sludges. This is an area of potentially unpredictable outcomes, since the extraction test is responsive to factors that are not focused on inherent properties of chemical concern for the safe assimilation in a land treatment system. Thus, the divergence between measures that reflect significant environmental concern and the basis of the test for that concern remains a major issue for debate.

For any waste classified as hazardous, the treatment, storage, or disposal requires a full scale Part B permit, which is issued at the Federal or State level. This permit contains a technological section and a series of less or non-technical issues that must be addressed. These latter concerns include the following:

1. liability insurance,

2. complex legal responsibilities, and

3. public hearings.

Unfortunately, with the quantum adverse response associated with hazardous waste and the equating of land treatment with land disposal, these non-technical facets often become the over-riding issues. These public concerns are legitimate, but those based on scientific inaccuracy and the absence of technical information would appear to be a disservice to society and those concerned with the environment. Land treatment of hazardous waste appears to have a large number of these gaps and differences with respect to technology and regulation.

The stages for addressing the technical issues are clearer and have been developed and refined to handle pre-existing as well as new land treatment systems. In a manual (Brown et al., 1980) and book by (Brown et al., 1983), the procedures for design of a HWLT site were developed. This is currently being revised to include evolved information (Sims,1985). These manuals have numerous flow-charts, and the reader should consult these for information. In addition, the connotation of hazardous waste requires the designer and permit writer to comply at this time with 18 other regulatory manuals, the total page length of which no one can probably define with accuracy. These are listed in Table 1. The comparision of Table 1 with the other uses and permit situations involving land treatment is an interesting commentary on RCRA and the national program for the management of any material classified as hazardous waste in the future. To the extent that hazardous waste regulatory criteria and the public perspective are

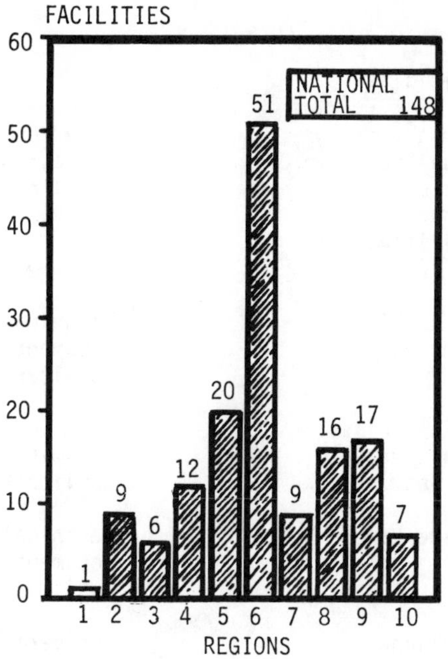

Fig. 2. Hazardous waste land treatment systems subject to part B application process as distributed by EPA regions, U.S.E.P.A., Permits and State Programs Division (through 9/30/85).

TABLE 1

RCRA PERMIT GUIDANCE DOCUMENTS - LAND DISPOSAL/LAND TREATMENT

Document Title	Description
1. RCRA Permit Writers' Manual for Ground Water Protection (40 CFR Part 264 F)	To be used by both permit writers and applicants
2. Permit Applicants' Guidance Manual for Hazardous Waste Land Treatment, Storage, and Disposal Facilities	Similar concept as the storage PA's guide; Will include model permit application; Emphasizes use of TRD series

Table 1 con't

Document title	Description
3. Permit Writers' Guidance Manual for Hazardous Waste Land Treatment, Storage, and Disposal Facilities (2 volumes)	Same concept as other PA's guides; Will include a model permit;
4. Solid Waste Leaching Procedure Manual	Technical resource document for permit applicants and writers
5. Soil Properties, Classification, and Hydraulic Conductivity Testing	Technical resource document for permit applicants and writers
6. Hazardous Waste Land Treatment	Technical resource document for permit applicants and writers
7. Draft RCRA Guidance Document: Land Treatment	For permit applicants; Presents specifications which would comply with the Part 264 Land Disposal Standards
8. Permit Guidance Manual on Hazardous Waste Land Treatment Demonstrations	Permit guidance manual for permit applicants and writers
9. Permit Guidance Manual on Unsaturated Zone Monitoring for Hazardous Waste Land Treatment Units	Permit guidance manual for permit applicants and writers
10. Permit Writers' Guidance Manual for Hazardous Wastes Land Storage and Disposal Facilities--Phase I: Criteria for Location Acceptability and Existing Applicable Regulations	For permit writers; Presents five criteria for acceptable location of storage and disposal facilities
11. Alternate Concentration Limit Guidance Based on S264.94(b) Criteria Part I	Information required in alternate concentration limit demonstrations

Table 1 con't

Document title	Description
12. Test Methods for Evaluating Solid Waste 2nd edition	Technical information on testing of hazardous wastes for all applications; Being updated on a regular basis; Current update package contains methods for use in ground water monitoring & incinerator performance measurements; In addition, hierarchial scheme for screening ground water for appendix VIII toxicants will be presented as ANPR
13. Liability Coverage: A Guidance Manual (Subpart H)	Information for permit applicants and writers on liability coverage for TSDs
14. Financial Assurance for Closure and Post Closure Care: A Guidance Manual	For permit applicants and writers; the main sourcebook on the Subpart H requirements
15. RCRA Personnel Training Guidance Manual	An early attempt at providing guidance to owners and operators of hazardous waste management facilities and their personnel
16. Closure and Postclosure Interim Status Standards (Subpart G)	For edification of permit applicants; Explains requirements for closure and post closure
17. Financial Requirements-Cost Estimates: Interim Status Standards (Subpart H)	Guidance on cost estimates
18. Model Permit	Companion to Permit Writers' Guide; Boilerplate language and modules for different types of permit conditions

easily transferred beyond the original intent (such as to municipal sludge), the phenomena of HWLT should be of concern to the professional engineering and scientific community.

An overview of the soils and technical stages involved in hazardous waste indicates that the concepts and procedures have maintained a close resemblance to those developed in the 1979-1983 period (Brown 1983, Loehr 1980, Overcash 1979). That is, the site testing, matching individual waste constituents to the assimilative capacity of soils, and the land-limiting constituent (LLC) concept are still largely followed. Therefore, a detailed review of those aspects is not included herein. Instead, certain of the differences and potentially incorrect facets of HWLT are described to focus on areas needing review. The following discussion is based on an inherent comparison of HWLT to that of the majority of other land treatment usages.

The structure of the design procedures for a HWLT contains a dominant reliance on field pilot-scale testing. While literature-based design is acknowledged, the practical result has been to require demonstration and treatability in the field. A whole series of issues are left unanswered, and the time/cost factor increases without recognition of alternate approaches. Identification and focus on the truly critical constituents is blurred by the need to address RCRA lists, acknowledgement of prior information is diluted, and a series of proof-of-a-negative-effect situations related to the land disposal analogy are created. These represent differences and issues raised by the reliance on field treatability and the aura of hazardous waste.

The judgment criteria for HWLT state and imply that the closure period is characterized by a return to background conditions. That is, after a design life has been achieved, the land application of waste is ended, and a series of regulated steps are undertaken to terminate responsibility for that area. This is referred to as closure. One of the criteria for satisfactory closure and correspondingly a short post-closure period is the return to background, presumeably the conditions of surrounding similar soil. It is and has been obvious that for metals such conditions could never be reached, even if the concentrations were well within acceptable ranges for such practices as municipal sludge land treatment. It also seems likely that background levels of slowly degradable organics may not return to control levels. Polychlorinated biphenyls and large - structured polynuclear aromatics may be examples. Both of these are ubiquitous, and the issue is not comparisons against background but comparisons against levels that might pose realistic concern. Thus, even for organics and certainly for many cationic inorganics, the standards for closure are inappropriate for HWLT, again a fallacy of equating such sites to landfill closure.

A third judgement criteria for a HWLT site is that there be no leaching of chemicals that pose a concern relative to listed RCRA constituents. This non-leaching perspective is sometimes difficult to reconcile with the leaching of a number of inorganic anions, which is inevitable in natural systems. It is anticipated that nitrate, chloride, boron, arsenate, and molybdate as example species would be judged against drinking water standards, thus allowing for significant leaching. However, the landfill perspective and the partial disregard for the occurrence of leaching within acceptable standards, which are a part of the perspective of hazardous waste, may pose substantial difficulties with natural leaching.

Within the statutes for managing hazardous waste by utilizing land spreading, there has been substantial emphasis on the capture and treatment of all runoff. The concept of runoff containment is substantively at odds with the regulatory practices for other wastes applied to land or for agricultural chemicals used in crop production. These other land-based systems have instead adopted best management practices as a realistic means of protecting the environment. The "hazardous" characteristics of waste are not directly a justification for runoff control, particularly when the hazardous definition centers on leaching within a landfill environment. Further, the potency of a number of hazardous waste constituents is less than that of some agricultural chemicals. In many cases, the RCRA chemicals could not exert environmental burdens of a fraction of the phosphorous or sediment lost in rainfall runoff. As an alternative, the emphasis of a valid technical approach should be on levels in runoff and whether any significant effect would result. The runoff collection philosophy should be modified on scientific grounds before the broad-scale use of such an approach heightens the divergence of HWLT within the overall technology of land application.

The above four areas pose substantive examples within the letter of the hazardous waste regulations and, importantly, within the interpretation, often at the State level, of the need to correct technical inaccuracies. These corrections will move HWLT nearer to the broad technology of or the application and treatment of wastes by the terrestrial system.

HAZARDOUS WASTE LAND TREATMENT RESEARCH

An overview of the areas of substantive HWLT research is appropriate in this paper. A major topic upon which new information is currently being generated is the field-scale demonstration of the treatability of various hazardous wastes. The specific wastes are wood treating, petroleum refining, and potentially explosive wastes. These efforts are an important contribution, which allow a particular industry to observe and review HWLT technology.

The monitoring aspect of HWLT as a subset of the monitoring for landfills remains an area of research. Unsaturated zone sampling has yet to be resolved. A third area of major research represents a significant advance in basic information for land treatment technology. Nearly 60 specific organic chemicals are to be studied under field conditions, when applied to soils as pure compounds and as chemicals present in a hazardous waste matrix. This effort is focused on a significant measure of the magnitude of a waste matrix effect, which will greatly expand the ability to use the available organic chemical assimilative information in the literature.

As a consequence of the equation linking land treatment and land disposal (or landfill), there will be substantive pressure to close these HWLT units, as a whole range of chemicals are banned from land disposal. Whether this pressure will actually lead to closure is unclear, but a fourth area of major research is on closure. A study on closure, including field-scale runoff measurements and surface soil changes, has recently been completed (Overcash et al., 1985). For the detailed results and conclusions, the reader is referred to the entire EPA report. The summary conclusions were as follows:

1. Field results at a closed refinery land treatment (LT) system closely paralleled the results obtained in a greenhouse simulation of closure using soil/waste mixtures obtained from the same refinery. This being only one comparative study, no firm conclusion can be drawn as to whether the greenhouse studies using soil/waste mixtures from additional refineries would adequately predict full-scale results at the respective locations. However, it can be concluded that greenhouse simulation studies, which have advantages in terms of cost and controlled environment, may be an important aspect of full-scale closure evaluation at a LT system; and at such time that results are statistically comparable, emphasis might be shifted to the greenhouse studies at a considerable cost savings. It can also be concluded that the results of this study for one refinery operation suggest that future comparative greenhouse/field studies will show the predictive usefulness of the greenhouse, leading hopefully to the emphasis being placed on greenhouse simulation with periodic field confirmation.

2. Based on two years of data collected at one land treatment closure site, closures having similar conditions of waste/soil and climate will have insignificant downward migration of the organics and heavy metals studied.

3. Based on field results at one refinery for 2 years, grass vegetation (as opposed to no vegetation) will improve runoff water quality by controlling migration of eroded particulate material contaminated with organic and inorganic constituents originating from the waste/soil mixture.

4. Based on greenhouse results representing four refinery land treatment systems and three native grasses, preliminary testing of soil/waste mixtures and different grasses is recommended to identify adverse effects that may result from incomplete germination and/or grass kill.

5. Based on controlled greenhouse studies of 2-years duration, the zone of soil/waste mixture for land treatment closure will exhibit essentially no change in total and solubilizable metals. The mixture will also exhibit an asymptotic decline with time to greater than background levels in oil and grease and in total organic carbon. The methylene chloride extractable (TCO+GRV) organics declined for one site but not significantly for the others; however, for all sites there was loss of aromatics and aliphatics and some increase in the polar fraction organics.

6. For the ten RCRA Appendix VIII polynuclear aromatic hydrocarbons (PNA's) analyzed in field and greenhouse studies, the expected levels in the surface soil/waste mixtures (similar to those investigated) will be $\leq$ 5 ppm. These levels may be expected to decrease over 2 years by $\geq$ 80%.

CONCLUSIONS AND RECOMMENDATIONS

At this time, the equating of land treatment with land disposal presents major philosophical and administrative consequences for the adoption of this technology. While unofficial recognition exists that HWLT is primarily treatment, the official weight of acceptable versus unacceptable practices in land disposal remain the dominant control. At a minimum, the public acceptance is severely reduced by the official land disposal categorization. The primary consequence in the short term will be increased incineration and a less than optimal economic situation.

The land treatment technical and professional community must clearly recognize that hazardous waste regulations have adopted a very different viewpoint of the practice of land application. This viewpoint is at present a very dominant perspective within State and Federal regulations and is at odds on several issues with other regulatory programs. Many anticipate that the banning restrictions for land disposal and other non-technical factors associated with the classification of a waste as hazardous will effectively reduce or eliminate HWLT. This would adversely affect one of the few low cost alternatives. However, the decisions on environmental impact of land disposal practices seem likely to be based on transport to off-site receivers, as predicted by mathematical models. These models allow for an extra module in which HWLT provides assimilation; therefore, acceptable practices would be defined. Further, it is likely that small demonstration projects will be able to demonstrate effectiveness and will be allowed to expand to full-scale. This will be a slow process, and the design constraints that evolve may significantly alter accepted practices. These changes are generally in the non-technical category.

For engineers and soil scientists, the challenge is to continue improving the technical base and to convey this information into the regulatory structure. This process is necessary to correct present technical inaccuracies and to assure that over-response or under-responses are minimized. The land treatment community must advocate that direct and indirect inaccuracies in the regulations be removed. The enforcement of reasonable regulations should be supported to correct abuses in the use of land treatment. The use of rigorous enforcement is likely to be an overall more cost effective national policy than regulation on the basis of worst case scenarios. Finally, all uses of land treatment i.e., hazardous waste through farm manures, should be viewed as a unified technology, so that the full weight of cost-effectiveness and environmental protection are more clearly evident.

REFERENCES LISTED

1. Brown, K. W. and Associates, Inc., 1980. Hazardous Waste Land Treatment. SW-874. USEPA, Office of Research and Development, Cincinnati, Ohio.

2. Brown, K.W., G.B. Evans, Jr., and B.D. Frentrup (eds.), 1983, Hazardous Waste land Treatment, Butterworth Publishers, Woburn, MA.

3. Loehr, R. C., W. J. Jewell, J. D. Novak, W. W. Clarkson, and G. S. Friedman, 1979, Land Application of Wastes, Van Nostrand Reinhold, Co., New York, N. Y.

4. Overcash, M. R. and D. Miller, 1981, Integrated Hazardous Management, Today Series, American Institute of Chemical Engineers, New York, N.Y.

5. Overcash, M. R. and D. Pal, 1979, Design of Land Treatment Systems for Industrial Wastes, Ann Arbor Science Publishers (now Technomics Publishers, Lancaster, PA.)

6. Overcash, M. R., W. L. Nutter, R. L. Kendall, J. R. Wallace, 1985, Field and Laboratory Evaluation of Petroleum Land Treatment System Closure, report for American Petroleum Institute and U. S. Env. Protection Agency, Ada, OK.

7. Sims, R. 1985, personal communication, Utah State University, Logan, Utah.

Cleanup on contaminated soils

Ronald C. Sims and Judith L. Sims, Utah Water Research Laboratory, Utah State University, Logan, UT 84322-8200

Uncontrolled or poorly managed disposal of hazardous wastes frequently produces large quantities of contaminated soils. An in situ treatment approach may be effective in eliminating or reducing hazards to acceptable levels and it may also be the most cost-effective alternative for remedial action.

In situ treatment technologies are used for treating contaminated soils until an acceptable level of treatment is achieved, and for protecting groundwater and surface water resources without physically removing or isolating the contaminated soil from the contiguous environment. Treatment refers to the processes of immobilization, degradation, and detoxification of specific target waste constituents in the contaminated soil system.

In situ treatment technologies generally can be classified into three categories: 1) immobilization, 2) degradation, and 3) attenuation. Immobilization techniques are designed to capture chemical constituents within the contaminated soil mass. Three major classes of immobilization techniques are adsorption, ion exchange, and precipitation. Degradation techniques may refer to biological, chemical, and photochemical reactions. Biological degradation refers to the use of soil microorganisms, primarily bacteria, actinomycetes, and fungi, for the metabolism of organic constituents, and some inorganic compounds, leading to the decomposition, detoxification, and mineralization of the parent compound(s). Chemical degradation techniques convert contaminant species through the processes of oxidation, reduction, precipitation, and polymerization. Photochemical degradation (photolysis) utilizes the action of ultraviolet radiation for the breakdown of organic constituents.

Attenuation involves the mixing of clean soils or other bulking agents into the contaminated soil to reduce contamination to an acceptable level. This approach may be used to reduce toxicity to soil microorganisms, or to reduce hazards due to human ingestion or direct contact, or due to potential uptake by vegetation and ingestion by animals, i.e., transport through the food chain.

In situ treatment of contaminated soils must be based on an understanding of factors and processes that determine the behavior of chemicals in soil systems. Specifically, an evaluation of chemical properties, biochemical processes, and environmental factors influencing the behavior and fate of chemicals in soils is required.

Factors affecting the success of in situ treatment

Site, soil, and waste factors

Before beginning in situ remedial actions to treat hazardous waste contaminated soils, relevant site characteristics must be identified and evaluated. Site characterization may also assist in evaluating how site modification or management could enhance the protection of human health. Soil characteristics that affect water movement (i.e., infiltration and permeability) and factors that affect contaminant mobility are the most important. The specific site and soil characteristics that need to be identified when assessing a site for in situ treatment as well as the site and soil conditions that may be managed to enhance soil treatment are identified in Table 1.

The properties of waste that affect the behavior and fate of chemicals in soil systems must be characterized because these properties directly affect how the waste will be treated (or assimilated). Factors important in determining the behavior and fate, and therefore the treatment pathways, of waste constituents in soil are listed in Table 2. For each chemical or chemical class, the information needed can be summarized as characteristics related to:

1. Potential leaching (e.g., water solubility, octanol/water partition coefficient, solid sorption coefficient).
2. Potential volatilization (e.g., vapor pressure, relative volatilization index).
3. Potential decomposition (e.g., half-life, degradation rate, biodegradation index).
4. Chemical reactivity (e.g., oxidation, reduction, hydrolysis potential).

Comparing the properties of the soil at a specific site with the characteristics given above permits an evaluation of the potential for 1) soil treatment and 2) off-site contamination.

Soil-waste processes affecting constituent behavior

Immobilization

The relationship between sorption (immobilization) of chemical constituents in soil systems (based on soil properties) and chemical class (based on chemical structure) is summarized in Table 3. Generally, nonionic constituents of low water solubility and cationic constituents have low mobilities and leaching potential. Acid constituents at neutral and high pH values are most easily leached from soil systems.

Understanding the relationship between soil water content and extent of sorption of hazardous chemicals provides the hazardous waste manager with a tool for controlling potential release and migration of constituents through the control of the leaching process. One commonly used isotherm

Table 1. Site and soil characteristics identified as important in in situ treatment (Sims et al., 1984).

Site location/topography and slope
Soil type and extent
Soil profile properties
boundary characteristics
depth
texture*
amount and type of coarse fragments
structure*
color
degree of mottling
bulk density*
clay content
type of clay
cation exchange capacity*
organic matter content*
pH*
Eh*
aeration status*
Hydraulic properties and conditions
soil water characteristic curve
field capacity/permanent wilting point
water holding capacity*
permeability* (under saturated and a range of unsaturated conditions)
infiltration rates*
depth to impermeable layer or bedrock
depth to groundwater,* including seasonal variations
flooding frequency
runoff potential*
Geological and hydrogeological factors
subsurface gelogical features
groundwater flow patterns and characteristics
Meteorological and climatological data
wind velocity and direction
temperature
precipitation
water budget

*Factors that may be managed to enhance soil treatment.

Table 2. Soil-based waste characterization (Sims et al., 1984).

Chemical class	
	Acid
	Base
	Polar neutral
	Nonpolar neutral
	Inorganic
Soil sorption parameters	
	Freundlich sorption constants (K, N)
	Sorption based on organic carbon content (K_{oc})
	Octanol/water partition coefficient (K_{ow})
Soil degradation parameters	
	Half-life ($t_{1/2}$)
	Rate constant (first order)
	Relative biodegradability
Chemical properties	
	Molecular weight
	Melting point
	Specific gravity
	Structure
	Water solubility
Volatilization parameters	
	Air/water partition coefficient (K_w)
	Vapor pressure
	Henry's law constant ($1/K_w$)
	Sorption based on organic carbon content (K_{oc})
	Water solubility
Chemical reactivity	
	Oxidation
	Reduction
	Hydrolysis
	Precipitation
	Polymerization
Soil contamination parameters	
	Concentration in soil
	Depth of contamination

Table 3. Leaching potential of chemicals in soil systems (Sims & Wagner, 1983).

	Chemical Class							
	Nonionic			Ionic				
	Water solubility			Basic		Cationic	Acidic	
Leaching potential	high	med	low	low pH	neutral pH		low pH	neutral pH
Low			X	X		X		
Medium		X			X		X	
High	X							X

that is useful in describing the immobilization of organic constituents in soil is the Freundlich isotherm:

$$S = KC^{1/n} \tag{1}$$

where
K and n are constants,
S = amount of chemical associated with solid phase, or the solid phase concentration, and
C = amount of chemical associated with the solution phase, or the solution phase concentration.

The Freundlich isotherm relates the solid phase concentration to the solution phase concentration at equilibrium conditions.

An important linear isotherm can be obtained from the Freundlich isotherm when n=1, i.e.,

$$S = K_dC \tag{2}$$

where K_d = the distribution coefficient.

The relationship between K_d, soil moisture content Θ, and percent adsorption of an organic chemical can be used to manage a soil system:

$$\text{percent adsorbed} = K_d/(K_d + \Theta) \tag{3}$$

The extent of sorption as a function of soil moisture content for different values of the distribution coefficient is illustrated in Figure 1. Optimization of cost effective and efficient treatment may require a compromise between optimum soil moisture content for biodegradation versus sorption. Careful control of soil moisture content will determine, to a large extent, the relative immobilization of a given set of chemical constituents identified at a remedial site.

Understanding the effect of different solid surfaces on hazardous waste-constituent immobilization provides a mechanism for rationally selecting additional sorbents for use in augmenting the natural ability of a soil system to immobilize hazardous chemicals.

Biodegradation
Quantitative aspects of microbial decomposition for hazardous constituents in soil systems relate directly to the time required for in situ cleanup of contaminated soils. Mathematically, the rate of biodegradation represents a sink term in organic transport models that can be used to predict potential groundwater contamination with respect to magnitude and type of contamination and the time factor for contamination (rate of transport). Mathematical models that include biodegradation and transport information can be used to rank chemical constituents with respect to potential mobility and, therefore, provide an approach for determining which chemicals at a contaminated site require the highest priority for immobilization treatment.

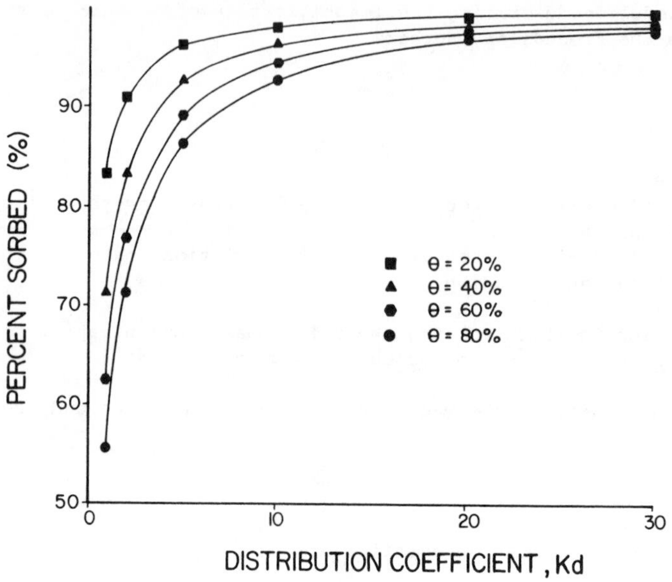

Figure 1. Extent of sorption as a function of soil moisture Θ and K_d.

Also, evaluation of the effect of soil amendments, including nutrients, carbon substrates, pH, etc., on biodegradation kinetics provides a rational approach to selection of amendments for use in augmenting the natural ability of a soil system to biodegrade hazardous constituents.

Volatilization and photodegradation

The major contaminant property affecting volatilization is the vapor pressure of the compound. Major environmental factors affecting the vapor pressure of a constituent are the soil/water and air/water partition coefficients that exist for the soil/water/air environment within a soil system. Additional complexity results if the contaminant(s) is added with an additional adsorbing fluid such as oil in refinery waste, where partitioning of the constituent between the oil/soil, oil/water, and oil/air phases would also be expected to affect the volatilization, as well as the immobilization of the constituent(s). A mathematical model developed by Dr. Thomas Short, Robert S. Kerr Environmental Research Laboratory, U.S. Environmental Protection Agency, includes these phases (oil, soil, water, and air) for the evaluation of the effect of biodegradation and immobilization on constituent travel through a soil system. A description of this model is presented in the Permit Guidance Manual on Hazardous Waste Land Treatment Demonstrations (U.S. Environmental Protection Agency, 1986).

Photodegradation of organic constituents may occur by two processes: 1) direct photodegradation and 2) sensitized photooxidation. The relative importance of photodegradation of constituents on or within a soil will depend upon its partitioning between the air/water/soil media within the soil system. Using photochemical reactions to enhance constituent degradation is an area of interest for in situ treatment of contaminated soils.

Monitoring of in situ treatment

To ensure that the objectives of in situ treatment are attained, a monitoring program must be established to: 1) ensure that the hazardous or toxic constituents of the waste are being degraded, detoxified, or inactivated as planned, 2) monitor degradation rates of degradable constituents, 3) ensure that waste constituents are not entering runoff or leachate water and leaving the area in unacceptable concentrations, and 4) determine whether adjustments in treatment management are required to maintain the treatment process.

A complete program would include the monitoring of 1) soil core and soil-pore liquid in and below the treatment zone, 2) groundwater, 3) runoff water, and 4) atmosphere. Constituents that should be monitored include those determined to be hazardous in the initial site/waste characterization study, as well as expected important degradation or transformation products. The monitoring program may also include substances needed for treatment, whether these substances are native to the soil or added as a treatment agent or amendment.

Application of in situ treatment

Uses of treatment technologies

Three major types of site scenarios are appropriate for utilizing in situ treatment technologies. These include:

- low residual levels of contamination in the hot or contaminated zone which remain on site following the bulk of contaminant removal;
- low level contamination around the periphery or in the transition zone of a landfill, impoundment, etc.; and
- large widespread contamination problems.

Examples of in situ treatment application include chemical spills, chemical residuals in industrial surface impoundments containing process water or sludge, soil contaminated by surface leachate, landspreading operations, and dredge spoil containment basins. In situ treatment techniques are applicable for soils contaminated with organic and/or inorganic constituents, as well as organic-inorganic complexes. Individual chemicals or complex mixtures in soils may be treated using in situ techniques.

In situ treatment techniques utilizing physical, chemical, and biological processes for the treatment of hazardous waste contaminated soil have been used at the laboratory scale, field bench scale, and field full scale. Treatment agents have been added directly to the soil-waste matrix to accomplish the objectives of treatment, while other techniques have augmented natural processes occurring in soils to accomplish treatment.

Examples of treatment technologies

Examples of in situ treatment technology groups are listed here and discussed in detail in succeeding sections:

Neutralization
Oxidation
Reduction
Precipitation
Bioreclamation
  Natural
  With bacterial augmentation
  With oxygen augmentation
Immobilization
Photolysis
Permeable treatment beds
Solution mining
Vitrification

Neutralization
In situ neutralization is the technique of applying or injecting substances into a contaminated site to neutralize the pollutants present and is highly waste specific. The process of neutralization is the interaction of an acid with a base. In a strict sense, neutralization is the adjustment of pH to 7, at which level the concentration of hydroxyl and hydrogen ions are equal. Since adjustment to pH 7 is not often practical or even desirable, the term "neutralization" may be used to describe adjustment of pH to values other than 7. Adjustment of pH may also be used to insolubilize certain chemical species or to control chemical reaction rates. A typical in situ neutralization process might include a well point injection system and chemical and feed systems (U. S. Environmental Protection Agency, 1982).

Neutralization has found wide application in the treatment of aqueous industrial wastes containing strong acids such as sulfuric and hydrochloric, or bases such as caustic soda and ammonium hydroxide. In situ neutralization may potentially be used to neutralize acidic or basic wastes at a waste disposal site or contaminated area. This technique would be most applicable to industrial waste disposal sites, since municipal landfills would continually be generating anaerobic decomposition products that would require neutralization over a long period of time. If the material is present as a solid and is constantly dissolving, neutralization would have to be performed repeatedly over a period of time. Therefore, this technique should be used for wastes that are completely dissolved or readily mobile. Also, the degree of effectiveness of in situ neutralization is difficult to determine (i.e., whether the reaction agent has contacted all of the wastes, especially if the wastes are buried). Heterogeneous wastes may require different treatments for different wastes, and treatment for one waste may be unsuitable for another. Safety and pollution control must be considered in the selection of particular reagents, because of the violence of the resulting reactions and the possibility of hazardous reaction products.

Oxidation

Chemical oxidation may represent a significant treatment process in a soil system. Oxidation, a process in which the oxidation state of a substance is increased by the removal of electrons or the addition of oxygen to the substance, may result in its transformation, degradation, and/or immobilization. Oxidation reactions may occur naturally in a soil system by clay-catalyzed reactions or by the addition of an oxidizing reagent to the soil/waste complex. Certain compounds (e.g., phenols, aldehydes, and aromatic amines) are more oxidizable in soils than others (e.g., halogenated hydrocarbons, chlorinated insecticides, benzene, and saturated aliphatic compounds). General characteristics of organic chemicals likely to undergo oxidation include: 1) aromaticity, 2) fused ring structure, 3) extensive conjugation, and 4) ring substituent fragments.

Natural soil catalysts promoting oxidation of constituents in soil systems include iron, aluminum, trace metals within layer silicates, and adsorbed oxygen. For oxidation to occur in soil systems, the redox potential of the solid phase must be greater than that of the organic chemical contaminant. Therefore, the half-cell potential of chemical contaminants needs to be below the redox potential (approximately 0.8 volts) of well-oxidized soil (Dragun & Baker, 1979).

More water-soluble compounds should be more readily oxidized in clay-catalyzed systems, because sorption to the hydrophilic clay mineral surface, which is the oxidation reaction site, precedes the oxidation process. Also, greater oxidation of chemical contaminants is expected in less saturated soils. To manage soil-catalyzed oxidation of organic compounds, control of soil moisture by drainage and/or addition and incorporation of uncontaminated clays may be used.

Another method of oxidizing organic wastes in soil systems is by the addition of oxidizing agents such as ozone or hydrogen peroxide to the contaminated soil. Ozone is an oxidizing agent which may be used to degrade recalcitrant organic compounds directly or to create an oxygenated environment for the enhancement of biological activity. Ozone either reacts directly with organic compounds or, in a free radical reaction, produces a hydroxyl free-radical intermediate. If the specific organic constituents present in contaminated soil are relatively biodegradable, ozone treatment may be effective as an enhancement of biological activity. However, if a large fraction is relatively biorefractory, the amount of ozone that will be required to treat the waste by chemical degradation will be a direct function of the organic materials, both natural and waste-related, present and will require an increased cost of treatment.

Hydrogen peroxide acts as an oxidizing agent in three major ways: 1) direct reaction with a substrate to form oxygen; 2) autodecomposition in the presence of a metal catalyst to form oxygen; and (3) degradation by UV light to form hydroxyl free radicals.

Both ozone and hydrogen peroxide are strong oxidizing agents and are nonselective. If their use results in the decrease of natural organic materials in soils, decreased sorption capacity for waste organics due to oxidation of natural organics may result. The addition of the oxidizing agents to contaminated soils may result in violent reactions with certain classes of compounds and increased mobility of some metals. The use of oxidizing agents may also affect soil hydraulic properties (e.g., infiltration rate), especially in structured soils.

Oxidizing agents may be applied in water solutions directly to the soil surface, injected into the subsurface, or applied through injection wells, depending on the depth and location of contamination.

Soil-catalyzed reaction has been verified in the field for several chemical classes, including s-triazines and organophosphate compounds, while the oxidation of other compounds have been verified in the laboratory. Oxidation of hazardous metals is usually not effective as a treatment method, because most metals (except arsenic) tend to be more mobile and/or toxic at higher oxidation states. Overall, for use at hazardous waste sites, this technology is at the conceptual stage.

Oxidizing agents have been used in wastewater treatment, but there is little experience with their use in terrestrial systems. Hydrogen peroxide is used in septic tank drainfields experiencing failure due to biological clogging.

Reduction
Chemical reduction is a process in which the oxidation state of an atom is decreased. Reducing agents are electron donors, with reduction accomplished by the addition of electrons to the atom. Reduction of chemicals may occur naturally within the soil system or by the addition of reducing agents.

Reduction mechanisms of organic compounds include hydrogenolysis, hydroxylation, saturation of aromatic structures, condensation, and ring opening. Chemical reduction may be accomplished using catalyzed metal powders and sodium borohydride. Chlorinated organics, unsaturated aromatics and aliphatics, and other organics susceptible to reduction are amenable to treatment by reducing agents.

Chromium and selenium are two metals which can be treated in soil systems by chemical reduction. Hexavalent chromium (Cr(VI)) is highly toxic and highly mobile in soils. Cr(VI) can be reduced to trivalent chromium (Cr(III)), which is less toxic and is readily precipitated by hydroxides over a wide pH range. Cr(VI) itself is a strong oxidizing agent under acidic conditions and will readily be reduced under natural conditions, without the addition of strong reducing agents. However, reducing agents such as leaf litter, acid compost, or ferrous iron and acidification agents such as sulfur may be added to enhance the conversion of Cr(VI) to Cr(III). After reduction, liming of the soil may be used to precipitate the Cr(III). The pH of the soil must be maintained greater than 5 to ensure the continued immobilization of chromium.

Hexavalent selenium (as selenate) is the dominant form of selenium in calcareous soils and is highly mobile. Elemental selenium and selenite (Se(IV)) are less mobile in soils. Se(VI) may be naturally reduced to Se(IV) or elemental selenium under acid conditions. Elemental selenium is virtually immobile in soils, while Se(IV) will participate in sorption and precipitation reactions. However, Se(IV) as selenite is an anion and its potential leachability increases with increasing pH. In situ treatment of selenium consists of addition of reducing agents to the contaminated soil, as with chromium, and the acidification with sulfur or other agricultural acidifying agents to pH 2-3. However, immobilization of most other metals requires pH values of 6.5 or above. Also, microbial degradation of organic waste constituents would be adversely affected at such low pH levels.

The addition of organic materials as reducing agents may have beneficial effects on soil properties (e.g., structure, water holding capacity, and reduction in soil erosion potential). However, the impact of the use of other reducing agents on soil organic matter, soil permeability, and mobility of metals is not known. The toxicity and potential mobility of reduced products are not known.

Two processes have been developed for the reduction and dehalogenation of PCBs and dioxins (Franklin Institute Research Laboratory, Inc., 1981; Mulle, 1981). Both employ a sodium-based chemical reagent to remove chlorine from PCB and dioxin molecules. The degradation products are generally non-toxic or less toxic than the parent compound.

Reduction of organic compounds has been used in wastewater treatment. Reduction of paraquat in soil systems with sodium borohydride and powdered zinc has been demonstrated in small-scale field plots (Staiff et al., 1981). Reduction of chromium and selenium has been accomplished in laboratory studies, but the status of the treatment technique at the field scale is conceptual. Sodium reduction and dehalogenation of PCBs and dioxins have been applied to the treatment of PCB-containing oils but are still in the developmental stage for soil decontamination.

Precipitation

Precipitation of metals occurs when the solubility product of the ions forming the precipitate is exceeded in the solution. Metals may be precipitated as sulfides, carbonates, phosphates, and hydroxides, with sulfide precipitates being the most insoluble for several metals (e.g., copper, cadmium, lead, mercury, and zinc), even at acidic pH levels. The extent of metal sulfide precipitation is a function of 1) pH, 2) type of metal, 3) sulfide content, and 4) interfering ions. A high salt content of the waste will reduce the theoretical extent of precipitation. Treatment may be accomplished by adding calcium or sodium sulfide. The use of sodium sulfide salts may, however, adversely affect soil permeability. The soil should be maintained in a reduced state to prevent the oxidation of precipitated sulfides to more soluble sulfate compounds.

To precipitate metals as other compounds, such as carbonates, phosphates, and hydroxides, substances such as calcium carbonate and treble superphosphate fertilizer may be added to the contaminated soil. If arsenic is present, however, the use of phosphate may result in the release of arsenate to the soil solution. For maximum treatment effectiveness, soil pH should be maintained at the appropriate level to maintain maximum insolubility of the metal precipitates.

Removal of metals from wastewater and river water has been demonstrated for sulfide and hydroxide precipitation, but has not been extensively tested in soil systems. The use of limestone as a barrier to retard the migration of metals from landfill leachates has been successfully tested at the laboratory scale (Fuller, 1978; Artiole & Fuller, 1979).

The kinetic aspects of precipitation and dissolution reactions involving metals in a soil system have not been well-characterized and may limit the effectiveness of this treatment technology. Chelating agents, other competing reactions (e.g., formation of soluble metal complexes), and salt content of the waste may also reduce the effectiveness of treatment.

Bioreclamation

Bioreclamation treatments are directed toward enhancing biochemical mechanisms for accomplishing the degradation, detoxification, and mineralization of hazardous organic and organic-inorganic complexes in soil systems. Both aerobic and anaerobic biological systems have been utilized to treat hazardous waste constituents in soil systems. Soil bioreclamation techniques have been used specifically in the petroleum, wood preserving, and pesticide industries.

Soil microorganisms, principally bacteria, actinomycetes, and fungi, are important in decomposition and detoxification processes. Therefore, treatments applied to the soil to enhance biological processes must not alter the physical environment to an extent that biochemical processes are inhibited. Environmental conditions generally conducive to soil microbial activity include soil temperatures between 50 and 60°C (Atlas & Bartha, 1981), soil water potentials greater than -15 bars (Sommers et al., 1981), pH values between 5 and 9 (Atlas & Bartha, 1981), and oxidation-reduction potentials between pe + pH values of 17.5 to 2.7 (Baas Becking et al., 1960).

The use of soil microorganisms has been demonstrated for the treatment of several classes of organic compounds, including pesticides, halogenated aliphatics, phthalate esters, monocyclic and polycyclic aromatic compounds, phenolic compounds, halogenated ethers, and polychlorinated biphenyls.

Specific types of bioreclamation that will be discussed include natural bioreclamation and reclamation using bacteria augmentation and oxygen augmentation.

Natural Bioreclamation

Natural bioreclamation utilizes the indigenous soil microorganisms for accomplishing treatment, i.e., degradation and detoxification of toxic and hazardous constituents. The two major pathways in soil for accomplishing treatment include aerobic treatment and anaerobic treatment. Natural bioreclamation may achieve complete treatment (mineralization) or may effect only partial degradation of constituents, resulting in the production of metabolic intermediates.

This technology also uses augmentation techniques for stimulating natural processes in soil systems in order to optimize or improve the extent and rate of treatment by the indigenous microorganisms. Augmentation techniques may include addition of an amendment (e.g., moisture, nutrients, organic matter, or analog enrichment), pH adjustment, or tilling.

Microorganisms, primarily bacteria, actinomycetes, and fungi, are most important in effecting treatment by biodegradation. Parameters affecting the biodegradation of constituents in soil systems are of two types: 1) those that determine the availability and concentration of the constituent to be degraded or that affect the microbial population size and activity; and 2) those that control the reaction rate.

Waste constituents that are potential candidates for natural bioreclamation include most organic compounds and organic-inorganic complexes. This technology has been studied at the laboratory scale and has been applied at the field scale for hydrophobic, hydrophilic, and halogenated organic constituents.

Natural bioreclamation technology is most advanced at the laboratory level, but has also been applied at the field-scale level. Field-scale applications include full-scale land treatment sites, site cleanup of soil contaminated with polynuclear compounds, and cleanup of creosote contaminated soil. There is also significant research being performed at the laboratory scale regarding the use of augmentation techniques for optimizing in situ natural bioreclamation.

Natural bioreclamation is economical, for it does not require expensive chemical addition or site modification. Bioreclamation is also low technology. It does not require sophisticated technology with high training requirements. Bioreclamation is associated with ease of application, as the technology for application already exists, i.e., utilization of common agricultural techniques, practices, and equipment. However, months or years may be required for significant or complete biodegradation of target constituents, and public relations may be impaired, for the public may want faster cleanup of the contaminated site.

Bioreclamation with bacterial augmentation

Bioreclamation with bacterial augmentation involves the addition of exogeneously grown microorganisms to a soil system. Bacterial augmentation may be used when the metabolic range of the natural soil microbiota is not sufficient to degrade specific target compounds or classes of compounds, or when sufficiently large populations are not available in the natural soil environment for effective and rapid biodegradation of constituents of concern. Frequently, the application of microbes to the soil is combined with other treatment techniques, such as soil moisture management, aeration, and fertilizer addition.

Methods for application are determined in consultation with the vendor of the microorganisms. The microorganisms may be applied in liquid or with a solid carrier. Depending upon the method of application, runon and runoff controls may be necessary. The ease of application depends on the trafficability and the depth of contamination.

Microbial inoculants are available commercially with broad ranges of metabolic capabilities. Experience in their use in soil systems contaminated with wastes is expanding.

Compounds or classes of compounds which may be degraded by mutant or selected bacterial cultures include:

* alcohols
* aromatic hydrocarbons
* ketones
* alkyl halides
* chlorinated aromatics
* nitriles
* amines
* esters
* phenols

This method may be most effective against one compound or closely related compounds.

This technology has been demonstrated in the laboratory and has been used in several full-scale soil decontamination operations. Case histories of successful treatment of chemical spills for contaminants including oil, ortho-chlorophenol, and acrylonitrile have been reported (Thibault & Elliott, 1979, 1980; Walton & Dobbs, 1980).

Bacterial augmentation allows for specific organisms to be cultivated for specific constituents and may be less expensive than excavation. However, a relatively long period of time may be required to complete treatment, and the public may desire that a site be cleaned up faster than it is possible with microorganism augmentation. Retreatments, especially after precipitation events, which may "wash out" the inoculum, or treatment augmented with chemical addition are often required. More information is also required on the ability of exogenous organisms to survive, grow, and function in the soil environment.

Bioreclamation with oxygen augmentation (air, ozone or hydrogen peroxide)

Augmentation of the soil/waste system with oxidants, such as oxygen, hydrogen peroxide,or ozone will generally improve the biodegradation of organic chemicals. Aerobic metabolism is more energy-efficient and microbial decomposition processes are generally more rapid under aerobic conditions. The use of hydrogen peroxide and ozone will also result in direct chemical degradation or partial degradation of chemicals which are more slowly biodegraded (more refractory constituents). Hydrogen peroxide has been used in conjunction with ozone to degrade organic compounds which are refractory to either oxidant individually.

Oxygen may be added to a soil by tilling the soil and/or draining the soil. Tilling and drainage of soil systems are easily accomplished and are currently practiced at full-scale hazardous waste land treatment sites. Oxidizing agents such as hydrogen peroxide or ozone may be applied in water solutions directly onto the soil surface, injected into the subsurface, or applied through injection wells, depending upon the depth and location of contaminant(s). Oxidizing agents are dangerous and require special treatment, for they may result in violent reactions with certain classes of compounds (e.g., metals) and may be corrosive to application equipment.

Ozone has been used in full-scale applications for reclamation of subsurface aquifer materials by increasing microbial activity due to the enhanced dissolved oxygen level in the soil solution (Nagel, 1982). Hydrogen peroxide has been demonstrated to cause an increase in microbial activity and to result in increased degradation of organic contaminants in soil/groundwater systems. Hydrogen peroxide is also used in septic tank drainfields experiencing failure due to biological clogging. Augmentation of soil with hydrogen peroxide or ozone may result in the oxidation of nontarget organic material and may result in a decrease in the sorptive capacity of the treated soil. Depending upon the amount of soil requiring treatment and the type and amount of organic material in the soil, the cost of treatment may be very high.

Immobilization: sorption, ion exchange, and attenuation
Both inorganic and organic wastes can be immobilized in soil systems by natural processes and by the addition of immobilizing agents. The immobilization of organics in soil may allow additional time for further treatment by biodegradation. However, long-term stability of immobilization agent/waste constituent complexes against decomposition or degradation is not yet known.

Immobilization techniques are designed to capture waste constituents within the contaminated soil mass. Three major types of immobilization techniques are sorption, ion exchange, and attenuation.

Sorption of a pollutant refers to processes which result in a higher concentration of the chemical at the surface or within the solid phase than is present in the bulk solution of soils. Adsorbed compounds or ions are in equilibrium with the soil solution and are capable of desorption. Soil sorption is perhaps the most important soil/waste process affecting immobilization of waste constituents. Leaching potential and the residence time in soil for constituents which undergo degradation are directly affected by the extent of immobilization. Processes included in sorption are: 1) physical sorption through weak atomic and molecular interaction forces (van der Waal forces), 2) specific adsorption exhibited by anions involving the exchange of the ion with surface ligands to form partly covalent bonds, and 3) chemisorption involving a chemical reaction between the compound and the surface of the sorbent. Sorption is most closely associated with the organic soil fraction. The sorption process is usually described by an adsorption isotherm, which expresses the relationship between the amount adsorbed on a solid and the concentration of solute in solution at equilibrium.

To utilize sorption as an immobilization technique, an assessment of the native sorption capacity should be made by conducting adsorption isotherms to see if the natural soil under natural soil moisture conditions is adequate for immobilization of the waste constituents. If additional sorptive capacity is required, sorbing agents such as activated carbon (for metals and organic constituents), agricultural products and by-products, sewage sludges, and other organic matter (for metals and organics), and tetren (for metals) may be added to the contaminated soil.

Ion exchange is a process in which certain minerals and resins in contact with a solution, particularly an aqueous one, release ions in preference for ions of another type present in the solutions. Ion exchange materials can be classified into two principal types: cationic and anionic.

Soil clay minerals and soil organic matter have high surface areas and are negatively charged, thus acting as effective cation exchangers. The cation exchange capacity (CEC) of a soil is defined as the number of milliequivalents of an ion that can be exchanged per 100 grams of soil on a dry weight basis. Some soils also have an anion exchange capacity, but usually this is not as great as the CEC. As with sorption, the native ion exchange capacity of the contaminated soil should be evaluated, and if additional capacity is required, such ion exchangers as additional clays, synthetic resins, and zeolites may be added to the soil.

Desorption or release of ions from exchange sites may adversely affect treatment effectiveness. Factors important in waste constituent release from sorbent and ion exchange materials include the amount of leachate (soil/waste ratio) and the amount of constituent contaminating the soil (soil/constituent ratio).

Immobilization of organic and inorganic waste constituents may also be accomplished by attenuation, which is the mixing of contaminated soil or wastes with clean soil to reduce the concentrations of hazardous components to acceptable levels. Either uncontaminated soil from the contaminated area can be mixed with the contaminated soil layer, or uncontaminated soil from an adjacent area can be used. Soils may also be purchased from local contractors, or pure soil materials (e.g., bentonite) may be obtained from commercial suppliers.

Most immobilization techniques have been evaluated at the laboratory scale and many have been used in wastewater treatment. Activated carbon has been used to immobilize pesticides and herbicides in soils.

Photolysis
Photodegradation is the use of incident solar radiation to accomplish photoreaction processes. Both direct photolysis (photoreactions due to direct light absorption by the substrate molecule) and sensitized photo-oxidation (photoreactions mediated by an energy-transferring sensitizer molecule) are possible under environmental conditions. Sensitized photoreactions are characteristically reactions involving photooxidation, resulting in substrate molecule oxidation rather than the substrate isomerism, dehalogenation, or dissociation characteristic of direct photolysis reactions. Photooxidation may aid in microbial degradation through the oxidation of resistant complex structures. Photoreactions are limited to soil surfaces due to light extinction within the soil system, but coupled with soil mixing, may be highly effective as an in situ treatment technique.

Utilization of the lower atmosphere as a treatment medium requires an analysis of both the photoreaction potential and the volatility of the compounds of interest. Volatilization and dilution alone are not considered as acceptable methods. An adequate assessment of the potential for use of photodegradation requires information concerning the atmospheric reaction rate of the compound and anticipated reaction products (especially concerning the formation of hazardous degradation products).

Photodegradable compounds include those with moderate to strong absorption in the 290+ nm wavelength range. Such compounds usually have an extended conjugated hydrocarbon system or a functional group with an unsaturated hetero atom (e.g., carbonyl, azo, nitro). Tetrachlorodibenz-p-dioxin (TCDD) (Crosby et al., 1971; Plimmer & Klingebiel, 1973), kepone (Dawson et al., 1980), and PCBs (Occhiucci & Patacchiola, 1982), have been treated with this method. Classes of compounds that usually do not undergo direct photolysis include saturated aliphatics, alcohols, ethers, and amines.

The rate of photoreaction is influenced by the nature of the light reaching the reaction medium, the adsorption spectrum of the reacting species of sensitizer, the concentration of reacting species, the energy yield produced upon light absorption, the nature of the media in which the reaction is taking place, and the interactions that occur between the contaminant and its surroundings.

Photolysis of soil contaminants may be enhanced in two ways: 1) by addition of proton donors, such as polar solvents, and 2) by enhancing volatilization leading to photodegradation (e.g., by increasing soil vapor pore spaces by drying of the soil system).

Several hazardous waste sites have been treated by enhanced photolysis using proton donors, including sites containing TCDD, kepone, and PCBs. Laboratory studies have demonstrated potential for degradation of other compounds.

Permeable treatment beds

Contaminated groundwater may be treated in place by constructing a permeable treatment bed that can physically and chemically remove contaminants (U.S. Environmental Protection Agency, 1982). The construction of a permeable treatment bed consists of excavating a trench to intercept the flow of contaminated groundwater, filling the trench with an appropriate material, and capping the trench. The trench should be designed long enough to contain the plume of the contaminated flow and deep enough to prevent groundwater from flowing beneath the bed. Groundwater pumped from the trench during construction will likely be contaminated and require treatment.

Materials which are not appropriate for use in a trench consist of those with a short useful life, high cost, and re-activation difficulties. Materials which have been identified as feasible fill materials include: 1) limestone or crushed shell for the neutralization of slightly acidic groundwater and the removal of certain metals (Cd, Fe, and Cr); 2) activated carbon, for the removal of organics; and 3) glauconitic greensands or zeolite, for the removal of metals. Permeable treatment beds may become saturated or plugged and, therefore, should be considered as temporary rather than permanent remedial actions. Also, desorption of hazardous absorbed materials to the renovated clean water flow may result in recontamination.

The state-of-the-art on the use of permeable beds to treat groundwater flow is at the present time conceptual. The use of the trench fill materials for treating contaminated groundwater in place has in most cases been tested only on a laboratory scale.

Solution mining
Solution mining involves the elutriation of waste constituents from a contaminated soil for recovery and treatment. The site is flushed with an appropriate flushing solution, and the elutriate is collected in a series of shallow well points or subsurface drains. The elutriate may be 1) recycled back through the soil for treatment by degradation, using appropriate application rates for controlled biodegradation; 2) treated to remove metals by precipitation and recycled back to the soil system; or 3) treated and disposed of to a different ultimate receiver (e.g., watercourse).

Solution mining may be used to remove both organic and inorganic waste constituents from a hazardous-waste contaminated site. Solution mining may also be appropriate for the recovery of mobile, oxygenated degradation products formed after soil treatment with chemical oxidizing agents. The technique may be particularly applicable if there is a high safety and health hazard associated with excavation.

Flushing solutions may include water, acidic aqueous solutions, basic aqueous solutions, solvents, and surfactants (U.S. Environmental Protection Agency, 1982, 1984). However, the solutions used for flushing may themselves be potential pollutants with toxic and other environmental impacts on the soil and water receiver systems. There may also be difficulty in determining whether the solvent has sufficiently contacted all of the waste constituents.

This technology is currently being studied at the laboratory level, with investigations being conducted to determine appropriate solvents for mobilizing various classes and types of waste constituents. The success of this treatment depends upon the availability of an appropriate inexpensive solvent that is available in large enough quantities to adequately treat the site.

Vitrification
In situ vitrification is a process for stabilizing and immobilizing contaminated soils by converting the contaminated soil into a durable glass-like form through melting by joule heating, which occurs when an electrical current passes through a molten medium.

Graphite electrodes are inserted vertically in the ground in a square array. Graphite is placed on the soil surface between the electrodes to form a conductive path. An electrical current is passed between the electrodes, creating temperatures high enough to melt the soil. The molten zone grows downward, encompassing the contaminated soil and producing a vitreous mass. Convective currents distribute the contaminants uniformly within the melt. The depth of contamination that can be treated is limited as the heat losses from the melt approach the energy deliverable to the molten soil by the electrodes. During the process, many organics are pyrolyzed, and gaseous effluents emitted from the molten mass are collected by a hood over the area and routed to an off-gas treatment system which scrubs, absorbs, and filters hazardous components. When power to the system is turned off, the molten volume cools and a block of glass-like material resembling natural obsidian is produced. Hazardous materials incorporated in the vitreous mass are not available for further release, except in direct proportion to vapor pressure, and in inverse proportion to solubility in molten gas. Any subsidence that may occur may be covered with uncontaminated backfill. Increased soil moisture is not a barrier to the use of vitrification, but does increase power requirements, which increases electrical energy costs, and run time, which increases the contribution of labor to costs.

Vitrification is being developed for stabilizing transuranic (TRU) contaminated wastes in place by the Pacific Northwest Laboratories, but is also being tested for application to unprocessed buried hazardous chemical wastes (Fitzpatrick et al., 1984).

As of November, 1984, the Pacific Northwest Laboratory had conducted 21 engineering-scale (laboratory) tests and 7 pilot-scale (field) tests and is fabricating a large-scale system for testing. Various hazardous, simulated hazardous, and organic materials were tested, including Co, Mo, Sr, Cd, Cs, Pb, Ce, La, Te, and Nd as nitrates; chlorides and oxides; and organic solvents such as carbon tetrachloride, tributyl phosphate, and dichlorobenzene.

Conclusions

In situ treatment of contaminated soils requires considerable information and understanding concerning site/soil/waste interactions. Available treatment techniques need to be carefully evaluated and selected based on this information and understanding. In addition, evaluating the success of any treatment or combination of treatments requires an effective monitoring program. Further information concerning in situ treatment techniques and use may be obtained from Review of In Situ Treatment Techniques for Contaminated Surface Soils, Volume I: Technical Evaluation (U.S. Environmental Protection Agency, 1984) and Volume II: Background Information for In Situ Treatment (Sims et al., 1984).

References

Artiole, J., & W. H. Fuller, 1979. Effect of crushed limestone barriers on chromium attenuation in soils. J. Environ. Qual. 8:503-510.

Atlas, R. M., & R. Bartha, 1981. Microbial Ecology, Fundamentals and Applications. Addison-Wesley Publishing Co., Reading, PA.

Baas Becking, L. G. M., I. R. Kaplan, & D. Moore, 1960. Limits of the natural environment in terms of pH and oxidation-reduction potentials. J. Geology 68:243-284.

Crosby, D. G., A. S. Wong, J. R. Plimmer, & E. A. Woolson, 1971. Photodecomposition of chlorinated dibenzo-p-dioxins. Science 73:748.

Dawson, G. W., B. W. Mercer, & C. H. Thompson, 1980. Strategy for the treatment of spills on land. In: Control of Hazardous Materials Spills, Vanderbilt Univ., Nashville, TN.

Dragun, J., & D. E. Baker, 1979. Electrochemistry. In: The Encyclopedia of Soil Science, Part I: Physics, Chemistry, Biology, Fertility, and Technology. Dowden Hutchinson, and Ross, Inc., Stroudsburg, PA, p. 130-135.

Fitzpatrick, V. F., J. L. Buelt, K. H. Oma, & C. L. Timmerman, 1984. In situ vitrification - a potential remedial action technique for hazardous wastes. In: Management of Uncontrolled Hazardous Waste Sites, Proc. Fifth National Conference, Hazardous Materials Control Research Institute, Washington, DC.

Franklin Institute Research Laboratory, Inc., 1981. The Franklin Institute chemical method for detoxifying polychlorinated biphenyls (PCBs) and other toxic waste. The Franklin Institute, Philadelphia, PA.

Fuller, W. H. 1978. Investigations of Landfill Leachate Pollutant Attenuation by Soils. EPA-600/2-78-158, U.S. Environmental Protection Agency, Cincinnati, OH.

Mulle, G. J. 1981. Chemical decomposition of PCBs in transformer fluids: The Acurex Process. Acurex Waste Technologies, Inc., Mountain View, CA.

Nagel, G. 1982. Sanitation of groundwater by infiltration of ozone treated water. GWF-Wasser/Abwasser 123(8):399-407.

Occhiucci, G., & A. Patacchiola, 1982. Sensitized photodegradation of adsorbed polychlorobiphenyls (PCBs). Chemosphere 11(3):255-262.

Plimmer, J. R., & U. I. Klingebiel, 1973. Photochemistry of dibenzo-p-dioxins. In: Chlorodioxins - Origin and Fate. Blair, E. H. (ed.) Am. Chem. Soc., Washington, DC, p. 44-54.

Sims, R. C., D. L. Sorensen, J. L. Sims, J. E. McLean, R. Mahmood, R. R. Dupont, & K. Wagner, 1984. Review of In-Place Treatment Techniques for Contaminated Surface Soils, Volume 2: Background Information for In Situ Treatment. EPA-540/2-84-003b, Municipal Environmental Research Laboratory, U.S. Environmental Protection Agency, Cincinnati, OH.

Sims, R. C., & K. Wagner, 1983. In situ treatment techniques applicable to large quantities of hazardous waste contaminated soils. In: Management of Uncontrolled Hazardous Waste Sites, Proc. Fourth National Conference, Hazardous Materials Control Research Institute, Washington, DC.

Sommers, L. E., C. M. Gilmour, R. E. Wildung, & S. M. Beck, 1981. The effect of water potential on decomposition processes in soils. In: Water Potential Relations in Soil Microbiology. SSSA Special Publication No. 9, Soil Sci. Soc. Amer., Madison, WI.

Staiff, D. C., L. C. Butler, & J. E. Davis, 1981. A field study of the chemical degradation of paraquat dichloride following simulated spillage on soil. Bull. Environ. Contam. Toxicol. 26:16-21.

Thibault, G. T., & N. W. Elliott, 1979. Accelerating the biological clean-up of hazardous materials spills. In: Oil and Hazardous Material Spills: Prevention-Control-Cleanup-Recovery-Disposal. Information Transfer, Inc., Silver Spring, MD, p. 115-120.

Thibault, G. T., & N. W. Elliott, 1980. Biological detoxification of hazardous organic chemical spills. In: Control of Hazardous Materials Spills. Vanderbilt Univ., Nashville, TN, p. 398-402.

U.S. Environmental Protection Agency, 1982. Handbook for Remedial Action at Waste Disposal Sites. EPA-625/6-82-006, Municipal Environmental Research Laboratory, U.S. Environmental Protection Agency, Cincinnati, OH.

U.S. Environmental Protection Agency, 1984. Review of In-Place Treatment Techniques for Contaminated Surface Soils, Volume 1: Technical Evaluation. Municipal Environmental Research Laboratory, U.S. Environmental Protection Agency, Cincinnati, OH.

U.S. Environmental Protection Agency, 1986. Permit Guidance Manual on Hazardous Waste Land Treatment Demonstrations. EPA-530/5w-84-015. Office of Solid Waste, U.S. Environmental Protection Agency, Washington, DC.

Walton, G. C., & D. Dobbs, 1980. Biodegradation of hazardous materials in spill situations. In: Control of Hazardous Materials Spills. Vanderbilt Univ., Nashville, TN, p. 23-29.

Use of soils to retain waste in landfills and surface impoundments

K. W. Brown, Soil and Crop Sciences Department, Texas A&M University, College Station, TX 77843

ABSTRACT

Soil lined facilities have been used extensively for the containment and disposal of waste liquids. Often slowly permeable natural clay-rich deposits were relied upon to retard the movement of liquids from landfills or surface impoundments. In some cases, remolded layers of soils with laboratory hydraulic conductivities of 10^{-7} cm s^{-1} or less have been constructed with the intention of retaining liquids. There is an increasing body of data which indicates that the hydraulic conductivity of both in situ clay deposits and recompacted clays may be greater than those measured on samples in the laboratory. In addition, these facilities have received a wide range of waste liquids with properties that differ greatly from those of water. In fact, most of the waste liquids which have been disposed in landfills are nonaqueous.

Water is well known for its ability to hydrate clay soils and cause them to swell, resulting in low conductivities. Many organic liquids are known to cause the interlayer spacing of smectitic clays to decrease from those which occur when the same clay is wetted with water. Thus, organic liquids could possibly cause hydrated clay-rich soils to shrink and crack, which could result in an increase in the conductivity of the soils intended to retain organic liquids.

A theoretical evaluation of the influence of dielectric properties of liquids on the thickness of the double layer between adjacent clay minerals suggests that the spacing should decrease when minerals are hydrated with liquids having dielectric constants lower than that of water. Most common organic liquids have dielectric constants considerably lower than that of water, suggesting that they should cause hydrated soil to shrink. X-ray observations of a smectitic clay mineral wetted with organic liquids confirmed that the interlayer spacings were less than those observed when the clay was wetted with water. Electrophoretic mobility studies indicated that organic liquids with low dielectric constants cause suspended clay to flocculate. Flocculation studies using dispersed clays indicated that smectitic, micaceous, and kaolinitic clays all flocculated rapidly when they were added to organic liquids, which had low dielectric constants and which were only sparingly soluble in water. The clays also flocculated when placed in a solution containing greater than 50% water soluble organic liquid. Observations of bulk samples of the three above-mentioned clays indicated that water wetted specimens swelled more than similar samples wetted with organic liquids.

Laboratory studies of conductivities using a range of organic liquids including both polar and nonpolar solvents, waste solvents, and commercial petroleum products indicated that the hydraulic conductivities of compacted soils to organic liquids were one to five orders of magnitude greater than those to water. Observation of the soils permeated with dye labeled organic liquids revealed the formation of platy structural units near the surface. The dye stains in the soil revealed that the organic liquids moved through cracks that penetrated the soil, which originally had a massive structure. Field test cell liners were constructed using three clays and two organic waste liquids. The conductivity measurements in the test cells confirmed the laboratory findings. The nonpolar solvent waste containing xylene which was used to permeate the 1.5 m square and 15 cm thick field test section of compacted clay, broke through many of the replications within two weeks. The acetone waste took as long as two years to break through the test sections; however, in the end, sections of each type of clay were also permeated by the acetone waste.

Field data and observations collected at active landfills and surface impoundments suggest that organic liquids have moved 10 to 1000 times faster than anticipated based on laboratory measurements made using water. Some of this increased mobility may be attributed to differences between laboratory and field conductivities, while the remainder is likely due to the impact of organic liquid on the properties of clay soils. There are now sufficient data available to provide a mechanistic explanation as to how organic liquids migrate rapidly through soils. These data suggest that organic liquids, which are only sparingly soluble in water or water soluble liquids in concentrations greater than about 50%, will dessicate clays causing them to shrink and crack. The liquids are then able to flow through the newly formed macropores in the soil much more rapidly than when the soils are wetted with water.

INTRODUCTION

Early waste disposal operations often relied on the ability of soils to retard the movement of water and associated pollutants via low conductivity, adsorption, and degradation. As long as wastes were not contaminated with the products of our modern industrial society, the slow rate of release of pollutants, which had low toxicity and persisted for only a limited time, likely did little damage to the environment. However, now that our wastes contain manufactured halogenated organic chemicals, solvents, and biocides, which are toxic and potentially mutagenic in very low concentrations, even small leaks from landfills or waste impoundments may be detrimental.

As our awareness of the possibility of pollutant migration through soils increased, clay-rich in situ soils were increasingly relied upon to contain waste leachates. If a natural body of clay into which a landfill or waste impoundment could be dug was not available at a particular location, then clay was imported and recompacted into pits dug into more permeable material to form a barrier to the migration of pollutants. Both in situ clay deposits and recompacted clay liners were often referred to as being "impermeable" despite the general awareness that all soils by their very nature contain pores and thus are permeable. Of course, if Darcian flow applies and liquids move through soil in a uniform wetting front, then it can be shown that the flux of liquids and the distance they travel in a given time may both be rather small, and thus, may not reach underground or adjacent water resources for hundreds or thousands of years. Typically, hydraulic conductivity measurement in clays intended for use in waste retention have been made on either undisturbed cores taken from the field or on soils which have been dried, ground, sieved, rewetted, and remolded in laboratory columns. Hydraulic conductivities of 10^{-7} cm s^{-1} or less have been typically accepted as being sufficiently slowly permeable to be acceptable for retaining contaminated liquids (EPA, 1984). Assuming Darcian flow, a hydraulic gradient of 1 and typical porosities, the flux and distances of movement can be calculated as shown in the second and third column in Table 1. The flux of liquid through a saturated clay liner with a hydraulic conductivity of 10^{-7} cm s^{-1} is 308.6 m^3 ha^{-1} yr^{-1}, which is obviously not

Table 1. Flux and distances traveled through soils of different hydraulic conductivities. (The calculations were done for a head of one).

Hydraulic Conductivity (cm s^{-1})	Flux (m^3 ha^{-1} yr^{-1})	Distance Traveled Assuming 30% Porosity (m yr^{-1})	Distance Traveled Assuming 1.5% Effective Porosity (m yr^{-1})
10^{-5}	30,865	10.5	210
10^{-6}	3,086.5	1.05	21
10^{-7}	308.6	0.10	2.1
10^{-8}	30.9	0.01	0.21
10^{-9}	3.09	0.001	0.02

a trivial amount of liquid being released into the environment particularly if it is contaminated. If the transport occurs as a uniform wetting front, the liquid would, however, penetrate only 0.1 m per year; thus, if the waste impoundment were 10 m from a water table, one hundred years would pass before the contaminated leachate would be expected to reach the water table. If, however, as suggested by the data of Bouma and Wosten (1979); Germann and Beven

(1981); and Horton, et al. (1985); only a small fraction of the total porosity may be effective in transporting liquids in the soil, and pollutants may move much more rapidly than they would if all the liquids in the pores were being displaced. The available data suggest that the effective porosity may be only 1.5 to 5% of the total volume of a well-structured soil. If we assume that 1.5% of the total volume is effective, we find that instead of moving 0.1 m per year, contaminants may move 2.1 m yr^{-1}, thus reaching a 10 m deep water table in slightly less than 5 years instead of 100 years.

If laboratory measurements underestimate the hydraulic conductivity of native soils in the field as suggested by the data of Ritchie et al. (1972) and Griffin et al. (1985), and of soil remolded in the field as suggested by the data of Daniel (1985) (Table 2); then, the potential flux of liquids may be underestimated by one to three orders of magnitude, as can be seen from Table 1. The observed differences may be attributed to alterations in the soil structure or truncation of larger conducting pores when samples are remolded or enclosed in permeameters for testing. Such underestimation of hydraulic conductivity could result in contaminated leachate migrating much more rapidly than would be anticipated.

Table 2. Differences between laboratory and field hydraulic conductivities as reported in the literature.

Laboratory Conductivity ($cm\ s^{-1}$)	Field Conductivity ($cm\ s^{-1}$)	Note	Reference
3×10^{-6} to 8×10^{-7}	3×10^{-5}	Houston Black clay soil	Ritchie et al. (1972)
2×10^{-7} to 4×10^{-9}	3×10^{-4} to 4×10^{-7}	4 depths 1 location	Griffin et al. (1985)
1×10^{-8} to 8×10^{-8}	2×10^{-4} to 3×10^{-6}	4 samples 1 location	Daniel (1985)

In addition to these difficulties with using clay soils to retard the movement of contaminated liquids, there is an increasing body of evidence which suggests that liquids dessicate clays and may cause them to shrink and crack forming secondary porosity which facilitates flow. The evidence for this phenomena and the practical implications will be reviewed here.

THEORETICAL CONSIDERATIONS

Double layer theory suggests that the spacing between clay minerals (H) should be proportional to the square root of the dielectric constant of the liquid (D) in contact with the clay as:

$$H = \sqrt{\frac{DKT}{8\pi\eta e^2 V^2}}$$

where K is the Boltzman's constant, T is the temperature, η is the concentration of electrolyte, e is the ionic charge, and V is the valence of the primary ion (Van Olphen, 1977). This relationship suggests that a decrease in the dielectric constant will result in a decrease in the basal spacing between clay particles. Many common organic chemicals and petroleum products have dielectric constants which are considerably less than that of water (Table 3), suggesting that water-wet clays subjected to organic liquids may shrink.

Table 3. Dielectric constants of common organic solvents at 20 to 25°C.

Solvent	Dielectric Constant	Source*
n-hexane	1.9	1
Cyclohexane	2.0	2
1,4-dioxane	2.2	1
Benzene	2.3	1,2
o-xylene	2.5	2
Di-ethyl ether	4.3	1
Chloroform	4.8	1
Ethyl acetate	6.1	1
Aniline	6.9	2
t-butanol	12.7	1
Benzaldehyde	17.6	1
Isopropanol	18.3	2
Acetone	21.0	1,2
Ethanol	25.0	1,2
Methanol	33.2	1,2
Nitrobenzene	35.4	1
Nitromethane	37.0	1
Acetonitrile	37.2	1
N,N-dimethylformamide	37.3	1
Dimethyl sulphoxide	47.3	1
Water	79.4	1,2
N-methylformamide	187.0	1

* 1 = Murray and Quirk, 1982
2 = Fernandez and Quigley, 1985

X-RAY EVIDENCE

Clay mineralogists have utilized organic liquids to adjust the basal spacing of smectitic clays for decades (MacEwan, 1948). More recent evidence of the influence of dielectric constant on the basal spacing of a calcium montmorillonite was presented by Murray & Quirk (1982). They gathered data from several sources relating the spacing to the dielectric properties of various liquids. Although, there are a few spurious results, their data indicate that spacing is least at low dielectric constants and increases to spacings similar to those of water-wet clay at dielectric constants greater than 40. Brindley et al. (1969) reported on the influence of dilutions of water soluble organic chemicals on the basal spacing of calcium mortmorillonite. They found that dilute solutions of acetone caused basal spacing to increase from those found with water, but as the solution became more concentrated, the spacing decreased in several steps until they shrank to distances less than those in water at dielectric constant less than 28.

X-ray data on the influence of dielectric constant on a smectitic clay (sodium bentonite) was reported by Brown & Thomas (1986) for dilutions of acetone and ethanol in water. The results shown in Figure 1 indicate that for both of these miscible organics, the clays exhibited increased basal spacing at dilute concentrations, but decreased basal spacing at dielectric constants less than 47 for acetone solutions and less than 28 for ethanol solutions.

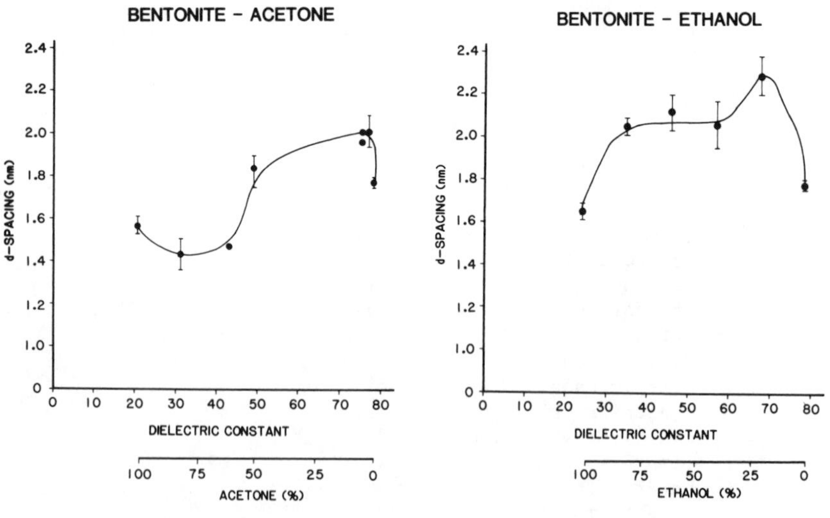

Figure 1. Basal spacings of bentonite clay equilibrated with water-acetone and water-ethanol solutions of various dielectric constants. Vertical bars represent one standard deviation (Brown and Thomas, 1986).

While x-ray data can only be used to evaluate the influence of dielectric constant on the interlayer spacing of smectitic clays, double layer theory suggests that the spacing between adjacent particles of smectitic and other clays will also likely decrease when they are wetted with chemicals having dielectric constants less than that of water.

ELECTROPHORETIC MOBILITY AND ZETA POTENTIAL

Either electrophoretic mobility, zeta potential or both may provide insight into the interaction between clay minerals and the liquid in which they are suspended. Data on the mobility of individual clay particles as a function of applied voltage may be collected by use of a microscope as described by Riddick (1961). Electrophoretic mobility is expressed as $\mu m\ s^{-1}$ per volt cm^{-1}, while the zeta potential, ζ is calculated as:

$$\zeta = \frac{\mu\ (D \times D_o)}{\eta}$$

Where μ is the electrophoretic mobility, D is the dielectric constant of the liquid; D_o is the permitivity of free space and η is the viscosity of the liquid. An abrupt decrease in electrophoretic mobility or decrease in the zeta potential indicates that clays have flocculated as a result of decreases in the spacing between individual clay particles. Data of Brown & Thomas (1986) summarized in Table 4 indicate that for the smectitic (bentonite), micaceous, and kaolinitic clays they evaluated, abrupt decreases in both parameters occurred at dielectric constants below 37 to 40 for smectitic clays subjected to dilutions of acetone in water and ethanol in water, and 26 to 35 for kaolinitic and micaceous clays in similar solutions. This data confirms the theoretical predictions that 1:1 and non expanding 2:1 clays also undergo decreases in spacing between particles, and as a result, flocculate below an apparent threshold dielectric constant.

Table 4. Dielectric constants at which the d-spacing dropped below 1.8 nm, the electrophoretic mobility was midway between zero and the plateau, the zeta potential was midway between zero and the plateau, and the apparent clay concentrations reached 0.5.

	Soil	Basal d-Spacing	Electro-phoretic Mobility	Zeta Potential	Apparent Clay Content	Average
Acetone	Kaolinite		31	35	31	32
	Mica		26	26	37	30
	Bentonite	47	37	41	38	41
Ethanol	Kaolinite		28	32	30	30
	Mica		31	30	33	31
	Bentonite	28	38	39	49	39

FLOCCULATION - DISPERSION TESTS

Another measure of the flocculated or dispersed state of a clay mineral in a liquid of interest may be made by observing the behavior of a dilute suspension of the clay of interest in a settling chamber as would be done in a standard particle-size analysis. Such observations made by Brown & Thomas (1986) are summarized in Table 4 for the same three clays and dilutions of the two organic liquids mentioned above. For all the systems evaluated, the curve of C/Co to dielectric constant was steep, indicating that the change from a dispersed system to a flocculated system occurred abruptly within a narrow range of given dielectric constants. The dielectric constants were read off plots of the data at which the apparent clay content decreased to 50% of the actual clay content. These results (Table 4) are in reasonable agreement with the dielectric constants at which the electrophoretic mobility and zeta potential indicated the clays had flocculated.

BULK VOLUME CHANGE

The data reviewed above suggest that the spacing between individual clay minerals will decrease rather abruptly at dielectric constants below values which range from 26 to 49 depending on clay minerals and liquid properties. It is now necessary to determine if the changes in spacing between individual particles will alter the bulk physical properties of the soil.

The impact of liquid dielectric constant on the shrinkage of illitic clays was investigated by Murray & Quirk (1982); kaolinitic soil by Green et al. (1983); and soils with the three mineralogies cited above by Brown & Thomas (1986). All three studies indicated a decrease in bulk volume as the dielectric constant decreases as can be seen in Figure 2 from Murray & Quirk (1982). The authors proposed a linear relationship as shown by the dashed line, however, the solid curve added here may better describe the relationship and indicates an abrupt drop in the amount of swelling at a dielectric constant of about 40, which is similar to that suggested by the electrophoretic mobility and flocculation data discussed above. Data of Brown & Thomas, 1986 indicate that the volume change of the smectitic clay was greatest, but that all three types of clay minerals they tested had smaller volumes when equilibrated with acetone or xylene than when they were wetted with water. This group of reports indicates that the changes in spacing between individual particles does have an impact on the bulk volume of the soil.

VISUAL OBSERVATIONS

Visual observation of the soil structure of the soils before and after permeation by organic liquid has provided qualitative information evidence of the impact of organic liquids on clay soils. Anderson et al. (1985a) reported that the surface of recompacted native clay-rich soils exhibited massive structure which changed to a platy-type structure following permeation with xylene.

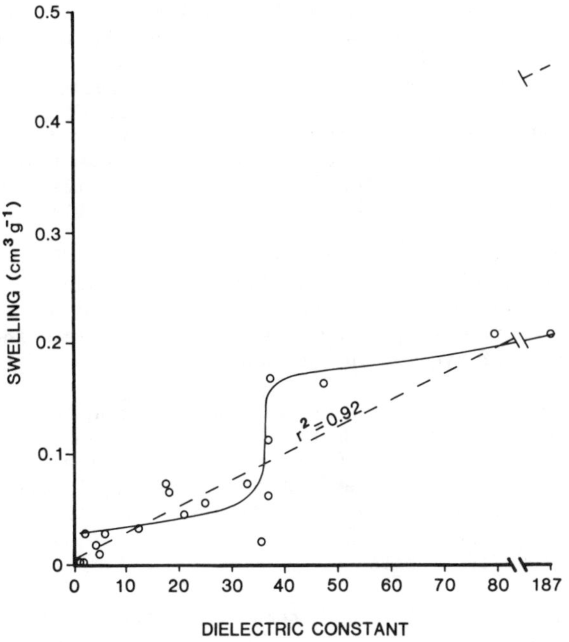

Figure 2. Swelling of Urrbrae B soil in various organic solvents as a function of dielectric constant. (After Murray and Quirk, 1982).

Similar changes were observed in the top 1 to 2 cm of three soils which were compacted into field test cells and permeated with a xylene waste (Brown, 1986). Xylene appeared to completely wet and alter the structure of the surface of the compacted soils. It is possible that the compaction on the surface oriented the clays immediately below the surface, and as organic liquids dessicated the clay, it caused them to collapse into platy structural units near the surface.

At greater depths, however, the xylene appeared to flow between structural units and did not completely saturate the individual structure unit. Below the surface few centimeters, vertical cleavage planes were coated with dye and paint pigments which had been added to the waste while cut surfaces were not dyed. Structural development was most evident in the kaolinite soil, less in the mica, and least in the bentonite. Evidence of the movement of waste acetone through the soils in the field test cells was less due to the absence of paint pigments in the waste, however, similar trends were still evident (Brown et al., 1986). Vertical cleavage planes were coated with dye indicating that large amounts of waste acetone passed through these pores. The cut surfaces smelled of acetone but were not dyed. Since acetone is miscible with water, it is likely that it penetrated more rapidly into the soil mass than xylene did.

The above visual observations are in agreement with observations of Barbee & Brown (1986), who reported that movement of dyed xylene through undisturbed well-structured soils left dye in narrow bands as a coating on natural ped faces, extending deep into the soil with no visible signs of dye within the peds.

The concentration of xylene in samples scraped from dyed ped surfaces were much greater than those from the center of dissected peds (Brown et al., 1986). While the concentration differences for acetone were also evident, they were not as great, likely because the acetone diffused much more readily into the soil matrix.

LABORATORY CONDUCTIVITY DATA

Hydraulic conductivity has been typically measured utilizing water or a dilute solution of calcium sulfate on remolded soils in the laboratory utilizing either fixed wall or flexible wall permeameters. Some authors have undertaken studies to compare the results of these two techniques, while others have used only one of the methods. Daniel et al. (1985) have suggested that while the different methods may give different results, the choice of method should be based on the use to which the data will be put. Fixed wall permeameters are useful for studying the impact of various chemicals on the hydraulic conductivity and the results are applicable to situations where lateral forces or excess overburden pressures are not likely to deform the soil closing fissures which result from the penetration of the organics. Triaxial units, in which pressure is applied to the flexible walls of the cores, result in the closing of pores which develop in the soil, and thus, such data is useful for design purposes where lateral forces are likely to be present. Even if such forces are present, however, some increases in permeability may be expected, since the forces may not be able to compensate for all the possible shift in pore-size distribution.

Much of the controversy about the use of one technique or the other has been focused on the possibility of sidewall flow. Acar and Seals (1984) has suggested that valid measurements can be made only in triaxial cells, otherwise, the results will be distorted by sidewall flow. If, however, there are no lateral forces to close the cracks which open, it likely makes little difference if the remolded clay cracks down the middle of the core or pulls away from the sides of the mold, since the resulting increase in leakage rate will be manifested in either case. In an attempt to determine the influence of possible sidewall flow on the calculated hydraulic conductivities, Anderson et al. (1985b) constructed a permeameter which had a sharp concentric ring in the bottom of the permeameter that separated the outflow in half, with the inner half of the cross section being collected from the center of the core and the outer half being collected from the reminder of the core and the sidewall flow. While the resulting measurements are not always exactly equal, they are often quite comparable demonstrating that organic liquids caused increased conductivity as a result of changes in the properties of the soil in the center of the core, as well as, along the sidewalls.

Typical data collected with reagent grade xylene, a non-polar solvent; and acetone, a polar solvent on the hydraulic conductivity of two soils are shown in Figure 3. The hydraulic conductivity of each of the soils to xylene increased sharply after about 0.2 pore volumes of liquid has been displaced from the core. The increase ranged from two to three orders of magnitude. The changes in hydraulic conductivity shown here and those discussed below are far greater than can be explained by the differences in density and viscosity between water and the organic liquids. For acetone, the initial response in all soils was a decrease in hydraulic conductivity. This is likely a result of acetone diffusing into the water in the soil, resulting in a dilute solution penetrating the soil in advance of the more concentrated acetone, causing the soil to swell as suggested by the x-ray data in Figure 1. At pore volumes

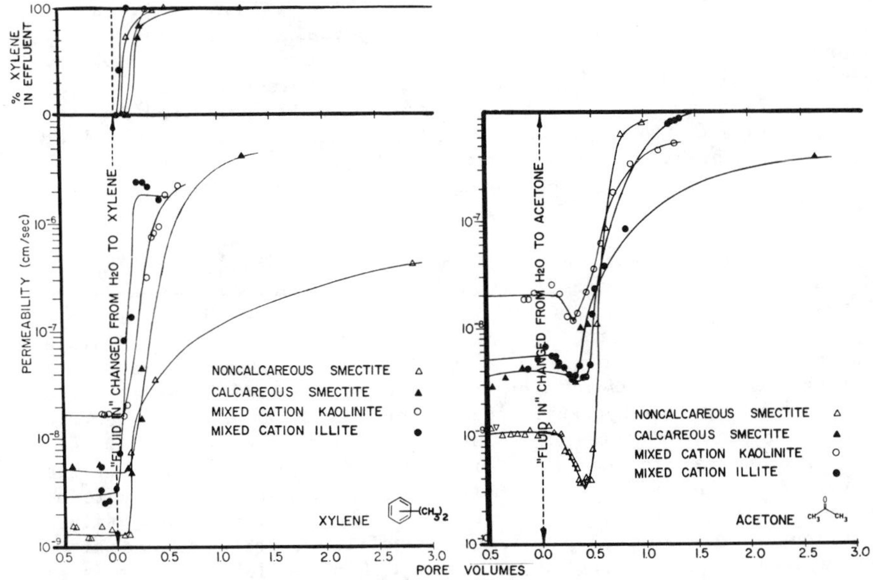

Figure 3. Conductivity curves for four soils treated with xylene (a), and acetone (b); treatment fluids introduced at pore volume = 0.0. (After Anderson et al. 1985a).

greater than 0.5, the hydraulic conductivity began to increase sharply, much as was seen for the xylene reaching levels which are again two to three orders of magnitude greater than those measured on the same cores with water (0.01N $CaSO_4$). The increase in hydraulic conductivity is likely a result of the concentrated acetone having penetrated the soil.

Much of the data available in the literature on the influence of concentrated organic liquids on the hydraulic conductivity of remolded clay soils is summarized in Table 5. The table includes data on many different soils which contain a variety of clay minerals which represent most of the soil mineralogies which would commonly be used for waste containment across the country. It includes data on fourteen chemicals and petroleum products representing a range of the organic liquids in common use. With very few exceptions, increases in hydraulic conductivity ranging from less than 1 to 5 orders of magnitude were observed. The exceptions include the trichloroethylene and acetone data by Green et al., 1981 which is very likely a result of not having run enough liquid through their soil. Although it is not possible to calculate exactly the pore volume of liquid they passed through their column, it is likely that their data were based on less than one half pore volume of liquid, and as can be seen in Figure 3, one would expect the hydraulic conductivity for acetone to still be decreasing. The reason why their conductivity to trichloroethylene decreased is not known. It is interesting to note, however, that Evans et al. (1985) reported similar initial decreases with carbon tetrachloride which were followed by increases in hydraulic conductivity of several orders of magnitude. When Green et al. (1981) evaluated their data, they chose to ignore the increased in hydraulic conductivity they observed for benzene and xylene and instead selected only low values. Their plot on the influence of dielectric constant on hydraulic conductivity is the reverse of that suggested by the theory reviewed above and is completely opposite that reported by other researchers. It appears that Green et al. (1981) simply did not run their experiment long enough to determine the impact of the concentrated organic liquids on their soil.

It is thus evident that most concentrated organic liquids, and in particular, those with low dielectric constants or form separate phases that either float on water or sink to the bottom of a water column will increase the hydraulic conductivity of soils. Dilute concentrations of water miscible organic liquids or concentration of sparingly soluble organic liquids would be expected to have little or no impact on the hydraulic characteristics because the dielectric constant of such mixtures or solutions would be near that of water. The data on the influence of dilutions of organic liquids on the hydraulic conductivity of compacted clay is summarized in Table 6. The data of Daniel and Liljestrand (1984) indicate that dilute solutions of methanol, xylene, and landfill leachate do not alter the hydraulic conductivity. Brown and Thomas (1986) investigated the influence of a range of concentrations of acetone and ethanol on the hydraulic characteritics of clay. They found that mixtures with concentrations greater than 75% acetone with dielectric constants greater than 35 were required before the hydraulic conductivities were altered. These results are in good agreement with the electrophoretic mobility and flocculation data for the same soil discussed above.

It, thus, appears that while concentrated organic liquids with low dielectric constants will increase the conductivity of soil, dilute chemicals or leachates will either not alter the hydraulic characteristics or may decrease the hydraulic conductivity.

FIELD TEST CELL DATA

In order to gather data on the impact of organic chemicals on the conductivity of remolded clay on a larger scale than can be achieved in the laboratory, Brown et al. (1986) made measurements and observations on field test cells. The cells 150 x 150 cm were equipped with a sand underdrain overlain by two lifts of compacted soils, which were flooded with either xylene or acetone wastes. Three soils with predominantly kaolinitic, micaceous, and smectitic clays were studied. Leachate was collected and used to calculate hydraulic conductivities. The xylene waste rapidly penetrated each of the compacted clays, with some yielding a separate organic phase which leaked through the 15 cm layer of clay within two weeks after the waste was placed in the cells. Average values for the final conductivity were all greater than the laboratory conductivities measured with water or the initial conductivities measured in the cells and are shown in Table 5. The xylene rapidly penetrated all three soils and the hydraulic conductivities continued to increase as the cumulative volume of leachate increased. For the acetone waste, the hydraulic conductivity initially decreased through about 0.5 pore volumes again. This likely resulted as the acetone first diffused into the water in the pores resulting in low concentration causing the soil to swell, as is suggested by the x-ray data in Figure 1 and the laboratory data in Figure 3. As the acetone permeating the soil became more concentrated, the soil began to shrink, the trend reversed, and the hydraulic conductivities at pore volumes greater than 0.5 increased with increasing pore volumes.

Once the soils had been permeated, they were removed from the cells and dissected for observation and sampling. The compacted soil liners were contained in a thick plastic envelope, and a plastic flap which was welded to the wall of the envelope was positioned between the two lifts of soils to minimize side wall flow. There was ample visual evidence as discussed above from the dye patterns that the organic liquids moved through channels in the center of the soil liner. In some instances, the dye patterns indicated that the liquid moved vertically through one lift, then spread horizontally just above the second lift until it migrated through a channel in the second lift.

Table 5. Summary of the laboratory and field measurements of the conductivity of compacted clay soils to water and concentrated organic liquids.

Permeant	Type of Study and Equipment	Type of Clay	K_{Water} ($cm\ s^{-1}$)	K_{Final} ($cm\ s^{-1}$)	Source
Methanol	Fixed wall permeameter	Kaolinite	6.0×10^{-8} to 8.0×10^{-8}	3×10^{-7}	Daniel et al. (1985)
	Flexible wall permeameter	Kaolinite	6.0×10^{-8} to 8.0×10^{-8}	1.0×10^{-7}	"
Carbon tetra-chloride	Flexible wall permeameter	Unspecified	4.0×10^{-9}	5.0×10^{-7}	Evans et al. (1985)
Xylene	Fixed wall permeameter	Noncalcareous smectite	1.4×10^{-9}	3.0×10^{-7}	Anderson et al. (1985a)
	"	Calcareous smectite	5.6×10^{-9}	4.0×10^{-6}	"
	"	Kaolinite	1.8×10^{-8}	2.0×10^{-6}	"
		Illite	3.5×10^{-9}	2.0×10^{-6}	
Heptane	Fixed wall permeameter	Noncalcareous smectite	1.5×10^{-9}	3.0×10^{-7}	"
	"	Calcareous smectite	3.6×10^{-9}	1.5×10^{-7}	"
	"	Kaolinite	1.9×10^{-8}	3.0×10^{-6}	"
		Illite	4.3×10^{-9}	5.0×10^{-6}	
Methanol	Fixed wall permeameter	Noncalcareous smectite	1.6×10^{-9}	1.0×10^{-6}	"
	"	Calcareous smectite	5.1×10^{-9}	4.0×10^{-7}	"
	"	Kaolinite	1.5×10^{-8}	4.5×10^{-7}	Anderson et al. (1985a)
	"	Illite	5.5×10^{-9}	5.0×10^{-7}	

Table 5 continued.

Permeant	Type of Study and Equipment	Type of Clay	K_{Water} ($cm\ s^{-1}$)	K_{Final} ($cm\ s^{-1}$)	Source
Acetone	Fixed wall permeameter	Noncalcareous smectite	1.1×10^{-9}	7.0×10^{-7}	"
	"	Calcareous smectite	3.5×10^{-9}	3.0×10^{-7}	"
	"	Kaolinite	2.0×10^{-8}	4.5×10^{-7}	"
	"	Illite	3.1×10^{-9}	8.0×10^{-7}	
Ethylene Glycol	Fixed wall permeameter	Noncalcareous smectite	1.4×10^{-9}	2.5×10^{-7}	"
	"	Calcareous smectite	4.7×10^{-9}	2.0×10^{-8}	"
	"	Kaolinite	1.6×10^{-8}	incomplete	"
	"	Illite	6.8×10^{-9}	incomplete	
Aniline	Fixed wall permeameter	Noncalcareous smectite	2.9×10^{-9}	2.5×10^{-7}	"
	"	Calcareous smectite	3.9×10^{-9}	9.0×10^{-6}	"
	"	Kaolinite	1.5×10^{-8}	3.5×10^{-7}	"
	"	Illite	3.9×10^{-9}	1.3×10^{-7}	
Acetone	Fixed wall permeameter	Smectite	3.6×10^{-8}	5.1×10^{-5}	Brown & Thomas (1984)
	"	Synthetically treated smectite	2.6×10^{-8}	1.4×10^{-6}	
	"	Mica	1.6×10^{-8}	2.5×10^{-7}	"
Xylene	Fixed wall permeameter	Smectite	3.6×10^{-8}	1.8×10^{-4}	Brown & Thomas, (1984)
	"	Synthetically treated smectite	2.6×10^{-8}	7.3×10^{-4}	"
	"	Mica	1.6×10^{-8}	1.0×10^{-4}	"

Table 5 continued.

Permeant	Type of Study and Equipment	Type of Clay	K_{Water} ($cm\ s^{-1}$)	K_{Final} ($cm\ s^{-1}$)	Source
Gasoline	Fixed wall permeameter	Smectite	3.6×10^{-8}	2.0×10^{-4}	"
	"	Synthetically treated smectite	2.6×10^{-8}	9.1×10^{-5}	"
	"	Mica	1.6×10^{-8}	6.2×10^{-5}	"
Kerosine	Fixed wall permeameter	Smectite	3.6×10^{-8}	1.5×10^{-4}	"
	"	Synthetically treated smectite	2.6×10^{-8}	9.1×10^{-5}	"
	"	Mica	1.6×10^{-8}	5.7×10^{-5}	"
Diesel Fuel	Fixed wall permeameter	Smectite	3.6×10^{-8}	5.2×10^{-5}	"
	"	Synthetically treated smectite	2.6×10^{-8}	4.5×10^{-5}	"
	"	Mica	1.6×10^{-8}	6.3×10^{-7}	
Motor Oil	Fixed wall permeameter	Smectite	3.6×10^{-8}	6.1×10^{-6}	"
	"	Synthetically treated smectite	2.6×10^{-8}	2.1×10^{-6}	"
	"	Mica	1.6×10^{-8}	9.5×10^{-7}	"
Acetone	Fixed wall permeameter	Kaolinite	1.1×10^{-8}	3.7×10^{-6}	Brown et al. (1986)
	"	Mica	1.5×10^{-8}	4.5×10^{-8}	"
	"	Bentonite	3.5×10^{-9}	1.5×10^{-7}	"

Table 5 continued.

Permeant	Type of Study and Equipment	Type of Clay	K_{Water} ($cm\ s^{-1}$)	K_{Final} ($cm\ s^{-1}$)	Source
Waste Acetone	Fixed wall permeameter	Kaolinite	1.1×10^{-8}	4.6×10^{-9}	"
	"	Mica	1.5×10^{-8}	2.4×10^{-8}	"
	"	Bentonite	3.5×10^{-9}	incomplete	"
	Field cell	Kaolinite	1.1×10^{-8}	7.7×10^{-8}	"
	"	Mica	1.5×10^{-8}	1.0×10^{-7}	"
	"	Bentonite	3.5×10^{-9}	3.4×10^{-8}	"
Xylene	Fixed wall permeameter	Kaolinite	1.1×10^{-8}	1.0×10^{-4}	"
	"	Mica	1.5×10^{-8}	2.2×10^{-5}	"
	"	Bentonite	3.5×10^{-9}	1.5×10^{-4}	"
Waste Xylene	Fixed wall permeameter	Kaolinite	1.1×10^{-8}	6.1×10^{-6}	"
	"	Mica	1.5×10^{-8}	6.4×10^{-6}	"
	"	Bentonite	3.5×10^{-9}	8.5×10^{-7}	"
	Field cell	Kaolinite	1.1×10^{-8}	1.1×10^{-6}	"
	"	Mica	1.5×10^{-8}	2.1×10^{-6}	"
	"	Bentonite	3.5×10^{-9}	1.3×10^{-7}	"
Xylene	Fixed wall glass column	Fire clay	1.4×10^{-8}	4.7×10^{-8}	Green et al. (1981)
Acetone	Fixed wall glass column	Ranger shale	3.8×10^{-8}	2.5×10^{-9}	"
Trichloro-ethylene	Fixed wall glass column	Ranger shale	3.8×10^{-8}	2.0×10^{-9}	"

Table 6. Summary of the conductivity of compacted clay soils permeated with dilute organics and water.

Permeant	Concen-tration	Type of Study and Equipment	Type of Clay	K_{Water} (cm s^{-1})	K_{Final} (cm s^{-1})	Source
Acetone	2.0%	Fixed wall permeameter	Kaolinite	1×10^{-8}	1.5×10^{-8}	Brown (1986)
	12.5%	"	"	1×10^{-8}	2.0×10^{-8}	"
	25.0%	"	"	1×10^{-8}	2.0×10^{-8}	"
	50.0%	"	"	1×10^{-8}	1.5×10^{-8}	"
	75.0%	"	"	1×10^{-8}	1.0×10^{-7}	"
	100.0%	"	"	1×10^{-8}	1.0×10^{-6}	"
Ethanol	60.0%	"	"	3×10^{-9}	5.0×10^{-6}	"
	80.0%	"	"	3×10^{-9}	8.0×10^{-6}	"
	100.0%	"	"	3×10^{-9}	2.0×10^{-5}	"
Methanol	5.0%	Flexible wall permeameter	Smectite Chlorite	1×10^{-9} 2×10^{-9}	2.0×10^{-9} 5.0×10^{-9}	Daniel & Liljestrand (1984)
Xylene	196ppm	Flexible wall permeameter	Smectite Chlorite	1×10^{-9} 2×10^{-9}	1.0×10^{-9} 1.0×10^{-9}	"
Solid Waste Landfill Leachate		Flexible wall permeameter	Chlorite	2×10^{-9}	1.0×10^{-9}	"
Liquid Waste Impoundment		Flexible wall permeameter	Chlorite	2×10^{-9}	2×10^{-9} to 3×10^{-9}	"

FIELD OBSERVATIONS

Although many instances of groundwater contamination, particularly with volatile organic solvents and refined petroleum products resulting from landfills, surface impoundments, back lot dumping, and leaking underground storage tanks are known to the regulatory community, there are few documented field observations which suggest that the impact of organic chemicals on clay soil discussed above have influenced the rate of movement of contaminants. One instance where some evidence is available is a landfill at Wilsonville, Illinois, which received some 88,000 drums of waste, many of them containing organic liquids. The site geology has been extensively investigated by Griffin et al. (1984, 1985). Organic liquids migrated from the trenches to adjacent monitoring wells much faster than would be suggested even by the *in situ* hydraulic conductivity data. A separate phase of organic liquids has been found in several monitoring wells. The facility relied on *in situ* clay deposits for a liner. Field observations indicated the presence of free organic liquids in the larger pores and fissures in the soil adjacent to the trenches. It is likely that the organic liquids expanded the preexisting fissures in the clay facilitating their movement toward the monitoring wells.

Another example of a case in which organic liquid moved through clay soil much more rapidly than water, occurred in a newly constructed underdrained surface impoundment in Oklahoma. The impoundment had a 180 cm thick compacted clay liner with a gravel underdrain which was used for treatment and evaporation of liquid wastes. Only a limited amount of leachate was collected during the first two years of operation. At that time, about 10,000 l of a chlorinated solvent which was denser than water was inadvertently discharged into the impoundment. While Darcy's Law calculations indicate that about 50 years should be required for water to penetrate the liner, the organic liquid began appearing as a separate phase in the leachate collection system several months after it was placed in the impoundment (Files of the Oklahoma Department of Health).

CONCLUSIONS

There is an increasing body of evidence to suggest that soils are less effective in retarding the movement of liquids, and in particular, organic liquids than had generally been assumed. It appears that field conductivity of *in situ* or recompacted clays are greater than those measured in laboratory permeameters, perhaps, because of structural differences or truncated pores which occur when samples are subjected to laboratory testing.

In addition, there is ample evidence to suggest that concentrated organic liquids and liquid organics which form separate phases when mixed with water increase the hydraulic conductivity of soils. Double layer theory predicts that liquids with low dielectric constants should cause decreases in the spacing between clay minerals. This prediction is substantiated by x-ray data, electrophoretic mobility and zeta potential data, and flocculation studies.

Data on the influence of organic liquids on the bulk volume of soil samples, as well as, visual observations indicate that permeated soils shrink, resulting in cracks and fissures which account for the increased hydraulic conductivities observed in laboratory permeameters, field cells, and at operating waste disposal facilities.

Thus, it is likely that many soil-lined facilities which have been used for the disposal or containment of concentrated liquid organic chemicals have or continue to leak more rapidly than would be predicted from laboratory hydraulic conductivity measurements in which water was the permeant. When soils are to be used to line containment facilities, their hydraulic characteristics should be determined using the liquid to which they are expected to be subjected.

REFERENCES

Acar, Y. B. & R. K. Seals. 1984. Clay barrier technology for shallow land waste disposal facilities. Hazardous Waste Vol. 1:2, pp. 167-181.

Anderson, D. C., K. W. Brown & J. C. Thomas. 1985a. Conductivity of compacted clay soils to water and organic liquids. Waste Mgmt & Res 3: 339-349.

Anderson, D. C., W. Crawley, & D. Zabcik. 1985b. Affects of various liquids on clay soil: bentonite slurry mixtures. In: A. J. Johnson, R. K. Frobel, N. J. Cavalli & C. B. Petterson (eds.). "Hydraulic Barrers in Soil and Rock". ASTM STP 874. Philadelphia, Penn, pp.93-103.

Barbee, G. C. & K. W. Brown. 1986. Movement of xylene through unsaturated soils following simulated spills. Water, Air, and Soil Pollution (in press).

Bouma, J. & J. H. M. Wosten. 1979. Flow patterns during extended saturated flow in two undisturbed swelling clay soils with different macrostructures. Soil Sci. Soc. Am. J. 43:16-22.

Brindley, G. W., K. Wiewiora & A. Wiewiora. 1969. Intercrystalline swelling of montmorillonite in some water-organic mixtures (Clay-Organic Studies XVII). The American Mineralogist 54:1635-1644.

Brown, K. W. & J. C. Thomas. 1984. Conductivity of three commercially available clays to petroleum products and organic solvents. Hazardous Waste, Vol. 1:4 pp. 545-553.

Brown, K. W. 1986. Review and Evaluation of the Influence of Chemicals on the Permeability of Soil Clays. Final Report to USEPA, National Environmental Res. Lab., Cincinnati, Ohio. Grants Nos. CR 808824030 and CR 811663010.

Brown, K. W. & J. C. Thomas. 1986. Mechanisms by which organic liquids affect the conductivity of compacted soils . (In review).

Brown, K. W., J. C. Thomas & J. W. Green. 1986. Field cell verification of the effects of concentrated organic solvents on the conductivity of compacted soils. Hazardous Waste & Hazardous Mgmt. (in press).

Daniel, D. E. 1985. A case history of leakage from a surface impoundment. In: Proceedings of the Symposium on Seepage and Leakage from Dams and Impoundments. ASCE, Denver, Colorado, U.S.A.

Daniel, D. E., D. C. Anderson, & S. S. Boynton. 1985. Fixed-wall versus flexible-wall permeameters. In: A. J. Johnson, R. K. Frobel, N. J. Cavalli & C. B. Petterson (eds.). "Hydraulic Barrers in Soil and Rock". ASTM STP 874. Philadelphia, Penn, pp. 107-126.

Daniel, D. E. & H. M. Liljestrand. 1984. Effects of landfill leachates on natural liner systems. A report to Chemical Manufacturers Association. Department of Civil Engineering, University of Texas, Austin, Texas. 86 pp.

EPA, 1984. Public Law 98-616. Hazardous and Solid Waste Amendment of 1984, Section 3004, Paragraph "O": Minimum Technologies Requirements, Nov. 9, 1984.

Evans, J. C., H-Y. Fang & I. J. Kugelman. 1985. Influence of Hazardous and Toxic Wastes on the Engineering Behavior of Soils. In: H. G. Bhatt, R. M. Sykes & T. L. Sweeney (eds.). Management of Toxic and Hazardous Wastes. Lewis Publishers, Inc. 121 S. Main Street, Chelsea, MI. pp. 237-264.

Fernandez, F. & R. M. Quigley. 1985. Hydraulic conductivity of natural clays permeated with simple liquid hydrocarbons. Can. Geotech J. Vol. 22, pp. 205-214.

Germann, P. & K. Beven. 1981. Water flow in soil macropores. I. An experimental approach. J. Soil Sci. 32:1-13.

Green, W. J., G. F. Lee & R. A. Jones. 1981. Clay-soils permeability and hazardous waste storage. Journal WPCF, Vol.53:8, pp. 1347-1354.

Green, W. J. G. F. Lee & R. A. Jones, and T. Palit. 1983. Interaction of clay soils with water and organic solvents: Implications for the disposal of hazardous wastes. Environ. Sci. & Techn, 17:278-282.

Griffin, R. A., R. E. Hughes, L. R. Follmer, C. J. Stohr, W. J. Morse, T. M. Johnson, J. K. Bartz, J. D. Steel, K. Cartwright, M. M. Killey & P. B. DuMontelle. 1984. Migration of industrial chemicals and soil-waste interactions at Wilsonville, Illinois. U.S. Environmental Protection Agency, Cincinnati, OH. EPA-600/9-84-007. pp. 61-77.

Griffin, R. A., B. L. Herzog, T. M. Johnson, W. J. Morse, R. E. Hughes, S. F. J. Chou & L. R. Follmer. 1985. Mechanisms of contaminant migration through a clay barrier -- Case study, Wilsonville, Illinois. U.S. Environmental Protection Agency, Cincinnati, OH. EPA-600/9-85-013. pp. 27-38.

Horton, R., M. L. Thompson & J. F. McBride. 1985. Estimating transit times of noninteracting pollutants through compacted soil materials. In: Proceedings of Research Symposium on Land Disposal of Hazardous Waste. EPA 600/9-85-013. U.S. Environmental Protection Agency, Cincinnati, Ohio. pp. 275-282.

MacEwan, D. M. C. 1948. Complexes of clays with organic compounds. I. Complex formation between montmorillonite and halloysite and certain organic liquids. Trans of the Farada Soc. 44:349-367.

Murray, R. S. & Quirk, J. P. 1982. The physical swelling of clay in solvents. Soil Sci. Soc. Am. J. 46:865-868.

Riddick, T. M. 1961. Zeta-potential: New tool for water treatment, Part II. Chem. Eng. July 10, 1961, pp. 141-146.

Ritchie, J. T., D. E. Kissell & E. Burnett. 1972. Water movement in undisturbed swelling clay soil. Soil Sci. Soc. Am. Proc. 36:874-879.

Van Olphen, H. 1977. An Introduction to Clay Colloidal Chemistry for Clay Technologist, Geologists and Soil Scientists, 2nd Ed., John Wiley & Soils, Inc.

Land disposal of wastes and groundwater protection: An overview of electric utility industry sponsored research

Ishwar P. Murarka, Electric Power Research Institute, Palo Alto, California 94303

Solid wastes are generated in all industrial, municipal, and domestic operations. Over 70% of the three billion tons of solid wastes produced annually in the United States are disposed of on land and in waters. The passage of the Resource Conservation and Recovery Act (RCRA) in 1976 and its subsequent amendments in 1980 and 1984 has heightened the public awareness of the need to properly dispose of wastes in ways that protect human health and minimize impacts on the environment. The biggest concern in land disposal of solid wastes is that harmful chemicals will contaminate groundwaters to unsafe levels.

Proper management of the disposal of industrial wastes requires that the composition, environmental releases, and fate of released chemicals be accurately predicted. Unfortunately, methods and data needed to model accurately the release and migration of chemicals from solid waste disposal sites do not yet exist.

RESEARCH SPONSORED BY ELECTRIC UTILITIES IN THE UNITED STATES

Electric Power Research Institute (EPRI), a nonprofit R&D organization financed by U.S. electric utility companies, began in 1982, a large interdisciplinary research project on solid wastes from the utility industry. The overall goal of this project (known as Solid Waste Environmental Studies [SWES]) is to develop methods and produce data needed for quantitatively predicting the release, transformation, transport, and subsurface environmental fate of chemicals associated with the disposal of utility industry solid wastes. The utility industry generates approximately 75 million tons of solid wastes annually through the combustion of fossil fuels. Because of this large volume of wastes, the industry has both a responsibility to protect groundwater and a considerable financial stake in meeting regulatory requirements as efficiently as possible.

DESCRIPTION OF RESEARCH IN SWES PROJECT

A brief description of research under SWES project in progress is given below. For more details readers should contact the author. CRC Press, Inc. is expected to publish a book edited by Murarka (1) which contains more detailed information on electric utility industry wastes.

- Leaching Chemistry Studies: This research focuses on understanding the geochemical processes that govern the dissolution of chemicals from utility solid wastes once they are placed in the disposal environment. The objective of laboratory and field experiments in this phase of the research is to predict on the basis of waste characteristics and environmental factors: (a) release rates, (b) release concentrations, and (c) persistence in the subsurface environment, of the chemicals of interest.

- Subsurface Transport Studies: Dispersion of dissolved chemical wastes in groundwater systems is influenced by properties of permeable media and the characteristics of the released chemicals. Field-scale experiments will measure solute dispersion rates for the saturated and unsaturated zones. Several tracers have been selected for use in these experiments. Results from these experiments will improve solute migration models and enable researchers to quantify the effects of physical processes on transport of released chemicals.

- Chemical Attenuation Studies: Interactions between unconsolidated porous media and the chemicals in solution commonly result in reduced movement of released chemicals. Therefore, a series of laboratory experiments will produce data on rates of precipitation and adsorption in the subsurface environment of: chromium, cadmium, arsenic, selenium, boron, vanadium, and zinc. These findings will provide the basis for modeling the geochemical reactions that alter the movement of released solutes.

- Field Sampling Methods: Laboratory experiments and field studies are being carried out to define the precision and accuracy of field methods for sampling groundwater quality, moisture distribution, water movement, and physico-chemical characteristics of waste, soil, and underlying geologic formations.

- Geohydrochemical Models Evaluation: Public domain computer codes for modeling migration of solutes in the unsaturated and the saturated groundwater zones have been evaluated. Among the code characteristics examined were: theoretical bases, assumptions, and representations (or algorithms) of the hydrologic and geochemical processes. Evaluation of newly released models will continue so that the SWES project will not duplicate already available results.

- Interim Assembling of Geohydrochemical Models: As a result of evaluation of available codes, it became clear that the research should focus on assembling an interim code that would integrate the already available knowledge of hydrologic and geochemical processes. On the basis of detailed specifications, a two-step coupled approach was chosen. FASTCHEM, the resulting code, will simulate dynamic of migration of chemicals released from utility waste disposal sites along one-dimensional pathlines derived from a two-dimensional flow-field with reaction chemistry calculated using equilibrium and kinetic equations.

- Geohydrochemical Models Improvement and Validation: Future research in the SWES project will improve the one-dimensional interim model by extending it to multiple dimensions and incorporating the detailed process representations and quantitative results developed in the SWES project. We plan to develop detailed data needed for complete validation of these models by making extensive and intensive measurements at selected waste disposal sites units at power plant sites.

AVAILABILITY OF SWES RESEARCH RESULTS

EPRI has published 29 reports on SWES research. EPRI publications are available for purchase to the general public through its Research Reports Center, P. O. Box 50490, Palo Alto, CA 94303. Contractors have published scientific papers, based on their work for EPRI, in the open literature. However, such papers are usually published a year or two later than the EPRI reports. EPRI conducts annual technology transfer seminars to disseminate results of its research.

REFERENCES

1. Murarka, I. P. (editor). Solid Waste Disposal and Reuse in the United States. CRC Press, Inc., Boca Raton, FL (in publication), 1986.

2. Murarka, I. P. General Description and Status of Solid Waste Environmental Studies Project (RP2485). Mimeograph Report, Electric Power Research Institute, Palo Alto, Ca., 1985. 28 pp.

The role of soil science in the utilization, treatment, and disposal of waste

Raymond C. Loehr, H. M. Alharthy, Centennial Chair and Professor of Civil Engineering, The University of Texas, Austin, TX 78712

The purpose of this presentation is to discuss the role soil scientists have in the management of wastes. One might ask whether soil scientists should be involved in such a mundane technical area as waste management. The answer is unequivocally yes - of course. The involvement is needed on a personal level, as responsible citizens, and on a professional level to conduct research, train students and advise regulatory agencies and industries. The concerns are large, the opportunities are good, and the challenges are intellectually stimulating.

The soil is a natural resource that is used by humans for food and fiber production and structural purposes. It also is used to assimilate the residues of human activity, such as human, animal, municipal, and industrial wastes. This assimilation may be random and unmanaged or it may be purposeful and managed. Soil scientists can and must play a much larger role in determining: (a) the soil assimilative capacity for specific wastes and chemicals and (b) how that assimilative capacity can be managed in a positive manner.

There are only three major "sinks" into which wastes ultimately are discharged: water, air, and land. Each has a specific assimilative capacity for the constituents that are discharged. Inadequate understanding of the water and air assimilative capacity contributed to the water and air pollution that has been obvious in the past decades. As the nature of the assimilative capacity of the air and water has been established and the environmental impacts have been better understood, water and air quality criteria and standards have been developed to protect these resources.

However, although knowledge about the soil assimilative capacity is increasing, it is less extensive than that for air and water. Reliable, quantitative predictions of contaminant movement in soil remain difficult. Such predictions require an understanding of the processes that control the transport, hydrodynamic dispersion, and chemical, physical and biological changes of the contaminants in the soil. Knowledge developed by soil scientists can help provide that increased understanding. Such knowledge is vital to the maintenance of human health, environmental quality, and the land and soil resource.

The assimilative capacity of the soil is poorly understood and the fact that it has been exceeded is increasingly obvious (NRC, 1984; Pye et al., 1983; Wood et al., 1984). Industrial waste lagoons, land disposal sites (landfills, pits, ponds, and lagoons), septic tanks and agricultural practices can be sources of groundwater contamination. Incidents of groundwater contamination have occurred in every state.

Unfortunately, standards and criteria to assess the degree of risk associated with the land application and disposal of wastes and to protect the public and the environment are not extensive. Except for perhaps drinking water criteria and standards, most of the available environmental criteria and standards have been developed to protect air and water, and not soil, from adverse impacts. Guidelines that will protect the soil resource and avoid exceeding the soil assimilative capacity are being developed at federal, regional, and state levels. Soil scientists should play an active role in developing and interpreting the scientific and technical information for these guidelines.

SOIL SCIENCE AND WASTE MANAGEMENT

Two basic concepts of soil have evolved through several centuries of study (Brady, 1984). The first, pedology, considers soil as a natural entity, a weathered product of nature. The second, edaphology, considers the soil as a natural habitat for plants and justifies soil studies primarily on that basis. These two concepts have permitted soil scientists to provide considerable service to humankind. However, the involvement of soil scientists with waste management requires a third concept or focus. That concept relates to consideration of the assimilative capacity of the soil for the residues of man.

This third focus does not require soil scientists to learn different principles or acquire new talents. This focus still requires knowledge of the basic physical, chemical, and biological characteristics of soils and of the properties of soil as they relate to food and fiber production.

What this focus does require is the application of this knowledge in a different direction, in what might be considered as "the other side of the coin" or "the other part of the story." Soil scientists need to be concerned with:

1. what happens to excess metals, nutrients and pesticides added to soils
2. what happens to liquids from septic tanks and surface impoundments
3. if there is a chemical spill, how much soil is contaminated and are there in-situ techniques that can remove or control that contamination
4. how clean is clean when contaminated soils are de-contaminated
5. what is the plant uptake of waste constituents
6. what are background levels of potential contaminants in soils
7. what are appropriate monitoring systems to detect contaminants in soil, soil-pore water and groundwater at land based waste disposal sites

These and related questions relate to: (a) the physics, chemistry, and microbiology of soils, (b) rates of transformation, transport and loss of contaminants and (c) the fate of aqueous and non-aqueous wastes in a mixture of soil, water and solvents. What other discipline knows more about:

1. the hydrological properties of soil.
2. sorption and desorption in soils.
3. metal mobility and immobilization in soils.
4. gas transport through soils.
5. microbial reactions in surface and subsurface soils.
6. soil ecology and the potential for bioaccumulation.
7. effect of waste constituents on soil structure.
8. nutrient availability.

Such knowledge is only illustrative of that which is needed to define and use, but not abuse, the soil assimilative capacity of soils for wastes and residues. Disposal of wastes and residues must be done in a manner that will protect human health and the environment.

A responsible scientist cannot say, "waste management is not my problem, I am only interested in the use of the soil for food and fiber" as some have said. Waste management is a problem for soil scientists both personally as a concerned citizen and as a professional. The advice soil scientists provide can mitigate or exacerbate soil and groundwater pollution problems.

A responsible soil scientist also cannot say, as some have, "the soil should not be used for disposal of municipal and industrial wastes." Municipal wastes result from human activities. Industrial wastes are a by-product of the quality of life we desire. Such wastes must be disposed of in an environmentally sound manner somewhere. There is no "away" into which the wastes can be tossed or placed. The world is a closed system and must be so considered.

Fortunately, most soil scientists do not have such an "ostrich" approach. The key involvement of soil scientists with waste management is to consider "the other side of the coin" and to develop ana apply fundamental knowledge so that the waste assimilative capacity of soils is known and properly used but not abused.

LAND APPLIED WASTES - MAGNITUDE AND TYPE

More and more attention is being given to the use of soil for waste disposal. This is a result of: (a) constraints being placed on other disposal options, (b) increasing quantities of residues (sludges and ashes) generated by conventional waste treatment processes, (c) the increasing number of conventional waste treatment systems, and (d) residues from the increasing population and from our desired standard of living.

Large quantities of wastes are discharged to the soil from homes, cities, and industries and many diverse approaches are used to apply and dispose of the wastes in the soil. Table 1 estimates the quantity from some of these sources. For other sources, there are no sound estimates of the amount that reaches the soil. The basis for these estimates and information on other sources are noted in subsequent paragraphs.

Table 1. Estimates of the Amount of Wastes applied to the Soil

Source	Amount	Comment
Septic Tanks	3.5 to 4 billion gallons per day	contains organics, nutrients, pathogens, metals and organic chemicals used in the home
Municipal Sludge	4.8 million dry tons per year	contains organics, nutrients, and inorganics and organics contributed by homes, industry and business
Hazardous Waste	113 million metric wet tons to surface soils and 67 million metric wet tons disposed of by injection wells	inorganic and organic constituents from industry

Septic Tanks - Over 17 million families in the United States use septic tanks or other types of subsurface seepage to dispose of their wastes. This is about 30 percent of the population.

Assuming that about 50 gallons per capita are disposed of in this manner per day, about 3.5 billion gallons of waste are introduced into the soil each day. In addition, there are business and commercial sources that use septic tanks. The wastes discharged to septic tanks contain human wastes, metals from corrosion products and inorganic and organic chemicals used in the home or business.

Municipal Sludge - The treatment of municipal wastewaters occurs in large publicly owned treatment works (POTWs) which remove as much of the suspended solids and dissolved organics as possible in order to meet the POTW discharge permit conditions. This results in large quantities of primary and secondary sludge that must be stabilized, handled, and properly disposed. Approximately 2000 pounds of dry sludge solids and about 12,000 gallons of liquid sludge are generated from each one million gallons of municipal sewage that is treated. This is the equivalent of about 0.2 to 0.3 pounds of dry solids and about 1 to 1.2 gallons of liquid sludge being generated each day by each person in the United States as a result of conventional wastewater treatment. These quantities can more than double if chemical precipitation processes are used to achieve greater pollutant removals. Such would be the case if, for instance, phosphorus removal is required.

In 1980, about 6.8 million dry tons of sewage sludge were produced in the United States (EPA, 1983). Of this amount about 70% was disposed of in or on land: 12% landspreading on food chain crops, 12% landspreading on non-food chain crops, 18% distribution and marketing (which ultimately is returned to the soil somewhere), 15% landfill, and 12% long-term lagooning or other. Thus about 4.8 million dry tons of sewage sludge containing organics and inorganics from homes, business and industries reach the soil of the nation each year.

The annual sludge production will increase as wastewater treatment requirements increase to provide further protection to surface waters.

Hazardous Wastes - Estimates of the quantities of hazardous waste generated by industry vary depending on the definition of hazardous waste that is used. The Congressional Budget Office (1985) has estimated that about 226 million metric tons (MMT) (wet weight) of hazardous wastes were generated in 1983 and that about 280 MMT may be produced in 1990. In 1983 about 113 MMT were disposed of using surface impoundments, landfills, and land treatment. All of this material reaches the surface soils. In addition in 1983, about 67 MMT were disposed of by injection wells to deep confining geologic formations.

Underground Storage Tanks - An underground storage tank is one or a combination of tanks, including underground piping, that has 10% or more of the volume beneath the ground. It has been estimated that about 100,000 such industrial and gasoline storage tanks are leaking and that another 350,000 are expected to leak within the next five years.

Agricultural Sources - Agricultural sources contribute fertilizers, pesticides, salts in irrigation water, and organic and inorganics in manures. Manures have been of concern primarily on lands adjacent to high-density animal production operations.

Land Based Waste Disposal Approaches - In addition to septic tanks, there are many other purposeful disposal approaches as well as spills that can apply wastes to soils. Table 2 summarizes some of these approaches. The total quantity of contaminants that enter and pass through the soil from these approaches is unknown but it is not trivial.

Summary - The list of sources that contribute chemicals to the soil can be extended. The quantities that are applied to terrestrial environment are large and the types of chemicals that are applied are diverse. Such quantities will increase as the population increases, industry expands and as other disposal sinks are constrained.

The soil represents not only an appropriate treatment and stabilization medium for many wastes but also an opportunity to manage wastes and recycle organics and inorganics with a minimum of adverse health and environmental effects. Deliberate application of wastes such as manure and municipal wastes to land has occurred for centuries.

Table 2. Sources of Contaminants That Enter The Soil

Purposeful Application

1. animal manure storage and disposal
2. pesticide applications and waste disposal
3. municipal and industrial wastewater and sludge land treatment facilities
4. food processing waste applications
5. septic tanks
6. mining waste disposal sites
7. fertilizer applications

Inadvertent and Unplanned Release

1. municipal and industrial landfills
2. waste impoundments and piles
3. municipal and industrial pits, ponds and lagoons
4. operating hazardous waste landfills
5. underground storage tanks
6. abandoned municipal and industrial land disposal sites
7. de-icing salts
8. urban runoff

Indirect

1. precipitation

The terrestrial environment can deactivate and stabilize many wastes as a result of the physical, chemical, and biological mechanisms in the soil. As long as the assimilative capacity of a specific site is not exceeded, the waste constituents are degraded, immobilized, and not transmitted through the atmosphere, water or food chain in amounts that are of concern to human health or the environment. If the assimilative capacity of the soil is not exceeded, there can be a societal benefit from the reuse of the water, nutrients and organics in the wastes. If, however, the assimilative capacity is exceeded, water contamination and the loss of the soil as a productive resource can occur.

The fact that there have not been more instances of water and soil contamination is testimony to the extensive assimilative capacity of the soil. This capacity must be better understood if greater incidences of contamination are not to occur and portions of these resources are not to be lost. Soil science is a key discipline in developing this understanding.

THE ENVIRONMENTAL AGENDA

In discussing the role of soil scientists in waste management, it is appropriate to identify whether, in spite of the large quantities of contaminants that are reaching the soil, the environment is still an important public concern. This evaluation considers whether there is likely to be a continuing market for the talents of soil scientists in this technical area.

The birth of the environmental movement in the United States dates from the mid to latter nineteenth century. This movement resulted in the creation of state and local boards of health, development of professional societies to address environmental issues and the passage of initial environmental legislation. State and local primacy in these issues continued well into the 1940's and 1950's. However, with the recognition that environmental problems did not stop at county or state boundaries and that protection of the environment was a broad public good, the federal government assumed a major regulatory role. As a result, we now have federal regulations controlling surface water pollution, air pollution, sludge management, hazardous wastes, and the production of safe drinking water.

Several patterns have emerged from this environmental activity. One is the largely reactive nature of environmental policies and decision making. Our environmental history is filled with examples of how environmental action, policies and regulation represent reactions to events such as: (a) fish kills in rivers, (b) the "death" of Lake Erie, (c) air pollution episodes in Los Angeles and Donora, Pennsylvania, and (d) most recently the increasing contamination of groundwater.

In spite of this generally reactive conditioning, there is increasing recognition that, while crises will continue to occur, long-term scientific and engineering knowledge is needed for sound regulatory decision making. Support for fundamental and applied studies is continuing and appears to be increasing. Such support for research related to the terrestrial ecosystem comes not only from traditional agricultural bases such as USDA and agricultural industries but also from non-traditional organizations such as the Department of Defense, other federal agencies and industries. Examples of the organizations supporting research related to soil assimilative capacity and the use of land as a waste management alternative are presented in Table 3. Soil scientists can play a role in the development of the needed knowledge through research on the transformations, transport, and fate of constituents entering the soil.

Another pattern is the acceptance of environmental protection as an approved governmental activity. Environmental protection now is considered as a core value of society. Public polls continue to confirm this acceptance. For example, during the 1981-82 economic slump, there was very little public support for relaxing environmental standards. Any political or governmental official who does not recognize this fact of life does so at his or her peril, especially if there is a significant environmental concern, such as a hazardous waste problem or groundwater contamination, in their area.

Table 3. Organizations supporting waste management studies that relate to the soil.*

1. Environmental Protection Agency

   - monitoring, prediction, aquifer cleanup and restoration, land disposal technologies

2. U.S. Geological Survey

   - geology, water chemistry, hydraulics, biology and geochemistry

3. U.S. Department of Agriculture

   - nutrients, pesticides, salinity, modeling, improved chemical disposal practices

4. U.S. Department of Energy

   - aquifer cleanup, monitoring, transport and fate, geochemical processes

5. U.S. Air Force

   - fate of solvents, waste disposal and accidental spills, dioxin contamination, metal mobility, in-situ degradation

6. U.S. Army

   - treatment methods for soil contamination, leachate control methods contaminant control and cleanup

7. Tennessee Valley Authority

   - disposal of power plant wastes, land application of wastewater treatment sludge

8. Electric Power Research Institute

   - fate of constituents in solid waste at utility disposal sites, geochemical models for predicting the release, transport, transformation and fate of chemicals applied to the soil

9. American Petroleum Institute

   - land treatment of petroleum wastes, transport of immiscible liquids in the subsurface

*adapted from EPA 1985a

A third pattern that is evolving is an understanding that pollution control is a cross-media issue rather than a single media issue. Pollutants are discharged into more than one media and few remain in a single media. Contaminants removed from one media, such as wastewater, are concentrated into a slurry or solid and can be discharged to the atmosphere by incineration or to the land using landfills as an ultimate disposal site.

To understand cross-media problems four broad factors need to be emphasized. The first is the initial creation and source of pollutants. The second is the fate of pollutants when controls are applied. The third is the physical, chemical, and biological transformations of pollutants after their release into the environment. The fourth are the various ways human and environmental receptors are exposed to pollutants. Of all the disciplines, soil science is the most central to an improved understanding of the last two factors especially when the soil is used as a waste management alternative.

Despite the many years of environmental concern and our increasingly sophisticated industrialized society, we have yet to establish criteria and policies to judge when environmental policies are successful. How clean is clean? What is an adequate margin of safety? The public is unwilling to tolerate low levels of risk from the environment even though they tolerate much higher risks as the result of personal habits and lifestyles.

There is no indication that environmental concerns will be a non-issue in the near future. All evidence suggests that environmental protection will continue to be a public priority. When this is combined with the fact that the soil will continue to receive emphasis as a waste management alternative for diverse wastes, there is every indication that there will be a "market" for soil scientists who have the knowledge and the interest in waste management efforts.

PERTINENT SOIL SCIENCE ACTIVITIES

There are many waste management activities in which soil scientists can be involved. Examples are indicated in Table 4 and summarized in the following paragraphs:

Table 4. Activities related to waste management and chemical application in which soil scientists can be involved.

Classification of Soils
Chemical Degradation and Loss
Microbiological Degradation and Transformations
Soil Physics and Chemical Mobility
Mathematical Modeling
Consulting
Research
Advisory

1. Classification - selection and characterization of soils and sites suitable or not suitable for waste disposal. Examples include the location of land treatment sites, septic fields, landfill sites and soils of low permeability for use as landfill liners and covers.

2. Soil Chemistry - evaluation of the chemical transformations, mobility and fate of waste constituents added to the soil. These include adsorption and desorption, hydrolysis, metal immobilization, and volatilization.

3. Soil Microbiology - determination of degradation rates and pathways of organics added to the soil. These include biodegradable organics, synthetic organics, bacteria and viruses. Important aspects include: (a) the influence of oxygen, nutrients, moisture content and co-metabolism, (b) measurement of microbial biomass in surface and deeper soils, (c) the toxicity of primary waste constituents and resultant by-products and (d) die-off of pathogens.

4. Modeling - mathematical description of the transformations, transport and fate of organics and inorganics added to the soil. Modeling is becoming increasingly important to: (a) determine the effect of waste loading or chemical application, (b) identify proper application rates to use but not abuse the assimilative capacity of the soil, (c) predict the environmental impact of land based waste management alternatives, and (d) determine the relative food chain, human, animal and avian risk associated with these alternatives.

5. Consulting - assistance with the selection, design, and operation of land-based waste management alternatives and assistance in determining the proper application rate of chemicals added to land, such as pesticides and fertilizers. Involvement can be with regulatory agencies, industries, consulting firms and private organizations. The capability of soil scientists in this regard is well recognized. For instance, in an EPA Permit Writers Guidance Manual (EPA, 1985), it is stated, "To adequately evaluate soil suitability and the ability to monitor at a given site, the permit writer should have a strong background in soil science, hydrology, or hydrogeology."

6. Research - development of data that can be used to determine the soil assimilative capacity for wastes and chemicals. Examples of the type of research soil scientists can undertake are noted in Table 5. The identified research is both fundamental and applied, can be done in the laboratory and the field, and is pertinent to the interests of many agencies and organizations.

7. - Advisory - providing pertinent technical advice regarding the application of chemicals and wastes to soils. Such advice can be: (a) through cooperative extension activities and at public hearings, (b) through review of documents such as regulatory

guidance documents and regulations, and (c) critique of field studies, engineering reports, proposed or actual chemical applications, and land-based waste management options. It is important to avoid a biased stance in providing such advice. Facts and not personal preferences are what is needed.

These activities are a natural extension of the current professional activities in which soil scientists are involved. What they require is that soil scientists emphasize "the other side of the coin" or "the other part of the story.

SUMMARY

After many years in academia and in working with government and industry, several enduring points are obvious. One is that change is happening all the time -- in the political, economic, and social sectors; in public concern; and in technologies. Recognizing and accepting change is a cornerstone of responsible science and engineering and is the stimulation that helps cause progress.

Another is that as professionals we must deal with a circular, not a linear approach to waste management. With the linear treatment approach, one keeps adding processes and chemicals to treatment system to achieve better effluent quality. One result is increasing quantities of sludge for ultimate disposal. A circular approach considers control of the waste at the source, treatment process steps and ultimate disposal such as the use of land for waste management. Originally the raw products (food, fiber, ores) came from the soil and it is appropriate to return the wastes and residues to the soil in an environmentally sound manner.

A third is that we make progress in a professional area when there is active and stimulating cross-communication between disciplines. For example, in my own discipline, environmental engineering, progress was accelerated when information related to chemical process theory, microbiology and biochemistry, and physical chemistry was incorporated. Similar examples can be provided in other disciplines.

A fourth is that you are not part of the solution, you may be part of the problem. These seemingly random points relate to the topic of this paper -- the role of soil science in the utilization, treatment and disposal of waste -- since they identify the challenge and the opportunity for soil scientists.

Changes are occurring that allow soil scientists to "consider the other side of the coin" and use their talents to determine the soil assimilative capacity for wastes and chemicals. The future belongs to those who can foresee the changes and find opportunities where others only see problems. It is important that soil scientists see the opportunities and be part of the solution rather than be part of the problem.

It is equally important that soil scientists participate in meaningful cross-communication with other disciplines on these issues.

Pertinent disciplines include environmental engineering, microbiology, chemistry, and law. These disciplines will benefit from the knowledge soil scientists can transmit. In a like manner, soil science research and education will benefit by the information from these disciplines.

As a soil scientist, you are part of a special group of concerned professionals who are making the world a better place to live. It is appropriate that soil scientists focus their talent on a real societal need, the utilization treatment and disposal of wastes. The problems exist, the public interest in environmental matters continues and the challenges and opportunities are real. Five or so years hence, when soil scientists are asked if they recognized the environmental changes, if they attempted to find better ways to use but not abuse the assimilative capacity of soils for wastes and chemicals, and if they reduced the public and environmental risk, I trust that many more soil scientists will be able to declare an emphatic yes.

ACKNOWLEDGEMENTS

The contributions of Drs. K. W. Brown, J. T. Ling, and T. F. Yosie in the development of this paper are gratefully acknowledged. The summary of research needs (Table 5) was adapted from a paper prepared by the author for a Conference held in September 1984 on Long-Term Environmental Research and Development. The Conference was sponsored by the National Science Foundation and the Council on Environmental Quality.

Table 5. Pertinent waste management research soil scientists can undertake.

NITROGEN

-- refine the knowledge of nitrogen behavior in soils that receive waste, especially with regard to ammonia volatilization, nitrogen mineralization, and nitrification.

-- determine nitrogen cycling in forest lands and determine the assimilative capacity of such soils so that valid criteria can be developed for waste utilization on forest lands.

-- develop management techniques for increased nitrogen removal or loss under field conditions when wastes are applied to soils.

METALS

-- evaluate the availability of waste applied metals in soil to plants and animals.

-- identify soil parameters other than CEC to establish annual and cumulative limits of waste-applied metals.

Table 5, cont.

PATHOGENS

-- conduct field monitoring studies on the survival and transport of of waste-applied viruses under varying climatic, hydrologic, and waste conditions.

ORGANICS

-- develop screening and definitive models that can be used to determine the synthetic organic compounds of environmental concern when added to the soil and the soil assimilative capacity for such compounds.

-- develop reliable and quantitative predictions of synthetic organic movement in soils under field conditions.

-- determine the degradation rates and half-lives of synthetic organics in soils under varying environmental conditions.

-- determine the partition (sorption) coefficients of synthetic organics in soils having different clay and organic contents.

-- verify and extend the apparent relationship between K_{ow}, K_{oc}, and K_s for different synthetic organics and different soils.

-- determine the bioconcentration of synthetic organics by soil invertebrates, by plants and by animals when wastes are applied to the soil.

-- develop a data base on K_{oc}, K_{ow}, K_s for different synthetic organics and different soils.

-- investigate the phenomena of "facilitated" transport of organic chemicals in soils and determine the factors that cause such transport.

-- determine the rates of volatilization of synthetic organics when wastes are applied to land and the importance of this loss mechanism.

-- develop protocols that can estimate the treatability of synthetic organics in wastes that are applied to the soil.

MANAGERIAL

-- field verification of predictive design -- Much can be learned from field investigations (essentially post-mortums) of existing sites, especially those at which groundwater contamination or other environmental problems have occurred. Knowledge of the causes of previous environmental problems will improve future use of the soil as a waste management alternative, better protect the land/soil resource and allow the soil assimilative capacity to be better understood.

Table 5, cont.

-- determine operational practices that can be used to maximize and extend the use of land treatment and disposal sites -- Such sites can receive different wastes over a period of time. Methods that can be used at a site to avoid environmental and health problems when conditions change or assumptions prove invalid should be determined.

-- develop monitoring protocols and practices to assure that the soil assimilative capability is not exceeded -- Land treatment and disposal sites are dynamic systems. Adequate monitoring procedures are needed to identify when the soil assimilative capacity begins to be exceeded. The type and location of monitoring equipment, the parameters to be monitored and the frequency of monitoring need to be determined.

REFERENCES

Brady, N.C. 1984. The nature and properties of soils. 9th ed. Macmillan Publishing Co., New York.

Congressional Budget Office. 1985. Hazardous waste management: Recent changes and policy alternatives. Congress of the United States, U.S. Government Printing Office, Washington, D.C.

Environmental Protection Agency. 1983. Process Design Manual - Land Application of Municipal Sludge. EPA 625/1-83-016, Municipal Environmental Research Laboratory, Cincinnati, Ohio.

Environmental Protection Agency. 1985a. Report on the review of the Environmental Protection Agency's ground water research program. Science Advisory Board, July, Washington, D.C.

Environmental Protection Agency. 1985b. Permit writers guidance manual for the location of hazardous waste land treatment facilities -- criteria for location acceptability and existing applicable regulations. Draft, August, Office of Solid Wastes, Washington, D.C.

National Research Council. 1984. Groundwater contamination. Geophysics Study Committee, National Academy Press, Washington, D.C.

Pye, V.I., R. Patrick, and J. Quarles. 1983. Groundwater contamination in the United States. University of Pennsylvania Press, Philadelphia.

Wood, E.F., R.A. Ferrara, W.G. Gray, and G.F. Pinder. 1984. Groundwater contamination from hazardous wastes. Prentice-Hall, Inc., Englewood Cliffs, NJ.

SNOWBALL EARTH

SNOWBALL EARTH

THE STORY OF THE GREAT GLOBAL CATASTROPHE THAT SPAWNED LIFE AS WE KNOW IT

GABRIELLE WALKER

BLOOMSBURY

First published in Great Britain 2003

Bloomsbury Publishing Plc, 38 Soho Square, London W1D 3HB
Published in association with Crown Publishers, New York

A CIP catalogue record for this book is available from the British Library

ISBN 0 7475 6051 X Hardback

10 9 8 7 6 5 4 3 2 1

ISBN 0 7475 6433 7 Export Paperback

10 9 8 7 6 5 4 3 2 1

Printed by Clays Ltd, St Ives plc

For

Rosa, Helen and Damian

CONTENTS

ACKNOWLEDGMENTS

For the past two years or so, I have been a Snowball Earth groupie. Wherever the story was told or challenged, at conferences, field trips, lectures and campsites around the world, I appeared with my notebook and questions. Some of the researchers I spoke to are mentioned by name in the book, while others are not. All gave generously of their time and knowledge.

Thanks are due first of all to Paul Hoffman, who spent countless hours talking to me. He drove me around the Namibian desert and the Boston marathon course, welcomed me into his field camp, his home and his lab and sent me a continual stream of information by e-mail. He sought no influence over what I wrote.

Thanks also to Dan Schrag, who first introduced me to the Snowball. His ideas have been central to the theory, and he shared them with me repeatedly and very willingly.

Many researchers helped me explore the various Snowball sites around the world. Thanks to Tony Prave, Mark Abolins and Frank Corsetti for introducing me to the rocks of Death Valley in California. Ian Fairchild worked hard to set up the Snowball Earth workshop in Edinburgh, after his field trip to the Garvellachs was sadly scotched by the outbreak of foot-and-mouth disease. Joe Kirschvink invited me on his field trip to South Africa, and again to his lab in Caltech. (Thanks for Joe stories also go to Dave Evans, Ben Weiss, Kristine Nielson, Tim Raub, Curtis Pehl and many other members of Joe's irrepressible student family.)

ACKNOWLEDGMENTS

Dennis Thamm delayed his holiday for a day to show me around Mount Gunson mine. Jim Gehling and Linda Sohl drove with me to Pichi Richi Pass in South Australia, and Jim then gave me an unforgettable tour of the Ediacaran fossils in the Flinders Ranges. George Williams sent me copious papers covering his work on the South Australian deposits, in spite of his distaste for the Snowball idea. Kath Grey and Malcolm Walter talked to me about stromatolites before I headed to Shark Bay in Western Australia to see the living rocks there for myself. Guy Narbonne showed me the fabulous forms of the Mistaken Point assemblages, along with Bob Dalrymple and Jim Gehling. On the same trip Misha Fedonkin told me tales of the Russian White Sea, most notably while we sat in a Newfoundland late-night bar, and Bruce Runnegar gave me a whole new perspective of molecular clocks and pizza-shaped fossils. Thanks also to Ben Waggoner for his description of the flies at Arkhangel'sk.

Nick Christie-Blick told me his stories over dinner in Nevada and English tea in Newfoundland. Linda Sohl described the sheep incident and the noise that frightened kangaroos make, while the two of us were squashed together in the back of a van in Newfoundland. I shared a truck for three days with Martin Kennedy and Tony Prave in Death Valley and grilled them both further in the more genteel environs of Edinburgh.

Roland Pease enlisted me to present a BBC radio programme on the Snowball and provided me with copies of the interviews that we recorded. Mark Chandler explained about climate models and Jim Walker about seasons. Oliver Morton pointed out the significance of "outrageous hypotheses". Brian Harland tolerated three separate visits from me to his Cambridge home and office, the first when he had only just emerged from the hospital, having suffered a shattered knee. Ian Fairchild and Mike Hambrey told me more about Brian's early work, while fascinating insights into Paul Hoffman's background came from Terry Seward, Peter von Bitter, Erica Westbrook, Sam Bowring, Jay Kauffman and Dawn Sumner.

As for my own background, if it hadn't been for John Maddox and Laura Garwin, who took me on as a rookie at *Nature,* I would never have known about the wonderful world of earth science. Many thanks are also due to my colleagues at *New Scientist,* especially Alun Anderson, who hired me to do the best job in the world ("go out and look for stories that you find fascinating") and Jeremy Webb, who taught me more about writing and editing than either of us realize. The book began as a feature article for *New Scientist,* one of many that came from my insatiable thirst for stories about ice. The U.S. National Science Foundation sponsored me on two wonder-filled trips to Antarctica, and the Canadian Department of Fisheries and Oceans conveyed me to the Arctic Ocean to catch my first, gripping glimpse of what happens when the seas themselves freeze.

I wrote mainly in London at the British Library, where the staff is marvellous—particularly in Science 2 South. The town of Kirkcudbright in Scotland provided another haven for a while. I didn't manage much writing there, but did plenty of thinking and made twenty-four pounds of procrastination jam. In California I wrote in the beautiful Sausalito Public Library. And Niles Eldridge at the American Museum of Natural History in New York kindly granted me a desk for a month of writing there. I often stayed beyond closing time and felt a privileged shiver each night that I walked past the dinosaurs in the dark.

Many people read the manuscript in whole or in part and made helpful comments and criticisms. Thanks for this go to Robert Coontz, Richard Stone, Sarah Simpson, Michael Bender, John Vandecar, Helen Southworth, Rosa Malloy, Diane Jones, Jaron Lanier, Dominick McIntyre, Jeff Peterson, Edmund Southworth, Paul Hoffman, Doug Erwin and Jim Gehling. Any errors are, of course, my own. Particular thanks go to David Bodanis, who was there from the beginning with his encouragement and insights. He would save me a coveted booth at the British Library café and spend lovely long lunches there helping me find a way through my story. The edits he suggested were often painful, but invariably right. From David I also learned the two most important secrets for book writing: write every day, and always have a map.

My agent, Michael Carlisle, has given me unstinting support and encouragement. Alexandra Pringle, my editor at Bloomsbury, was full of enthusiasm from the beginning. I owe more thanks than I can say to Emily Loose, my editor at Crown, who encouraged me to write the stories I truly wanted to tell, with her intelligent suggestions on structure and her knack of teasing details out of me that I'd been tempted to gloss over. At one point, I read of a famous writer who challenged his editor about the importance of details over structure. God is in the details, the writer said. No, God is in the structure, his editor replied. When I asked Emily about this, she answered without hesitation. "God," she said firmly, "is in both."

I suppose I must have talked a lot about the Snowball over the past couple of years. Thanks for tolerance and encouragement go especially to Helen Southworth, Diane Martindale, Barbara Marte, John Vandecar, Jaron Lanier, Dominick McIntyre, Jonathan Renouf, Karl Ziemelis and Christine Russell. Thanks to Rachel Rycroft and

Barbara Nickson for setting me on what—to my surprise, though probably not to theirs—turned out to be the right path after all.

Above all, thanks to my family: Rosa, Helen, Damian, Ed and Christian. They are a rock that never cracks or crumbles, no matter how bad the storms.

Out of whose womb came the ice?
And the hoary frost of Heaven, who hath gendered it?
The waters are hid as with a stone
And the face of the deep is frozen.

—*Book of Job*

SNOWBALL EARTH

PROLOGUE

Boston, 20 April 1964

The weather was foul. Flurries of snow turned to wet sleet and slapped at Paul Hoffman's face as he warmed up on Hayden Rowe, just around the corner from the marathon starting line. The wind, he noticed, was gusting from the northeast. Not good. At six foot one, Paul made a large target for a headwind, and this one would be in his face for most of the twenty-six miles and three hundred and eighty-five yards to come.

Paul was tall for a distance runner, too tall, but at least he had the right sinewy build. He was lean and gangly and had just turned twenty-three. He'd always been athletic, but running suited him better than any sport he'd ever tried. Especially distance running. Paul liked his own company, and he didn't function particularly well in a team. He enjoyed the long solitary hours of training, and liked the feeling of pitting himself against the world.

Apart from his height, there wasn't much to distinguish him from the rest of the runners: his dark hair was neatly cropped, short back and sides, parted on the left; he had hollow cheeks and a long thin face. Though he was a star in his local athletics club, Paul was a nobody in these circles. He was anonymous, wedged in towards the back of the pack.

That was about to change.

On this chilly morning, Paul was feeling excited and nervous in equal measure. He had never run a marathon in his life, not even in training, yet he had decided to start at the top. Boston is the oldest city marathon in the world, the race of legends. If you are a long-distance runner, this is the race that matters. Normally you'd build up to it, try a few lower-profile marathons, and work out your pacing. But Paul Hoffman had never been particularly interested in preambles. Up in his native Canada, he had pushed himself for mile upon lonely mile over the hills of the Niagara escarpment, and afterwards, as he neatly inscribed the latest times and distances in his running log, he calculated how his efforts would translate into a full marathon. Every time, he was looking for one figure: six minutes a mile. That was the mental cut-off. If he could sustain that pace for the whole length of a marathon, he'd have a most respectable time of just under two hours forty. As soon as Paul figured he could make the pace, he sent in his entry forms.

Now, on this cold, filthy day, the doubts were kicking in. What if he had guessed wrong? What if he ran out of resources shy of the finish line? How hard could he push himself at the beginning? *He'd never run the distance.* Even the crowd was unsettling. Paul had never seen so many spectators at a race. It was Patriot's Day holiday—the day commemorating the "shot heard round the world" that started the American Revolution—and despite the nasty weather, the citizens of Massachusetts were out in voluble force.

The race began at noon. From the outset, the course sloped downhill in a long looping curve, begging Paul to stretch out his legs and increase his speed. He resisted the impulse. "Keep breathing," he told himself. "Concentrate. How am I feeling? How am I *really* feeling?" His legs seemed fresh, and he was pacing well.

PROLOGUE

The headwind was blowing obliquely from the right. Some of the runners ducked behind each other to block it. But Paul kept open road ahead of him. If he ran behind someone, his stride felt cramped. He'd rather face the wind.

The field stretched out, and the leading pack pulled out of sight. Paul made his way steadily up to the second group, which contained six other runners. They swept into the town of Natick to a wave of applause from the crowd. "You're looking good," someone bawled at him. "Keep going!" A slender clock tower stood to the left, its hands pointing to 12:56. Paul had been running now for precisely fifty-six minutes. Natick was around the ten-mile mark, and at six minutes a mile the time should have been closer to one o'clock. "We're going too fast," Paul shouted to another runner alongside him. Neither of them slowed down.

On the other side of town, the group entered a road lined with bare winter trees and screaming fans. Thousands of students had poured out of the venerable Wellesley college buildings on the right. This was the Wellesley gauntlet, a tradition Paul could well have done without. A wall of screams and shrieks tore at his ears, as he shook his head irritably and ploughed on.

Soon the volume of the shouts dropped and he could recognize words again from the spectators. "Paul! You're fifteenth!" Fifteenth place. The ranking stunned him. This was better than he'd ever imagined.

There were four runners immediately ahead of him now. That meant there were ten people out of sight in the leading group. Ten people. Ten trophies. If Paul could outstrip the rest of his group and overtake just one of the leaders, he'd be in the top ten. From his first-ever Boston Marathon he would take home a trophy.

PROLOGUE

The road suddenly plunged downhill. At the bottom was Newton Lower Falls, and the first of the uphills. Paul knew he was good with hills. He took his chance, stretched out his stride, and pushed the pace, leaving the other four behind.

In . . . out . . . in . . . out . . . breathing was getting harder. No more holding back now. This was flat out, as fast as he could force himself to run. The yelling was increasing in pitch. "You're eleventh!" "There's no one behind you!"

Six miles to go, and Paul was entering the unknown. He knew his body would soon run out of sugar reserves and start burning fat. Sugar gives instant energy. But fat is a long-term energy store, not designed for hard, sustained effort. When you start burning fat, you hit the wall. Your legs feel suddenly like lead and your arms can no longer pump. This happens at different times for different people, but it's usually something over two hours. That's why marathons are so hard. Paul had no idea how long his sugar reserves would last.

Another hill appeared up ahead, the site of one of the Boston Marathon's most famous dramas. In 1936, reigning champion Johnny Kelley caught up here with a young Narragansett Indian, Ellison Myers "Tarzan" Brown, who had been leading the field. Kelley patted Brown's shoulder as he reached him. "Good race," the gesture said. "Goodbye." Brown responded with an instant surge. It was textbook running. When someone has just expended supreme effort to catch up with you in a marathon, they need to coast along for a while. Take off again, and you break their spirit. Also, in this case, their heart. Brown went on to win the race while Kelley limped in fifth. And the rise was immediately dubbed "Heartbreak Hill".

Paul was on his own now, with no runners in sight ahead or

behind as he set off up Heartbreak. At the brow, he passed two exhausted runners who had dropped back from the leading group. They were toast. He felt a surge of elation. That means ninth place! I'm in ninth place!

His legs were still working, there were no signs of cramp. But his body was screaming at him. All his concentration was taken up with ignoring the pain.

He turned left on to Beacon Street, heading into town on the long, straight stretch before the final few twists. Far up ahead blazed a Citgo sign marking the twenty-five-mile point. He was striving to reach it but the sign seemed no closer. Please let it be over soon. No runners in sight. Just the crowds yelling. "Paul! You're ninth! Paul! Keep going!"

When he finally reached the Citgo sign, a paltry rise in the road nearly finished him off. Somehow he forced himself to run on and make the right turn on to Exeter Street. He flicked his eyes back as he turned and saw another runner coming up hard behind him. Ahead, he caught his first glimpse of the leading group, and he summoned one last huge effort. He began to gain on the leaders, but the runner behind was gaining on him. In the end, all ran out of road. Paul crossed the finish line in ninth place, to a roar like he'd never heard before.

In his first marathon, Paul had run *two-twenty-eight-oh-seven.* He'd hoped for something like two hours and forty, a good ambitious target. But he'd wildly exceeded that. His marathon time was world class, less than fourteen minutes short of the world record. If he wanted to, he suddenly realized, he could make this his life.

Paul's summer was already planned. He was training to be a field geologist, and intended to go to the Arctic. The Geological Survey of Canada had offered him a place on a field party at

Keewatin, west of Hudson Bay. He'd been out in the field before, but this was the first time he would be a senior assistant, allowed to do independent mapping of the geology. To join the Survey full-time was Paul's dream, and this summer job could be the first step.

But everything suddenly seemed different. This was an Olympic year, so should he try for the Canadian Olympic team? Should he abandon his fieldwork for the summer and go to the games in September? He was already too late to have a serious chance of winning the Tokyo Marathon, but perhaps he should put his geology on hold and devote the next four years to training for Mexico City in '68.

He drove back to Canada that night, stiff-legged and aching, statistics spinning in his head. He'd run two-twenty-eight-oh-seven, an average of five minutes thirty-nine per mile. The world record was a five-minutes-ten pace. With intense training, could he shave twenty-nine seconds per mile from his time? Could he ever hope to match the world record? If he went to the Olympics, would he be in contention? Was he a world-beater? Could he *win?*

Back and forth he went in his mind, but he always came up against the same wall. Five minutes ten a mile. That was fast. He could run at that pace for two miles, five miles even. But not ten. And if he couldn't run ten miles at that pace now, even intense training wouldn't help him sustain it for a full marathon. He could run a good race in Tokyo, but he wouldn't win. Whichever way Paul looked at it, he couldn't see a chance at the Olympic gold.

That made the decision for him. If he couldn't win at the Olympics, he wouldn't go. He'd find another route to glory. Come the summer, he was camping in the Canadian high Arctic, beginning the painstaking process of piecing together the stories hidden deep in its rocks.

ONE

FIRST FUMBLINGS

This is an extraordinary time to be alive. Look around you, take in the intricate complexities of life on Earth, and then consider this: complex life is a very recent invention. Our home planet spent most of its long history coated in nothing but simple, primordial slime. For billions of years, the only earthlings were made of goo.

Then, suddenly, everything changed. At one abrupt moment roughly 600 million years ago, something shook the Earth out of its complacency. From this came the beginnings of eyes, teeth, legs, wings, feathers, hair and brains. Every insect, every ape and antelope, every fish, bird and worm. Whatever triggered this new beginning was ultimately responsible for the existence of you and everyone you've ever known.

So what was it?

SNOWBALL EARTH

Paul Hoffman, part-time marathon-runner, full-time geologist, and obsessive, intense seeker of glory, thinks he knows. He believes he has finally struck science gold. Now a full professor at Harvard University and a world-renowned scientist, he has uncovered evidence for the biggest climate catastrophe the Earth has ever endured. And from that disaster, according to Paul, came a remarkable new redemption.

Shark Bay shows up from the air as a snag in the smooth coastline of Western Australia. Five hundred miles north of Perth, it lies just at the place where tropical and temperate zones rub shoulders. The area around the bay is a powerful reminder of how far we have come since primordial slime ruled the world. It is full of varied, vivid life.

This is one of the few places in the world where wild dolphins commune with humans, every day, regular as clockwork. At 7:00 A.M. each day a park ranger dressed in khaki uniform emerges from a wooden hut to focus a pair of binoculars on the horizon. Perhaps half an hour later, he'll spot the first dolphin fin. Somehow the word immediately spreads. Where there were only four or five people on the sandy beach, suddenly fifty or sixty appear.

Three harassed rangers do their best to marshal them into an orderly line. Everyone will get a chance to see the dolphins. No one will be permitted to touch them. No one must go more than knee-deep in the water. Another ranger deftly diverts the enormous wild white pelicans away from the beach by flicking on a water sprinkler. The birds flock around with gaping jaws—in this desert landscape, fresh water is irresistible.

The dolphins and their calves arrive. One of the rangers, a

wireless headset amplifying her voice, wades up and down in front of the spectators, introducing the dolphins ("This is Nicky and Nomad, Surprise and her calf Sparky") and reciting useful dolphin facts. The crowd surges into the water, like acolytes seeking a Jordanian baptism, their expressions beatific.

The dolphins are the crowd pullers—more than six hundred of them live here. But Shark Bay is also famous for the rest of its wildlife. The bay contains more than 2,600 tiger sharks, not to mention hammerheads and the occasional great white. The tigers show up in the water as streamlined shadows up to twelve feet long; often they are skulking beside patches of sea grass in the hope that dinner will emerge in the form of a blunt-nosed, lumbering grey dugong. Dugongs, or sea cows, are supposedly the creatures behind the mermaid myths, though I can't see it myself. They are too prosaic, placidly chewing away at the end of a "food trail", a line of clear water that they have cut, caterpillar-like, through the fuzzy green sea grass. They're exceptionally shy and rare, but here, among the largest and richest sea-grass meadows in the world, are a staggering ten thousand of them—tiger sharks notwithstanding.

Then there are sea snakes, green turtles and migrating humpback whales. And just a little to the north, where the tropics begin in earnest, lies Coral Bay—one of the world's top ten dive sites. *Come and dive the Navy pier! See more than 150 species of fish!* Also sea sponges and corals, brilliant purple flatworms, snails and lobsters and shrimp. And the vast, harmless whale sharks, the world's biggest fish. And on land there are wallabies and bettongs and bandicoots, emus and kangaroos and tiny, timid native mice.

There's everything in this region, from the wonderful to the plain weird. Evolution has been tweaking, adapting and inventing new forms of complex life for hundreds of millions of years, and here in Western Australia it surely shows.

But this is also a place where you can travel back in time, to see the other side of the evolutionary equation—the simplest, most primitive creatures of all. They come from the very first moments in the history of life, just after the dust from the Earth's creation had settled. And when these first fumblings of life appeared on Earth's surface, their form was exceedingly unprepossessing. Throughout oceans, ponds and pools, countless microscopic creatures huddled together in a primordial sludge. They coated the seafloor, and inched their way up shore with the tide; they clustered around steaming hot springs, and soaked up rays from the faint young sun. Dull green or brown, excreting a gloopy glue that bonded them together into mats, these creatures were little more than bags of soup. Each occupied a single cell. Each had barely mastered the rubrics of how to eat, grow and reproduce. They were like individual cottage industries in a world that had no interest in collaboration or specialization. They were as simple as life gets.

Although these primitive slime creatures have now been outcompeted in all but the most hostile environments, a few odd places still exist where you can experience the primeval Earth first-hand. The acidic hot springs of Yellowstone National Park, for instance, or Antarctica's frigid valleys. And here, in Western Australia, where countless microscopic, single-celled, supremely ancient creatures are making their meagre living in one small corner of Shark Bay: a shallow lagoon called Hamelin Pool. The pool's water doesn't mix much with the rest of the bay, and it's twice as salty as normal. Since few modern marine animals will tolerate so much salt, this is one of the last refuges of ancient slime.

THE SIGN pointing to Hamelin Pool is easy to miss, even on the desolate road running south from Monkey Mia. On the second pass I finally spot it, turn left, and bump along a sand track with scrubby bush to either side. For this first visit I avoid the restored telegraph station with its tiny museum and tea shop, and head straight for the beach. I want to experience primordial Earth without a guide.

There's an empty car park of white sand, with wattles and low-slung saltbushes clinging to the surrounding dunes, and a path threading through the bushes towards the sea. Though I've come to find the world's simplest creatures, the complexities of life are everywhere. From one of the bushes a chiming wedgebill incessantly reiterates its five-note melody. From another, a grey-crested pigeon regards me unblinkingly. The shells of the beach crunch underfoot; they are tiny, bone-white, and perfectly formed, and the bivalves that grew them are eons of evolution ahead of the simple creatures that I'm seeking. I step on to the boardwalk, which stretches like a pier out into the water. Each weathered plank of wood contains row upon row of cells that once collaborated in a large, complex organism. Signs on all sides show pictures of the slime creatures with smiley faces and cheery explanations of their origin. Flies buzz infuriatingly around my head, landing on my face to drink from the corners of my eyes. Black swifts swoop between the handrails, and butterflies the colour of honey, with white and black tips to their wings. Time travel is harder than it looks. The modern world is right here even in Hamelin Pool, and it's stubbornly refusing to leave.

I retreat to the telegraph station to plead with the ranger for permission to leave the boardwalk and wade out into the pool. He hesitates and then relents. "Go along the beach to the left," he says. "Don't step on the mats. Be careful." The mats he's talking about are one of the signs of primeval Earth. They are slimy conglomerates of ancient cyanobacteria, and they grow painfully slowly. At the beginning of the last century, horse-drawn wagons were backed into the sea over the mats, to unload boat cargo. A hundred years later the tracks they left are still visible as bare patches in the thin black sludge. An injudicious footprint here will last a long time. I promise to watch my step.

I return to the beach and this time walk carefully towards the water's edge. More striking than the ubiquitous patches of sludgy, foul-smelling bacterial mats are the "living rocks" in between. These strange denizens of Slimeworld are everywhere, an army of misshapen black cabbage heads marching into the sea.

The ones highest up the shore are now nothing more than dead grey domes of rock, shaped like clubs, perhaps a foot tall. They once bore microbial mats on their surfaces, but these have long since shrivelled, abandoned by the receding water. Closer to the Pool's edge the domes are coated with black stippling that will turn to dull olive green when the tide washes over them. Most of the stromatolites, though, lie in the water, stretching out as far as I can see. Between them the sand is draped with black-green mats of slime, and chequered with irregular patterns of sunlight as the waves ripple overhead. I wade up to my knees among these strange formations, basking in the sunshine. There is nobody else in sight.

The living rocks of Slimeworld are called "stromatolites", a word that comes from the Greek meaning "bed of rock". Though the interior of the stromatolites is plain, hard rock, their outer layers are spongy to the touch. Here on the surface is where the ancient microbes live. They're sun-worshippers: by day they draw themselves up to their full filamentous height—perhaps a thousandth of an inch—soak up the sun, and make their food; by night they lie back down again. The water that surrounds them is filled with fine sand and sediment stirred up by the waves. Gradually this sand rains down on the organisms, and each night's bed is a fresh layer of incipient rock. The stromatolites are inadvertent building sites; the sticky ooze that the organisms extrude acts as mortar and the sand acts as bricks. Every day, as the microbes worm their way outward, another thin layer of rock is laid down beneath them.[1]

It's a slow process. Stromatolites grow just a fraction of an inch each year. The ones in Hamelin Pool are hundreds of years old and would be astonishing feats of engineering, had they been created by design. For these micro-organisms to erect a stromatolite three feet high is like humans building something that reaches hundreds of miles into the sky, and scrapes the edges of space. I wade a hundred yards, two hundred yards offshore, and the slope is still so gentle that the deepening is barely perceptible. Mercifully, the flies and butterflies have dropped back, and the birdsong is out of earshot. At last I begin to feel that I've travelled back to life's earliest days.

Hamelin Pool's mats and stromatolites look utterly alien, but they were once ubiquitous. Time was, this scene of stromatolites and stippled microbial mats would have greeted you everywhere you went. Forget dolphins and wallabies. This is how the Earth looked for nearly three and a half billion years. The imprints of the stromatolites and their mats show up still wherever sufficiently ancient rocks poke through to the Earth's surface. I've seen them in Namibia, in South Africa, in Australia and California. They are sometimes dome-shaped like these in Shark Bay, sometimes cones, sometimes branching like corals. There are places where you can walk among ancient petrified stromatolite reefs, rest your feet on their stone cabbage heads, and see where they have been sliced through to reveal rings of petrified growth. And you can run your fingers over fossilized mats, which give rock surfaces the unexpected texture of elephant skin. This slime used to be everywhere, and now it's almost nowhere.

How did we get from there to here? This is at once a simpler and more powerful question than it seems. Of course, life took many separate evolutionary steps on its way from stromatolites to wallabies. It had to invent eyes and legs and fur and feet, and everything else that distinguishes marsupials from slime. But there was one particular step that was more important than all the others, one that made all the difference.

The step was this: learning to make an organism not from just one cell, but from many. Though the first microbes on Earth were woefully unsophisticated, they did gradually learn new tricks to exploit the planet's many niches. But they all still had one thing in common. Each individual creature was packaged in its own tiny sac, a single microscopic cell. Then at some particular point in Earth's history, everything changed. One cell split into two, then four. From that time onwards, organisms could be cooperative, and above all their cells could specialize. There could be eye cells and skin cells, cells to make up organs and tissues and limbs.

For life, this was the industrial revolution. Forget the old cottage industries. Now you could have factories with production lines. Parcelling out tasks and specializing is always more efficient than trying to do everything yourself. And there are some things, wallabies for instance, that can only be made with a massive collaborative effort.

In just the same way, when organisms developed the ability to become multicellular, they gained a world of possibilities. Your body is made up of trillions of cells. Every hair is packed with them. You shed skin cells whenever you move. Your blood cells carry energy around your body, to feed the organs made up of still more cells. This multiple identity is the one criterion that's vital for any complex creation. Every dolphin and dugong, every shark, pelican and wombat depends for its existence on that crucial leap from one cell to many. This was the point when simple slime yielded its pre-eminence to the complex creations that heaved their way out of the sludge and started their march towards modernity.

But why did it take so long? The Slimeworld lasted for almost the whole of Earth history. Let's put in some numbers. Our planet had been around for 4 billion years before the first complex earthlings emerged from the ooze. That's nearly 90 per cent of Earth's lifetime.

Four billion years is an insane amount of time, almost impossible to contemplate. There have been many attempts to capture this spread of time in ways that we can comprehend. If the history of life on Earth were crammed into a year, slime would have ruled through spring, summer and autumn, continuing well past Halloween into the beginnings of winter. If it were squeezed into the six days of creation, slime ruled until six o'clock on Saturday morning. If it stretched over a marathon course, slime would have led the field past the twenty-three-mile mark.[2]

But my favourite image is this one, borrowed from John McPhee.[3] Stretch your arms out wide to encompass all the time on Earth. Let's say that time runs from left to right, so Earth was born at the tip of the middle finger on your left hand. Slime arose just before your left elbow and ruled for the remaining length of your left arm, across to the right, past your right shoulder, your right elbow, on down your forearm, and eventually ceded somewhere around your right wrist. For sheer Earth-gripping longevity, nothing else comes close. The dinosaurs reigned for barely a finger's length. And a judicious swipe of a nail file on the middle finger of your right hand would wipe out the whole of human history.

Stephen Jay Gould set the discovery of these vast stretches of Earth time in a long line of findings that put humans firmly in our place.[4] Galileo, said Gould, taught us that the Earth isn't the centre of the universe. Darwin, that we're just another kind of animal. Freud, that we're not even aware of most of the things going on in our own heads. And geologists have now discovered that the Earth reached late middle age before we were so much as a glimmer in its eye.

Though we humans are certainly complex, also clever, perhaps even the highest form of life that Earth has so far produced, we're nothing like the most natural earthlings. Measured in units of staying power, Earth's first, most primitive experiment in life was also its best. With simple individual cells, nothing complex, nothing flashy, each creature out for itself, life had found a supremely winning formula. Why should it ever change?

That's the question that has plagued complex, clever, thinking, adaptable humans since they first uncovered this bizarre history of life. Earth looked set to stay locked in slime for ever. Why did complex life appear at all, and why did it wait to emerge until that one point in time, just a few hundred million years ago, nearly at the end of the marathon, somewhere near your right wrist, late in the Earth's middle age?

To answer this, Paul Hoffman has seized on an idea that was first proposed sixty years ago, and was then dropped, half-heartedly resurrected, and dropped again several times over the intervening years. There's nothing half-hearted, however, about the resurrection Paul has now effected. He's marshalled new evidence, restored and amalgamated old ideas, and employed fierce argument to persuade the people around him. According to Paul, life's richness, diversity and sheer overwhelming complexity arose from a mighty catastrophe. It's called the "Snowball Earth".

First came the ice. It crept from its strongholds at the North and South Poles, freezing the surface of the ocean, spreading gradually over the Earth's surface. A blue planet inexorably made white.

Individual crystals of ice first appeared in the sea like tiny floating snowflakes. They were smashed together by wind and waves, their fragile arms broken and their debris turning the seawater into a greasy slick. The surface thickened and froze into a thin transparent layer. As this layer thickened, it grew grey and then opaque from salt and air bubbles that filled its inner voids. In some places the greasy ice congealed into large round pancakes, with raised edges like giant lily pads, where they bumped and smashed against each other. And, a nice touch this, the fresh young sea ice grew a coating of frost flowers, each one the size and shape of an edelweiss.

Sea ice bends. Unlike freshwater ice, which can shatter like glass, the ice that forms first on the surface of the sea is elastic. When you try to walk on it, your legs unexpectedly buckle. But as it thickens it becomes reassuringly firm, like solid rock.

Though sea ice is grey when it first forms, it whitens year by year as its brine drains back into the sea. Even grey young ice is often dusted with white snow. But a frozen ocean is far from monochrome. Gashes of open seawater, created as the pack ice is ripped apart by wind and weather, expose the deep turquoise roots of the floating sea ice. And the dark ocean reflects in the clouds, streaking them the colour of a bruise. "Water sky" this is called, and polar sailors have long used it as a clue for where to point their ship next as they navigate perilously through the pack.

Where waterways have frozen over, the ice is smooth and level. Where the edges of an old water wound have been cauterized together again, untidy piles of ice blocks are an astonishing bright blue. Ice cracks suddenly like a whip. Sometimes pack ice groans and creaks as the wind crams floes together or prepares to break them open. But for the most part, the frozen polar oceans are shrouded in silence—eerie and absolute. There is no scene more alien on Earth.[5]

For perhaps a few thousand years, the white menace stole unheeded towards the equator. Earth's primitive life-forms had neither the eyes to see the encroaching ice nor the wit to fear it. Most of them lived their dull lives in a band around the Earth's waist, and as the ice advanced steadily from the far north and south, they bathed unconcerned in the warmth of their shallow, equatorial seas.

An occasional storm might have whipped up waves near the

shore. Perhaps the surf tore at the rubbery microbial mats that coated the seafloor and sprayed nearby rocks with scraps like soggy chicken skin. Stromatolites built up their stone reefs, layer by microscopic layer. Geysers blew. Rain fell. The sun shone again. There was no hint of the devastation to come.

But when the ice reached the tropics, its slow creep became a sprint. In a matter of decades, it engulfed the tropical oceans and headed for the equator.

Ice spread out from shallow bays and grew first a skin, then a carapace over the oceans. It clung to the beaches and scraped the mats on the seafloor. In some places this shell was still thin enough to crack and seal again. In others it was thousands of feet deep.

For a few hundred, perhaps even a few thousand years after the oceans were capped with ice, the land remained bare. But ice began to accumulate, gradually, in the thin air of mountain ranges, creating great frozen rivers that flowed down to fill the surrounding valleys. In the end, the whiteout was complete. Earth's surface looked like the frigid wasteland of Mars, or one of Jupiter's ice-covered moons. Sunlight bounced off the bright surface and was dazzled back into space. The mercury hit a staggering minus 40 degrees C. (Or it would have, except that at those temperatures mercury itself would have frozen.) There was little wind or weather of any kind. Clouds, by and large, disappeared, save perhaps for tiny ice crystals high in the atmosphere, which scattered sunsets into strange, lurid colours, blue and green, rimmed with vibrant pink. No rain fell and little snow. Every day brought silent, unremitting cold.

The Snowball wasn't just another humdrum old "ice age" like

those from more recent eras. The events we call ice ages were merely brief cold blips in an otherwise fairly comfortable world. There was ice in New York then, but none in Mexico. If you were in northern Europe during one of those ice ages, you shivered. But if you were in the tropics, you scarcely noticed.

Instead, Paul's Snowball was the coldest, most dramatic, most severe shock the Earth has ever experienced. It was the worst catastrophe in history. For perhaps a hundred thousand centuries, Earth was a frozen white ball, desolate and all but lifeless.

To the microbes, the Snowball must have seemed like the end of the world. Some survived, of course—they must have, or we wouldn't still see them today. Perhaps they huddled for warmth around undersea volcanoes. They might have survived near hot springs, or found fissures and cracks in the sea ice where the sun's rays could slip through. But for many, perhaps most, the Snowball was disastrous.

Eventually the Snowball empire began to founder. Volcanic gases gradually built up in the atmosphere, trapping the sun's heat and turning the air into a furnace. After millions of years of stasis, the ice finally succumbed, melting in a rapid burst of perhaps just a few centuries. Temperatures now soared to 40 degrees C. Intense hurricanes flooded the surface with acid rain. Oceans frothed and bubbled, and rocks dissolved like baking powder. Earth had leapt out of the freezer and into the fire.

There was at least one more of these Snowball-inferno lurches, and there may have been as many as four. But at the end of them all, after the last of the Snowballs and its attendant hothouse finally faded, some 600 million years ago, came the most important moment in the history of evolution. The rocks that appeared immediately afterwards bear fossils showing the first stirrings of complex life. Out of the ice and the fire that followed had come the complexity that we see around us today.

THIS IS Paul Hoffman's vision, and he is enchanted by it. Most other geologists are horrified. Accept his story, they say, and you have to reconsider everything you thought you knew about the workings of the world. Geologists are taught from an early age that the Earth is a slow and steady place. The past looked pretty much like the present. Change happens only very slowly, nothing is terribly extreme. True, there have been a few occasions where they have been forced to admit, somewhat grudgingly, that this picture falls short. The idea that an asteroid came from space to wipe out the dinosaurs was once derided, but is now widely accepted. OK, the argument goes, so the occasional extraterrestrial calamity can rock the Earthly boat. But broadly speaking, the geological picture of Earth's history is a settled, safe, comfortable one.

Compare that to Paul's picture of the Snowball. A global freeze. A planet that looked more like Mars than home. Ice *everywhere.* And then a sudden lurch from the coldest to the hottest that the Earth has ever been. Every way you look at it, his Snowball stretches the bounds of decency. It's as extreme and catastrophic as they come.

Small wonder, then, that the Snowball has become the most hotly contested theory in earth science today. Paul Hoffman, though, is resolute. He is the chief champion of the theory. By argument, evidence, and brute force of personality, he is determined to win over the unbelievers.

Paul is an obsessive man espousing an extreme theory. If he is proved right, we'll have learned something important about where we all ultimately come from. But there's a darker side to Paul's theory. He has uncovered behaviour in our planet that's unsettling in the extreme. If his vision is true, Earth can experience sudden lurches in climate that are more violent, and deadly, than anyone had ever imagined, and such catastrophes may well happen again.

TWO

THE SHELTERING DESERT

In the autumn of 1994, Paul Hoffman was back in Boston, nearly thirty years after he'd won his marathon trophy there. Though he'd continued to run marathons in his spare time, Paul had spent most of the intervening years sticking to his geological guns. He had acquired that most essential of accoutrements for the male geologist, a beard. His hair was more unruly these days. Thick and white like a goat's, it sprang up in surprise from a high forehead that was lined from too many days spent outdoors. Now fifty-three years old, he wore a pair of round wire glasses and was widely considered one of the top geologists of his generation. He had been elected a member of the prestigious National Academy of Sciences, had won countless awards, and written classic academic papers. He was back in Boston not as a callow nobody running the marathon, but as a full professor in the Department of Earth Sciences at Harvard University.

Paul had made it, then, into the ranks of world-class science. But still he wasn't satisfied. A cloud hung over him that he was desperate to shake off. After thirty years of fieldwork in the high Canadian Arctic, Paul had been abruptly forced out. He had picked a fight with the head of his home institution, the Geological Survey of Canada, and paid a high scientific price. A high emotional price, too. He had felt more at home working in the far north than anywhere else in his life, and now he was banned from returning there. When the blow first struck, he felt humiliated and lost. Now, two years later, he was arriving at Harvard with as much to prove as ever.

About this time, Paul's alma mater, McMaster University, contacted him as part of a survey of distinguished alumni. They asked what he'd like to be remembered for, and Paul replied without hesitation. "Something I haven't done yet," he said.

PAUL HAS been fascinated by minerals since he was nine years old. Next door to his elementary school in Toronto was the Royal Ontario Museum, and as a child he used to haunt the place. On Saturday mornings the museum held field naturalist classes, and Paul signed up with enthusiasm. The first year, the class studied butterflies. The second, fossils. But in the third year, Paul found himself studying minerals. They were perfect. It suited the atavistic urges of a young boy to acquire sparkling, shiny crystals of hornblende, quartz and fluorite, to hoard them and examine them, to try to obtain one of *everything*.

There were plenty of samples to be discovered in the rocks around Toronto, and always the chance of a new crystal, a rare crystal, a bigger, better sample than one Paul already had. And then the bargaining would begin. What did you find? What do you have that I haven't got? What have I got that you're dying for? Perhaps I'll trade you.

Paul's life became a treasure hunt. At first his mother would drive him and fellow members of the mineralogy club to their sites, but as they grew older they went unaccompanied. They scoured the public records in Toronto for locations of old, abandoned mines, then set off on camping trips to find them. Or they travelled to existing mines, where they charmed the workers and won permission to poke around the waste dumps.

Once they visited a quartz mine where a single large cavern was crammed with spectacular crystals, some milky, some as clear as water, some as long as your arm, and all gleaming like ice. From a mine in Cobalt, northern Ontario ("the town that silver built"), Paul brought back thin plates of native silver embedded in bright pink cobalt salts. During one morning of careful searching you could find ten or twenty ounces of silver among the rocks that had been tossed into the waste there. In the dump at a uranium mine, he found black cubes of uraninite set in a mass of pink and white calcite; also chunks of purple fluorite housing spectacular yellow needles of uranophane—calcium uranium silicate. Both of these uranium minerals are radioactive. Paul and his friends bought cheap Geiger counters from a scientific supply store in Toronto. They held the Geiger counters up to their finds and were thrilled by the staccato crackles that emerged. They weren't afraid of the radioactivity. As long as you're careful, as long as you don't spread the dust on your toast in the morning, you'll be fine.

Every fine weekend, Paul would head off to another mineral

site. He loved being outdoors. Continuing to collect minerals avidly throughout his teenage years, he evinced no interest in dating girls or following fashion. Instead he traded samples with the museum mineralogists, and swapped stories with the scientists at the University of Toronto. In minerals, Paul thought he'd found his métier.

But in 1961, during his freshman year at McMaster University in Hamilton, they turned out to be a major disappointment. The study of minerals mostly happened in a lab, it seemed, where you spent your day leaning over a desk, measuring the distances between spots on photographic film. Paul wanted to be outdoors, back on a treasure hunt. He wandered from the Mineralogy Department along to Geology, which sounded like the next best thing. Were there any opportunities for the summer? They sent him to the Ontario Department of Mines in Toronto, where the austere director, J. E. Thompson, looked him over and decided to take him on board. "Take the overnight train to Sioux Lookout on May tenth," Thompson told him. "Bring a good pair of boots."

Sioux Lookout was a tiny town surrounded by the ribbon lakes and dense forests of northern Ontario. Paul took both the train and the boots and soon found himself on a bush flight out into the wilderness. The lake shores were gorgeous, but the interior was a treacherous, forbidding place of dense bush and swampy ground. To reach the outcrops of rock hidden among the trees, you had to take a compass bearing and then fight your way through the undergrowth, counting paces to see how far you had travelled. The four members of the field party lived out of two canoes. Each morning they struck camp, stowed their tents and gear in the canoes, and then paddled on to a new site.

It was a bad year for forest fires, and sometimes the smoke grew so thick that the researchers could scarcely breathe. And then, several weeks into the trip, David Rogers, the party leader, felt a crippling pain in his gut, which he quickly realized must be appendicitis. There was no point in waiting for the weekly supply plane. Paul stayed with Rogers while the other two paddled north through the night for help. Eventually their route intersected with a railway line, and they managed to flag down a train. An intercontinental train takes a long time to stop, even after the driver has seen two young men frantically waving from the bush, and has slammed on his brakes. The driver and his precious radio finally came to a halt several miles down the track. A hasty call summoned a bush plane to pick up the patient and whisk him away to a hospital. Geologists are tough. Three days later, sans appendix, David Rogers was back in his canoe, in the field.

That summer, Paul spent more than four months canoeing and traversing and mapping the rocks. This, he felt, was the life. He'd camped before, plenty of times. He'd even been in northern Ontario with his parents, on holiday. But this was camping for *work*. Every day there was a new site to explore, every day a new set of rapids to run. Paul Hoffman was hooked. Geology seemed to be exactly what he was looking for, and when the next summer came around, Paul was eager for more. But this time he wanted to go somewhere different. Sioux Lookout was great in its way, but it wasn't the true North. Paul was hankering after remoteness. What he really wanted was the Arctic.

Scratch a geologist and, under their skin, almost invariably, you'll find a romantic. They will often be gruff about the landscape they work in. They are usually matter of fact about the rocks and how they interconnect. But try asking why they've chosen to spend their lives working in this particular place or on that particular terrain, and that's when the stories start to slip out.

When Paul was eight years old, just before he started with his mineral obsession, he heard a CBC radio drama about the last trip of the Arctic explorer John Hornby, an eccentric Englishman who had lived precariously in the Canadian Northwest Territories during the early 1900s. Hornby was quixotic, even by the extraordinary standards of the place and time. His eyes were an intense, piercing blue, and he refused—for luck—to travel with any man whose eyes were brown. Though his hair and beard were wild, he spoke with a soft, expensively educated accent. He was barely five feet tall, but his toughness was legendary. Once, so the stories go, he trotted for fifty miles beside a horse. Another time he ran a hundred miles in twenty-four hours, for a bet.

Hornby used to boast that all he needed, for a trip of any length, was a rifle, a fishnet and a bag of flour. He would take absurd risks, venturing into the barren lands again and again with scarcely any provisions. Finally, in 1926, he pushed the odds too far. He decided to spend winter in the remote Thelon River valley, a few hundred miles south of the Arctic Circle. With him he took his eighteen-year-old nephew, Edgar Christian, and an Edmonton man, Harold Adlard. Sometimes rowing, sometimes portaging, the party would take their hefty, square-sterned canoe across Great Slave Lake and eastwards to the Thelon River, where they would build themselves a log hut for the winter.[1]

The timing was crucial. This far north, winter would be excessively harsh. By November, thick ice would coat the lakes and rivers, and deep snow would smother the hut and its environs. After that, there would be little wildlife at large, and few opportunities for hunting. Yet Hornby, true to form, was taking few provisions. His entire plan relied on gathering meat from migrating caribou as they passed the Thelon River on their way south for the winter. If he missed the caribou, all would be lost.

Hornby, however, seemed to feel no urgency. He left several notes en route, stuffed into tins and marked by stone cairns. "Travelling slowly," one reported laconically. "Flies bad." And in another, left around 5 August: "Owing to bad weather and laziness, travelling slowly. One big migration of caribou passed."[2] By the time Hornby's party reached their wintering site sometime in October, most of the caribou had gone.

The party's attempts to stave off hunger grew increasingly desperate. They managed to trap a fox here, a hare there, sometimes a few scrawny Arctic ptarmigans. By early December, Hornby was reduced to digging up frozen blood from the site of an old caribou kill. It made, Christian wrote in his diary, "an excellent snack". Every day the party set traps. Every day now the traps were empty. "Got nothing but damned cold," Christian wrote on 18 February. And on 23 February, "this game of going without grub is Hell". Soon they were pounding old bones to squeeze out any sustenance, and scraping hides for fragments of meat.

By now all three were far too weak from hunger to attempt escape. They were hundreds of icy miles from the nearest humans, and in their poor condition, that distance might as well have been thousands. Edgar Christian remained touchingly optimistic in his diary. "We can keep on till caribou come North and then what feasting we can have," he wrote on 26 March. But Hornby died on 17 April, and Adlard on 3 May. Christian himself finally succumbed to hunger at the beginning of June, just days before the caribou were due to return. Two years later the Royal Canadian

Mounted Police discovered Christian's diary and the three bodies. Christian had laid Hornby and Adlard side by side and covered them as best he could. His own body had fallen from its bunk and broken on the floor. The silver watch in the breast pocket of his shirt had stopped at 6:45.

When Paul heard a dramatized radio production of Christian's diary, this extraordinary story struck a chord. He had just seen John Mills's portrayal of the doomed explorer Robert Scott in the stirring adventure movie *Scott of the Antarctic.* Scott embarked with a small band of followers for a daring adventure at the opposite end of the world. He had hoped to conquer the South Pole for England, but his expedition, too, was disastrous. When he and his men arrived at the Pole in January 1909, they were horrified to see a Norwegian flag already flying there, courtesy of their arch-rival, Roald Amundsen. The air at the Pole danced with tiny crystals of ice, "diamond dust", which cast bright rings of light in halos around the sun. But Scott's mood was black. "Great God!" he wrote in his diary. "This is an awful place and terrible enough for us to have laboured to it without the reward of priority."[3]

There was worse to come. On the return journey, Scott and his men gradually succumbed to the appalling weather. First one perished; then another famously walked out to his death in a blizzard. Finally, Scott and his remaining two companions starved to death, trapped in their tents by another blizzard, just eleven miles from a food depot.[4]

Like young Edgar Christian, the polar adventurers left diaries and letters from which *Scott of the Antarctic* quoted liberally. Scott's was particularly rousing. "Had we lived," John Mills's Scott intoned stentoriously at the end of the movie, "I should have had a tale to tell of the hardihood, endurance and courage of my companions which would have stirred the heart of every Englishman."

Scott and Hornby embodied the tragic heroes of fairy tales. Something about their stories tugged at the young Paul Hoffman. The two became confused in his head. He pictured Scott vainly seeking out caribou while icebergs crashed around him in the Canadian lakes. His eight-year-old mind retained only the haziest of details from these tales, but the romance of the planet's icy extremes took firm hold. One day, he'd decided, he would go north for himself.

So, in his sophomore year at McMaster, he began to ask around. The Arctic, he was saying. How can I get to the Arctic? For that, it turned out, he needed to approach the Geological Survey of Canada, an august, government-funded institution that sends geologists prying and poking at rocks in the remotest, most inaccessible locations. Paul took himself off to Ottawa, to sign up. Two months later he had won a place at a Survey field camp on the borders of Great Slave Lake in the Northwest Territories, just a short canoe ride from where John Hornby had suffered that last bitter winter.

Paul nearly blew it. Only three days after he arrived, he was horsing around, practising shot-put and discus using the rocks from thereabouts. One false move later, he had sent a discus of Yellowknife slate slicing through the tent belonging to the field party's leader. Fortunately, its owner was not yet in residence. A hasty but meticulous stitching job and a surreptitious switching of the tents got Paul off the hook and allowed him to stay for the rest of the season. That was all it took. The Arctic drew Paul as nothing ever had before. He was to return almost every year for the next three decades, until his contact with the Arctic—and the Survey—was unexpectedly and bitterly severed.

In some ways the appeal of the North was immediate and obvious. Fieldwork in the high Arctic had the three things that mattered to Paul more than anything. His work was an intellectual pursuit, it involved strenuous physical labour, and it happened in a place that was as beautiful as Paul had ever experienced. But the wildlife, or more particularly the insect life, would have dampened the enthusiasm of most. True, there were three magnificent weeks in June when the ice was still breaking up on the lakes and the place was heaven on Earth. These were warm, sunny, peaceful days when anything seemed possible. But then the flies came.

The mosquitoes appear first. They are big and noisy and desperately annoying. They insert hypodermic needles into your skin, and the moment they bite, you can feel it. A few weeks after the mosquitoes come the black flies, smaller but more devious. They are master miners. They carve out a cavity in your skin, injecting you first with anaesthetic to prevent you noticing. The anaesthetic they use is a nerve poison. If you get a few hundred black fly bites quickly enough—within an hour, say—you begin to feel the effects of the toxin. You feel nauseous, can't concentrate, and lose your bearings. You struggle to hold a line of argument in your head.

Spend long enough in the Arctic, and you will develop your own definition of a bad fly day. According to Paul, a bad fly day is when you can hit your arm once and find a hundred corpses in your hand. On bad fly days, mosquitoes whirr and whine around your head in a dense claustrophobic cloud. Black flies crawl everywhere on your clothes and skin, and into every crevice. To avoid inhaling them, you have to breathe through your teeth. If you run your hand through your hair, it comes back greasy and bloody. At the end of a bad fly day, you empty your pockets of globs of dead and half-dead flies. They have crept up your wrist, down your neck, under your belt, down the top of your boots. On bad fly days you soak yourself with industrial-strength Repex, the repellent of choice. Repex doesn't keep the flies away, but it stops them from biting. It lasts two to three hours. On bad fly days you don't have to be reminded to reapply.

In the Canadian Arctic, between the fine few weeks of June and the return of winter in late August, every day that is not freezing cold or blasted with wind is a bad fly day.

And then there are the bears. The first time Paul encountered a grizzly, he had been out all day on a long traverse, walking twenty or thirty miles. He was heading back to camp around 11:00 P.M., walking north fast, straight into the setting sun, his baseball cap pulled down low to block the dazzling sunlight. Suddenly the bear appeared under the brim of the cap, coming for him at full speed. The animal was backlit, its body in shadow but surrounded by sunlight. The ends of its hair shone silver, and foam and saliva were spewing from its mouth in glistening arcs. All Paul could see was a bear-shaped halo of light and foam.

Man and bear stopped in their tracks and stared. Paul remembers thinking, *Stand still. Don't move. If it charges, fall on your right side and protect your right hand.* Paul's right hand was precious, his drafting hand, the one he used to draw his meticulous geological maps. But the bear didn't charge. Paul made the slightest movement to the right, and the bear turned and raced off to the left, to where her two cubs were waiting on a small knoll. She cuffed her cubs and hustled them away. A few seconds later the foaming, glistening vision had vanished.

After that, Paul kept a pair of running shoes beside his sleeping bag while he was in his tent at night. If something pawed at the side of the tent, Paul would throw a shoe to shock it, and then rush out to scare it away. There wasn't much danger if you were awake and could frighten the bear off. The real trouble was if a bear came to the camp while you were away. A black bear or a grizzly could tear a camp apart trying to find food, and that would be disastrous. If your tent was destroyed, you were at the mercy of the flies. All day long you were fighting flies. You had to have a refuge from them at night, or you'd go mad.

The only way out was to shoot any bears that persisted in returning to the camp. Paul had to shoot three bears over the years—two black and one grizzly. He hated every time. He was shocked how much red-blooded damage you could do with one little squeeze of a trigger. The grizzly was the worst. When it came into camp, it made an angry beeline for the helicopter. It had probably been buzzed by some idiot joy-riders and was out for revenge. That was hardly the bear's fault, but the helicopter was too precious to risk. As Paul loaded the rifle, he felt sick. Afterwards the same helicopter slung the grizzly's body back out into the bush.

Paul knew that he couldn't afford to let the flies or bears get to him psychologically, so he never did. Gradually he got used to them both. After a few weeks of building up tolerance, he found that new fly bites didn't swell so much. And if you could ever see beyond the buzzing, whirring, whining clouds that enveloped you, the landscape was vast, empty and gorgeous. There were no trees to block the skyline, just mile upon mile of rounded rocks and the boggy Arctic vegetation known as muskeg. Air and light both had a clarity that Paul had never experienced before. During the fleeting summer months of his field season, when the outer vestiges of winter melted briefly, there were ponds and pools and lakes of water everywhere. The ground squelched underfoot. The only sound came from the nesting birds, loudly defending their soggy territories and raising their young. Even they quietened down at night, although the midnight sun still shone then. All day long the sun was low on the horizon, and at midnight it reached its lowest point. Then the sunlight slanted most steeply of all, and the shadows were dramatic and long.

The short summer and continuous daylight put everything into overdrive. Eggs hatched into fledglings and then grew into birds that were ready to leave their nests in a matter of weeks. Flowers appeared in the scrapings of soil between rocks and among the spongy mosses and lichens of the muskeg and then vanished again almost immediately. Summer after summer Paul returned to the Arctic, now a fully-fledged geologist for the Survey. He strode out his rock contacts, mapped his terrains, noted down the rock types and their structures. He worked sixteen to eighteen hours a day. There was no sign that any other human had ever set foot there. Paul felt that he was master of the landscape.

The rocks he was working on were among the oldest in the world. They came from a catch-all time that geologists call the Precambrian, because it led up to the Cambrian period—which heralded one of the most significant changes in the history of the Earth. Naming time slices by what comes afterwards is a peculiar geological habit. More peculiar than ever in this case, because the Precambrian is much more than just a slice of time. The Precambrian lasted 4 billion years, covering nearly 90 per cent of Earth's entire history.

And yet, to geologists, this has long been considered the

Earth's Dark Age. Plenty may have been happening, but nothing was recorded for posterity. The rocks of the Precambrian are like the history books of Europe's medieval Dark Ages—a blank. What was missing? Fossils. Geologists rely heavily on fossils. One rock can look much like another, and to find out when exactly it was formed you need to look at the creatures that are locked inside. The Cambrian, roughly 550 million years ago, is the time when serious fossils first appear in the rocks. If you look at a section of rock from the beginnings of the Cambrian, you start to see real creatures with legs and teeth and armour plating, and you see changes in the fossils over time. In more recent rocks, dinosaurs appear and then vanish, making way for the fossils of mammals, fish and birds. Each has its own season and time, and each dates the rock that houses it. Fossils provide a ready-made timescale. They are like clocks left frozen in the rock. Every slice of geological time that comes after the Cambrian can be divided into periods and eons, according to the creatures that lived then.[5]

But before the Cambrian there were no fossils to speak of. And the few algae and the simple, single-celled creatures of Slimeworld that did bequeath their forms to the rocks stayed more or less the same for billions of years. Because of this, the rocks of the Precambrian just merge together into one long, undifferentiated mass. This was the geological Dark Age because there was simply no way to tell one time period from another.

Look at a standard geological timescale, a poster pinned on to every geologist's wall, and you'll see the Precambrian squeezed into a tiny, unimportant-looking box at the bottom. "This squashed period contains almost all of Earth history," the legend ought to say, "and yet we know almost nothing about it."

Paul was fascinated by the Precambrian. He felt sure that this long, mysterious period of time must contain important secrets about how the world works, and he dearly wanted to find them.

The first project Paul embarked on was trying to discover whether the continents behaved in the same way in the Precambrian as they do today. On geological timescales nothing stays still—not even continents. Over millions of years, continents skip and skate over the Earth's surface, some crashing together to throw up mountains, others ripping apart to create ocean basins. Paul wanted to know if this had always been true, even in the Precambrian. And if so, was the dance of the continents a minuet or a jitterbug? Were their movements carefully orchestrated, or a random bump and grind?

Gradually, Paul began to piece together the way the plates that would become North America moved during the Precambrian. Rather disappointingly, they looked just as random as in more recent times. They were clearly dancing a jitterbug, not a minuet. Still, he put his results together with geological maps from all over North America and began to trace exactly how the continent had formed. He discovered that most of the formation took place in a short, frenzied burst of activity around 2 billion years ago, when seven small plates crashed together and stuck in place. After eight painstaking years of researching this tale, Paul published a massive synthesis, which he called "United Plates of America".[6] The research required two skills: careful attention to detail and the sort of mind that can synthesize countless arcane facts into one overall, compelling picture. Nobody else in the world could have written it.

LIFE WAS good, even away from the rocks, during the long Canadian winters when Paul was forced back southwards to analyse his data and kick his heels. He was still running, and he had a new obsession to add to his life: music. As a teenager, Paul's attention had been caught by modern classical music, but in his junior year at college he was introduced to African-American music: modern jazz and pre-war blues; Ornette Coleman, John Coltrane, Eric Dolphy. He collected recordings voraciously throughout the seventies. Soon he had a thousand records, then two thousand and more. His opinions were characteristically forceful. Miles Davis and John Coltrane? Overrated. Charlie Parker and Dizzy Gillespie? Fabulous. They're the real musicians, the ones who deserve the credit they never fully get. And Louis Armstrong. His care with notes! His extraordinary musicianship! People were put off by Armstrong's stage persona. They thought he was an Uncle Tom. But Armstrong knew what he was doing. Every note, every rhythm was as precise as they come. Billie Holiday, a singer with true soul. Ella. Yes, she had a fantastic voice. Yes, great technique. But she was never compelling as a musical artist. She never connected emotionally.

Paul began to host a radio show, which was aired live on Wednesday evenings from nine to eleven. He played an eclectic mixture of jazz, blues, gospel, country-and-western, all from his own record collection. He talked about the history of the music, the particular idiosyncrasies of the musicians, the merits of different recordings. He talked about how to listen to the music, what to like, what not to like. His show developed a cult following, and Paul loved it.

Also, much to his surprise, Paul began to share his life with a woman he had known for years. He had first met her in the sixties at the home of his mentor at the Survey, a geologist called John McGlynn and his wife, Lillian. Erica Westbrook was a friend of the McGlynn family. She was often at the house when Paul visited. He hadn't particularly noticed her back then, nor she him: she was a scornful teenager when Paul was a driven young college student.

But things were different in 1976, when Erica offered to sublet his house in Ottawa for the summer while he was away in the field. Paul had just turned thirty-five. He had never even had a girlfriend. His lifestyle wouldn't allow it. He spent too much time out in the field, and when he wasn't in the far north, he lived for running, and music. He was, he had always felt, too self-focused to have time and attention for a family. Erica was tall, an inch or so taller than Paul himself. She had long, thick, black hair, a generous smile, and a habit of casting amused sideways glances. This time around, she found Paul intriguing. She laid a bet with a girlfriend about which of them would succeed in seducing Paul. Erica won.

Still, Paul didn't particularly see a future in the relationship. His attention remained focused entirely on geology in the North. Erica's response was drastic. She took a plane to Yellowknife and spent a long, fraught week in the Northwest Territories, in Paul's field site, in Paul's home turf. That was the only place she felt she could count on his attention. She spent the week arguing passionately. She wanted the relationship. She wanted Paul. Once again, she won.

It was never going to be easy. Erica was sociable and warm. She went on to work as a palliative care nurse. She was a people person. Paul was utterly focused on his work. Once, Erica's resolve nearly cracked. There had been a snowstorm in Ottawa and the garage roof had fallen on top of Paul's car. Paul's precious car. A shiny red Lotus Elan that he had bought to compensate himself when an injury left him temporarily unable to run. When Erica saw the roof and the car that morning, she realized something that drove her crazy. Paul had already left to go to the Survey. He must have walked past the garage. He must have noticed the roof. He had done nothing about it. The Lotus was *his car,* but yet again he had left everything to her. He hadn't even mentioned it. Erica raced back into the house and dialled her mother-in-law's phone number.

Dorothy Medhurst, Paul's mother, has always been a formidable woman. Paul describes her as a whirlwind. Everyone else describes her with very healthy respect, bordering on awe. She is tall and strong and passionate. She is an artist. She is uncompromising. At eighty-eight, she now lives alone in an isolated cabin thirty miles from Toronto. The cabin has no electricity, no telephone, and no running water. Dorothy prefers living that way. All of her children were raised to think for themselves, to embark on projects, to stay outdoors, not to be home until the streetlights were coming on. When Paul cried as a baby, Dorothy would put him in his crib out under the tree. "If you're going to cry," she told him, "go cry to the mosquitoes." Paul can still trace the pattern of those branches in his head. The home Paul grew up in was not a cuddly, touchy-feely one. There were no soft furnishings. The wooden floors were decorated with field lines for ball games. The walls were festooned with paintings. Paul called his parents by their first names. You judged people not by their blood connections but by their talents, and how they used them.

Even as an adult, Erica was rather afraid of Dorothy. But still, on that snowy day in Ottawa, she dialled the number and blurted out her frustrations. Dorothy listened thoughtfully. When Erica had finished, this is what she said: "I agree. It's not normal behavior. But you have to decide now if you're prepared to put up with it. Because it's not going to change." This was excellent advice. Erica knew immediately that Dorothy was right. Paul wasn't going to change. She had known from the beginning that he was focused and obsessive and intense. That was his strength as well as his weakness. It was the source of his charisma and also the thing that made her want to scream at him. If Erica wanted any part of Paul, she realized that she had to take all of him. She stayed.

Erica was a big influence on Paul. He sought her advice, and she helped to temper his ferocity. If she had been around on 6 July 1989, Paul would probably still be at the Survey, and would probably never have heard of the Snowball Earth. But she wasn't. She was away, and when Paul decided to let fly, there was no one to caution against it.

He had received an essay that enraged him. Ken Babcock, the new head of the Survey, had sent the essay to all employees. Babcock was a political appointee, and he had no truck with the academic-style freedoms of the Survey researchers. He criticized everything that had gone before. This isn't a university, he said, it's a service to our clients in government and industry. Researchers at the Survey felt that he talked like a bureaucrat, not a scientist. In his essay, he told them to "get back to basics". They should focus on the practical needs of government and industry rather than on esoteric academic research.

The essay, entitled "The Search for Excellence", infuriated many of the Survey scientists. They seethed at the implication that their work was deficient in some way because they were driven by academic curiosity. How dare Babcock suggest that their work was irrelevant just because there wasn't an immediate payoff? Many of them despised Babcock and his bureaucratic ways. They believed that turning the Survey into some kind of glorified consultancy would destroy its fine reputation. But they all held their peace, except Paul. Paul couldn't help himself. He wrote a memo to all his colleagues, taking issue with Babcock's entire stance. That might have been all right, had his penchant for sarcasm not prompted him to add a caustic rider at the end of the memo. "The search for excellence at GSC [the Survey]," Paul declared, "should begin at the top."

Paul's memo inevitably found its way into the offices of the local newspaper, the *Ottawa Citizen*. The *Citizen*'s report was immediate and gleeful. "Top Survey Scientist Rebukes Boss", the headline declared. "Controversy Rages at Elite Government Agency."[7] Paul very properly refused the paper an interview. Babcock, however, did give an interview, in which he pointed out rather sourly that in the private sector Paul's memo would have been grounds for dismissal. "He is truly one of our outstanding national earth scientists," Babcock told the *Citizen,* which then told the rest of Canada. "I suspect that his knowledge of the world of politics and management is less well developed."

Privately, Babcock was furious. He had been personally attacked by a subordinate and now the whole world knew it. His backhanded compliment in the newspaper was a sure sign that he wanted Paul out, but Babcock didn't sack Paul—he couldn't. Instead, over the next few years, Paul felt that he was becoming a nonperson. His funding requests were refused, and he was passed over for all privileges. He was even turned down when he requested unpaid leave to teach for a semester in the United States. Nobody was *ever* turned down for unpaid leave. Paul had spent many semesters away before without difficulty. He began to realize that he'd have to leave, but what he didn't realize at first was that this would also mean leaving the Arctic. Wherever Paul went in Canada, he quickly discovered, he would be unable to get funding to finish any work he had started under the Survey's umbrella. Paul's memo cost him more than he'd ever dreamed.

Today he makes light of it. "I left as I arrived," he declares, "fired with enthusiasm".[8] But at the time he was humiliated, bewildered and hurt. And what hurt more than anything else was being barred from his beloved Arctic. He desperately wanted to finish the work he had started there. He wanted to be back, mapping the terrain, hiking across the bleak landscape of the barren land, a place that felt more like home than anywhere else on Earth. For the second time in his life, Paul was walking away from something precious to him. He'd done it with the chance of Olympic glory, and now he was doing the same thing with the Arctic. This time, as before, he responded the only way he knew how. If he couldn't go back there and finish his work, he'd find something better. He'd find a new problem to solve, a new route to glory. He'd find something new to be remembered for.

But where should he go? Harvard University offered him a haven for his academic base, and he moved there gladly. But he needed a new field site, one with exposed rocks from the right time, the Precambrian. The rocks had to be fairly easy to reach logistically, and yet it was important they hadn't been excessively studied already; there was no point going somewhere that had already been picked over by other geologists. Paul needed somewhere fresh, a place where a great story was just waiting to be unearthed.

He toyed with one or two possibilities. Kashmir, perhaps, in northern India. Or maybe China would work. Then he found the perfect candidate. South West Africa had just become Namibia, having won its independence from South Africa two years earlier, with none of the presaged bloodshed. For decades before independence, scarcely any geology had been done there by outsiders, thanks to the military occupation by the South African Defence Forces. But newly independent Namibia was beginning to open up to the outside world. And most of the country was taken up with a vast, empty desert, full of exposed Precambrian rocks. They were younger than the rocks Paul had worked on before. Rather than 2 billion years old, they were more like 6 or 7 hundred million years old. That put them closer to the end of the Precambrian, closer to that strange point in time when fossils suddenly appeared out of nowhere. Perhaps they might even hold some clues about why life had suddenly lurched away from the simple world of primordial slime into the complexity that we see around us today.

Paul had other reasons to feel pulled towards Namibia. His father's brother, "Izzy", had lived and worked there. A few times during Paul's childhood, Izzy had travelled to Toronto full of tales, and the young Paul's eyes had shone. Namibia had been on Paul's list for decades. Africa, too. After geology, Paul's other obsessions were jazz and athletics. Africa had consistently supplied the masters in both departments. Namibia won on all sides.

PAUL HAD to start again from scratch in Namibia. He didn't even know where the best rocks would be, or which places he should concentrate on. He pored over aerial photographs of the terrain, and tried to pick out likely rock outcrops, looking for ones that he could drive to on bush tracks, or reach with a short enough hike from a possible camping ground; ones, too, where the rocks seemed to be slightly tilted, so he would be able to walk from layer to layer, up and down, back and forth in geological time, without having to scale a vertical cliff face. The balance was delicate, though. If this tilting had been accompanied by too much bucking and rippling of the Earth's crust, the rock layers would be too complex to interpret.

In June 1993, armed with a list of outcrops to visit, Paul set off for Namibia. The contrast with Canada was stark. There were, mercifully, no flies in the desert. But there were also no long, slanting shadows. Sunlight in Namibia glared fiercely overhead. The dark rocks would soak up morning sunlight, and for the rest of the day heat would pour relentlessly back out of them. At noon, when Paul wanted to find some shade after hiking and measuring for hours, there were no shadows to be seen. There was no midnight sun. Summer or winter, the days were frustratingly short, and an impenetrable darkness would fall abruptly each afternoon at 5:45.

There were also more people than Paul had ever worked among. Even in the desert, driving along a bush track, he would suddenly come upon a village of round mud huts clustered around a tall, rickety windmill that pumped water from the local well. Paul quickly learned to take a "landing fee" with him. The front seat of his truck was perpetually wedged with bags of sugar and tobacco to offer to the locals. He learned his first halting words of Afrikaans, how to say "please" and "thank you" and ask directions. Around those villages, it was easy to get lost. The ground had been grazed bare of dried grass and there was only baked mud, rutted with myriad tracks from animals and carts and, occasionally, the tyres of vehicles like Paul's dusty white Toyota. Everyone wanted to help. When Paul said thank you, the villagers would reply "Pleasure!" in a lilting, cheerful tone. If there was nobody to ask for directions, Paul would sometimes have to cast back and forth, trying this track and that until he finally found one that seemed to be heading the right way. Sometimes the tracks disappeared completely, and Paul had to divert down narrow gullies, setting his vehicle pitching and yawing over the rocks, three wheels on the ground, one in the air.

He had never worked using a vehicle before. In Canada, everything was by planes and helicopters, boats and boots. In Africa, he had to learn how to cut across a dried-out riverbed without getting trapped with his wheels spinning helplessly in the soft sand. The tricks, he discovered, were to let air out of the tyres until they were half-flat and could grip the loose surface more easily, and never, ever to touch the brake in mid-sand.

Gradually the memories of Canada began to recede, and Paul found himself relishing the harsh aridity of the African landscape: the sweeping valleys, the narrow, winding canyons and the disdainful kicks of the springboks that bounced out of the way of the Toyota. Though he would never deviate from his geology for anything approaching a tourist activity, he grew to enjoy seeing African wildlife in the wilderness, where it belonged. Sometimes as he drove he would see ostriches, their short tails bouncing as they jogged through the bush. He saw giraffes with their black velvet eyes and absurdly long lashes ("the most beautiful eyes in the world"); grumbling warthogs and baboons, herons and bustards and African grey parrots whose monotonous "waaah, waaah" sounded like a whining child. In the air there was a flash of yellow as a southern masked weaverbird emerged from its dangling sack of dried grass and mud. On the ground, termite nests towered, with their turrets and tubes and demonic spires, all vivid red from the rusty Namibian soil.

And there were rocks and rocks and more rocks, all unstudied and enticing. North of Windhoek rose the great pink granite intrusion of the Brandberg, Namibia's highest mountain, flanked with flat-topped, chocolate-coloured hills. These were the remnants of a plume of hot rock that had risen up from inside the Earth some 133 million years ago, when South America and Africa were last conjoined. The plume had flooded angry lava on to the plains of both continents, helping to rip them apart and open up the South Atlantic Ocean. Even the soil thereabouts was magma-dark, barely covered with a pale blond beard of grass. There were no bushes or trees, just squat *Welwitschia mirabilis* plants, with a woody root from which sprouted two flat, flailing leaves. Each plant is miraculously long-lived. Its leaves grow slowly and steadily, corkscrewing around each other for hundreds, perhaps even thousands, of years.

Farther north still, the Precambrian outcrops emerged from beneath the volcanic floods. When these rocks formed, more than 600 million years ago, Namibia was covered with a broad, shallow sea that left behind sandstones and mudstones, pink carbonates and dark grey shales. Peering closely at these rocks, Paul found the thumbprint whorls of the ancient stromatolites that had inhabited the Precambrian shores; he found sand dunes, beaches and lagoons, all now petrified and awaiting his notebook and hammer. And where the ancient seafloor once dipped towards a western ocean, barren, rocky hills stretched for mile after mile, their layers in places magnificently buckled and twisted into vast folds that dwarfed the tiny Toyota as it jolted along the canyon floors.

Paul was entranced. Such geological riches, yet scarcely any of them had been studied. Surely among all these outcrops he would find something important, some intriguing new insight into the history of the Earth.

He intended to continue studying the bump and grind of continental motions, just as he had in Canada. He was used to working alone or with just a few students in tow, but for his first field season in Namibia he brought along another Precambrian expert, Tony Prave, a researcher from New York. Tony is a wisecracking Italian American, a jobbing geologist in his late thirties who works hard and stays out of the limelight (and hence, broadly speaking, out of trouble). With his thick, dark, shoulder-length hair and bronzed face he could easily be mistaken for a Native American. His accent, though, is pure Hollywood mafioso. He has a wide, charming smile and slightly wary eyes.

Tony had spent most of his career working in Death Valley in California. He got to know the Precambrian rocks there by heart. But though he loved Death Valley, he jumped at the chance to go to Namibia with Paul. Paul was a famous field geologist. This was, Tony felt, the opportunity of a professional lifetime. Throughout that season, Paul, Tony and two graduate students moved from camp to camp and outcrop to outcrop in the remote Namib Desert. They mapped, climbed, hiked and studied, walking up gullies and down valleys, musing, interpreting and learning to understand Namibia's deepest history.

Paul affectionately called Tony "Pravey". The two of them got along brilliantly. They were both opinionated, both robust in their arguments, both fascinated by the rocks. Tony found Paul's methods exhilarating. Out on the outcrops, Paul's mind was like a steel trap. "Why? *Why?* What's it mean?" Paul would ask, rapid-fire, when Tony reported an observation. And then, as they headed to a new outcrop: "What would you predict? We're going over there now. What would you predict, Pravey?" Back at the camp, the two of them would stay up talking until eleven or twelve o'clock at night. Unlike the Arctic, where you could do geology at any hour, Namibia had long nights of enforced absence from the rocks. These were times to sit around the campfire and drink whisky and bandy opinions back and forth. It didn't seem to matter that they sometimes disagreed, that Paul adored baseball, for instance, while Tony hated it. They delighted in each other's company, and Tony basked in Paul's warm approval.

Eventually, inevitably, trouble started. Paul has never found friendship easy. He's charismatic, but also self-focused and intense. As a child, his relations with his fellow mineral collectors were civil rather than warm. In athletics, even when he was part of a team, he raced alone. And many of the geologists with whom he once worked closely are now scarcely on speaking terms with him. People like Tony. Tony and Paul no longer collaborate, or go on field trips together. They are no longer friends.

The problem arose towards the end of the field season, when Tony began to disagree with Paul's interpretation of the Namibian rocks. The issue was an arcane geological one, involving the details of exactly when Africa collided with South America. During the Precambrian, a narrow ocean separated these continental behemoths—they wouldn't actually hit until sometime in the Cambrian. But Tony became convinced that Africa was nonetheless beginning to sense the coming collision, and that its rocks had begun to buckle and bend in response. Paul, on the other hand, maintained that there was no sign in Namibia's Precambrian outcrops of the impending pile-up.

This disagreement started to sour their relationship, and by the time they returned to Namibia next season, the early warmth between them had ebbed away. Now, in the talk around the campfire, Paul was sarcastic about what he called "the Pravey hypothesis". "Oh, so what does the great Pravey say is going to happen tomorrow?" Tony remembers him asking. "What does the great hypothesis predict?"

"Paul's very competitive," Tony says now. "He's one of these people where if you take three steps, he has to take four. If you've mapped ten square kilometres, he has to map eleven. He's always got to be that little bit better, that little bit more intense." Tony had started to see the sharp side of this competitiveness, and he didn't like it.

The final straw came when Paul returned from a day's work mapping a narrow canyon. He arrived in triumph. "The Pravey hypothesis is dead," he declared when he re-entered camp. He had, he said, gathered evidence that conclusively disproved Tony's interpretation. The next day Tony hurried to the site, now dubbed by Paul "the Canyon of Contention". And there, among the rocks that Paul had mapped, Tony saw a whole jumble of fault lines. The rock layers had been mashed up beyond measure, long after they had formed. You couldn't use them to prove or disprove anything.

Tony saw red. He launched himself back to the camp, and confronted Paul head-on. They were eyeball to eyeball. Tony swore at Paul. Paul swore back. They began to yell, spittle flying from their mouths in their fury and making arcs in the air. The two students who were also on the trip looked on in horror. Then one of them decided to intervene. She was diminutive but tough, and she wedged herself between Paul and Tony. "You should be ashamed of yourselves!" she told them both. "Stop it!"

Her intervention worked. The shouting stopped, and Paul stalked off to his chair where he sat silently, staring into the darkness. The camp was subdued that night, and shortly afterwards Tony left Namibia.

Relationships among geologists are intense. By its nature, geology involves travelling with your colleagues to remote places, working long, hard hours in sparse conditions, living on top of one another and away from other people for weeks on end, having little contact with the outside world. Think of submarine crews, or Antarctic explorers. Think of throwing obsessive, opinionated people together in places that they can't easily leave. Their personalities become magnified. They bond or they break. Paul in particular has had plenty of fights like the one he had with Tony. Stand-up, screaming fights. He reacts furiously when confronted, and he holds nothing back. He will rage one moment, and ten minutes later act as if nothing had happened. But those on the receiving end are slower to forget. His reputation as a brilliant geologist has been tempered over the years with his reputation as a hugely difficult character. The people who still work with him are the few who know how to handle him. He can be rude, sarcastic and unpleasant. He's dismissive. He often makes people feel small. He knows this. He even makes a joke of it sometimes. "Everyone's entitled to my opinion," he'll say. And then, "Gosh, I'm awful. I don't know how I'd react to me."

And yet, when he compliments people, they feel good. They feel special. There is something about Paul that makes you want his approval. I have met former students of his who are now established geologists, tenured professors with great careers in top universities. They have all this, but they *still* care desperately what Paul thinks and says. If you ask them why, they shrug helplessly. "I don't know," they say. Tony talks about the time before his fight with Paul as their "honeymoon", and the time after as their "divorce". Even now, years later, he gets a look of frustrated pain when he talks about their fight. Even now, he says this about Paul: "That man is so charismatic, if he'd been born two thousand years ago, he could have been Jesus."

PAUL SHRUGGED off his fight with Tony. He returned to Namibia the next season and the next, with a fresh batch of students to help map and measure and interpret. After that first flush of excitement, though, the outcrops were beginning to look disappointing. He had wanted to measure the timing of the continental shifts, but the rocks in Namibia turned out to be useless for accurate dating. Still, Paul couldn't shake off the feeling that these outcrops held some important secret. He felt he was big-game hunting, but for what?

Then something began to nag at him.

Everywhere Paul went in Namibia, he spotted signs of ancient ice. He would be hiking up a gully and suddenly he would see a huge white boulder embedded in the grey siltstone. Siltstone is formed from an ancient seabed. Over time in the ocean, a fine rain of sediment lands gently on the seafloor and is gradually converted to rock. But a boulder had to be brought in separately from the shore. Something must have carried it out into the ocean and then flung it overboard. There were no ships in the Precambrian, and certainly no creatures capable of flinging anything. The culprit had to be icebergs. The boulder, a "dropstone", must have fallen from a melting berg up on the sea surface.

That wasn't all. As Paul looked more closely, he would see a medley of rocks appearing in the siltstone. Not a single boulder now, but countless pebbles and stones, all shapes, sizes and colours; fractured and rounded; pink, brown, tan, white and grey; granite basement, quartzite and carbonate. This mad jumble had somehow become bound up in the fine grey silt. Like the boulder, these rocks were interlopers. Something had gathered them up from mountains and gullies throughout Namibia. Something had bulldozed them down to the shore and on into the silty sea. The mix of multicoloured rocks stretched in every direction for hundreds of miles. Only one agent was capable of transporting so many different kinds of rock over such large distances: ice.

Paul recognized these ice-signs immediately. His mother's fireplace in her Canadian cabin was held up with two great chunks of pale stone packed with ice-borne pebbles. Paul had seen them every weekend and throughout the summer as a child.

He had also known for years that rocks like these show up all over the world. They can be found in the Americas, Asia, Europe—in fact on every single continent. And they all date from one particular point in time: the mysterious end of the Precambrian, just before the first real fossils appeared, just before life went complex and the Earth changed for ever. Paul had known all this even when he was working in Canada, but he had never really thought much about it. His mind had been fully occupied with his work on the shifting of continents.

Now, though, faced on all sides with the Namibian ice rocks, Paul started to wonder just why they were there. Why they were *everywhere.* You expect to see ice at the North and South Poles. But to find signs of ice on every continent seemed extraordinary. And the ice rocks in Namibia came with an extra mystery. They appeared in the middle of rocks that had clearly been formed in warm, tropical waters. What was ice doing in the tropics? And why did it appear there at that crucial moment in the history of life? Was it a coincidence? As he probed and pondered over the Namibian ice rocks, Paul grew more and more intrigued. He was haunted still by the sense that Namibia held some extraordinary story, just waiting to be told. Could these strange rocks be the key? Forget the motions of ancient continents. Now all he wanted to know about was ice.

What Paul didn't know yet was that the ice rocks brought nothing but trouble. For decades they had been grabbing the imagination of geologists without revealing their secrets. There was always some reason why ice in many of these places simply had to be impossible. Until now, everyone who had tried to explain the ice rocks had faltered. On the way, though, they uncovered clues that would prove vital for the Snowball story.

THREE

IN THE BEGINNING

"Polar exploration is at once the cleanest and most isolated way of having a bad time that has ever been devised."[1] So wrote Apsley Cherry-Garrard, one of Scott's companions from the doomed South Polar expedition. It wasn't just the cold, or even the danger, that made early polar travel miserable, but the sheer physical effort of trudging over the snow for day after day, dragging *everything* on a sledge behind you. Henry "Birdie" Bowers, one of Scott's strongest and toughest men, called this the most backbreaking work he had ever come up against. "I have never," he said, "pulled so hard, or so nearly crushed my inside into my backbone by the everlasting jerking with all my might on the canvas band around my unfortunate tummy."[2]

These painful endeavours weren't confined to adventurers. If you were a geologist in the 1940s with a penchant for studying rocks in icy places, man-hauling was essential. There were no helicopters, or snowmobiles. To reach remote, unstudied outcrops in the centre of any white wasteland, you had to load up all your equipment—tents, food, cooking gear, fuel—harness yourself to the sledge, and *heave*.

For Brian Harland, a geology professor from Cambridge University, the effort was worth it. Brian is famous for his precise probing into the Earth's past. He put together the definitive "Harland timescale", a chart of neatly coloured rectangles that divides geological time into its separate periods, each with its own ascribed date and span, and which graces the walls of geology departments around the world.

But he is also renowned for his Arctic geology. From the beginning of his career, he was drawn to the rocks of the remote Arctic, sure that he would find extraordinary geological secrets half-buried beneath the ice. He was right. By scouring the scarce rocks of the far North, Brian discovered the first traces of a global glaciation. He was the grandfather of the Snowball.

Brian's fieldwork, though, was never easy. In August 1949 he was leading an expedition over the ice fields of Svalbard, a frozen archipelago several hundred miles north of Norway and east of Greenland. He and his four companions had been away from base for days, dragging all their supplies with them. Now their route back led up a dauntingly steep slope of ice. If hauling on the flat is bad enough, uphill it can seem nearly impossible. Still, Brian had decreed that at the brow of the hill they could stop and camp; there would be food, warm drinks and rest. The five geologists duly buckled up and began the long, hard pull. Their heads were down, their attention fully focused on gaining the top of the slope. They had no idea of the disaster that was about to strike.

IN THE BEGINNING

* * *

THIS WAS Brian's second visit to Svalbard. Eleven years earlier, in 1938, he had been there as a young graduate, part of a brief student expedition. Svalbard's rocks had immediately intrigued him. They were among the oldest in the world, and many had formed in the Precambrian, that long Dark Age of the Earth. Brian realized that these rocks could provide a rare window into this ancient, mysterious time. But they were also remote and inaccessible, covered for the most part with a thick blanket of ice. In just a few places, dark, conical mountaintops and ridges of rock poked out above the snow. Brian had been intrigued by these outcrops. He'd caught glimpses of great rocky cliffs bearing giant folds and faults. Where did the folds come from? How had the mountains formed? What could they reveal about the workings of the world?

On the '38 expedition, there had been little chance to find out. The expedition had other priorities. Brian was the only geologist among a group of geographers. There were ice-forms to study, and maps to make, and not enough time for everything. And then the Second World War had intervened. But now Brian was back, thirty-two years old, a fully-fledged Cambridge academic running his own show. This time he could decide for himself where to go and what to study. The geology of the islands was a blank, and he was determined to fill it in. He wanted to understand every outcrop, every layer of Earth's prehistory.

This was Brian's first time as expedition leader, and he felt the responsibility keenly. He was a slight man with pinched features and a nervous disposition. Brian planned, some people said, to excess. Most of his time beforehand was spent worrying over details. For every problem he had a contingency. When it came to samples, notebooks, photographs, all the paraphernalia of a geological expedition, Brian's numbering systems were complex, consistent and legendary. Everything had its own alphabetical or numerical code. Every item of equipment slotted neatly and clearly into the overall plan.

The expedition food was chosen for high calorific content rather than taste. There was margarine, processed cheese, sugar, oats, biscuits, chocolate and a fatty mix of dried meat called pemmican, all the same items that had sustained Antarctic explorers like Scott and Shackleton just a few decades earlier. This simple, efficient and egalitarian diet had appealed strongly to Brian's utilitarian instincts on the '38 Svalbard expedition, and he saw no reason to change it. (In later years he would bow to the necessity of supplementing the dull basic rations with spices, delicacies and other extras, but he never really approved.)

He had, however, learned one important lesson. In 1938, everyone had been ravenous. The rations had been designed for Antarctic expeditions using dog teams. Man-hauling required much more energy than the food provided, and there had never been enough to eat. When you're constantly hungry, staying warm becomes more and more difficult. At night your dreams are laden with food. You fantasize about medieval feasts and sweetshops and huge, rich desserts. And when you wake, you have to force yourself to harness up to a heavily laden sledge while your stomach is gnawing and your limbs feel weak and tired. The food on Brian's '49 expedition might have been dull, but he made sure it was plentiful.

Brian's watchwords—a legacy, perhaps, of his Quaker upbringing—were fairness, order and efficiency. He had already set in place the rules that were to govern his expeditions for the next forty years. No hoarding of food was allowed. Rations were divided evenly, and your portion belonged to you until midnight. Anything you hadn't eaten by then reverted to the general pile. Also, you were strictly forbidden to bring any additional delicacies secreted in your pack. What one member of the expedition ate, everyone ate. You could break these food rules if you chose, but only furtively and with a guilty aftertaste. Few people tried. Brian was scrupulously honest, and his attitude somehow spread.[3]

He also judged people firmly by their dedication to the task in hand. To be part of his expeditions meant abandoning any perceived status or sense of entitlement. His was an Edwardian value system. Would you volunteer, were you willing, had you put in the necessary effort to prepare? (When I first met Brian, years later, I had to lay down all my academic credentials before he would speak to me. He wanted to know about my degree, my doctorate, how much research I had already done. When he was finally satisfied that I deserved his time, he was promptly generous with it.)

Brian believed that a person's work should speak for itself, and he abhorred the notion of pushing himself forward. Take the naming of geological features. In those early days of exploration in Svalbard, many researchers gaily named the places they discovered after themselves and their friends. To immortalize themselves, they chose magnificent mountains, giant rivers of ice, great macho structures. But although Brian would become the world's leading authority on Svalbard, you'll struggle to find his name on the maps. Eventually, you may spot one small smudge, close to the summit of the ice cap, bearing the name "Harlandisen". Brian's students think this is hilarious. An isen is a rather nondescript patch of ice, usually found between more interesting places. Even so, Brian is embarrassed by the accolade. Ask him how the name came about, and he will blush faintly and mumble that the Norwegians insisted.

Brian's students loved him. They followed his codes strictly and with loyalty. On his '49 expedition he had brought along eleven students from Cambridge, split into different groups for maximum efficiency. Several parties had already investigated the coastal regions, tooling along the fjord-ridden coast in sixteen-foot open whaleboats, which Brian had christened *Faith* and *Hope.* The rest had taken the largest boat, an eighteen-foot dory called *Charity.* ("It's a biblical reference," Brian says. "You know. 'Faith, hope and charity, and the greatest of these is charity.'") Brian had bought *Charity* for seventeen pounds. She was a marvellous boat, big, wide and solid as a rock, with space enough for a ton of equipment. She had borne the third party to an encampment at the tip of Billefjorden, in the northwest of the main island. From there, Brian led a small party of four students, the "Northern Survey", out into the unknown territory of Ny Friesland. The plan was simply to map the rocks and begin to understand what was out there. Though these rocks came directly from the time of the Snowball, Brian as yet knew little about them.

The first few days were good ones. In clear weather the party sledged and skied, measured angles, surveyed the landscape, made sketches and took carefully numbered photographs. All around them, the dark brown tips of mountains and rocky cliffs poked through skirts of ice. Snow dusted every dip in the rocks. And flooding down every gully and alongside every cliff were Svalbard's great glaciers.

Glaciers are giant bodies of ice, with a texture like a strange combination of rock and river. They are solid, like the ice cubes in a refrigerator, and form out of snow the way rock is made from soft mud or sand. If mud falls consistently down on to a seafloor, its grains will eventually squeeze together and solidify into rock. Snow does the same thing. Individual snow crystals are gorgeous works of six-sided filigree. But if they pile up over time, these crystals begin to amalgamate. They squeeze up against one another. Their delicate arms smash and break and weld together. They trap pockets of air, meld into a hoary substance called firn, and then gradually solidify into hard, white ice. And then the ice begins to move. Like water it flows downhill, but at a magisterial, glacial pace. Glaciers don't just fill valleys; they create them. Flowing ice may be slow but it's inexorable, and a glacier can carve through solid rock.

For polar travellers, glaciers make great highways; but they come with hazards, too. When ice flows, it splits into deep fissures and cracks. Snow then drapes these crevasses, hiding them from the unwary. Break through one of these snow bridges, and you will find yourself plunging into the cold blue heart of the glacier. Usually you can avoid this hazard, since snow bridges often reveal themselves as tell-tale dips in an otherwise smooth surface. Usually, but not always.

Around a week into the expedition, Brian's party was travelling down Harkerbreen Glacier to investigate the rock cliffs on either side of the ice when the good weather abruptly deserted them. Thick clouds descended all around, until they could scarcely see the way ahead. They sledged gamely on, but the weather hampered all their efforts to investigate rocks, and Brian began to feel nervous. With only two days' rations in hand, he decided to try a new route back. If they could reach the wide sweep of Vetaranen Glacier, to the east, the way would be easy even in cloud.

To be safe, Brian decided to scout out the route ahead. With him he took one of the students, Chris Brasher. Chris was just twenty years old, but he was a fit and accomplished mountaineer. (He was also a most talented athlete. Five years later he would be one of the two pacemakers who propelled Roger Bannister to the first sub-four-minute mile. Two years after that he would win his own glory with Olympic gold in the 3,000-metre steeplechase.) Leaving the other three behind, Brian and Chris found a tributary glacier that snaked upwards and eastwards toward Vetaranen. They climbed doggedly up the steep ice slope, always checking for the dips in the snow that marked the presence of a crevasse. But the surface seemed innocent.

Back with the rest of the team, Brian directed operations. The way ahead was worryingly steep, but once over the slope everything should be easier. They would take a sledge at a time, starting with the heavier of the two, the Nansen. Nansen sledges are wonderful inventions, still used by polar explorers today. Their wooden parts are lashed together with hide, making them lithe and flexible enough to snake over bumps in the ice. At twelve feet long, they're also a good protection against crevasses. Even if you break through a snow bridge, the sledge will usually span the gap and act as a safety anchor, allowing you to climb back out again.

The five geologists attached their harnesses to the heavily loaded Nansen and began to plod their way up the glacier. Step, heave. Step, heave. They had almost reached the top of the slope.

Then the ground vanished from under them.

The sledge and the nearest two people plummeted immediately into a vast cavern of ice. One, two, three, the others followed, whipped backwards on their harnesses through a huge hole in the snow. The foremost man came last, his ski catching on the surface and tearing away from his foot as he fell.

Seconds later, all five found themselves miraculously alive, sprawled forty feet below the surface. Through bad fortune, they had broken through a wide, thick snow bridge, wide enough that the sled was no protection, and thick enough that it was invisible at the surface. But through good fortune, the bridge fell with them, so that all five had come to rest on a soft cushion of snow. And another piece of good luck: though the chasm continued down for hundreds of feet, the entire team and their sled had landed on a wide ledge of ice. There was only one casualty. Brian felt a pain in his right ankle and discovered that he couldn't stand on it. (He didn't want to claim any great injury. He later wrote that it seemed to be "slightly broken".[4])

Inside a crevasse, the temperature is many degrees colder than at the surface. Quickly your nose hairs and eyelashes are coated with a fine hoarfrost. Your face becomes numb, and begins to show white patches of incipient frostbite. The only light comes in feebly from the snow hole far above you, or as a blue gleam from the cavern's walls. For the next eight hours, Brian was forced to stay put in this ghostly glow while the four uninjured students began the rescue operation. First they crawled along the ledge until they found a place where it sloped up to the surface and a natural hole allowed them to climb back out. Back down the slope then, to where the second sledge held spares of everything, including ropes, a testament to Brian's meticulous contingency planning. Piece by piece, the students hauled every item of equipment up through the snow hole to the surface. They did their best to haul Brian out, too, but it proved impossible. Because the cavern's walls curved away from the dangling rope, Brian couldn't reach them to steady himself, and he swung and spun uncontrollably. Eventually he was lowered back down. He strapped on skis and shuffled slowly and painfully along the ledge to climb out the way the others had.

Outside, the cloud was still thick and low, heavy with the threat of snow. Brian and the team camped, on half rations, and considered their options. They were at least two days from their nearest food depot, and four or five days from base. The route they knew involved a steep downward slope and another long, heavy pull upwards again. But at least this way was definitely safe, and they resolved to take it. Broken ankle notwithstanding, Brian had no intention of being pulled along on the sledge. Each morning he would strap on his skis and begin a long, lonely shuffle over the snow. His right ankle was useless. He had to use his ski pole to point the ski in the right direction. Behind him the four students would finish their breakfast, pack the gear, and haul the sledges along in Brian's trail. Around midday they would catch him, and stop for lunch. Then they would continue on into the distance, leaving Brian to trudge painfully along in their tracks. By the time he arrived at the night camp, food was already prepared, tents were pitched, and he could fall into his sleeping bag.

After five days they finally reached base. *Charity* bore Brian back around the coast to Svalbard's main town, Longyearbyen, where he was ordered straight into the hospital. His broken ankle had finally earned him a "hot bath and excellent care", which, he later wrote, made him "the envy of the others" since they had to return to the privations of the field.

One of the other parties from the expedition, it turned out, had also fallen foul of the hazards of Arctic travel. As Brian later put it, *Hope* was all but lost. Boat and crew had to be rescued from mid-fjord by a rubber dinghy. But Brian had known all along that Svalbard wouldn't yield its secrets readily. His contingency plans had been effective. His students were eager for more. And the various field parties had only scratched the surface of the data to be had. By the time Brian reached Cambridge again, he was already planning his next trip. He insists still that he wasn't drawn by the romance of the place. What pulled him back to Svalbard, he says, were the *stories*. He wanted to understand what the rocks could tell him. He didn't yet know that the rocks of Svalbard held a more extraordinary secret than he'd ever imagined. Nor did he know what trouble that secret would cause him.

BRIAN HAD been pleased with much of the organization of that first venture, but during his next few expeditions to Svalbard he was continually testing possible improvements. He began to build up his equipment, buying a whole new set of Nansen sledges to distribute among different field parties. Even though the Nansen hadn't protected his party from the ice cavern, such wide snow bridges were rare, and in every other respect the sledges had been great. He even found a handy source of Nansens back in England—buying several from the film set of *Scott of the Antarctic,* the movie that was just about to catch Paul Hoffman's young imagination, across the Atlantic in Canada.

Brian also began to realize that self-sufficiency and self-reliance were the keys to operating in Svalbard. Anything he left to someone else carried the risk of failure. Materials that had to be shipped north every season could be lost in transit. Relying on someone else for transport by sea could mean hanging around for days by the quay. Gradually, Brian established a base for himself in Svalbard, where he could store goods over the winter. He set up mechanical and electrical shops there. He bought covered motorboats that could sail safely around the coast even in the choppiest of seas. His expeditions became like guerrilla raids. Every summer his geological parties swarmed over the ice of Svalbard. They set up survey stations and measured outcrops; they climbed cliffs, collected samples, and steadily filled in the blank map of Svalbard's geological history.

And as Brian returned to the islands time and again, one particular feature of the rocks began to bother him. In many of the outcrops that poked through Ny Friesland's sheath of ice, a strange red stripe stood out against the pale yellows and browns and greys around it. Up close, the stripe was a chaotic mix of reddish boulders and rocks, all shapes and sizes, bound together in a background of fine silt.

Brian had known this pattern all his life. He'd grown up in Scarborough, on the North Yorkshire coast of England, and as a child he would collect the alien stones that studded the cliffs there—the famous "boulder clays". The boulders embedded in Yorkshire's sea cliffs were an unmistakable signal that ice had been on the move. They had arrived fresh from Scandinavia, where glaciers had dragged them off the land and dumped them into the North Sea.

Glaciers don't just glide serenely over a surface; they grind into it. A glacier scratches and scours the bedrock with the boulders it drags along. It bulldozes yet more rocks ahead of its advancing ice front. The surface of a glacier can be littered with debris that has tumbled off the steep cliffs at the sides and is carried along with the slowly moving ice. Eventually the glacier will spill into the sea. Perhaps it will break off into chunks of iceberg that gradually melt and deliver their load of rock debris to the seafloor as individual "dropstones". Perhaps it will simply offload its rocks just a little way offshore. Glaciers dump similar rock jumbles on land, but land deposits tend to be eroded away by the action of wind and rain, and most of the really old ice rocks that have survived around the world were formed in the protected environment of a shallow sea.

That's exactly what had happened to create Brian Harland's mysterious red stripe, and he was baffled by it. Why should he be troubled by signs of glacier deposits in frigid Svalbard? Because he already knew that in Precambrian times, Svalbard was very much warmer than it is today. He was sure that when the rocks of ancient Svalbard formed, conditions there were positively tropical.

They must have been, he reasoned, because most of the rocks in the outcrops he was studying were tropical. They were carbonates, pale grey and yellow rocks made of the same stuff as seashells. These rocks, though, were born before shells even existed. Unlike the chalks and limestones formed in more recent times from the crushed shells of sea-creatures, these carbonate rocks had nothing to do with the presence of living things. Instead, they had formed from a purely chemical process in Precambrian seawater, and then rained down on to the seafloor to be compacted into rock.

And here's the important point: this process happens only in warm seas. Cold water clings to its carbonate; only warm water releases it. That's why carbonate platforms hold up the sunny islands of the Caribbean. You'll find them beneath the Great Barrier Reef, and throughout the islands of Indonesia. And you'll also find them on either side of Brian's icy red stripe.

What's more, in the carbonates below his red stripe, Brian found oolite, a strange type of rock made up of tiny spheres that are squashed together like petrified caviar. This bizarre texture is also utterly characteristic of tropical climates. Six hundred million years ago, the islands of Svalbard had clearly been hot. Finding signs of ice among oolitic carbonate rocks was bizarre, like watching a glacier march across Barbados.

Brian began to investigate further. How about northern Norway? There, too, he found Precambrian carbonates interrupted by a layer of ice rocks. Greenland? The same. Now he began to pore over published papers, marking out anywhere in the world where geologists had mapped ice rocks in the Precambrian, the mysterious geological period that was devoid of distinguishing fossils. They were everywhere. Every single continent had the clear traces of ancient ice.

And then a whisper started in Brian's head. Perhaps the ice was global. Perhaps it had been everywhere in the world. At first his main interest in this was an arcane geological one. If the ice really had once been everywhere, the ice rocks it had left behind could be a Precambrian global marker. This might be a way of matching time-slice to time-slice for rocks all around the world, shedding light on the Earth's otherwise obscure Dark Age.[5]

Then Brian realized something else, something much more important. This ice came just before one of the most dramatic periods in Earth's history: the great evolutionary explosion that created complex life. Perhaps the ice was the trigger. It might explain why the Earth moved from the Dark Age to the Age of Enlightenment. Brian knew that biologists couldn't explain this breakthrough. There were simply no theories that made sense. And he realized that he might now have the answer. A climate change as big as the one he was proposing—surely that would be enough to shake the Earth from its slimy idyll, and jump-start the true beginnings of biodiversity. Excited, Brian marshalled his arguments about this "Great Infra-Cambrian Glaciation" and set about writing them up.

BRIAN HARLAND wasn't the first person to propose a global ice age. A Swiss researcher named Louis Agassiz had done so nearly a hundred years earlier. Agassiz suggested that ice had run rampant around twenty thousand years ago, much more recently than the Precambrian. In this he was partly right—ice had indeed stretched beyond its polar bounds then. But Agassiz's ice age was nowhere near as extensive as he'd imagined.[6]

Agassiz came upon the idea of prehistoric ice in the 1830s, while studying the geology of the Alps. He surmised that parts of the Rhone valley had been carved out by ice that had long since melted, and found boulders transported far beyond the existing fringes of the Alpine glaciers. He then began to discover that other parts of the world also showed signs of extensive ice. Scotland, for instance, bore carved valleys similar to those in Switzerland, but no glaciers remained there. Putting the evidence together, Agassiz proposed that there had once been a mighty ice age with glaciers stretching from the North Pole down to the Mediterranean. Then he grew more ambitious. The ice, he declared, had been everywhere. He even claimed to have found glacial traces in the Amazon rain forest.

Agassiz was the first ice champion. In the end he succeeded in convincing a sceptical world that ice could *ever* stretch beyond its present polar bounds. Thanks largely to him, we now know that the polar caps wax and wane on timescales of a hundred thousand years or so. When they stretch to their largest size, the world enters an ice age. The most recent one, which finished just eleven thousand years ago, is also the most famous, the time of woolly mammoths, mastodons and sabre-toothed tigers. And the receding ice heralds an "interglacial", the warm time between ice ages that we are experiencing today. Though researchers still argue about what causes ice ages, most believe that they are driven by subtle changes in the amount of heat reaching the Earth as it wobbles in its orbit around the sun.

But Agassiz's ice ages were nowhere near as extreme as the Snowball. In the last of his ice ages, the sheet of ice that coated northern America reached only as far south as New York. Another ice sheet blanketed northern Britain and Scandinavia, but scarcely made it into mainland Europe, let alone as far south as the Mediterranean. Some pack ice spread outwards from Antarctica into the Southern Ocean, and New Zealand felt the chill, but that's as far as it went. Ice didn't get anywhere near the equator. Other than an occasional iceberg, most of the oceans were ice-free. Global white-out? It wasn't even close.

Agassiz had a religious reason for overblowing the extent of his ice age. He felt it provided concrete proof that a providential God intervened in Earth's processes. God, so Agassiz thought, had deliberately introduced the ice to wipe out all previous creatures and leave an empty and bountiful stage to be occupied by His chosen new race: mankind.

This part of Agassiz's theory tumbled in the 1850s, when fossil finds demonstrated without question that most species had survived the ice age, and that the ice couldn't possibly have been global. Agassiz's many critics had feared the implications of his dramatic white-out, and they were delighted by this development. Earth was still, after all, a well-mannered and temperate place. Its climate might oscillate over time, growing a little warmer at times, a little colder at others. But nothing too bad had really happened, nor anything too extreme.

In the hundred years that followed, several people had noticed the much more ancient ice rocks of the Precambrian. Geologists had mentioned them in passing. Sir Douglas Mawson, the renowned Antarctic explorer who knew a thing or two about ice, had spotted signs of them in South Australia, and knew that they could be seen around the world. "Verily," he said in an address to the Royal Geological Society of Australia in 1948, "glaciations of Precambrian time were probably the most severe of all in earth history; in fact the world must have experienced its greatest Ice-Age."

But nobody had run with it. After Agassiz's embarrassment, who would want to declare that ice could ever have been global? Who'd put forward such an extreme idea, and risk exposing himself to the inevitable ridicule? Someone, perhaps, who was happy to swim against the tide, who took things at face value, who was self-reliant, unconcerned about how other people viewed him and made his own rules about status.

Brian Harland had grown increasingly convinced that the Precambrian was a time of global ice, one that was vastly more dramatic than the recent puny ice ages. By 1963 he had prepared all his arguments and set off for an international conference in Newcastle, in the northeast of England. He was ready to put his idea before the world.[7] But he had, as it turned out, picked an unfortunate time to champion the ice rocks. They were about to go crashing out of fashion.

This was largely thanks to the efforts of John Crowell, a geologist at the University of California, Santa Barbara. John enjoyed going to England. Though he lived in California, he had been seconded to the Admiralty in London during the Second World War. His background was in geology, but he'd retrained in meteorology for the war effort. He was one of the three scientists who predicted the height of the waves, both surf and swell, that would be experienced by troops landing on the Normandy beaches. Now, on his way to that same conference in Newcastle, he had a paper to present that had arisen indirectly from those London experiences.[8] He had accumulated evidence that most of the so-called ice rocks around the world had nothing at all to do with ice.

Through his work in the Admiralty, John had become fascinated by the behaviour of the sea just beyond the shoreline. In particular, he had started to investigate a set of undersea canyons close to the California coast. Most geologists assumed that the canyons must have formed on land, and then been flooded at some point when the sea level rose. But John discovered that the undersea world itself was a violent place, and that the canyons had been carved out by massive mudslides.

The canyon floors contained a jumbled mess of rocks, sand and stones that had been carried along on the back of the sliding mud. John realized that this looked just like the supposed leavings from a glacier. For him, the mixed-up rocks that had been called "glacial" for decades were nothing more or less than the effect of underwater mudslides. How to explain the ubiquity of these rocks? Simple. You get mudslides everywhere.

John's idea caught on quickly. He'd written a few papers in the late 1950s, and many researchers were assimilating his ideas. Trends come and go in geology as in everything else. John's mudslides were the hot new thing—ice rocks, laughably old-fashioned. When Brian tried to talk about his global ice in Newcastle, his colleagues were scornful. Didn't he know about the new findings? Why was he still harping on about an interpretation that had been so clearly superseded?

On the bus back from the conference centre, Brian found himself sitting next to John Crowell. He told John about his glacial rocks in Svalbard, and about his idea of the Great Infra-Cambrian Glaciation. John's response was kindly, and almost unbearably infuriating. They're not really made by ice, John said. Tell you what. Why don't I get some funding to go to Svalbard? Then I can check out how your rocks really formed.

Brian ground his teeth. He knew his rocks. He also knew what he was doing. He didn't need someone else to go and check his work.

In truth, it's not too hard to distinguish ice rocks from those created by mudslides. When ice is on the move, the rocks that it carries are often scarred and scratched with lines all pointing in the same direction. You can tell from the types of rocks whether they have been transported from long distances, or originated close to shore. In your deposit you might find a boulder that appears alone and has gently distorted the lines of the mud around it. Such objects obviously fell on to the seafloor from a melting iceberg overhead. Brian knew what signs to look for, and he knew he was looking at the effects of ice. But nobody would believe him, and John Crowell's new theory stood.

Over the next few decades, Brian published more and more careful descriptions demonstrating that the Precambrian rock jumbles around the world had been created by ice.[9] John Crowell, meanwhile, travelled to all seven continents, scrutinizing the rocks, gradually ruling them in (though, ironically, he never went to Svalbard). In the end, John and the rest of the world conceded.

Brian, the man perpetually ahead of his time, turned out to have been right all along. "He was willing to take a flier," John says now, rather ruefully. "And he turned out to be more correct than us sceptics."

But that still wasn't enough for other geologists. There was more trouble for Brian on the horizon, in the form of an old theory: continental drift.

IN THE early 1960s, earth science was being shaken to the core. Before then, most people had believed that the continents were fixed in place. Afterwards, almost everyone believed that they shifted. Compared to this, any worries about the ice rocks seemed minor. For geologists, the safe, comfortable ground beneath their feet was suddenly moving. Everyone was talking about it.

Plate tectonics, as the theory became known, was the new manifestation of an old idea. Back in the early 1900s, the German meteorologist Alfred Wegener had already made the disturbing proposal that continents moved around the surface of the Earth.

Wegener was a man blessed with intense curiosity about the world around him. Though studying the atmosphere was his day job, he found it hard to resist almost any earthly mystery that came his way. The planet, Wegener felt, was teasing him with its secrets. Once, after a journey to the Arctic, he wrote about the marvel of the northern lights, and how tantalized he felt by them. "Above us . . . a powerful symphony of light played in deepest, most solemn silence above our heads, as if mocking our efforts: Come up here and investigate me! Tell me what I am!"[10]

Part-time astronomer, geologist, adventurer, he made many expeditions to the polar regions, and even dabbled in hot-air ballooning. (At the age of twenty-six, he set the world record along with his brother, by staying aloft continuously for fifty-two hours.) Though he made important discoveries about the physics of the atmosphere as well as the wayward behaviour of continents, he was repeatedly rejected for professorships of regular universities, mainly because he refused to confine his research to a single academic area.[11] He wanted to know everything at once.

Wegener came up with the notion of continental drift around Christmas of 1910. He was looking idly at a map of the world when he was struck by how snugly the coastlines of Africa and South America fitted together. They looked like two pieces of a jigsaw puzzle. He suddenly wondered whether they had once been part of the same continent, and had only later drifted apart. Intrigued, Wegener began to find other evidence around the world that disparate continents had once been connected. There were ancient floodplains of volcanic lava in Africa and South America, which matched like two halves of a coffee stain. There were animal fossils of exactly the same types and mixtures on both sides of the Atlantic. In place after place, the geology or the fossils matched uncannily between continents that were now far apart. He concluded that the continents must surely move.

The idea was bold, intriguing—and widely derided. Most of the geological world immediately rejected it. They had been taught from birth that the Earth's surface was safely fixed in place, and Wegener's alternative made them extremely uncomfortable. Wegener didn't help his case by coming up with a preposterous mechanism to explain how the continents moved; he mistakenly believed that they ploughed through the Earth's solid crust like an icebreaker, and at breakneck speed. It also didn't help that he was a meteorologist rather than a "proper" geologist, and yet was putting forward geological evidence to support his claims.[12]

But Wegener didn't give up. He pushed and harried and accumulated evidence, directing tireless energy and determination into trying to prove his point. But before he could convince the world that he was right, his curiosity finally killed him.

He died during an expedition to Greenland in 1930. The plan was to establish a station high on the summit of the ice cap, where a few researchers would spend the bitter Greenland winter. Through months of isolation and darkness, they would study everything: the wind, the weather, the stars, the auroras, the snow, the ice. But the expedition encountered problems from the beginning. Unyielding fields of pack ice stranded Wegener's ship off the coast of Greenland for an agonizing thirty-eight days. By the time he finally broke through to reach land, the summer was half over. Though he sent off his advance party to set up the central station, known as Mid-Ice, he was already worried that there would not be enough time to supply it fully for the long winter ahead.[13]

Eventually, but very late in the season, Wegener set out for Mid-Ice himself. He took a team of hired Greenlanders and fifteen dog-sledges weighed down with provisions. The conditions were appalling, and after one hundred miles of blizzard and intense cold, the hired hands rebelled. Wegener ploughed on with just two companions. By the time the three of them staggered into Mid-Ice, on 30 October, the temperature was fifty below zero, one of the party was badly frostbitten and they had no supplies left to deliver.

The situation was desperate. Mid-Ice had scarcely enough food and fuel to supply its two present occupants through the winter. There was no way that all three newcomers could stay as well. Wegener celebrated his fiftieth birthday on 31 October, and, taking Greenlander Rasmus Willumsen with him, he set back out again on to the ice.

The details of what happened next are sketchy, pieced together from the scant clues that Wegener left behind. Around 160 miles from base, it seems that he abandoned his own dog-sledge, and started to ski alongside Willumsen's. He always skied fast. "The journey must never come to a standstill," he had often told his companions on earlier expeditions. "The natural pace of the dog-sledges is the normal speed to which everyone else must adapt himself." Fine words for a young man. Not such a good idea for a fifty-year-old, in half-light and bitter cold, on a surface that had been whipped up into solid waves of ice by the driving wind. At some point during this frantic, lung-bursting dash for safety, Wegener suffered a heart attack and died. His body was found neatly buried in snow, the grave marked by a cross fashioned from his skis. Of his diary, and of Willumsen, no trace has ever been discovered.

Wegener knew the risks involved in his science. "Whatever may happen," he had written before the expedition, "the cause must not suffer. It is the sacred thing which binds us all together. It must be held aloft under all circumstances, however great the sacrifices may be. That is, if you like to call it so, my expeditionary religion. It guarantees, above all, expeditions without regrets."

The German government offered to send a battleship to retrieve Wegener's body for a state funeral. His wife refused, and so his body still lies somewhere deep in Greenland's ice cap.[14] One day perhaps it will find its way to the sea, encased in an iceberg. If so, when the ice eventually melts, Wegener's remains will be gently deposited like a geological dropstone on the seafloor.

* * *

WITH WEGENER'S death, continental drift lost its advocate and the idea foundered. But a few scientists still held the candle for him and his idea. And to anyone obsessed with both geology and the Arctic, Alfred Wegener was the perfect hero.

Brian Harland had certainly always loved continental drift, though he says that's because of the idea, not the man. The theory had been proposed in 1912, five years before Brian was born, and he first heard about it as a schoolboy. Delighted, he gave a talk about it to his school. By then, most geologists had discounted the theory, and his teachers were unimpressed. At Cambridge, continental drift brought Brian more trouble. Cambridge was one of the main centres of dissent about the idea, and whenever Brian mentioned it, nobody wanted to know.

But Brian wasn't particularly bothered. He believed in letting the facts tell their own story, and to him everything seemed to point towards continental drift. The eastern region of Greenland, for instance, had rocks that looked just like those in the east of Svalbard, several hundred miles away. But the ones in between, those of western Svalbard, were completely different, even though they were obviously from the same time period. Brian was sure that western Svalbard must have been a separate chunk of continent that had wandered north and thrust itself between the other two places. He was sure that Wegener was right.

Eventually, once again, everyone else caught up with Brian. Most of the world's rocks contain tiny magnetic particles of iron that act like compass needles, pointing towards magnetic north. Unlike compass needles, though, these particles can't swing around at will. When the rock hardens from its original soft sediment, they are frozen in place, and if the continents haven't moved since then, all the magnetic particles should still point north. But in the early 1960s, the tentative new science of rock magnetism began to reveal a surprise. Rocks on different continents had "compass needles" that pointed toward different "norths". The only explanation anyone could think of was that the continents had changed their geographical positions since the rocks first formed.

Then physicists began to get involved, and suddenly everyone seemed to have new evidence showing that the continents moved. Strictly speaking, the new theory of plate tectonics was different from Wegener's. Continental drift suggested that the continents themselves were moving. We now know that the Earth's entire surface is broken up into plates that shift around, some of them bearing continents on their backs. But still, in essence, Wegener's idea was vindicated.

It's ironic, then, that this vindication put yet another spoke into Brian's Snowball wheel. If the continents truly did move, then there was a much easier way to explain Brian's ice rocks than the outrageous idea of global ice. Everyone knew that the poles were cold and the equator was hot. So each continent must simply have drifted over to the polar regions to collect its ice, and then wandered away again.

Of course, Brian had already thought of that. He was a *champion* of continental drift. As he'd made clear in his papers about the ice rocks, he'd already tried to fit all the continents together in a huddle around the pole. But there was simply no way to do it. However he arranged his geological jigsaw, he couldn't cram all the continents into the polar regions. Some were always left out in the sun.

But plate tectonics was now on everyone's lips, and to many geologists, moving continents could explain everything. Brian realized that he had only one option. He had to prove that at least one of the continents was near the equator when the ice formed.

This, he figured, would be tantamount to proving that there had been a global freeze, since it's extremely hard to freeze the equator without freezing everything else, too. Our sun's rays come to us untrammelled, in single-minded parallel lines. At the equator, they strike the equator more or less full on. At the poles they always hit at an angle. Shine a torch directly on to a piece of paper, and you'll see a neat round circle. Now tilt the torch, and the circle will grow and distort into an ellipse—the same amount of light spread out over a greater area. The same thing happens on our spherical planet. Directly overhead at the equator, the sunlight is fiercely intense. But the further north or south you go, the more it spreads.

The upshot of our celestial geometry is this: it's easy to freeze the polar oceans and to make glaciers in Alaska and Antarctica, even at sea level where there's no thin mountain air to help. But the closer you get to the tropics, the harder it becomes to make ice. If the temperatures had somehow dropped low enough to freeze the equator, everything else must also have frozen.

To see if the equator really had been frozen, Brian decided to adopt the same technique that had been used to vindicate Wegener: rock magnetism. Many rocks come with their own magnetic birth certificate, because they adopt the local pattern of the Earth's magnetic field. This has a classic, characteristic shape. Stick one end of a piece of wire into the top of an orange, bend the wire over, and push the other end into the bottom of the orange. That will give you some idea of how the Earth's magnetic field looks. It shoots straight upwards at the poles and passes horizontally over the equator. If you were standing near magnetic north, the field would pass through your foot, say, and up out through your head. If you were standing near the equator, the field would pass through you horizontally, across your hips, waist and shoulders.

When they're young and soft, rocks are still impressionable; they can take on the stamp of the Earth's field. The magnetic particles they contain line up the way the Earth's field does. As the rock is compressed and hardens, these particles are fixed in place, and the direction they point in tells you where they were born. If their field is vertical, they were born near the poles. If horizontal, they come from the equator. The rock magnet is weak, to be sure, much weaker than one that you'd stick on your refrigerator. But it is just measurable.

Brian decided to try to find out if his rock samples from Svalbard and eastern Norway had fields that were horizontal. He built a new instrument, so sensitive that it could detect magnetic tremors from the elevator as it rose and fell in its shaft, fifty yards away. He measured sample after sample. At first he was thrilled. The Svalbard rocks showed a horizontal field—just what you'd expect if they'd formed near the equator.

But he couldn't really be sure. The weak magnetic field in the rocks might have been altered in the hundreds of millions of years since they'd been created. In the 1960s, rock magnetism was still in its infancy. There were no sophisticated techniques to rule this out. If the field had been altered after the rocks formed, there was no way that Brian would be able to tell.

Then came a devastating blow. The university authorities built a car park outside the lab. Any further magnetic experiments would be hopeless. Brian was already working at the limits of the available technology. The rock magnets were so weak, and the instruments to measure them still so crude, that any slight changes in the field around them would wreck the results. And now the magnetic field in Brian's lab changed every time a car entered or left.

He published his findings,[15] but he always knew that he hadn't made his case. Nevertheless, he continued to investigate the geology of Svalbard, organizing more than forty expeditions in all. Eventually he put his findings into a prize-winning book, the definitive geological guide to the archipelago, which contained 500,000 words and took him five years to write.[16]

Brian has never stopped working—he's not the retiring type. Even in 1990, at seventy-three, he was still studying Svalbard's rocks by day, and sleeping at night in a tent pitched on the shore. Thanks to their newly protected status, polar bears had grown bold by then, so the camp was surrounded by tripwire attached to a device that would fire blank cartridges to frighten off any marauders. But Brian was characteristically unfazed about this danger. The wire, he felt, was just a damned nuisance—set off many times by stumbling people, but never by bears. His final expedition was in 1992, but he has not stopped working on the Svalbard rocks in his collection. Now, at eighty-five, he still goes into his office in Cambridge every day.

Though Brian never succeeded in proving his Great Infra-Cambrian Glaciation, he was always obstinately convinced it was right. To bring the idea out of the cold, however, would require much more evidence. The next step would be to demonstrate without question that ice had been present at the equator. That would take an unusual scientist, someone with an eye for problems that were a long way out of the ordinary. Someone who was also a world expert in magnetism.

FOUR

MAGNETIC MOMENTS

Joe Kirschvink adores magnets. You might say he's irresistibly drawn to them. In fact, that's just the sort of crummy joke that Joe would probably make. Even now, in his late forties, he loves to act the clown, with his bright button eyes, and brows that periodically shoot high into his forehead. He's compact of build, and fizzes with energy and ideas. His unruly hair and neat moustache are the colour of wet sand.

Joe is a professor at the California Institute of Technology, an august institution that lies among the villas of Pasadena in southern California. Caltech professors are hard-nosed people. It's one of the most fiercely competitive academic establishments in the world, filled with some of the most gifted scientists. They work long hours, know how to sell themselves, guard their patches jealously, and make sure they stay ahead. You don't often come across a Caltech professor like Joe, who constantly describes his own ideas as "nutty", and invites you to call him a nut. "Honestly," he says. "I don't mind."

In truth, Joe Kirschvink is one of Caltech's most brilliant brains. His strength lies in his ability to look at old problems in a new way. He delights in topics that other scientists shun, ones that have a whiff of the weird about them. Joe often does his work away from the scientific spotlight, but he tends to make the kind of discovery that swings the spotlight over to him. And then he moves on to something else. His motto could be "never dismiss, never assume". In his introductory geology class, he has each student write a "nut" paper, in which they have to consider an offbeat hypothesis, ideally one that has been ridiculed by the scientific establishment, and then describe how they would rigorously test the idea. His students love it.

Joe first began experimenting with magnets when his father inadvertently wrecked the family microwave with an exploding golf ball. (He had been told that warmer golf balls travelled further and was attempting to heat the ball up quickly.) Joe got the machine parts to play with. Later, as an undergraduate at Caltech, he used his genius with magnets to good effect in a tradition called "stacks". At one point in their final year, all the seniors are supposed to create elaborate locks for the doors to their rooms and challenge junior classmen to break in. Joe's particular one is legendary. From the outside, the door was blank—there were only a few magnets and some written clues. But on the inside of the door, Joe had rigged up a series of magnetic switches that had to be tripped in exactly the right order to open the door. Standing outside, the hopeful lock-breakers had to move a magnet to different points in the blank door using only the written clues as a guide. Each wrong move was punished with a loud blast of "The Ride of the Valkyries". The stack proved too inventive even for Caltech students. Nobody managed to break the lock and claim the two gallons of ice cream waiting inside.

Around the same time, in the mid-1970s, Joe began to experiment with naturally occurring magnets. While on a trip to Australia, he heard that "north-seeking" magnetic bacteria had been discovered in Massachusetts. These creatures had achieved a clever evolutionary trick. Bacteria usually find their food in the depths of ponds and puddles, so they have an incentive to know which way is down. In the northern hemisphere the Earth's magnetic field lines slant downwards, so for the bacteria "north" equals "down" equals "food". To find food, they simply clamp their internal magnets on to the Earth's slanting field lines and slide down, like firemen down a pole.

Joe was in Australia when he heard about these strange magnetic bacteria. He knew that in the southern hemisphere, the Earth's field lines pointed the opposite way, so that "north" there was "up". What, he wondered, happened to southern bacteria? He immediately rushed off to find likely looking grubby ponds and pools of water, and snared a few bacteria. Using the magnifying glass and magnet that he always carried with him, he found that Australian bacteria swam consistently south. Their inner magnets were upside down compared with those of their northern cousins.[1]

The Aussies loved it. Joe found himself unexpectedly on the front page of the *Canberra Times,* brandishing a beaker of bacterial sludge from the Fyshwick Sewage Works. He knocked the Ayatollah Khomeini off the headlines. His south-seeking bacteria became celebrities. Joe set them up at a Canberra University party so geologists could watch them swimming backwards and forwards as he flipped a magnet below them. One partygoer, peering over his shoulder, asked if they liked beer. Joe promptly applied a drop of Foster's lager to one side of the beaker, and then flipped the magnet to make that side "south". The bacteria galloped toward the spreading yellow liquid, but as soon as they tasted it, they turned tail. Australian bacteria apparently do not like beer. Later, Joe tested the northern bacteria in his hometown of Phoenix, Arizona. The American bacteria showed no inclination to turn tail at the Foster's-water interface. They swam directly into the beer, and promptly perished. "They died happy," says Joe. American bacteria, unlike Australian, have no idea when to call it a day.

Joe was just fooling around with the beer, but these experiments had an underlying seriousness. He became intrigued by the way the Earth's magnetic field affected the creatures on its surface. At Princeton as a graduate student, he discovered tiny, pure magnets in the brains of honeybees and pigeons and proved—to everyone's astonishment—that the creatures use these in-built magnets to navigate.[2] This was a classic case of finding a scientific basis for ideas that had previously been ridiculed. Pigeon fanciers had long believed they should not race their birds when there was a magnetic storm. Beekeepers were convinced their charges had an innate sense of direction. Nobody else believed them. Joe put science where the myths had been. His findings, made while he was still a student, are now described in even the most staid of textbooks.

Joe also found magnets in the brains of fish, whales and even humans.[3] Thanks to Joe, it's now known that we all carry tiny, built-in magnets around in our heads. These magnets may even help humans navigate, although Joe never managed to prove that we use our magnets the way bees and pigeons do.

Joe's obsession with magnets even extends to the names of his children. His wife, Atsuko, is Japanese, and his sons are called Jiseki, which means "magnetite" in Japanese, and Koseki, which means "mineral". Jiseki came first, in 1984. As the firstborn son, with a lineage that traced back through Atsuko to the Japanese imperial family, the child had to be given a distinctive and meaningful name—one that would be approved by the temple monk back in Japan. This approval hinged crucially on how the name looked when written down. One night, after many fruitless suggestions, Joe asked his wife what "magnetite" would look like. She wrote down Jiseki. The name had a curious shape, like two lightning bolts next to one stone over another. Atsuko called her mother in Japan, who immediately took the name to the temple. It passed every test that the monk set.

Koseki, "mineral", appeared two years later. Though his name didn't pass quite as many temple tests, it was still perfectly acceptable, and everyone agreed that it was the natural follow-up. And if there had been a third child? This is what Joe says about the name that would have been next. "As well as magnetite and minerals, I work on meteorites. The Japanese word for 'meteorite' is 'inseki'. Time to stop."

Now Joe lives in Pasadena, but Atsuko and the two boys live across the Pacific, in Japan. Joe goes there for three months every year. Atsuko was very unhappy in California, and Joe told me rather sadly that this was the only arrangement that seemed to work: "I've discovered that if you spend a hundred per cent of your time with someone, and they're only twenty-five per cent happy, then life is miserable. But if you spend twenty-five per cent of your time with them and they're a hundred per cent happy, life is much better."

Perhaps that's why Joe is so close to his students. He treats them as family. He gives them responsibility, but tempers this with endless support. They know he would go the limit for them, and they love him for it. They say he's "generous", "modest" and "brilliant". Everyone has a "Joe story" to tell. Here's a sample. Joe, they say, speaks two languages—"English" and "foreign". He travels the world doing fieldwork, and after every new field site, his "foreign" becomes more confused. Once, at a petrol station in Baja California, Joe wanted to thank the elderly Mexican man who had pumped his petrol and cleaned his windscreen. Joe had just returned from collecting samples in Russia and, confused for a moment about which branch of "foreign" he should be speaking, said, *"Spasibo"*, which is Russian for "thank you". The Mexican petrol attendant beamed with delight. *"Pozhalsta,"* he said. "You're welcome." This particular Mexican turned out to be a Russian immigrant. It could only happen to Joe.

All of Joe's research ideas involve magnets of some kind, and there's usually some kind of controversy involved. But those are the only connecting points. Wherever the magnets are, in biology, geology, chemistry, or astronomy, you'll find Joe. He's suggested a way that animals might use magnetic fields to sense imminent earthquakes. He's even worked on evidence for alien life in a meteorite from Mars. (This last work was done in an ultra-clean lab to avoid any earthly contamination. The sign outside says: "This is the Door to the Planet MARS. Only the purest in Heart, Mind and Body may enter here.")

Still, Joe doesn't seek out controversies for their own sake, and he's no contrarian. He just enjoys delving into areas where his unusual way of thinking can resolve disputes and mysteries. He's like a curious and energetic child, blown by the wind into following now this idea, now that, whatever catches his attention. There's a downside to this. Joe claims he does things for fun, not for recognition, and that's just as well. Not many fellowship and prize committees appreciate researchers who work in so many different fields and in such wildly imaginative ways. Joe's students whisper about this angrily, sure that he has been passed over for honours that they feel he deserves. But his critics would say that he spreads himself too thin; he deals in too many different areas, they say, and doesn't follow them through.

When the idea of a global freeze caught Joe Kirschvink's attention in the 1980s, he betrayed both his greatest strength and his greatest weakness. He made what's probably the biggest, most imaginative leap of anyone involved with the theory. Without his insights, the Snowball idea would almost certainly have foundered. But then came the weakness: he didn't follow through. He came up with the ideas, put together the picture, and then moved on to where the wind next blew him.

JOE'S PERSONAL Snowball saga started one day in 1986 when he received a manuscript from a prominent geological journal. The paper was on a topic right in Joe's area of magnetic expertise, and the journal wanted to know whether they should publish it. Not surprisingly, since the world of magnet fanatics is a small one, he knew the researchers who had written it. George Williams, then from the Broken Hill mining company near Adelaide, Australia, and Brian Embleton from Sydney, had been studying a small outcrop in South Australia. The outcrop was full of stones dropped by icebergs, and the other tell-tale signs of a deep freeze. And the researchers wanted to figure out where the ice rock had been born: near the poles, where you'd expect, or near the equator?

Remember that the magnetic field frozen into a wandering rock at birth tells you everything you need to know about its place of origin. If the field is vertical, the rock was born at the poles. If horizontal, then it came from the equator. And according to the paper that had appeared on Joe's desk, the ice rocks from South Australia had fields that were almost as flat as they come. This place, the researchers claimed, had been ice-covered within a few degrees of the equator.

Joe was unimpressed. This work suffered from the same limitations that Brian Harland's had. For the approach to work, the researchers needed to prove that the field was frozen into the rocks *at the moment of their birth,* something that Brian had never managed. There were several ways in which the field from the original homeland might have been subsequently wiped out and printed over. Heating rocks to high enough temperatures erases their magnetic memory—like leaving a credit card on a radiator. Water flowing through pores in the rock can also deposit new magnetic minerals there, which adopt whatever the field direction happens to be as they settle in place. For any of these reasons, Williams and Embleton could have been reading a fake birth certificate.

As Joe read carefully through the manuscript, he realized there were ways the necessary proof could be found. Times and techniques had moved on since Brian Harland's day, and there were now ways to tell whether a field had been reset. In the paper he was reading, Williams and Embleton hadn't included these crucial tests. Joe recommended rejection.[4]

But something about this manuscript had piqued Joe's ever-lively curiosity. Could there really have been ice at the equator? Joe thought he already knew of a cast-iron reason why the Earth simply couldn't freeze this way. Snow and ice are dazzling. They reflect sunlight. A shiny white Earth would send the sun's rays bouncing back into space. So if Earth ever got into that state, Joe thought, it should never be able to get out of it.

That much had been first suggested back in the 1960s. Just about the time when Brian Harland was investigating his Svalbard rocks, a young Russian climatologist, Mikhail Budyko, was playing with this idea: what would happen if you let ice run riot on the Earth? Budyko set up a simple model in which ice started off at the poles, but could grow as it wished, and then let the model run.

The result horrified him. White ice at his model's poles reflected sunlight, making the Earth a little colder. Because temperatures were colder, more ice grew, which reflected more sunlight, and so on. The ice in Budyko's model grew and spread and grew and spread until it became unstoppable. When white ice reached the tropics, it tipped over a threshold and the entire Earth froze.[5]

This was the "ice catastrophe". After it, there would surely be no way back. If the Earth had ever frozen over like this, its shiny white surface would have reflected sunlight back into space. That, Budyko felt, would be a disaster. The planet would cool catastrophically, and he thought that the ice could never melt again. Once you entered the ice catastrophe, he decided, there would be no way to escape. Earth would have been doomed to spin through space, frigid and lifeless.

Obviously that didn't happen. So, equally obviously, Budyko reasoned, Earth must never have frozen over completely. Budyko, and everyone else, concluded that his ice catastrophe must never have taken place. He seemed to have provided yet another reason why the Snowball could not have been.

Two decades later, in the 1980s, Joe Kirschvink knew all about Budyko's ice catastrophe. Everybody did. And he also knew its corollary: the Earth can't freeze. If modellers came up with a white Earth in their experiments, they simply threw those results away.

So what about this new evidence from Williams and Embleton? Perhaps, Joe felt, he should probe a little more. An Australian geologist whom Joe knew happened to be travelling to Williams and Embleton's particular part of South Australia at about that time, and Joe gave him a compass. "Pick me up a sample or two," Joe said. "Nothing fancy, just hand samples. But check their orientation when you chip them off the outcrop."

THE AUSTRALIAN outback has many guises. There's the famous dry red centre that houses Uluru and Alice Springs and the weird red mining town of Coober Pedy, the world's biggest producer of precious opals. The summer's heat is so fierce there that half of its meagre population lives below ground in mud "dug-outs", and its post-apocalyptic scenery is the darling of movie producers. But there's also a subtler wilderness, several hundred miles south of Coober Pedy, on the way back to Adelaide, and civilization. There lie the Flinders Ranges with their dusty valleys and row upon row of rounded mountains. They are softer than the red centre, their colours more muted. And they have had many parts to play in the Snowball story.

The main route to the Flinders winds through a narrow valley, at the foot of a sharply pointed mountain called Devil's Peak. Here, at Pichi Richi Pass, is where Joe's rock samples came from.

To get to the outcrop, you climb over a small wire fence on to a patch of bare, rocky ground scattered with tufts of dry grass. Winter's the season to go there—in a southern summer, the heat would be unbearable. Even in March, the temperature quickly climbs to 90 degrees F. or more. At least there is merciful shade to be had among a stand of eucalypts with their peeling bark and graceful bone white trunks, and the odd desert oak with its black spiny fruit shells still clinging to the bare branches.

Through the trees, a hillside slopes down to the floor of a dry gully. There are no trees on the hill, just patches of golden grass, and at its feet lies a jumble of mud-coloured rocks with a strange pink tinge. The rocks are stranger still up close. All of them are shot through with rhythmic dark lines, as if they had been painted in neat, careful parallels.

The lines, though, pre-date artists by hundreds of millions of years. They are the last remnants of ancient tides. Once this area was underwater, just offshore from an estuary. The tide flooded in to land, carrying with it a slurry of fine sand. As the tide ebbed, sand was washed back out to sea and deposited gently right at Pichi Richi. In and out, ebb and flow, as sand landed periodically on mud, the rhythmic patterns built up. In the end they solidified, becoming sets of regular dark lines in a pale reddish mudstone. Such rocks are called "tidal rhythmites", and they are very rare. A few good waves will destroy the pattern completely before it can harden into rock. But somehow the seafloor thereabouts was protected from waves, and the layers survived.

George Williams, the researcher whose paper had caught Joe's eye, had already used the rhythmites to work out exactly how long days lasted when the Earth was young. One day hasn't always passed in a little over twenty-four hours, as it does now.

Since the whirlwind days of its youth, the Earth has been steadily slowing on its axis and days have been getting longer. Those short early days are frozen into the rhythmic patterns of Pichi Richi's rocks. From the monthly tidal cycles that he measured there, Williams figured out the number of days in a Precambrian month, and the number of months in a year. When his rhythmites were still mud and sand, a little over 600 million years ago, a year lasted thirteen months, and a day less than twenty-two hours.[6]

But for the Snowball story, the rhythmites have a more important role to play: their neat inscribed lines show clearly where the rock slumped and folded as it was forming. Joe planned to use these folds for a definitive test of whether the rock's magnetic field came from its birthplace.

Imagine a slab of something flexible—an eraser, say. Now take a pen and draw horizontal lines along the side of the eraser, so it looks a bit like a layer cake seen from the side. If you bend the eraser into an arch, the lines will bend too, following the curve of the arch. However, if you first bend the eraser, and then draw lines horizontally across it, the lines won't follow the shape of the arch at all. They'll cut right through it. A fold test works the same way. If the magnetic field in the rock formed before it folded, the field lines will follow the curve. But if they were overprinted later in the rock's history, they will cut through the curve, ignoring its shape completely.

So Joe took a slab of rock from Pichi Richi whose dark tidal lines had clearly folded into an arch, and began to investigate. Did the lines follow the curve, showing they were original, or did they cut through it, showing they were overprints?

A graduate student made the painstaking measurements and then presented Joe with the results. Joe was astonished. The lines did seem to follow the curve. This changed the odds dramatically in his mind. Perhaps the flat equatorial field really had come from the same time as the ice.

And then he had another idea.

On a field trip to Canada, he had noticed that among the Precambrian rocks there were thick red layers of ironstones. That was mysterious. Ironstones belonged to a time much earlier in Earth's history, when oxygen first appeared in the atmosphere. Before there was any oxygen, the seas were full of dissolved iron that had come from the Earth's interior, pouring out of underwater volcanoes and deep-sea vents. But as soon as the air became oxygenated, the ocean's iron literally rusted. It turned into solid iron oxide and was sprinkled on to the ocean floor to become the layers of ironstone you can see in ancient rocks today. All that makes perfect sense. But the ironstones then stopped. Since the air was full of oxygen, dissolved iron could never build up in seawater the way it once had, and there were no more ironstone layers.

Until, that is, one bizarre iron blip a few billion years later, the blip that produced the ironstones Joe saw in Canada. They appear elsewhere in the world, and always in the same geological time period—just towards the end of the Precambrian, just before the first complex animals emerged from the slime, just around the time of the mysterious ice deposits. And then, shortly afterwards, they vanish again. The very late Precambrian is the only time other than that very early period when ironstones appear in the whole of Earth's history. The question is why.

Joe realized that he might now have the answer. Maybe the ice was the cause. If the oceans had frozen over, perhaps seawater had been cut off from the air long enough to accumulate lots of dissolved iron from underwater volcanoes. If the ice then melted and exposed all this iron to the air, then—boom!—it rusted again, and a new set of ironstones was born.

Everything was making sense. The magnetics from Pichi Richi seemed to show that there was ice at the equator. That's the hottest place on Earth. If the equator freezes, everything else has to freeze, too. And now the ironstones provided independent evidence that the oceans had frozen over at the same time. Four years before Paul went to Namibia for the first time, Joe was looking, suddenly, at total white-out.

He was thrilled, but also troubled. He wanted to believe in a global freeze, but what about the contrary evidence from Budyko's ice catastrophe? The Earth couldn't possibly have frozen over, or it would have remained frozen for ever. And yet Joe had seen the evidence for equatorial ice with his own eyes, and measured it with his own instruments. In science, as in life, when theory conflicts with evidence, it's usually the theory that's wrong. But try as he might, Joe couldn't think why. The ice began to haunt his dreams. He found himself tossing and turning at night, waking up in a sweat, thinking, "Did the Earth really *do* this?"

And then suddenly he had it. He conjured up a way out of the ice catastrophe. The melters of the Snowball, the evaders of the ice catastrophe, were volcanoes. Then as now, Earth was scattered with volcanoes, which periodically spilled out molten rock and heat. The Snowball wouldn't have stopped them. They could erupt perfectly happily, even under ice—as they do today in Iceland.

The lava from these volcanoes wouldn't itself have been enough to melt the Snowball. But when volcanoes disgorge their lava, gas comes too. Curling plumes of gas rise from the sides of an active volcano. An eruption can fling great clouds of gas high into the atmosphere. Gas bubbles up from hot vents beneath the sea. And one of the main gases to come from the heart of a volcano is also a villain in the world today: carbon dioxide.

Carbon dioxide, CO_2, is the gas that threatens us all with global warming. Every molecule of carbon dioxide traps a little heat. The more CO_2 you have in the sky, the more heat you trap. The effect is, famously, like that of a greenhouse. Carbon dioxide lets sunlight in, but prevents the Earth's body heat from escaping, providing a very effective and cosy way of warming a planet up.

And here's what Joe suddenly, thrillingly realized. Each volcanic eruption would pour a little more CO_2 into the sky, and gradually the greenhouse would heat up. Carbon dioxide would build up in the air, and wrap the Earth in a blanket of warmth. This blanket would trap more and more heat, and after millions of years the heat would finally melt the Snowball.

The idea is even cleverer than it sounds. In normal times, carbon dioxide doesn't usually build up in the air like that. Volcanoes are erupting all the time, but the Earth usually has a sort of built-in thermostat operated by rainwater, which strips out the excess CO_2. When rain falls through the air it picks up CO_2 and, as a result, becomes slightly acidic. The acidified rain lands on rocks and reacts with them chemically, handing over its load of dissolved CO_2. By this mechanism, any excess CO_2 is scrubbed out of the air and locked away inside a new rocky matrix. Over millions of years the Earth ends up more or less in balance, never too hot or too cold.

But if the planet freezes over and its rocks are blanketed with ice, the CO_2 thermostat switches off. Now there is all give and no take. Volcanoes keep giving out carbon dioxide, but the ice-covered rocks can no longer soak it back up again. Left unchecked in this way, the greenhouse effect of the CO_2 will build and build until it's ten times, even a hundred times, what we have today. With our puny little efforts to pollute our atmosphere by burning oil and coal, we'll maybe double the amount of carbon dioxide. Over millions of years the runaway volcanoes would have sent CO_2 levels spiralling beyond the wildest imaginings of any oil conglomerate. The Earth's atmosphere would have turned into a furnace.

The aftermath of the melting would have been hell on Earth. Dante, says Joe, would be proud of it. For tens of thousands of years—until the excess carbon dioxide was finally locked back up in the rocks, and the furnace switched off—any slime creatures that had survived the freeze would have found themselves scorched. These would be prime conditions, perhaps, to weed out all but the few, and to set the stage for the emergence of a whole new kind of life.

Joe's idea was ingenious. So what did he do with it? Embark on a round of conferences and lectures presenting it to his peers? Publish the idea in an acclaimed scientific journal? Not exactly. Though the academic world runs on conference presentations and published papers, Joe did little of either. The graduate student who made the measurements on the Pichi Richi fold presented the results at a single academic meeting,[7] but never wrote them up. Joe put together his thoughts about the Snowball in a tiny two-page paper, in which he gave his idea about the volcanoes just a few paltry sentences.[8] The paper took four years to come out in an obscure book, a vast monograph read only by the highly committed.

Joe still can't explain fully why he didn't grab his brilliant idea and run with it. Years later, when he read Paul Hoffman's work, he was chagrined. "Damn it. Why didn't I think of that?" But by then he had moved on to other things, tugged by his graduate students into their areas of interest. Perhaps he simply didn't have enough confidence in his deep freeze. It was one more "nutty" idea among the rest.

Still, Joe had made the strongest case to date that there was once ice at the equator. He had envisaged a worldwide freeze-over, and found a way out of the worst problem the idea faced. In a way, he had put the story together. What was needed now was evidence that he was right. Was there any sign left on Earth of his global super-greenhouse? Could anyone really show what had been happening to life in the oceans when the Snowball was in progress, and during its heated aftermath?

Meanwhile, Joe contributed another essential item to the story. Brian Harland had called this global freeze the Great Infra-Cambrian Glaciation. Joe was made of snappier stuff. He remembered as a child moving from Arizona to Seattle, where he endured three miserable freezing winters. At first the wet Seattle snow bemused him, but he quickly learned to pack it around a rock to make a snowball that had maximum impact. What was Harland's Infra-Cambrian Glaciation but snow packed around a rocky planet? Joe rechristened the glaciation with the name it has borne ever since: Snowball Earth.

He also did one more thing, something that would prove crucial. At a conference in Washington, D.C., in 1989, Joe found himself chatting with Paul Hoffman over dinner. Paul was still working in Canada at the time. He hadn't even encountered the Namibian ice rocks then. But as they ate, Joe cheerily told Paul about his latest crazy theory. That night he planted a seed in Paul's mind, and when Paul found crucial Snowball clues in the ice rocks of Namibia, the seed began to germinate.

FIVE

EUREKA

When Paul Hoffman became obsessed with the Namibian ice rocks, he sensed they would eventually reveal to him some important story about the history of the Earth. Throughout the 1990s, he returned every year to study them. He was where he wanted to be, driven by a new mission, as happy now in Africa as he'd ever been in the Canadian Arctic.

Each season he went eagerly from outcrop to outcrop, making every moment of his precious field time count. There was no such thing as a rest day. Any days not spent on the outcrops were for driving between them. Paul would often hike up and down hills, over rocks and down stream cuts and gullies. He went out fast and hard. He would work his way up through the sequence, noting what type of rock was there, how the rock formed, whether it came from a delta, a river, a deep ocean floor, how it related to the rocks above and below. He would put a sheet of clear Mylar over an aerial photograph and draw in the rock type with a sharp pencil. Sometimes he stopped to measure the rocks in a particular section. He took out a folding carpenter's ruler and climbed up the section, recording its thickness yard by yard, noting any peculiarities in his waterproof yellow notebook (waterproof to prevent the figures smudging from sweat—there's no rainfall in the Namibian desert during the dry season). He developed his own shorthand of neat hieroglyphs for sandstone and mudstone, diamictite and carbonate, and the sharpness of the contacts between them.

This was standard field geology. The first task in a new field site is always to build up a detailed picture of the whole geological terrain. Nothing geologists discover holds any weight amongst their colleagues unless they fully understand, and can describe, the context in which it is found.

Sometimes Paul collected samples. He would take a geological hammer and smash chunks off the rock face. There's something about holding a geological hammer that makes you want to hit rocks. Weigh one in your hand, and you'll find yourself itching to whack something with it. Still, to carve a hand sample into just the right shape for your pocket requires considerable skill. Paul is very, very good at it. If his geology career went awry, he could make a living as an ornamental rock chipper.

Perhaps he wants a sample of a particular structure in the rock, or something that shows exactly how the contact looks between two different rock types. He holds an unwieldy chunk of rock in one hand and chips away at it casually. Thwack, smack, smack, and all the useless bits miraculously fall off, all the right bits stay behind. The trick apparently lies in choosing the right sample, finding the flaws that will dictate where the rock breaks, and then hitting it at the right angle in the right place, with just the right amount of force. When there's anyone around and a sample to be had, Paul can't resist showing off his skills. It's infuriating to watch him. Also oddly inspiring. You want to rush off furtively and practise. You want to do it as well as he does.

Every day Paul would stay out on the outcrops as long as he could—too long sometimes. He often had to race back to camp before the heavy Namibian night fell. The surface of the rocks was rough, like sandpaper. If he tumbled on them in the darkness, his skin would be shredded. The nights were cold as well as dark. Paul's field season spanned the Namibian winter, when the days were still hot, but evening temperatures slipped quickly down into the thirties. Back in camp, the next race would be to build up the wood fire in a sandy hollow, and then to huddle around perched on the cooler boxes that serve as chairs.

Preparations for dinner take place by the light of a headlamp. Peeling and chopping is on a trestle table, spread with a garish plastic sheet. (And I mean garish. Think enormous pink and purple flowers, shocking even by torchlight.) There are usually fresh vegetables in the cooler, red peppers say, or beans, which can be a little frayed at the edges if it's been a couple of weeks since the last resupply. They go into a pot over the fire, along with onions and garlic and tins of fish, mussels perhaps, crabs, shrimps or tuna. And then there's rice or pasta or potatoes. Paul's wife, Erica, was astonished the one time she saw him cooking for himself in the Canadian Arctic. At home, Paul is determinedly hopeless at domesticity. Even in the field, the more culinary of his students wince sometimes at his vagaries. Paul puts cucumber into stews. Once a student trained in Italian cooking caught him adding ginger to spaghetti sauce. After a hard day in the field, though, you're ready to eat anything. You'll soak up the food with dark, heavy bread and sip Namibian beer, "the best in the world". And then, in the darkness, you rinse off your plate or bowl carefully, sharing the miserly dishwater allowance. Paul has his own utensils in Namibia, his own red plastic bowl and enormous coffee cup, white with a dark rim. He jokes about his possessiveness, but nobody else touches them.

Water is the really scarce commodity. Namibia has hundreds of miles of coastline, but not a single year-round river. There are plenty of river channels, and water can run in them briefly during the wet season, when it's too hot to work. But by the time Paul arrives in Namibia, all the rivers are dry. He has to carry his water with him in giant plastic barrels crammed into the back of the Toyota. Water sets the limit on how long he can stay in the field before he has to go into town for a resupply. It is reserved strictly for drinking and cooking. Washing is banned. You're even supposed to swallow the water that you use for brushing your teeth rather than waste it by spitting it out. The lack of water is a blessing, of sorts. Namibia contains plenty of dangerous wildlife, but most shun the dry regions; lions, cheetahs, rhinos and leopards all rely on open waterholes to survive.

Still, the desert has hazards of its own. Late one afternoon Paul was driving down the dried-out bed of the Ugab River. The light was starting to fade, and he flicked on the truck's headlights. He was starting to feel worried. In Namibia, darkness falls quickly, and it would soon be too late to find a decent campsite.

Paul wound his way hurriedly down the canyon on the sandy river floor, dodging the rocks and branches swept there by an old flash flood. Up ahead his lights picked out a thick black log, maybe nine feet long, lying in the sand. The Toyota could handle that, no problem. But at the last minute Paul swerved around it, striking what might have been a glancing blow. The log had seemed to twitch as he passed.

He was intrigued. He slowly backed up, craning his neck to see the scene illuminated by his white tail-lights. The log had vanished. No, it was standing up, and heading towards the vehicle, fast. It was chest high, four and a half feet above the ground, just about the height of the Toyota's open window. Now Paul could see that it had curious yellow rings the length of its body. It was a zebra snake, a western barred spitting cobra. It had spread its black hood angrily around its face and it loomed unnervingly large in the wing mirror. Paul remembers wanting to laugh. This was like the *T. rex* scene from *Jurassic Park*. "Objects in the mirror are closer than they appear."

But he also knew that zebra snakes were deadly. The toxin would quickly paralyse his muscles, and shut down his breathing. He had no serum as an antidote, since serum has to be kept cool and Paul had no refrigerator. Without immediate artificial respiration, he would suffocate. If someone pumped his lungs with their own air constantly while he was rushed back along these twisting canyons, in the dark, out along the bush tracks to the nearest village and then on and on to a town that perhaps had a hospital, he might survive without too much brain damage. Zebra snakes don't even need to bite you. They are called spitting cobras for a reason. Normally they are excessively shy, but when aroused they can spit their cytotoxic venom six feet or more. This one was clearly aroused, and Paul hastily rolled up his window.

The snake book in the passenger door of Paul's Toyota contains many lurid pictures. Alongside the featured snakes from southern Africa, you can see the human effects of their venom:

rotten arms, legs and hands, attached to bodies with pained, hopeless faces; limbs and torsos with puncture points surrounded by skin that is black, blue, yellow, swollen, pitted and blotched. "Don't read the snake book," Paul says to every newcomer, to first-time field workers and naïve young graduate students. "It will only give you nightmares." Everybody immediately opens the book and stares.

You are told, when you first come to Namibia, never to unroll your sleeping bag until the very last minute, just before you climb in. Each morning, when you wriggle out of the bag, you immediately bind it into a tight bundle. Everybody knows about the sleeping bag left unrolled at Khorixas rest camp by an unwary student, about the zebra snake that slid inside during the day and was there waiting for him when he retired to his tent. He survived, just, since he was relatively close to town. When you're camping out in the remoter parts of the Namibian desert, you don't need to hear this story twice.

Paul refuses to be worried by the Namibian snakes. He says that they're shy, rare and usually more than happy to avoid him. In all his years working in Namibia, he's encountered only one other serious snake—a black mamba, the most aggressive and deadly of all the ones in Paul's gruesome book. Even then, Paul didn't see it himself. And the geologist who did disturb it from its rock barely had time to gasp before the snake had vanished.

Thanks to the lack of open water, there's little else to worry about. Except, that is, for the desert elephants. They can dig for water. By using their tusks to create wells and waterholes, they can survive in some of the slightly less parched parts of the desert—the Huab River, for instance, towards the Namibian coast, where water is relatively easy to come by. Though the riverbed is dry, ground water lies not far below, and the surroundings are unusually verdant. Pungent African lilies poke up through the sand, surrounded by unexpected pockets of green. There are spiky euphorbia bushes and stands of mopane trees and twisted acacias.

The elephants in the Huab needn't be a problem. Paul has often camped there. Unlike, say, Canadian bears, you can easily discourage elephants from visiting camp. Though the riverbed, which makes a perfect campsite, is also a highway for elephants, they prefer to take the shortest, easiest route on their nocturnal journeys. Pitch camp on the wide part of a bend, and they will generally leave you alone. Even when they occasionally do wander around a field camp at night, they tend to be respectful and disturb nothing. The next morning, you simply wake to see their giant prints in the sand—ridged ovals the size of serving platters, or snowshoes, XXL. Paul isn't bothered by elephants any more than he is by snakes, or than he was by the flies and bears in Canada. Though they're dangerous, he says, elephants also tend to be shy. But they get angry when aroused. When I went to the Huab, I quickly found this out.

THE TUSKS were the first thing I saw—short, white and wicked. Then the rest of the elephant's head took shape against a backdrop of dusty acacia leaves. Tiny eyes set in a creased, anxious forehead. Ears thrust outwards, making the great head seem monstrous. This, I dimly recalled, was the elephants' universal warning signal. "Ears back: good. Ears forward: very bad." I froze.

African elephants are immense creatures—they weigh up to six tons and stand some eleven feet tall. They're fabulous, viewed from the window of a safari truck. But between me and this tusker lay just fifty yards of bare, scuffed sand, flanked with thorn bushes. I had found no trace of my companions in two hours of hard hiking. The camp was further still—direction unknown. I was lost and alone. I was also—now—in big trouble.

This predicament was partly my fault, but not entirely. Paul Hoffman had invited me to see his field site in Namibia, and I'd been counting on him to see me safely through. A few hours earlier he had laid out the plan for the afternoon. Five of our party were to squeeze into a vehicle and negotiate the wide sweep of the Huab River's dried-out bed, while the remaining four of us hiked to meet them across a sandy basin, carpeted with thorn scrub and occasional groves of acacia trees. We all knew where we were heading. Paul had pointed out a rich red outcrop of rock, standing against the skyline a few valleys away.

So I was frustrated but not unduly alarmed when I realized that—characteristically and without warning—Paul had charged off into the bush to begin the hike some minutes before, without checking to see who was following. His two harassed field assistants, more accustomed than I was to this habit of his, had apparently grabbed their packs and plunged after him. Hastily, I picked up my camera and ran, calling as I went. No sign. Back I went to the vehicle to check the final destination. But the vehicle had gone.

Of course, I could have walked back to the camp—a mere thirty minutes or so away—and spent the afternoon in craven contemplation. But I could see the outcrop quite clearly from where I stood. Nothing was stirring—even the air was still. I couldn't resist. This was my chance to show Paul how well I could manage alone in the bush. (Why did I want to? I don't really know. Paul has that effect on people.)

The day was beautiful. Though the rains were long gone, there were still green patches of grass, crowned with a silver-gold sheen where the tips had begun to dry. Between the grass and twisted black thorn trees were bare patches of sand spattered with mustard-coloured mosses. I even felt a frisson of delight when I came upon old elephant dung—cannonballs of dried grass and mud that marked a network of tracks winding through the tough, thorny scrub of the river valley. "Elephants make the best paths," Paul had told us, and it's true that the trails they blaze are easy to follow—wide, sandy and obstacle-free. My chances of spotting an elephant were remote, but at least I could use their tracks, taking my bearings from the jagged outcrop of rocks on its distant hillside.

After more than an hour of hard walking, I finally reached the dry Huab riverbed. There on the sand were two sets of vehicle tracks, but both looked old even to my inexperienced eye. There was no sign of human footprints. The elephant track cut diagonally across the river, heading in the direction of the outcrop. I followed it.

I was hot and tired by now, and increasingly discouraged. But then, in the distance, I saw a distinct brown shape—an elephant—loping across the sand. Enchanted, I stared as it crossed the river and began to climb the far slope. I followed as closely as I dared, watching in delight as it found shade under a large tree and began lazily twitching its ears. Then, planning how I'd boast about my sighting to the rest of the crew, I marched on.

My new enthusiasm didn't last. The outcrop was getting no closer and I had still seen no human signs. Relief when I heard shouts of "Oi! Oi!" evaporated when I realized they weren't human voices but those of baboons, barking warnings from up on the hillside. In the distance I saw another elephant crossing the river, trunk trailing in the sand. Now my close-up on wildlife seemed much less enchanting. The air was cooler, and the afternoon was drawing in. Suddenly everything was stirring. With growing unease, I waited until the elephant was out of sight, and then continued cautiously along the riverbed.

Then I heard the roar. The quintessential lion sound. The noise you'd make at the zoo to tease kids. I reasoned with myself. Lions are rare in the Huab. Unlike the elephants, they need standing water to survive, and this place is far too dry. They're common in Etosha National Park, far away to the northeast, but I'd be very unlucky to find one here. Perhaps I just imagined it.

Right on cue, the roar came again, from the dense patch of scrub directly ahead. This scrub lay right beside the escarpment that sloped up to the outcrop, and to find my companions I had to walk past it. I squared my shoulders and strode ahead, pinning everything on finding the van, finding my team, finding something safe. Along the escarpment, around the side of the outcrop, not running, not smiling, I squinted up into the sunlight, searching for signs of human life. There was no one there.

I panicked.

Suddenly all directions looked equally alien. A civilian in geologists' territory, I'd foolishly kept all my attention on the outcrop up ahead, and taken scant notice of the landscape through which I was passing. Looking back now, I could see no landmarks that I recognized. Blindly I plunged into the bush, and the thorn branches tore at me as I fought my way past. A full ten minutes passed before I forced myself to stop and try to think clearly. I had to find the camp before daylight failed. But how?

Hansel and Gretel. Though I'd left no white stones as markers, all I had to do, I realized, was follow my footsteps back. Here in the bush I'd trod mainly on springy grass, and there were almost no prints to follow. But I could head back to the escarpment, and up till then I'd mostly walked on sand. I resolved to find my footprints and follow them back exactly the way I came, no guesses, no shortcuts. And I began to calm down. At the foot of the escarpment I found the first clear footprints. More deep breaths, and I headed back along the elephant track.

There was no warning. Suddenly the large, angry elephant appeared, blocking my path, scarcely fifty yards ahead. Ears outstanding, it clearly wanted me out of the way. Still, I had a mad impulse to take a photograph. I resisted. These beasts scare easily. Three months earlier a Namibian man had been trampled to death just north of here, when he surprised an elephant and then tried to run for it. Climbing trees doesn't help. Make no threatening gestures. Slowly and carefully get out of the way.

Behind me was no go—that's where the elephant was heading. And I had no desire to get closer. I turned to the side and walked out into the riverbed. The great head turned too, and watched, thoughtfully, ears still spread, as I slowly, steadily, began to cross the river. In the open sand I felt even more vulnerable. If it charged, what then? Don't think, just walk.

I reached the other side, turned, headed homewards. The elephant hesitated, and then it too continued on its way. Stopped, stared and started again, this time rubbernecking. That mighty creature and I walked past each other, on opposite sides of the river, heads turned, each watching the other's every move. I have no idea how long I had been walking before I was finally clear. But by the time I heard human shouts and found the real outcrop—the one that I'd passed inadvertently, hours earlier—I was exhausted. Now that the fear had gone, I was furious. Paul knew I was a neophyte who had no idea where we were heading. How could he have abandoned me like that? How could he be so self-absorbed?

Paul greeted me with a cheerful smile. There was still time to see the rocks, he said, but I was in no mood to admire them right then. Why had he left without checking that I was there? Why hadn't he stopped when he realized I wasn't with his group? Confront Paul at your peril. Criticize him head-on, and his temper will flare. How dare I, he responded. It was my responsibility, not his. I should have let him know I was following him. (But this had been his plan, not mine. And who else would I have been following? How else would I have reached the outcrop?)

I stalked off to look at the rocks. They were beautiful, a delicate rose colour, and as I watched the sinking sun spill on to them, I tried to calm myself down. What's the point of letting it get to you? You know what Paul's like. He dishes this out to everyone. It's nothing personal. By the time I returned, Paul was wreathed in smiles again. He congratulated me on having acquired the day's best campfire story and offered to show me, as we walked back, the baboon skull he'd found on the way. He was charming and I was soothed. And I'd had my first taste of the strangely mixed experience of working with Paul Hoffman.

PAUL SAW the ice rocks everywhere he went in Namibia. He sought them out, and they intrigued and confused him. The rocks had formed in a Precambrian shallow sea, and though each outcrop was different, all bore the distinctive signs of ancient ice. Some contained lone boulders that had been dropped by icebergs floating overhead. Some contained the mad jumble of rocks and stones that had been scraped off the nearby land by glaciers, and bulldozed into the sea. Occasionally these jumbled rocks bore scrape marks where the ice had dragged them over the ground.

Some deposits were hundreds of metres thick, while others were just a thin skin. Many were also capped by a mysterious layer of carbonate rock, which had often turned pink, or perhaps ochre, with the touch of wind and weather. Paul, like Brian Harland before him, was baffled by this. Carbonates usually show up in warm water, in the tropics, but these appeared immediately after ice. And the contact between the glacial rocks and the carbonates was always knife-sharp, as if there had been some sudden, dramatic change from ice to tropics, from cold to hot.

After the summers mapping the Namibian ice rocks, Paul spent his winters puzzling over them back home. At the end of each season he brought samples to Harvard, and stared at the rocks in his office. He crushed them and measured them in his lab, all the time wondering what story they had to tell. Then one day, he remembered that odd conversation he had had with Joe Kirschvink years earlier about his "nutty" Snowball Earth idea.

Off to the library Paul went, to look up Joe's work on the Snowball. There was little enough, just that one short paper buried in a vast, obscure book. Paul read the paper and was gripped. He quickly dug out Brian Harland's research, and the magnetic work by George Williams, and Mikhail Budyko's papers about the ice catastrophe. He couldn't get enough of the Snowball. This was a story indeed. But each of the proponents of the Snowball had dropped it, one by one, starved by lack of evidence. What if Paul could provide the evidence? What if his Namibian rocks held the clues to this extraordinary catastrophe?

Now Paul started looking directly for Snowball evidence, not in the ice rocks themselves, but in the carbonates that bracketed them below and above, geologically before and after. Geologists have many possible tools for extracting stories from stones, and one of the best involves measuring the ratio of their isotopes—heavy and light versions of the elements they contain. Carbonate rocks, for instance, contain different isotopes of carbon. There's a lightweight version called carbon-12, and a heavier one called carbon-13. Comparing the ratio of the two can usually tell you something about the seawater that the rocks were formed in. And when Paul looked at the lab results from his Namibia samples, he was astonished. They had a bizarre carbon isotope signature, one that he had never seen before, with much less carbon-13 than he had anticipated.

Usually, ocean water and the carbonates that it produces are both rich in carbon-13. The ratio gets skewed to the heavier side because of the activity of living creatures. Bacteria in the ocean need carbon to grow, and carbon-12 is their favourite flavour. They grab carbon-12, and leave carbon-13 behind. Think of a box of red and green jelly beans. As you gradually pick out the red ones, the rest of the box will start to look more and more green. The same thing happens with carbon isotopes in seawater. When bacteria grow and grab carbon-12, the seawater ends up with proportionately more carbon-13. This seawater carbon is then bound up in carbonate rock.

So, when life is flourishing, the carbonate rocks formed at the same time have the skewed heavy seawater ratio. That's what was so strange about Paul's rocks. For carbonates, they were extraordinarily light. Before the Snowball, and for what looked like a long time afterwards, life apparently wasn't active at all.

Paul felt this was important, but he couldn't figure out what it meant. He was more baffled still by the "cap" carbonates that came after the ice. These are the same rocks that show up all around the world. Brian Harland had seen them in Svalbard. They stretch for miles in Australia, Canada, almost everywhere that the glacial rocks appear. And that's peculiar. One of the first things you learn in geology is that the Earth is emphatically not one big layer cake. Sure, individual regions might end up with layers of different rock types, cut through by rivers the way a knife cuts through a cake. But the rock layers are still different in different places. Take a snapshot of the rocks forming today on Earth, and here's what you might see. One place might have sandy seafloors or beaches that eventually solidify to produce sandstone. Somewhere else might be in the act of producing mudstone. Perhaps some volcanoes spew out their lava to cover another region with black basaltic rocks, and elsewhere you might find rocks that have been pummelled and transformed by the inner churnings of a mountain belt. The Earth is a very big, very patchy place. You simply don't get single events that blanket the entire planet with one type of rock. Period.

So where did these cap carbonates come from? Everywhere the ice rocks appeared, the caps seemed to be. And the ice rocks showed up on every single continent. Why? *Why?*

The caps also contained strange textures. The strangest were in the rock outcrops around the Huab River. There, set in a fawn-coloured cliff of carbonate, Paul found brown tubes resembling burn marks made by long, thin pokers. From a distance they were dark vertical stripes marring the cliff face for hundreds of feet. They weren't just on the surface of the cliff, either. Where chunks of rock had broken off, you could see more tubes marching on into the interior. In some places the breaks had sliced horizontally through the tubes, exposing them as a neat array of dark circles, each the size of a penny. The tubes looked like the regimented burrows of a highly organized worm colony, but there were no worms in the Precambrian. Paul was baffled by them.

Across the valley from the tubes, a student of Paul's found something just as strange. He had climbed up a steep ridge to inspect the carbonate outcrop. At the top was a plethora of huge rose-coloured crystals. They stood out against the pale carbonate rock around them, looking like giant splayed paw prints set into the vertical rock face. Or like the kind of feathered fans that Victorian ladies carried to the opera, though some of them were as tall as Paul himself. When Paul first saw these fans, he was astonished. He thought at first that they were fossils of some kind. They looked almost like giant clamshells. But no clams existed in the Precambrian, nor did any other creature that could make shells like these. The fans had to come from some bizarre physical process. But what?

Crystal fans, tubes, ice rocks, strange isotopes. Paul was increasingly convinced that all this evidence added up in some way, and would somehow yield crucial clues about the Snowball. He tried continually to make sense of all these features. He visited and revisited the strange carbonate formations to collect samples, to note and map and muse.

By the end of 1997, Paul was feeling frustrated. He had now spent five years doing fieldwork in Namibia, and he still had no research papers to show for it. He finally decided to write a paper about the carbon isotopes, even though he didn't yet fully understand them. Over Christmas and on into January, he perfected his paper, which was destined for a small journal.[1] He talked about the glacial rocks, and the strange isotopes in the carbonates that bracketed them. He talked about, and discounted, several possible explanations for the ubiquitous ice. And then—right at the very end—he suggested that Joe Kirschvink's ideas about the Snowball might provide a possible explanation. For once in his life he was being cautious. Not by choice, though. He was truly wrestling with the Snowball conundrum.

Until, that is, another new player entered the scene—a young colleague of Paul's, Dan Schrag. At the time, Dan knew nothing about the Snowball idea. He knew nothing about Brian Harland's work, or Joe Kirschvink's. He knew nothing about Precambrian rocks or Namibian geology. But there was one thing he knew plenty about, and that was carbonates.

DAN SCHRAG is Paul's best friend at Harvard. They look almost like partners in some comedy routine. Paul is in his sixties, tall, thin, shock-headed and white-bearded. Dan is in his thirties, short, plump and blond, with thin hair and a smooth round face. Dan's manner is smooth, too. He is supremely sociable. People are his thing—his networks stretch through every scientific field. He has hordes of friends. Many of them are hot young scientists like him, but there's a smattering of other types, too—artists, designers, people he knew at school. Every year Dan rents a house somewhere beautiful with five particular college friends. Spouses and children come, too. When a child is born to one of the group, there is a complex gifting scheme. Each group member provides the child with three favourite books to be read now, and two hundred dollars to be used later. The money is invested in a college fund, earmarked for entertainment purposes only.

Dan doesn't take anyone with him to these college reunions. He hasn't met the right girl yet. Instead he throws himself into his work, with sharp eyes and a quick wit and the veneer of arrogance that often comes with high intelligence. Friends say he is warm and generous; enemies say he is calculating. Everyone says he is brilliant. Dan has just won a MacArthur Foundation "genius" award: half a million dollars to spend as he wishes. He has decided to use the money for building a science retreat near the ocean on Cape Cod. The house will have skylights, an inglenook, a huge kitchen, plenty of rooms where Dan and his many friends can gather, cook, talk science and think.

Princeton gave Dan a professorship when he was only twenty-seven. He moved to Harvard four years later and was quickly, precociously, given tenure. Normally you'd work up to a place like Harvard. Normally it would take years to get tenure there. You'd expect to be in your forties, maybe, to have lots of research years behind you. Not Dan. He'd already published seminal papers by the time he hit thirty-four. By then, he had more ideas than he knew what to do with.

Dan loves ideas. He loves bouncing them around, tasting them, testing them, and seeing how they might work. He loves having intense conversations about them. Especially with Paul. Late at night, before he leaves the Geological Sciences Department and heads for home, Dan calls in on Paul. Even at eleven o'clock, or midnight, Paul is invariably still in his office. Dan sends his dog, Max, on ahead. Max, an amiable black Akita, knows the way. He ambles through Paul's outer lab, turns right and wanders into the office, walks up to where Paul is sitting at his desk, and thrusts his nose into Paul's hand. Dan stands outside and listens. He can tell Paul's mood by the tone of his grunt.

What follows is usually intense. Dan often ends up staying in Paul's office bouncing ideas back and forth for one hour, two hours . . . until the early hours of the morning. Sometimes the discussions become heated, but Dan isn't afraid of Paul. Heated arguments don't bother him much. "Everyone who's ever worked with Paul really closely has had a catastrophic falling-out with him," Dan says. "I know they're all watching me, waiting for the hatchet to fall. And yes, Paul and I fight. I don't just mean those tongue-lashings—I get those ten times a day. I mean violent, stand-up, screaming fights. 'You are the scum of the earth!' But I don't think I am the scum of the earth, so that's OK. Many's the time I've vowed never to speak to Paul again. But I keep coming back for more, because I like him."

"You know why Dan and I work so well together?" Paul often says. "Because friction creates heat."

Paul and Dan depend on each other. Though Dan is a professor in his own right, he still loves getting Paul's attention. He offers his ideas to Paul the way an eager puppy might. Once, when I was walking down the street with Dan at a conference in Edinburgh, we spotted Paul up ahead. It was late in the evening, and Paul was strolling hand-in-hand with Erica. Dan raced to catch up with them. As we all stood rather awkwardly outside Paul's hotel, Dan poured out his latest idea in a torrent of words, excited, watching all the while for Paul's response. Something to do with how many kilotons of carbon are bound up in Paul's carbonate rocks. Erica turned to me, looking amused. "How many kilotons of this have *you* had?" she murmured.

And Dan is Paul's conduit to the outside world. He's a Paul antidote. He deals with people smoothly and easily, soothing the feathers that Paul invariably ruffles. In some ways their relationship is uneven. Dan has plenty of people he can talk to about his ideas, but Paul doesn't have many friends. Perhaps because of this, Paul was at least as nervous as Dan about the tenure process. Paul was safe at Harvard. He was tenured. He could stay as long as he liked. But if Dan hadn't been awarded tenure, Paul would have lost his closest friend.

Over the months that Dan was being considered for tenure, Paul grew increasingly nervous. He gave evidence to the committee in Dan's favour. He tried to gauge how the decision might go, and spent most of decision day pacing up and down the department's corridors. Four minutes after Dan received the congratulatory phone call, Paul appeared in his office and collapsed on the couch in an exhausted, relieved heap.

Dan was the perfect person to ask about the Snowball rocks. The key to the story surely lay in those strange carbonates with their fans and tubes and weird structures. The ones that had somehow formed in the Snowball's aftermath, when the ice had finally gone. Carbonates come from tropical oceans, and Dan is an expert in them. He is an oceans-and-carbonates man.

Unlike Paul, Dan doesn't collect his samples from dusty deserts. Instead he tips himself off the back of a boat, usually in some gorgeous tropical location. The tropics are the Earth's heat engine, Dan says, because the sun's rays fall most intensely there. They are the key to understanding the Earth's climate.

Finding climate records from the tropics isn't easy. In polar regions the ice caps are like time capsules. Every year snow falls, and with it come dust and other chemical clues to the climate *du jour*. Gradually these layers build up, are buried, and turn into ice. When researchers drill down into the ice, they can uncover a climate record going back for millennia. In Antarctica, for instance, the ice cap is so thick that the layers at its base are more than 400,000 years old. There are bubbles of ancient air in the ice to be measured, too. Want to know precisely what the prehistoric atmosphere was like? Look no further. Here in this bubble is a whiff of air that was last breathed by *Homo erectus,* and was trapped and buried long before the Neanderthals ever appeared.

In temperate zones there's not much ice to be had. But researchers can at least study the thickness of tree rings. They don't even have to chop the trees down. They carefully bore into the side of the tree and pull out a thin core of wood, about the size of a drinking straw. And then they just need to count and measure. A thick ring? That was a wet summer. A thin one? Must have been dry. If they pick big enough, old enough trees, researchers can build up a climate record going back hundreds of years.

Dan, however, takes a different approach. His obsession is the tropics, where ice caps occur only on high mountains, and trees don't tend to grow annual rings because one season is pretty much like another. Instead he seeks out the climate records hidden inside giant corals. Like trees, corals lay down a new growth ring every year, but their rings are made of carbonate rock rather than wood. Measure the growth rings year by year, and you can learn how the climate of the tropical ocean has changed. If the coral is big enough and old enough, its rings can take you back hundreds of years. Find an ancient, fossilized coral, and you might even learn about the tropical climate from thousands of years ago.

When Dan dives for samples, he carries along his scuba gear, lift bags and a huge, one-hundred-pound drill. Even underwater, that's heavy. You dive in pairs. One person kneels on the coral with the drill; the other holds on to the rest of the equipment. The noise from the drill is so overpowering that you can hardly hear yourself breathe. When you're drilling, the fish keep well away. A fine powder gradually emerges from the edges of the drill hole, floats over the coral surface, and then disappears into the surrounding water. You stop every so often to add an extra length of tube above the drill bit. You wear thick knee pads to keep your wetsuit from being cut to pieces by the coral's spiky surface. Sometimes there are strong currents. You have to wear extra weight belts and try to anchor yourself on to the coral, and the sensation is like trying to drill the road during a hurricane. Sometimes the coral is deep, say sixty feet or so below the water, and you have only about an hour of air to land on the coral, drill a core, heave the drill up again, and race back to the surface. But if the coral is shallower you can take your time, enjoy the scenery, sneak some moments when you've finished drilling to wedge yourself in among the corals and sponges and watch the fish go by.[2]

Though Dan had never worked on rocks as old as Paul's, he had spent plenty of time studying corals and unlocking their carbonate secrets. And his insights into Paul's mysterious Namibian carbonates were about to prove crucial.

HARVARD, SUNDAY, 15 FEBRUARY 1998

PAUL COULD still feel a slight ache in his legs as he sat at his desk in Harvard. The Boston Marathon was coming up in a couple of months, and he had upped his training levels. Yesterday he had run almost twenty-two miles. Now, though, he was back in his office. He had worked through the evening and it was getting late, but he still didn't feel like going home.

That's when Dan wandered in, with a friend who was giving a seminar at Harvard the next day. "Come and meet Paul," Dan had said, over dinner. On a Sunday night? At this time? Oh yes, no problem. Paul will be there.

Dan wanted to talk to Paul about his Namibia paper. Paul had handed it to him a couple of days earlier. He'd scanned it, and felt annoyed because the interesting part seemed to be buried right at the end. As soon as Paul asked him about the paper, Dan waded in. You want to know what I think? This stuff about the Snowball is fascinating! You can't bury it! You can't just put one little sentence in about the implications. You need to think about this idea more. What does it *mean*?

Paul didn't need a second invitation. He had been hoping Dan would bend his brain to the Snowball story. So Paul told Dan about the strange cap-carbonate rocks and about how unexpectedly light the carbon isotopes were. He told him about the weird textures: the huge, graceful crystal fans, and the wormlike tubes.

He also told him the whole story as it stood then. That the Earth had frozen over, top to toe, pole to pole. That during this Snowball, volcanoes had continued to spew their greenhouse gases over the frigid Earth. That over millions of years the planet's atmosphere had become scorching. That this super-greenhouse catastrophically melted the ice. That the greenhouse gases stayed around afterwards for tens of thousands of years, blasting the Earth into a hothouse until they finally subsided.

Dan listened carefully. He retreated into a corner to think, while Paul and the friend politely chatted. When Dan is concentrating on a problem, he goes silent. His eyes dart around, focusing on something out of view. He often bites his bottom lip. Then, when he comes upon an answer, his eyes fire up. He immediately, eagerly, blurts it out. Wait! I've got it! I can explain the carbonates!

His idea was brilliant.

The planet was in stasis, cryogenically preserved. A thick layer of ice covered the oceans. Great glaciers crept and ground their way over the rocky surface of the continents, slowly pulverizing everything in their path. Ice bred more ice as the shining white surface repulsed sunlight, locking Earth in the mother of all winters.

So it was, and so it would always have been, but for volcanoes that poked above the ice or squatted on the seafloor. They erupted, as they always have, and each eruption spewed out ash and lava and—above all—carbon dioxide gas. Gradually, slowly, this volcanic gas built up in the air, wrapping the Earth in a blanket of warmth. And in the end, fire conquered ice. Drip, drip, came the first sounds of change, then trickle, then flow, then flood. Then meltdown. The ice vanished, and Earth went from icehouse to hothouse in a geological instant. So much was Joe Kirschvink's vision.

But now comes the new part, the part that suddenly dawned on Dan one Sunday night in Harvard. That hothouse, he realized, was like the tropical heat engine gone mad. The ice had gone, but the heat that melted it remained behind, on full blast. Dry, scorched air sucked up moisture from the oceans and whirled it into storm clouds. Hyper-hurricanes raced around the Earth's surface, flinging their watery burden back on the ground in torrents. And that burden was no longer just water. The air was filled with carbon dioxide. Whatever rain passed through it turned immediately to acid.

What did the acid rain fall on? An inviting layer of powder. Over millions of years, glaciers had ground the continents' rocks into dust. Ground-up material is always easier to react with. Think how much faster sugar dissolves when it's not bound up in a lump. In the post-Snowball world, that combination of ground-up rock and torrential acid rain was a chemical factory waiting to happen. Rock dust and acid met, mated and were swept off into the sea. They set the waters fizzing and foaming, creating a Coca-Cola ocean.

And then a new snowstorm began, this time underwater. All around the world, the post-Snowball ocean turned milky with flakes of white. They poured down on to every inch of the ocean floor. From the chemistry of acid rain and rock dust had come a massive outpouring of carbonate, which blanketed the entire planet. The flakes squeezed together, and hardened and turned into rock. They were the cap carbonates. This was Dan's idea. The cap carbonates, he said, arose directly from the intense, bizarre conditions that had rescued the Earth from its Snowball.

Dan and Paul both pounced on the idea and began to probe it. Did it work? Could it explain Paul's other conundrums? First, the strange tube rocks and rose-coloured crystal fans. Both could have come directly from the ocean's effervescent fury. The tubes might have formed when bubbles of gas shot upward inside the fast-forming carbonates. The crystal fans might also be some weird by-product of this frantic fizzing. In acidic hot springs like the ones at Yellowstone, you often find fan-shaped crystals, their arms radiating outwards with the sheer pace of precipitation.

Next, the rapidity of the change. The contact between carbonate and glacial rocks was always knife-sharp. That's just what you'd expect if the carbonate formed immediately after the ice melted.

What about the isotopes? Remember that the rocks showed a light, "lifeless" signal both before the Snowball and then for a long time afterwards. To explain what happened before was easy. Living things in the oceans pick out the light carbon atoms, the "red jelly beans", and leave the heavier carbon behind for the carbonates.

Paul had already suggested that before the Snowball, life's pace was probably slowing down as a reaction to the growing ice. Fewer living things meant less pickiness, and more light carbon left around to be bound up in the carbonates. That, Paul felt, was why the carbonates grew steadily lighter as the ice approached.

But afterwards was trickier. Life must have rebounded quickly after the ice melted, but the light, lifeless signal continued on in the carbonates for tens of thousands of years. Perhaps the "light" signal in the aftermath of the Snowball had nothing to do with whether or not life was flourishing. This intense formation of carbonate rock would swamp any normal signal. Carbonates were madly precipitating everywhere. They would be grabbing so much of the ocean's carbon, both light and heavy, that it wouldn't matter any more how picky the bacteria were being. It's as if you were sedately choosing red jelly beans from the pile, when a greedy cousin came along and snatched handfuls of the lot, both green and red, faster than you could eat any of them yourself. The post-Snowball carbonates were light because they swamped the signal from the bacteria. It all made brilliant sense.

Back, forth, back, forth. Whatever piece of evidence Paul could think of, Dan managed to fit neatly into his scheme. When they assumed a Snowball, all the pieces fell into place. The carbonates and the isotopes weren't mysteries any more. They were just what you'd expect. They were *predictions* of this new, improved Snowball theory.

The two of them grew more animated and excited. Try it this way and that. Look from every possible angle. The more they probed, the more Dan's idea really did seem to tie everything together. Brian's ice rocks, Joe's volcanoes, Paul's isotopes and cap carbonates, all added up into one elegant story. It was intoxicating. This, at last, was what Paul had been seeking. He could scarcely believe his luck.

Dan didn't leave until nearly three. After he had gone, Paul sat in his office, staring at his computer screen. At 3:04 A.M. he sent Dan an e-mail. The subject line was "funk in deep freeze" (the name of a jazz album). The message said, "Muchas gracias for tonight. I needed it badly. Thanks for the kick in the ass."

The next day, Dan was back. He'd been thinking all night about the Snowball. He wanted to work with Paul on a new paper. And he wanted to send it to *Science,* one of the world's most prestigious and highest-profile journals. This was to be their first direct collaboration, and Paul was delighted. For the next few weeks Paul and Dan wrote and rewrote and discussed and argued. They haunted each other's offices. They fired ideas at each other, stopping wherever they happened to meet. Students going to classes often had to step over them as they sat in the stairwells of the Geology Department, thrashing out the latest details of their theory. Here's something I just thought of! Hey, I've just made another connection! They both describe this period as the most exciting of their lives.

In science, good luck can be as important as good judgement. When Brian Harland first came up with the Snowball idea, he was too far ahead of his time. But Paul Hoffman became obsessed by the ice rocks in Namibia at the perfect moment. The Snowball stage was set when he and Dan experienced their eureka, and the key criticisms that had long dogged the Snowball idea were already solved. There was no ice there? Virtually everyone now believed Brian's argument that the rocks were formed from the action of grinding glaciers, and dropped by overhead icebergs. There was no way out of the ice catastrophe? With his super-greenhouse, Joe

Kirschvink had found a way that the Earth could go into a deep freeze and still recover. All the continents had been huddled around the frigid poles at the time the ice appeared? Thanks to the magnetic records from the Flinders Ranges, everyone now believed that in at least one place, in what is now the South Australian outback, there was ice within a few degrees of the equator.

For Paul and Dan, it could hardly have been more perfect. Their predecessors had each, individually, solved the arguments *against* the theory. Now they themselves had produced evidence *for* the theory. Paul's isotopes were evidence that large numbers of living things perished before the Snowball, and Dan's carbonates were evidence for the super-greenhouse that came after the ice.

Coming up with the Snowball story—understanding how the new evidence fitted in with the work that had already been done—took a particular combination of capabilities. Paul had the deep knowledge of Precambrian geology, the long years of fieldwork in Canada and Namibia. Dan had the understanding of how oceans work.

Telling the story would require yet another important combination, but this time of personalities rather than knowledge. Paul had the vehemence, the stubbornness, the single-minded obsession. Dan had the social network, the grace, the names and phone numbers of smart, imaginative scientists in many different disciplines. Unlike Brian Harland, Joe Kirschvink, or any of the other people who had worked on the glacial rocks in the past fifty years, Paul and Dan were prepared to run with the Snowball idea. Having put the theory together, they wanted to taste it, test it and spread it around. Paul in particular. This idea felt different from any others he'd been involved in. This was his chance, finally, to make a world-class difference to the way we all understand the Earth.

SIX

ON THE ROAD

AUGUST 1998

Paul and Dan's Snowball Earth paper was published in *Science*.[1] There was an immediate flurry of interest, and the next step was to turn that flurry into a storm. Dan began calling his friends, and Paul took to the road. In the autumn of that year Paul went from one institution to another, purveying the good news. He was a very talented speaker, giving persuasive lectures that were both clear and impassioned. He had never, he said repeatedly, been so convinced that something was right.

Science is often messy. When you're judging any new theory, it's rarely as simple as yes or no, right or wrong. This is particularly true in geology. Reading the messages hidden in rocks is a craft, and different researchers invariably notice different things.

Theories in geology can rarely be accepted or dismissed out of hand. Even the ones that turn out to be broadly right often need to be massaged and modified, and given the initial benefit of the maybe.

But when a big new idea hits the scene, there's almost always a pattern of polarization. Though a few researchers keep a genuinely open mind, others immediately entrench either into pros or cons. These vehement souls will fight, criticize, and try to pull one another down. The survival of an idea can depend as critically on the quality of the rhetoric as on the robustness of the data.

That's exactly what happened when Paul went on the road. The more he promoted his idea, the more other researchers reacted against it. Many of them did so *because* Paul was promoting his idea so vehemently. He made no secret of his fervour. What he did, relentlessly, was force his opponents to face the Snowball theory, in a continual stream of public lectures and private seminars, papers, comments, reviews, e-mails and faxes, on stairwells at conferences, over lunch and around the campfire. He worked to get influential scientists on board and—yes—to squash those who disagreed. Sometimes it seemed as if he was trying to achieve the Snowball Revolution by the sheer force of his energy.

He even divided the science world into Snowball "believers" and "non-believers". When he was accused by one bitter critic of founding the "Church of the Latter-day Snowballers", Paul found the comment merely amusing—partly, I think, because he had a great riposte: "Someone once asked Charlie Parker if he was religious and he said, 'Yes, I'm a devout musician.' Well, it's the same for me. My approach to geology is that I'm a religious fanatic."

And Paul's relentless advocacy of the Snowball brought him continual criticisms that he lacked that most precious of scientific commodities: objectivity. His response was to liken himself constantly to one of his heroes. Alfred Wegener, the German meteorologist who had first championed continental drift, and who had perished at the age of fifty on the Greenland ice cap.

Paul saw plenty of parallels between his Snowball idea and Wegener's theory. Like the Snowball, continental drift (or, more strictly speaking, its later incarnation as the theory of plate tectonics) can explain many disparate puzzles under one elegant umbrella. As soon as you allow the continents to move, many other things follow. Where continents separate, they produce oceans. Where they collide, they make mountains. Where plates rub against each other, they can stick and suddenly slip, rumpling the Earth's skin with an earthquake. New seafloor is formed along the hitherto mysterious great ridges that run through the centres of the oceans like giant backbones. Old seafloor disappears by plunging down trenches at the edges of continents. Volcanoes form in the crust above these trenches. When the wet seafloor plummets into the Earth's interior, water creeps up into the overlying rocks, encouraging them to melt and spill their hot load on to the planet's surface. One idea explains all.

Also like Paul, Wegener was vilified as much for his *approach* as for his ideas. Wegener offended his opponents by the very way he reported his research. In his book *The Origin of Continents and Oceans,* he described his initial insight as an "intuitive leap". Intuition, many geologists felt, had no place in science. And there was worse to come. When Wegener discovered that fossils from South America uncannily matched those found in Africa and that the geology matched eerily too, he performed, he said, a "hasty analysis of the results of research in this direction in the spheres of geology and palaeontology, whereby such important confirmations were yielded that I was convinced of the fundamental correctness of my idea."[2]

He made a "hasty analysis"? He was convinced that he was correct? These sorts of comments deeply disturbed geologists, who felt that Wegener was far from dispassionate about his ideas. "My principal objection to the Wegener hypothesis," thundered one critic, "rests on the author's method. This, in my opinion, is not scientific, but takes the familiar course of an initial idea, a selective search through the literature for corroborative evidence, ignoring most of the facts that are opposed to the idea, and ending in a state of auto-intoxication in which the subjective idea comes to be considered objective fact."[3]

Another critic, Bailey Willis, said that Wegener's book describing the theory of continental drift gave the impression of having been "written by an advocate rather than an impartial investigator". (Willis wrote a paper[4] about Wegener's theory in 1944, which he titled "Continental Drift, Ein Märchen" [a fairy tale].) Joseph Singewald claimed that Wegener had "set out to prove the theory . . . rather than to test it" and accused him of "dogmatism", "overgeneralizing", and "special pleading".[5]

Ever since Wegener's vindication, proponents of a controversial idea in geology like to align themselves with him, and Paul does this a lot. Of course, there was nothing to prove conclusively in 1912 that Wegener was right, and he might have turned out in the end to be wrong. But still, the Wegener story is a cautionary tale to all geologists not to dismiss extraordinary ideas out of hand.

"Good ideas, when they're young, they're vulnerable," Paul said to me at a conference in Reno. "They're a pain in the ass, so you want to trash them. But the danger is the old ideas that everyone has got comfortable with. With a new idea, you have to cultivate it and let it grow and see where it takes you, and if you do, I think you'll learn faster where it's wrong than if you stomp all over it."

A few weeks later he sent me the following quote from Mott T. Greene, a biographer of Wegener:

> Throughout the entire course of the debate [about Wegener's theory] neither his supporters nor his detractors seemed to have the clear grasp of a theory which comes from having read it carefully. The reason for this is a kind of guilty secret: most scientists read as little as they can get away with anyway, and they do not like new *theories* [Greene's emphasis] in particular. New theories are hard work, and they are dangerous—it is dangerous to support them (might be wrong) and dangerous to oppose them (might be right). The best course is to ignore them until forced to face them. Even then, respect for the brevity of life and professional caution lead most scientists to wait until someone they trust, admire, or fear supports or opposes the theory. Then they get two for one—they can come out for or against without having to actually read it, and can do so in a crowd either way. This, in a nutshell, is how the plate-tectonics "revolution" took place."[6]

Paul was clearly convinced this was true of the Snowball. He was certainly doing his best to force other geologists to face up to the theory. But it wasn't just Paul's *style* that made his colleagues object to the Snowball theory. There was another reason why many people found the theory discomfiting and even dangerous: it was a theory that required its proponents to think what to geologists was unthinkable. The Snowball Earth was different in almost every characteristic from the planet we see today. Accept Paul and Dan's theory, and you have to imagine our home planet behaving like Mars or Europa or some other alien place. That was more than enough to make many geologists shiver.

The problem was that the Snowball—as Paul described it—violated a key geological maxim called "uniformitarianism". This rule was first articulated in the eighteenth century, when geology in its modern scientific sense was born, and all geologists learn it at their mother's knee. It says that the present is the key to the past. The general assumption behind this rule is that the same things happening in the world today have been happening throughout Earth's history. Uniformitarianism is generally a good rule of thumb.[7] Try to explain baffling evidence from the past by invoking changes in the way the world worked, and you risk straying into the world of mysticism and magic rather than accessible, empirical science.

But there are some phenomena that don't show up in the everyday world, and yet are no less scientifically valid for all that. Uniformitarianism encountered one of its most serious challenges in the 1980s, when Walter and Luis Alvarez succeeded in convincing most—if not quite all—of the scientific world that the dinosaurs were killed when Earth was struck by a giant asteroid.[8] Their theory was, at first, most unpopular. Invoking some outside celestial agency for the dinosaur extinctions contravened the law of uniformity; it was like attributing it to an act of God rather than to some ordinary and perfectly explicable Earthly process. But then researchers found a huge crater from exactly the right time, off the coast of modern Mexico. Just because you can't see mighty asteroids hitting the Earth and destroying untold species of animals and plants today, the message ran, that doesn't mean it never happened.

And there are plenty of other reasons not to trust a simple reading of the world around you. You could do very sensible and careful experiments in the everyday world and end up thinking that time flows smoothly, that rulers are the same length for everyone, and that clocks tick at the same rate regardless of where you are or how fast you're moving. All of these assumptions are wrong. On large enough or small enough scales, the world doesn't work like that at all. Clocks can tick more slowly or quickly, time comes in packets, objects can be in two places at once, and the faster something is travelling, the shorter it gets.

All of these things were discovered in the early part of the last century, when relativity theory and quantum physics shattered our comfortable connections between direct experience and natural laws. There's a reason our intuition is often wrong: we evolved that way. In our normal lives we don't deal with relativistic or quantum scales. Nor do we deal with vast geological timescales.

And that fact has not escaped Paul and Dan. They both have a habit of talking about the Snowball as an "outrageous hypothesis". This is a nice touch, with an instant resonance for all geologists. The phrase comes originally from William Morris Davis, who, like Dan and Paul after him, was a professor of geology at Harvard. In 1926, inspired by the extraordinary happenings in physics at the turn of the century, he wrote a famous paper entitled "The value of outrageous geological hypotheses". Here's what he said:

> Are we not in danger of reaching a stage of theoretical stagnation, similar to that of physics a generation ago,

> when its whole realm appeared to have been explored? We shall be indeed fortunate if geology is so marvelously enlarged in the next thirty years as physics has been in the last thirty. But to make such progress, violence must be done to many of our accepted principles. And it is here that the value of outrageous hypotheses, of which I wish to speak, appears. For inasmuch as the great advances in physics in recent years and as the great advances of geology in the past have been made by outraging in one way or another a body of preconceived opinions, we may be pretty sure that the advances yet to be made in geology will be at first regarded as outrages upon the accumulated convictions of to-day, which we are too prone to regard as geologically sacred.[9]

Davis *wanted* people to take risks in geology. And he was sure that any important new theories that stood a chance of invigorating the study of rocks would be outrageous. They would fly in the face of our intuition, just as the new theories of physics had changed everyone's assumptions about how clocks and rulers behaved.

That poses a particular problem, because geology is as much art as science. After geologists have painstakingly assembled all the evidence that rocks have to offer, they still need a certain amount of intuition for the interpretation. With physics or chemistry, you can test different mechanisms one by one. But geologists are fond of saying that their experiment has already been done. They can't rerun the Earth with slightly different conditions and see what happens. Instead they often have to use their instincts.

And Paul and Dan are convinced that when it comes to the

Snowball, you can't trust your instincts. This was a world that didn't obey normal rules. "The Snowball is a different planet," Dan says repeatedly. "You can't judge it by the same criteria we use today." Instead you have to trust the evidence, however strange it appears to be. And you have to be able to interpret it by thinking out of your skin.

This is something that suits Paul very well. He has spent his life running counter to convention. His passion for music began as a teenager when he became gripped by atonal twentieth-century classical music—the sort of music that breaks all the rules. Paul loved it precisely because it sounded so different. "We were brought up to challenge everything. Conventional wisdom was bound to be wrong, and so if you were unconventional at least you had a small chance of being right," he says repeatedly.

Paul's family bears that out. At the age of nine, his sister Abby cut her hair, called herself "Ab", and joined the local boys' hockey team. She played a ferocious left defence for an entire season before she was picked for the all-stars and her sex was discovered. The story was plastered all over the Canadian press. She was featured in *Time* and *Newsweek.* That was when Paul started calling her Miss Canada. Abby went on to win Commonwealth Gold in the eight-hundred-yard sprint, represented Canada at four Olympic games, and became a famously outspoken member of the International Olympic Committee.

But there are many occasions when imagination is no substitute for experience. Geology is all about weary legs and backpacks weighed down with rock samples. It's about looking at the world you see around you, whether as a record of times past, as an exemplar of the present or as a predictor of the future. To be a geologist is to be rooted in the real world, to go with what you know.

And to some of the people listening to Paul's lectures and seminars, the Snowball was going too far. How could the Earth possibly behave in such an extreme way? He wants oceans that freeze over completely, even in the tropics and at the equator. An ice age that lasted for millions of years. A planet that then plunged from the coldest temperatures it had ever experienced into an intense hothouse within just a few centuries. Carbon dioxide levels hundreds of times higher than have ever been seen in the geological record. Rock weathering rates like nothing on Earth today. How could anyone ever accept a theory that was so far out of the box?

The more Paul pushed, the more vehement many of his critics became. Particularly a certain geologist from New York called Nick Christie-Blick.

NAMIBIA, JUNE 1999

A FLEET of trucks swept off the concrete forecourt of the Safari Hotel in Windhoek and began the long trek north. This was phase two of Paul Hoffman's Snowball mission. After his intensive programme of lectures, seminars and presentations, he now brought a selection of his peers out to Namibia to see the Snowball rocks for themselves. Among them was Nick Christie-Blick, a professor of geology at Columbia University's Lamont Doherty Earth Observatory. Nick was unsympathetic to the Snowball, annoyed by Paul's combative style and dismayed by the implication that the Earth had behaved quite differently in the past. When Paul invited Nick on the field trip, he expected a certain amount of trouble. "I knew Nick would be a pain in the ass," he told me later, "because he always is." What Paul didn't realize was that the field trip was about to turn Nick into the Snowball's Chief Unbeliever.

Geology is an intensely personal science. It's not enough to study a sample of rock that someone has carried home, or even to see photographs that they've taken. In this mind-twisting game of constructing a three-dimensional jigsaw from rocks that have been bent, thrust over one another, eroded away or buried, the context can be everything. Show me just how sharp the contact is between the two rock types on the outcrop, and how far it extends before it disappears from view. Where exactly in the cliff face did these measurements come from? How accurate are your sketches? How detailed are your maps? Geologists usually trust their own field data completely, but are much more reluctant to place reliance on data from places they've never seen for themselves. Plate tectonics pioneer William Menard put it well: "Some earth scientists believe in God," he said, "and some in Country, but all believe that their own field observations are without equal, and they adjust other data to fit them."[10]

That's why field trips make up an important part of how geology is done. You do the field *work* on your own, or with your few closest collaborators, for month after lonely month of mapping, sampling, walking out contacts. When you're back home again, you might write a scientific paper describing what you've seen and adding whatever interpretation you see fit. But the real test comes when you take your colleagues out to look at the outcrops you've been working on, so they can judge the rocks for themselves.

So Paul had brought along experts in Precambrian geology from around the globe. He had arranged everything, even paid for the trip out of his own precious grant money. This, he felt, was the one way to convince his fellow geologists that the fledgling Snowball theory was sound.

Yet Nick Christie-Blick grew more antagonistic every day. Nick, like Paul, is a field geologist. He patrols the world's rock surfaces for a living, measuring and probing and hammering his way through millions of years of history. He lives in the United States, but is thoroughly British. He drinks tea, speaks softly with a clipped Home Counties accent, and has the peculiarly English habit of carefully enunciating some words in a sentence, and then unaccountably rushing and garbling the rest. His short dark hair curls slightly over his forehead and around his temples. He has an engaging, self-deprecating smile. His face is clean-shaven and square, and his build is muscular. He's more football player than long-distance runner. He prides himself on his fitness.

The world of ancient geology is a small one, and Nick and Paul Hoffman have known each other for years. They first met on a Christmas expedition to the Grand Canyon in 1974, when Paul was thirty-three and already a well-established geologist, and Nick was a callow young graduate student of twenty-one, fresh off the boat from England. The North Americans on the trip were kind to their young English colleague. They lent him extra clothes at night when his sleeping bag proved woefully inadequate. They included him in their arguments about geology, bebop and baseball. They called him "Blick".

Paul in particular made a big impression. Even back then, Paul had a reputation as a very talented field geologist who was both a "doer" and a "thinker". Nick was eager to learn from him. But the memory that stayed with Nick most strongly came on the day the group was climbing out of the canyon up the Kaibab trail. Though the climb was steep, Nick wasn't particularly worried.

He'd spent the previous three years rowing for his college in Cambridge. He was strong and fit, an outdoors type well used to holding his own on arduous hikes. He was also used to being first up the mountain. That's why he remembers so keenly the moment when a lean, spare Paul Hoffman overtook him from behind. The trail was as steep as it gets in the Grand Canyon, but Paul was almost running. There was no catching him. As the rest of the group watched in astonishment, he tore up the hill and disappeared. Paul swears he didn't do this for effect. "I just climb fast," he said, shrugging. But effect it certainly had. "He left us for dust," Nick told me. "It was certainly impressive." And he threw back his head and laughed.

At first, Nick hadn't intended to gun for the Snowball theory. He was already involved in too many other arguments. In a way, you could call Nick a professional critic. He is famous for picking away at the threads of theories until he finds a detail that unravels the whole thing. But that kind of criticism takes time and trouble. Nick was forever coming across theories he disagreed with—some were big overarching ideas like the Snowball, others were arcane details of rock behaviour. He wouldn't have the energy to try to disprove them all. "Life is too short," he told me once. And if the Snowball had been fated to disappear back into cosy obscurity, Nick would probably have stayed out of it.

His opposition had grown, though, after Paul turned up at Nick's home institution to give a Snowball lecture. Paul arrived late. He had been caught—ironically enough—in a snowstorm driving down to New York from Boston. One of his tyres had been cut by jagged ice on the road, and he'd had to replace it in the nasty driving sleet at the roadside. The folks at Lamont Doherty had almost given up on Paul when he finally arrived.

Lamont's genteel buildings are set on a lovely old estate just outside New York. It's one of the world's top places for studying the way the Earth works, and it is famous for being fiercely competitive. Scientists there are actively encouraged to comment on and criticize other people's work. Interactions are forthright and robust. Paul knew well that if you take a new idea to Lamont, you'd better be ready for a fight.

So Paul went into Lamont with all guns firing. And as Nick listened to the talk, he became increasingly incensed. Paul used words like "panacea" and "triumph", the sort of words that bring Nick out in a rash. He hates big ideas that purport to explain everything. In his view, they are invariably wrong. And the people who advocate them nearly always end up sweeping inconvenient details under the carpet. That's why details matter so much to Nick. The world, he says, is complicated and the only way to explain it is by laboriously piecing together small parts of each individual puzzle. When you start talking about panaceas, he says, that's the first step towards donning blinkers and losing all sight of what's really out there.

Nick takes his obsession with details out into the field. He has been known to stand up on a cliff top with an ironic grin, throw out his arms, and say: "Hallelujah, come on down, all you believers!" He heard Billy Graham say that once, inviting the converts down to the stage, and it struck him as the perfect metaphor for geologists reverently making their way down to the rock face and the precious clues about the Earth that lay therein. But in reality, Nick is more of a Doubting Thomas. When he goes to see the rocks for himself, he has to put his hands in the wounds. He has to see the processes for himself. Only when every single question, however small, has been fully answered and every doubt satisfied, does he allow himself to believe.

And on Paul's field trip to Namibia, that attitude proved disastrous. Between Nick and Paul there couldn't have been a more dramatic clash of personalities and styles. Nick wanted to find holes in everything. He argued incessantly about every outcrop and every interpretation. Paul, on the other hand, didn't want to know. From the first day and the first outcrop, it became obvious to Nick—and everyone else on the trip—that Paul didn't really want to hear alternative interpretations. Paul had convened the field trip to persuade people, not to hear his theory criticized at every turn. The more Paul refused to listen to Nick's criticisms, the more determined Nick became to find fault.

Nick, in confrontational mode, can be truly infuriating. I first met him in the departure lounge at Las Vegas airport, months after Paul's field trip. By then, Nick was implacably opposed to the Snowball. His first words to me were, "Snowball Earth is dead." He didn't say, "I disagree with some aspects of this theory," or "I think there are certain problems with the interpretation." He said it was dead. The airport was full of geologists on their way to a conference, and many of them were buzzing with the Snowball idea. It was manifestly alive and kicking. But rather than pointing that out, I replied that I'd be interested to hear why he believed that, and mentioned that I too had visited Paul's field site in Namibia. Nick curled his lip. "Oh," he drawled scornfully, "so Paul's taking *tourists* to the field now, is he?" That was the most damning thing he could think of to say. Later he apologized. Though Nick is exasperating in the heat of battle, he can also be humorous and pleasant when he backs off. He said he had just spent the past few days arguing with a long-standing adversary about how exactly to interpret some rock arcana. He had, he explained ruefully, "come out punching".

Paul's response to this kind of behaviour, though, was equally infuriating. Paul has a habit occasionally, if you've said something that he doesn't want to hear, of simply erasing it from the airwaves. He might do this with anything he doesn't want to comment on—an anecdote about someone you know and he doesn't, an opinion he disagrees with, an emotional experience that he can't connect to. When you say one of these things, he doesn't react in any way. He just pauses until you've finished, and then continues with whatever he was talking about before. Rather than ignoring your comment in some pointed way, he behaves in every sense as if you simply didn't say it. Sometimes, talking to Paul, I've caught myself wondering if I really did say something, or just spoke it in my head. It can be unsettling, but it's also relatively infrequent.

But with Nick, Paul began to do this constantly. Many of the other scientists on the trip started to feel uncomfortable. This behaviour seemed just as bad as Nick's continual carping. During a field trip, up on an outcrop, you're *supposed* to discuss things. Dan, the "people person", did his best, trying to engage Nick in just the sort of discussion that Paul was eschewing. But it didn't help. Nick was scornful of Dan's interventions. Dan wasn't a field geologist, and he knew little about rocks as ancient as these. He was no substitute for Paul, and Nick had no qualms about saying so.

On 22 June, just over a week into the field trip, the convoy reached a spectacular outcrop in the northwest of the country. As with many of the best outcrops, Paul had stumbled across this one almost by chance. Two years earlier he had been exploring a dried-out gully, trying to trace the uppermost part of the glacial rocks. The gully was moderately hard going, choked with hefty pale grey boulders. As Paul made his way laboriously up the main channel, a graduate student named Pippa Halverson ducked off down an innocuous and apparently uninteresting side channel. A few minutes later Pippa reappeared. "You might want to come and look at this," he said.[11]

"This" turned out to be an outcrop that took Paul's breath away. To reach it, he and Pippa scrambled down the side channel and then turned the corner that was hiding it from view. The rocky ground rose steeply up ahead, but the surface was so pitted by the action of wind and weather that their boots stuck to it like glue. There was little vegetation, just a few scrubby bushes and arthritic trees, one with a gleaming bark like thickly smeared cream. And rising up on the left was a sheer cliff face that was the embodiment of the Snowball story.

The base of the outcrop, later called "Pip's rock" in honour of its discoverer, was crammed with ice-borne rocks of every shape and colour. These were the so-called dropstones that had been delivered to the seafloor by ancient icebergs. White and pink and tan and orange, they stood out spectacularly against the dull mudstone. They were the unmistakable sign of ice. But that wasn't all. Around halfway up the cliff, the scene abruptly changed. Suddenly the mudstone transformed into a pinkish carbonate that contained no interlopers, no boulders, no signs of ice at all. Below this knife-sharp edge between rock types, the Snowball was in full force. A fleet of icebergs floated on an ancient sea, discharging the rocks they carried into the soft mud on the seafloor. Above this edge, everything had changed. The ice had melted, the sea had boiled with carbon dioxide gas and a milky carbonate rain. In spectacular fashion, these rocks had captured the transition between icehouse and hothouse. This one outcrop encapsulated everything about Paul and Dan's story.

Paul is immensely proud of Pip's rock. When he takes you there, he can't resist building up a sense of drama. He climbs on ahead over the giant boulders of the dried-out gully that leads you to the rock, and as you round the corner to see the cliff face, he is already there, ready to gesture towards the outcrop with a triumphant flourish. And Dan then calls for all present to doff their caps in mock respect for the rocks. Nobody is allowed to use their geological hammers to pry rock samples from the surface. Paul has decreed that this particular outcrop must remain pristine.

Geologists often have their sacred places, the ones that hold the key to their ideas. I've seen researchers take off their shoes when walking on rock surfaces. Though they say it's because they don't want to risk damaging the outcrop, it is a strangely reverent gesture.

Some of the most famous outcrops are sites of pilgrimage, for which geologists compile "lifetime lists". One place on every geologist's list is Siccar Point in southern Scotland, where the father of all geology—an eighteenth-century gentleman farmer named James Hutton—first learned about the Earth's great age. Before Hutton, the prevailing theory of how rocks appeared on the surface of the Earth was called neptunism. This theory held that the Earth was once covered completely with a single vast ocean. Each layer of rocks was formed in the ocean, the most primitive ones first, and the more recent ones last. Eventually the ocean dried up and the rocks have remained the same ever since. This idea fitted beautifully with the biblical notions of creation, Noah's flood, and an Earth that had existed—according to the most literal interpreters of the Bible—for just a few thousand years.

After Hutton, this biblical interpretation was swept aside in favour of a new, more rational approach. Hutton realized that rocks had been created at different times and in different ways. Some were laid down on the floors of ancient oceans, others created by volcanic eruptions, and others still by the erosion of mountains, whose pulverized rocks spilled into nearby valleys to create new layers of geological history. And, crucially, the Earth cycled through these processes. What had once been an ocean floor could be thrust upwards to become a mountain, then be eroded into a valley, and eventually be flooded to become an ocean once again. The Earth's surface was continually created and eroded away and re-created again, in a process which Hutton famously said had "no vestige of a beginning—no prospect of an end". Through his insights, Hutton had laid the foundations of the notion of deep, unfathomable geological time. His dear friend, the professional mathematician and amateur geologist John Playfair, described the vision of geological eternity thus: "The mind seemed to grow giddy by looking so far into the abyss of time."[12] Of the perception that the Earth had existed for dizzying eons, the late Stephen Jay Gould said that "all geologists know in their bones that nothing else from our profession has ever mattered so much."[13]

But even the rational Hutton obtained his inspiration from religious convictions. In his farming days, Hutton noted that soil was created when old rocks were eroded away, with the debris carried ultimately off to sea. If this were the only process allowed, all of Earth's land would ultimately erode away and there would be nowhere left for mankind to live. Since Hutton believed that a kind and loving God had created the world expressly for the benefit of its human occupants, he reasoned that there had to be another process that rebuilt the Earth's surface and kept it comfortably habitable. That's how he developed the idea that seafloors could become mountains, and that volcanoes could create new land to replace the land that had washed into the oceans.[14]

Hutton's arguments about God's motivations would hold no weight in modern geology, but they show that science is muddier than it seems, and that scientists' ideas and inspirations can come from unexpected sources. What distinguishes science from pseudoscience is not whether your theory originated with some particular conviction about how the world works, or whether you feel an emotional attachment to it. What matters is the evidence you find to support it, and whether you are ultimately prepared to accept that it could be wrong. Perhaps it's appropriate, then, that students of geology flock to the site of Hutton's original inspiration with a most irrational reverence. They go for the sheer pleasure of witnessing first-hand the rocks that inspired it all.

It must have been obvious to everyone on the field trip that Pip's rock was sacred to Paul. And Nick was exasperated. This outcrop wasn't informative so much as photogenic. That, Nick felt, was why Paul was revering it. It was just showmanship. Nick marched up to the rock face and looked for some fault to find.

He found something almost immediately, in the stones that Paul claimed had been dropped on to the ancient seafloor from icebergs floating overhead. You can tell when a stone has come from an iceberg because it deforms the soft mud that it lands in. Instead of lying flat, the layers of mud immediately beneath the dropstone are squashed downwards.

But this should only happen to the sediment *below* the stone. If you peer at the cliff face and see lines of sediment deformed above as well as below the boulder, that's a warning sign that the stone may never have been dropped at all. Instead, it probably rolled down a subterranean slope. And then when the soft sediments were squeezed around the hard boulder they bent around it, top and bottom. Crudely put, distortion below an embedded boulder implies a dropstone, while distortion both above and below implies something else. That's what Nick was looking for at Pip's rock. He moved along the rock face peering at the boulders until he found one to be suspicious of. Look! he shouted in triumph to anyone who would listen. There's deformation above this one as well as below. That's compaction! Look. That shows that some of the boulders didn't come from ice.

Some of the boulders didn't come from ice. But even Nick accepted that many of them did. In other words, this was a pointless criticism. Pip's rock had been created in the presence of ice. Even if a few of the boulders it contained were not dropstones, the rest of them clearly were. Nick wasn't claiming that there had been no ice when the rocks were formed. He was simply pointing out a slightly different interpretation for a few patches of the cliff.

Didn't Nick realize that if he criticized so pointlessly, it would drive Paul mad? "Yes," Nick told me later. "But that's too bad. The point of the field trip was to have people come and take a look. He knew what he was getting."

By the end of the trip, the battle lines were drawn. On the long flight home from Johannesburg to New York, Nick wrote Paul an eight-page, single-spaced e-mail detailing all the arguments and criticisms that he felt hadn't had a proper hearing on the outcrops. In the opening, he was the soul of politeness:

> Once again many thanks to you and all those involved in the excellent excursion. . . . I very much appreciate the opportunity to see these fascinating strata first hand, and also the effort you made to obtain financial support for the trip.

But it didn't take long before Nick was launching into detailed criticisms that seemed almost calculated to enrage. He sent the message to Dan, Paul and everyone else who'd been on the trip, and quite a few people who hadn't been—something that Paul later claimed had been done purely to blacken the Snowball's name among people who hadn't seen the rocks for themselves. To Nick's chagrin, although a few of the recipients wrote back to say thanks for the insights, nobody took up his invitation to engage in further discussion. Paul outwardly ignored the e-mail, and inwardly seethed.

The following spring, Nick taught a graduate class at Lamont about the Snowball, and followed up with an e-mail to all the students outlining his criticisms, and warning them that Paul was "a great salesman". Inevitably enough, the e-mail found its way to Paul, who was seriously stung. What he objected to most of all, he said in a heated message to the course's organizer, was the way Nick seemed to want to pass him off as an ideas person who paid no attention to detail. This, he declared, was patently untrue:

> Anyone who doubts I have the ability and the will to walk the extra mile at the end of a long day to get the facts right should try me out in the Boston marathon some year.

Nick had no intention of running against Paul in a marathon. But he did decide to take the battle to enemy territory. In September 2000, Nick went to the Massachusetts Institute of Technology, just down the road from Harvard. He had chosen a deliberately provocative title for his seminar: "The Snowball Earth Hypothesis: A Neoproterozoic Snow Job?" The convener insisted on the question mark. Nick hadn't wanted it in. During the lecture, Nick's main criticisms centred around how Paul and Dan were presenting the Snowball idea. It was a cottage industry, he said. A bandwagon. Paul and Dan were in the audience, and both were furious.

Nick and his fellow contrarians are as important for scientific progress as the people whose new ideas they challenge. This process of putting up and knocking down can be one of the best ways to find out whether a theory really holds, whether parts of it need to be massaged, or whether the whole idea should be dropped.

Still, the heat of the Snowball interchanges had its inevitable effect. The aftermath of the "Snow Job" seminar was exactly what Nick must have predicted. There was no more talking with Paul. Nick had defined himself as an enemy of the theory, and there was no going back. Now it was time for real scientific challenges to take over from the rhetoric.

SEVEN

DOWN UNDER

Paul Hoffman had woven together the Snowball story's different strands. His theory was new, but it also rested firmly on observations and ideas from the past, especially those of Brian Harland and Joe Kirschvink. With his carbonates and isotopes, however, he and Dan had provided the first evidence that the Snowball might be right.

Now the theory was about to face its first serious challenges. They came, like so much in this story, from the rocks of South Australia. After Joe had performed his magnetic magic on Flinders rocks in the 1980s, many other geologists had been back to poke at the outcrops, and prize out more of their data. By the time Paul and Dan published their Snowball theory, there was already a stockpile of Australian data just waiting to be mined. And from that data, two tests of the Snowball quickly emerged.

Since Nick Christie-Blick had become the main focus of much of the Snowball resistance, it's appropriate that one of those tests came from his own research group. Ironically, though, the result didn't conflict with Paul's theory at all; quite the contrary. Working in the Flinders, a student of Nick's had already uncovered some evidence that turned out to be greatly in Paul's favour. She had addressed the issue of timing. For Paul and Dan's explanation to work, the ice had to last many hundreds of thousands or even millions of years, long enough for carbon dioxide to build up in the atmosphere and set the conditions for the global layer of carbonates to form. So were they right? Just how long did the Snowball last?

Bennett Spring, South Australia, 1995

Linda's arms were already starting to ache. But she couldn't let go. If the drill doesn't go in straight, the core is ruined and you have to start again. Water, milky with rock dust, was spurting out of the hole and spraying her jeans from the thigh down; they were already clammy and soon would be sodden. She shifted angle awkwardly, trying to rest her right arm on her knee. Even through ear plugs, the noise was deafening.

To her right and left ran a dried-out stream cut, its steep walls casting her into shadow. Though Australia was on the fringes of winter, the temperature had crept into the seventies. The sky was scattered with bright cirrus clouds. Thirty yards to the right, splashes of brilliant green rushes marked a sluggish spring—the only water in this parched country for miles around. Linda had surprised a grey kangaroo there that morning; it had made its characteristic "shhht" sound of alarm and thudded hastily away. Here and there along the stream were stately red river gums, leaves still green, grey bark peeling to reveal the silver trunk beneath. One good reason not to camp in the stream cut—you never know when a gum tree will decide to shed a branch.

Linda's camp was a hundred yards back, up on the flat, surrounded by dry grass and nondescript scrubby bushes, their leaves a dusty grey-green. This was the centre of the Flinders Ranges, a few hundred miles from the rhythmites of Pichi Richi Pass. The ground was a rich golden brown, and the limestone hills around the plain were low, their slopes gentle.

Her vibrant red Suzuki truck, with its soft top, clashed nastily with this muted countryside. But the rest of the camp blended. A pale wooden table with folding legs, a hefty plastic water carrier that held 25 litres, and a tidy pile of twigs and small branches collected in advance for the night's fire. (Always collect wood from gum trees—the few cypresses and pines leave a sticky black goo on the bottom of your pans.) A half-dome tent, Outback colours, green and grey, carefully upwind of the fire.

The bulky HF radio Linda had left in her truck. That had been a waste of limited space. It was a loan from the mines department and was supposed to be working, but all she could hear was static. So much for communicating with the outside world. The only human contact she'd had for a week had been two days earlier when, bumping her way in the Suzuki past the main buildings of the sheep station, she'd stopped and asked permission to put holes in the rocks. The men at the station had stared at her, baffled. Why should she want to? Why should they care?

Linda Sohl, a girl from the Bronx, was finally beginning to feel at home in the Outback. This was her third field season, she was nearly halfway through her Ph.D., and things were looking good. She was twenty-eight years old, plump, pretty, with neatly manicured fingernails, tiny silver hoops in her ears, soft brown hair cut and flicked into symmetrical waves that framed her face. She had enormous deep brown eyes. In different circumstances, without the safety goggles and the field gear and the rock drill, you'd probably take her for a sensible older sister. She'd seem cautious, level-headed, often reserved. But she had an air of resolve about her and—on occasion—the unmistakable look of a dreamer.

Dreams of adventure had taken her two years earlier from her well-paid but dull job at a New York publishing house and propelled her back to college. Her parents were horrified. A Ph.D.? In geology? Where's the security in that? Still, in spring 1993, Linda had arrived in the office of Nick Christie-Blick, whom she'd met at a seminar.

Nick had invited Linda to visit his lab, in the leafy New York suburbs, and Linda was feeling hopeful. She knew how hard it was to get into a Ph.D. programme at such a prestigious place as Lamont Doherty, especially as she hadn't taken the conventional route straight from college. Still, Nick had expressed interest, and that was a good sign. Linda didn't know quite what to expect—an interview, perhaps a tour of the lab. She was determined to make the best possible impression.

No need. Nick had already decided to take her on. When she walked through the door at nine-thirty in the morning, he had spread out various geological maps of South Australia and he immediately began suggesting different field areas. "Do you have a passport?" he asked her. "We need to get you a visa." Several weeks later they were driving together through the Australian Outback.

DOWN UNDER

She'd never camped before, not even with the Girl Scouts. She'd never even seen a sheep before, not close up at least.

A few days after they set out, Linda and Nick were in the car park of a small motel, transferring gear from one vehicle to another. Back and forth they went, arms full, past a pickup truck loaded with rusty old farm equipment, bales of wire, and a sheep—lying on its side. Linda was fascinated. She stopped and stared. It didn't blink, didn't seem to be breathing. It was absolutely still. Its eyes were a dull light brown colour. She thought they looked like the blank, dead eyes of a stuffed teddy bear. She put her head closer. Suddenly, without warning, the sheep blinked and turned its head. Linda yelped and leapt backwards. "Nick! The sheep's alive!" Damn, damn, damn. Not the way to impress your brand-new adviser, before you've even made it into your first semester.

Nick was unfazed. He bought her a tent, a lantern, a set of pans. He took her to various possible research sites in the Flinders Ranges. He helped her to choose which rocks she wanted to work on for her Ph.D., showed her how to hook up the radio, and left her there.

At first the silence was the worst. Then the unexplained bumps in the night. Kangaroos would hop silently by until they hit a gravelly streambed and suddenly wake her with their crash. Reclusive emus would come out at night to skitter on the stream pebbles, left, right, left, right, like people running. Once there really were people, whooping their way along a remote bush track, shining bright searchlights and firing guns. The air was full of bullets and agonized screams. Linda stayed in her tent. Later she discovered it was part of an authorized cull of the feral alien animals, the cats and foxes that bedevil the Outback.

But for the most part the Outback turned out to be benign and beautiful. For annoyance value, there were just the idiotic galahs, brightly coloured parrots that would periodically launch themselves en masse from a tree, shrieking wildly and dive-bombing the site. And the bugling cockatoos, and the oversized Australian magpie, actually a relative of the crow, but with an enchanting, lyrical warble that echoed hauntingly among the gum trees, and almost made up for the rest. The birds were useful in a way. No need for an alarm clock when you have an ear-shattering dawn chorus every morning, 6:45 sharp.

Linda had never intended to test the Snowball idea. She was supposed to be studying some carbonate rocks that had formed long before the ice. But, having worked among the ice rocks for the past two years, she couldn't help being intrigued by them. She'd heard about Joe Kirschvink's magnetic work and the more detailed studies by George Williams. And she decided to see whether she too could detect the faint magnetic traces in the rocks.

A few weeks into her 1995 trip, Nick came out to help. He hadn't encouraged Linda to work on the ancient ice, but he was curious to know how her results would turn out. Nick, the details man, was always pestering and probing. In the field he could drive you mad. He doesn't—ever—let something lie. As he and Linda drilled and washed and carried samples back and forth, he asked her incessantly about the precision of her magnetic measurement. To Linda, it simply wasn't the most important issue. Details often matter, but that particular detail would come out in the wash. But Nick was consumed by it. A few weeks after he left to return to his own field site, Linda went into the small town of Hooker to collect her mail, sent to her care of "general delivery". Among her letters was one from Nick, via his field camp several hundred miles to the north. The page was covered with careful diagrams and a detailed analysis of the factors affecting the precision of Linda's measurements. He just couldn't let it go.

At the end of the season, Linda headed back to New York to analyse her samples. The work was exhausting and painstaking. The magnetics machine was already booked up for the days and evenings, so she worked on it through the nights for six weeks. But in the end she hit pay dirt. Though all the rocks she collected had looked more or less the same, pale and reddish from the magnetic minerals they contained, some of them had a field pointing northwest, and in others the field was southeast. To Linda, that meant only one thing. These ice rocks contained a record of ancient magnetic reversals.

A magnetic reversal occurs when the Earth's magnetic field spontaneously swaps directions, so that magnetic north becomes magnetic south. Our planet's magnetic field is generated by molten iron sloshing around in the Earth's core, and the flipping must somehow be related to changes in this deep, hot liquid. But the details remain a mystery. What we do know is that this bizarre event takes place roughly once every few hundred thousand years. If you could slow down your frenetic human lifespan until it matched the stately passage of geological time, you could study a compass and watch these flips in action. Flip, and the needle would swing halfway around the dial to point south; flip, and it would point north again; flip, back to south. The magnetic minerals trapped in Linda's rocks had behaved exactly like that geological compass needle. Each ancient flip had been frozen in.[1]

These flips are useful to those who study the ancient Earth, for two reasons. First, if they appear in a sequence of rocks, the magnetic field there has to be original. Neither subsequent heating of the rocks nor the influx of new magnetic material can produce this alternating pattern of fields. By finding reversals in the Flinders ice rocks, Linda confirmed Joe's discovery that ice had been present near the equator.

More important, Linda had also found evidence that the ice was extremely long-lived. Her rocks contained perhaps as many as seven flips. If magnetic reversals in the Precambrian happened roughly as often as they do today, Linda's ice rocks had to span at least a few hundred thousand years, and probably several million. Here was the first tangible evidence that the Snowball glaciations really were the longest ever known. It seemed as though they had lasted quite long enough to build up huge amounts of carbon dioxide in the atmosphere, and trigger the events that Paul and Dan envisaged.

Paul was thrilled when he heard about Linda's reversals. The Snowball theory had survived its first important challenge—the test of time. Nick was perfectly aware of the irony that his own student's work had ended up supporting the Snowball, but he was also proud of Linda. She had made an intriguing observation, providing a potentially important clue about the conditions that prevailed during the glaciations.

Still, just because the ice had stuck around at the equator for an inordinately long time, that didn't mean the rest of Paul and Dan's theory was correct. Nick and Linda were both still determined to show that the overall picture was nothing like Paul and Dan's dramatic total freeze-over.

The next serious test of the Snowball, however, wasn't from Nick's camp, though it did come with another dose of irony. Once again the challenger had worked in the Flinders Ranges, and provided key evidence that was later incorporated in Paul's theory. But this wasn't Linda, or anyone who worked with her. The new challenger was George Williams, the Australian who had instigated the research that caught Joe Kirschvink's attention back in the 1980s. George had worked in the Flinders Ranges for decades. He's the person who studied the tidal rhythmites there, and whose work laid the foundation for all the magnetic work that followed. He had done plenty of further magnetic work there, too, and he was utterly convinced that there had been ice at the equator. But he was also convinced that this had nothing to do with a global Snowball. George had an alternative explanation: the Earth, he said, had tipped over on to its side. Sound crazy? Well, he had evidence that seemed to prove it, and that was soon giving Paul Hoffman sleepless nights.

PORT AUGUSTA is a small, grim coastal town a few hundred miles north of Adelaide, just west of the Flinders Ranges. Its street signs proclaim it to be the "Gateway to the Outback". At Port Augusta the rough stuff begins. Though the roads to the south are smooth and civilized, many of the northern ones are little more than dirt tracks.

From there, the driving instructions to Mount Gunson Mine are simple enough. One paved road goes north, the Stuart Highway. Take it. After a hundred miles or so, turn right. Yes, this will be the only right turn. You can't miss it. No, really, you can't possibly miss it. There's nothing else there.

Most of the trade between Port Augusta and the Australian interior is plied along the narrow Stuart Highway by terrifying, thundering "road trains". These are linked caravans of two or maybe three huge trucks pulled by just one engine, whose driver has probably been up all night, fuelled only by greasy steak sandwiches "with the lot" (onions, cheese, tomato, bacon, fried eggs, you name it, all oozing with ketchup and evil yellow mustard) from the occasional wooden roadhouses. The trucks rock your car with shock waves as they rip past you. They are death to kangaroos, and grey-furred flesh periodically smears the road, picked at by huge, black, wedge-tailed eagles. Even a full-sized kangaroo won't make much of a dent in a road train. But if you're in a regular car, you'd better not drive at dusk or dawn when the 'roos are on the move. Evolution hasn't had time yet to equip them with road sense. They'll lurk in the scrub by the roadside and then leap into your path without warning. Hit one, and it could well be the death of you, too.

There's little else to distract you during the drive, just the flat, empty, featureless scrub of the Australian desert. Eventually a slightly battered sign points to a dirt road on the right. These days not many vehicles take this turning. In the 1980s, Mount Gunson was a flourishing copper mine, but now most of the operations have closed down. A few workers' huts remain, with their peeling girly posters. Also some yellow warning signs—SULPHURIC ACID. CORROSIVE!—next to rusting pipes almost the same colour as the dusty red soil. But the rest of the site is scarcely different from the surrounding desert, its landscape scattered with drab grey saltbushes and Flinders Ranges wattles, fast-growing, fast-dying, dried-out skeletons rubbing up against the vivid bottle-green of the living. When the miners finally pull out completely, the only sign of the once-frenetic mining will be the great open pits carved deep into the ground. Now that the mining has ceased, water is no longer pumped out of these pits, and soon they will become lakes. Then, there will be no more chances to see the geological signs that caused Paul Hoffman such anxiety.

DOWN UNDER

The Northeast Pit, cut into the Cattlegrid ore body, bears the features that seemed to challenge Paul's Snowball. This great hole in the ground, a thousand feet long and almost as wide, sinks down to a flat floor. The sides are steep and layered, dark quartzite near the floor of the pit, paler sandstone above, and then, in the topmost layer, the ubiquitous iron-rich red of the soil. You crunch along the floor of the pit on delicate yellow nodules crusted with white crystals of gypsum. Patches of soil are slotted with short parallel lines, where passing kangaroos have dug in their heels. The sun is invariably dazzling, glaring off the sandstone. Squinting even through sunglasses, you have to be close to the walls before you notice anything different about them. But as soon as you pick out one of the strange structures in the pit's steep walls, you start to see them everywhere.

They are wedges of sandstone, six feet tall and triangular, like a row of massive shark's teeth. Or perhaps witch's teeth, since they are stained green by the flow of copper-rich water. The quartzite they sit in has been smashed and broken like a jumbled pile of bricks, but the wedges themselves, and the rock layers above them, are made of smooth, flowing sandstone. They line the walls of the pit. As you walk along you see first one, then another, then a whole row of them, strung as on a necklace. Occasionally you see the outline of a wedge or two above the main row. To an untrained eye they look bizarre. But to geologists they're classic. They're textbook. Anyone can tell you what they mean.

Sand wedges are the clear signs of a climate pattern called freeze-thaw. Here's how they form. First, freeze. The temperature drops quickly, and in response the ground shrinks and cracks into regular polygonal shapes, like the mud cracks that form in the bed of a dry lake. Into the cracks blow sand and dust. Next, thaw, and the cracks are kept open by the presence of the sand. Next, freeze again, and the cracks open wider, with more sand tumbling in. Eventually the sand that has wedged into the cracks solidifies and turns into sandstone. In Mount Gunson's Cattlegrid pit, the broken-up quartzite is the ground that was repeatedly cracked by freezing and thawing, and the shark's teeth are slices through the solidified sand wedged into the cracks.

So Mount Gunson must have suffered from repeated episodes of freezing and thawing. Why should that be a problem? Mount Gunson was an island then, surrounded by a frigid ocean. Over to the east are the ice deposits of the Flinders Ranges, where fleets of icebergs dropped their load of stones and boulders into a shallow sea. Freezing the ground thereabouts should have been easy.

But to get sand wedges doesn't just require cold temperatures. It also takes warm ones, followed by sudden, repeated temperature drops. Sand wedges need cycles of freeze and thaw—in other words, *seasons.* The problem is that Mount Gunson was close to the equator when the ice was present. And at the equator, seasons simply don't happen.

We have seasons because our planet is tilted. If the Earth remained bolt upright in its progress around the sun, there would be no such thing as summer and winter. All year round, every place on Earth would experience the climate that its local position deserved. Close to the equator, where the sun is fierce overhead, the climate would be hot. Close to the poles, where the same amount of sunlight spreads over a larger area, the climate would be cold. January or June, there would be no difference.

But the Earth's tilt makes life more interesting. Superimposed on the overall pattern of climate—hot equator, cold poles—is a seasonal shift. In January the southern hemisphere is thrust out towards the sun. Australians and South Americans head to the beaches. Antarctica basks in the midnight sun, and temperatures there can stay above freezing for days. By June, halfway again around the Earth's annual orbit, the northern part of the Earth receives more than its share of sunlight. Now Antarctica is shrouded in permanent darkness, and northerners take their turn in the sun.

The equator is the only place on Earth to escape this annual cycle of hot and cold. No matter which hemisphere is grabbing its extra share of sunlight, the equator feels it, too. Equatorial regions muscle in on everyone else's summers.

So how to explain the Mount Gunson sand wedges? Cast-iron magnetic evidence says that Mount Gunson was at the equator when the ice came. Nobody doubts this. But the wedges seem to show seasonal changes. What gives?

George Williams thinks he knows. Today the Earth's axis isn't tilted by much, just twenty-two degrees, a sixteenth of a full circle. But, says George, what if it used to be much more tilted? Perhaps, in "Snowball" times, the Earth had tipped over on its side, by something closer to a full quarter-circle.

If so, everything we now know about the climate would be turned upside down. Poles and equator would swap characteristics with the sun blazing directly overhead at the poles, and spreading feebly out at the equator. And this, George felt, would nicely explain two of the main conundrums of the ice era without requiring a frozen Earth. The Arctic and Antarctic regions would be balmy, and the equatorial regions frozen, which would explain the Australian evidence for ice at the equator. What's more, the sun would now be wreaking seasonal changes on to the frozen equator. That would explain the sand wedges. If George is right, the Snowball simply didn't happen.

George had been talking for decades about the Big Tilt.[2] But until Paul came along, nobody was particularly listening. When George heard about Paul's Snowball, however, he went immediately on the offensive. In a magazine called *The Australian Geologist,* he published a ten-point criticism outlining exactly why he thought Paul was wrong and the Big Tilt was right. (George called this criticism "Has Snowball Earth a snowball's chance?"[3] Paul's immediate response was wryly titled "Tilting at Snowballs".[4])

The Big Tilt itself, after all, faced plenty of problems. For one thing, it couldn't explain nearly as much as the Snowball purported to. The Tilt accounted only for the equatorial ice and the strange sand wedges. It said nothing about the cap carbonates, the isotopes or the ironstones.

More seriously, as Paul quickly pointed out, there was no easy mechanism for righting the Earth. Tilting the planet in the first place would have been quite straightforward if it had happened early enough. When the solar system was born, there were plenty of planet-sized chunks of rock flying around the place, crashing into each other like giant pinballs. Most scientists believe that an almost-planet the size of Mars smashed into the young Earth, creating our moon from the debris. A collision like this could easily have knocked the Earth over on to its side. But in the relatively sedate time of the past few hundred million years, what could have righted the Earth to the more gentle tilt that it bears today? By the end of the Precambrian, any builder's rubble from the early solar system was long gone, and there's no other easy way to move the Earth back upright.[5]

Of course, saying we don't know how the Earth could have

righted itself doesn't prove George's theory to be wrong. Until recently nobody knew what exactly was causing the continents to drift, though they were still moving for all that. But the Big Tilt had a bigger problem, one that Paul seized on gleefully. The evidence for glaciation comes in one particular time slice, right at the end of the Precambrian. So George couldn't rely on the creation of the moon to knock the Earth over. Instead he had to propose that something else had suddenly tilted the Earth around 700 million years ago, billions of years after the formation of the moon. And that something else again had caused the Earth to right itself abruptly again a couple of hundred million years later. That's an even bigger stretch of the mechanism problem.

But there still remained the vexing issue of the sand wedges. How did a place that was within perhaps ten degrees of latitude of the equator experience seasonal changes? Paul did his best. He came up with a possible explanation involving glaciers that alternately surged forwards and drew back, sometimes insulating the ground they covered from the bitterly cold air and enabling it to thaw a little, sometimes exposing it for another freeze cycle. Not many people were persuaded by this, though. It sounded rather too much like special pleading.

Salvation came from a different source, and showed that even Paul had been too conditioned by how the Earth works today. The key turned out to be how different the Earth's seasons would have seemed on a planet that was blanketed in ice.

By 2001, veteran climate researcher Jim Walker, from the University of Michigan, had become intrigued by Paul's Snowball theory, and began tinkering around with a simple climate model to try to figure out what the weather would have been like. He picked the most extreme of Paul's conditions—a globally frozen ocean—and set the model on its way. Day to day in Jim's model, the weather was rather boring. Nothing much changed. There were no travelling storms and no temperamental weather patterns. At every point on the surface of Jim's model planet, one day was pretty much like the next. The Snowball must have been a little like Mars, he says. Apart from the occasional dust storm, the whole Red Planet just settles down into a placid, predictable weather pattern. "The wind there always blows from the same direction at four o'clock in the afternoon."

But to Jim's astonishment, the seasons on his model Snowball were an entirely different matter. They were exaggerated, larger-than-life versions of the seasons we are familiar with. At any one point on the frozen surface, there was a *huge* temperature difference between winter and summer, much bigger than we see today.

Why? Well, our modern wet and windy Earth has an ingenious built-in mechanism to guard against extremes. If the climate everywhere simply depended on the direct overhead sunlight, the tropics would be much warmer than they are today, the higher latitudes would be much colder, and the seasons would be very much more marked. Instead, the Earth's oceans damp down the sun's ardour.

All through the summer, the oceans soak up sunlight. Unlike land, the oceans are pretty much transparent, so the sun's rays can penetrate deep into the interior. Also, the ocean's currents keep water moving. Warm surface water is replaced by cold water rising up from the deep to take its place in the sun. The oceans work like a vast storage heater: they absorb heat throughout the summer, and then slowly release it during the winter. That's why seasonal changes are so much more extreme in the middle of continents than they are at the margins. Places like Nebraska or central Siberia are too far from the ocean to benefit from its summer cooling and winter warming.

But on the Snowball, all that would have changed. According to Jim's model, the whole place would have been like Nebraska or Siberia. With ice covering the oceans, there would have been no more of this gentle amelioration of the seasons. And here's the key point: that argument would apply right down to the tropics. Even places just a short distance from the equator would have had exaggerated seasons. The Flinders Ranges of South Australia would have experienced a temperature difference between winter and summer of perhaps 30 degrees C. And though the annual average temperature would have been bitterly cold, summer temperatures could even have crept up above freezing for a few brief months of the year. Bingo. There's your freeze and thaw. There's your explanation for the sand wedges.[6]

George Williams still believes in his Big Tilt, but most other researchers have begun to edge away from it. The Snowball, it seemed, had survived yet another test. But Linda and Nick were still working on finding an alternative idea, one that was less wacky that George's and less extreme than Paul's. Nick's final e-mail to Paul, back when they were still—just about—speaking, had contained this jaunty assurance: "In parallel with an effort to develop good tests, I also accept your challenge to seek a better hypothesis." And that's exactly what Nick did. He and Linda teamed up with an Australian researcher and longtime collaborator of Nick's, Martin Kennedy. And by the end of 2000, it was beginning to look as if they were on to something.

EIGHT

SNOWBRAWLS

"How can *anyone* look at these deposits and *still* be talking about a Snowball?" Martin Kennedy thumped his fist against the dashboard in frustration. He was in a truck with several other geologists, heading north towards the long, thin ribbon of Death Valley in California, and as the ground fell further and further below sea level, Martin's blood pressure was rising.

Martin is tall, slender, half Australian and half American. He is thirty-eight. He's intense and ornery, but this is coupled with bursts of humour that make him unexpectedly good company.[1] He can launch into a sudden fury, and then pull himself out of it just as quickly. (During a brief period working at Exxon, he underwent the regulation psychological tests, which determined that he was a "red" person, categorized as highly aggressive. He was, he says, disappointed. "I was kind of hoping to be blue-green.") His hair is short, brown and slightly curly. The corners of his eyes are crinkled with laughter lines. Martin has boyish features, a button nose and thin lips that can make him look engaging or petulant, depending on circumstances. He has a pathological aversion to authority.

Martin hadn't meant to pursue a career in geology. Initially he had intended to run a farm in Australia, where "you can live by your own hand". He would probably have done so if the purchase hadn't fallen through. Though he's now based in the relative civilization of the University of California's Riverside campus in Los Angeles, he's happiest in the Australian Outback, where he's done fieldwork for many years. He trusts himself, resents interference, and instinctively challenges received wisdom.

Martin had become increasingly outraged by the claims that Paul and Dan were making for their Snowball theory, and was now a key player in Nick Christie-Blick's stated quest to find an alternative.

In November 2000, he had joined a small field trip that was visiting some Snowball rocks in Death Valley. The trucks headed north through the valley, hugging the east side of the central saltpan. At first the sand was hazy with low, olive-green saltbushes, but the vegetation gradually disappeared, and by the time the trucks reached the sluggish saline pools of Bad Water, the lowest point below sea level in the western hemisphere, there was just bare sand streaked with salt.

Lining the sides of the valley were mountain peaks, sometimes jagged, sometimes rounded. From a geological perspective, all the mountains are recent upstarts. Around 13 million years ago, a mere snip compared to the Snowball timescales, this part of the Southwest was fairly flat, dipping off the back slope of the Sierra Nevada. Then the land began to stretch. As the crust thinned, cracks opened in its surface. Some parts fell downwards to make long, thin valleys, while others were thrust upwards into mountains. The whole area today is riddled with these stretch marks. And thanks to the overhead thinning, volcanoes sprang up from the deep and spilled their magma on to the Earth's surface. Many of the mountains left from that turbulent time are piled with lava, the rocks painted in a desert palette of burnt sienna and ochre, chocolate, Venetian red and tan.

The Snowball rocks that Martin had come to see hailed from a time long before the Earth's erratic bucking threw up these mountains and opened these valleys. They come not from 13 million years ago, but 600 million years ago and more, eons before dinosaurs roamed North America, in fact before anything roamed anywhere. When the Snowball gripped the Earth, the only living things hereabouts, and anywhere else in the world, were those tiny, single-celled sacs of chemicals bound together by extruded slime.

Though Death Valley is full of Native American trails, it wasn't seen by white men and women until the mid-nineteenth century, when the first pioneers passed through on their way to hunt for California gold and glory. Its moniker is unfair. The valley wasn't so dangerous, even in the early days, if you could find the waterholes. Few people died here, though many had miserable crossings. The landscape is bright with sand and salt and sun. And in spite of the generations of geologists who have picked over the valley in the past century, it is still a place of geological mysteries. Like the migrating boulders of Racetrack Playa, a dried-out lake bed to the north and west of Bad Water. The floor of this valley is strewn with rocks of every size: pebbles, stones and vast boulders weighing seven hundred pounds or more. And, inexplicably, these rocks move. Try to catch them at it and they will just lie there, solid, immovable and innocent. But between visits from the geologists who carefully plot their positions, these restless boulders somehow skid along the valley floor, twisting, turning, zigzagging, and leaving grooved trails behind them. There have been many attempts to explain this bizarre Death Valley phenomenon. Some people think the culprit is an intense gust of wind occasionally funnelling through the valley; or a sudden rainfall that coats the mud with a slippery sheen; or thin sheets of ice that lift the boulders and make it easy for them to slide. Nobody really knows.

The trucks turned west now, where a wide sweep of open land sloped toward the sand dunes in the centre of the valley. Except for the small pools at Bad Water, the valley floor was bone-dry. But up over the western mountains, occasional snow clouds hovered like pale will-o'-the-wisps, trailing off at their base where the snow evaporated before it ever hit the mountain surface. As Martin's truck wound gently up the road through Emigrant's Canyon to the mountain pass, a few fat flakes of snow smacked the windscreen.

Over the pass, walls of rock stretched upwards at the roadside, speckled with coloured rocks and boulders. This was the Kingston Peak diamictite, a direct remnant of the Snowball. The mismatched mélange of rocks had tumbled into the ice-covered seas of Snowball time and left behind a deposit more than two miles thick. Glancing at the rocks, Martin began to get annoyed. "How could you get all this in the final days of the Snowball?" he demanded, gesturing out the window. "If you believe Paul's Snowball, all this was deposited in a thousand years or so. You just can't get sedimentation rates like that!"

This was part of an ongoing argument. In recent months, many people had been wrestling with the question of when exactly the ice rocks had formed. Was it only at the end of the Snowball, or were they continuously created throughout the previous several million years that it gripped the Earth? It mattered because of thick deposits such as these. Everyone agreed that they were created when glaciers gathered up rocks and dragged them to the sea, or when icebergs melted on the water and dropped their load of debris on to the mud below. But in the first incarnation of Paul and Dan's Snowball idea, this couldn't have happened during the long, cold millennia of the Snowball. When the oceans are frozen over, icebergs can't break off and move around, because there's nowhere to move to. And with frozen oceans, it's extremely hard to get ice on the land. To make a land glacier you need snow, and to make snow you need some patches of open water to deliver the moisture. At first Paul and Dan believed that the Snowball would have been cold, dry and dead, with no snow, no glaciers and no icebergs until, perhaps, the very final days when the ice began to melt back.

Then how to account for two vertical miles of rock? Martin was right—you couldn't get such thick deposits from such a short time. But Paul's Snowball idea had evolved since he first started working on it. Now he was allowing for some glaciers to form and move even while the oceans were still frozen. He had realized that the wind could erode ice from the sea surface, transport it to land, and deposit it there. Though the process of creating glaciers would be painstakingly slow, during the Snowball there was no shortage of time. When you have millions of years to play with, it's not so hard to create a river of ice, inch by inch.

Martin knew about this argument, but he wasn't impressed. He switched to another complaint. "What about the cap carbonates?" he demanded. "To make them in the short time that Paul and Dan want, you'd have to have weathering rates a thousand times faster than today's. It's impossible!"

Then, unexpectedly, he grinned. "I feel like I'm getting riled here, and I shouldn't be," he said. "Look, to be honest, I hope the Snowball's right. It's a beautiful idea. But I just don't like the way they're ramming it down our throats. I feel . . ." He hesitated, searching for the right word. "I feel violated."

Martin had first met Nick Christie-Blick, the Chief Unbeliever, back in 1993 on a field trip in central Australia, which had been organized by several senior Australian geologists. Martin was in a foul mood. Even though the group had travelled to his field site, the place he had spent his Ph.D. mapping, Martin had been the last to hear about it.

Geologists can be very protective of their field sites. They spend months there, often alone or with only a few others for company. They leave their bootprints on the soil, and the marks of their hammers on the outcrops. Day after day they climb cliffs and hike through gullies, walking out the contacts between rock types. They learn how every rock and stone is related. In their heads and their field notebooks they gradually assemble the complex, four-dimensional jigsaw that tells them the area's ancient history. And they don't just become experts in the rocks; they often also develop a physical, almost proprietary connection to the landscape. If you plan to visit someone else's field site, the first thing you'd better do is call them.

Martin was no exception. He had grown to love the austerity of his research site, a day's drive east of the remote central Australian town of Alice Springs. He loved the vivid red colours and the vast, empty proportions of the landscape. He loved jumping into his beat-up old Land-Rover and bumping along the aboriginal trails that took him into the heart of the bush. He mapped alone. And he knew those rocks better than anyone else on the planet did.

But on the field trip in 1993, nobody had called Martin. He was still only a student, and the status-obsessed organizers had arranged everything without involving him at all. It was as if someone brought a field party tramping through your backyard without warning or explanation. Martin was furious. As the trip progressed, he made himself more and more obnoxious. He challenged everything, did his best to humiliate the leaders by pointing out their errors, and one night over dinner he brought one of them close to tears. (Eight years later this highly eminent geologist can still barely bring himself to mention Martin's name.) The more people tried to slap Martin down, the more belligerent he became. Nick was intrigued. Here was someone else who constantly challenged and harried. What's more, he was often right. When Martin argued about the rocks, he did so with little tact but plenty of intelligence. As soon as the trip ended, Nick started talking to Martin about how the two of them could collaborate.

Shortly afterwards, Martin took Nick out to see some other Australian rocks that he'd been working on, just outside Adelaide. Now Nick was the abrasive one. Throughout the trip, he was incessantly challenging and infuriating. He argued every point. "How do you know this rock isn't the same as *that* rock *there*?" By the end of the trip, Martin felt as if he'd been put through a wringer. But he also realized that his mapping had been tested as fully as it could ever be. When Nick is finally satisfied with the picture you paint him, you know it must be right. "Nick's very bloody-minded." Martin says. "He's a contrary person, and that's his value in the world. I've learned a lot from him. I don't speculate any more. I don't just let my lips flap."

Nick and Martin have worked together on and off ever since. They cherish the collaboration. And in annoyance as much about

Paul Hoffman's blowhard style as his scientific substance, they have become firm allies in the anti-Snowball game. Now Martin figured that he had managed to cook up a real alternative to Paul's theory. He was in Death Valley to check out the evidence for this latest challenge to the Snowball.

Daylight had almost gone when the truck finally arrived at the Noonday dolomite, a pale tan slope of rock that marked the cap carbonate overlying the glacial deposit. The rocks were shot through with dark, thin vertical lines that looked like worm burrows, though they had formed long before worms were invented. They were exactly like the tube rocks in Namibia that had so baffled Paul.

And there were other strange structures in the rock. The ancient mud layers were occasionally buckled into chevrons, and the insides of the chevrons were filled with very fine-grained cement. That was confusing. In geology, cement appears when a gap somehow opens in the rock after it has first formed. When the mud was first hardening into rock, something happened to create an internal space, like a crack in the middle of a brick wall. At some time later, fluids pouring through the rock deposited cement, which filled the gap.

But a chevron-shaped gap is strange. To make a chevron, you need to squeeze the rock until you rumple it into a fold. To make space for cement, you need to stretch the rock until a gap opens up. What process on Earth can both squeeze and stretch at the same time?

Martin thought he knew. He jumped back into the truck, eyes shining triumphantly. The signs in the Noonday dolomite were just what he'd been looking for. He was assembling evidence that the carbonate rocks that capped the Snowball deposits came not from some mighty weathering in the inferno that followed the ice, as Dan had suggested, but from an entirely different phenomenon, one that didn't need an inferno, didn't need a total deepfreeze, didn't need any of the things that, to Paul and Dan, made up the essence of a Snowball world. All Martin's model needed was for a strange substance, a chimera, half-ice and half-fire, to be scattered throughout the Snowball world.

Martin's chimera is called methane hydrate and is made of tiny ice cages, with molecules of methane—natural gas—trapped inside. Methane hydrate is amazingly abundant in the world today. Together with other gas hydrates, it harbours more than twice as much carbon as all the known natural gas, oil and coal deposits on Earth. It looks like dirty ice. It's often smelly, too, giving off a whiff of rotten eggs from the sulphurous activity of bacteria that are typically found in the sticky mud alongside. It's very unstable. Today methane hydrate survives only in the frozen Arctic soil, or in the high pressures that exist beneath the sea. Raise the temperature, or bring a lump of hydrate up to the surface, and it will rapidly disintegrate. You can hold it in a gloved hand and watch the ice disappear, fizzing with pops but no crackles. Strike a match, and the gas coming off it will ignite, and then burn with a reddish flame. All that's left behind at the end is a muddy pool of water.[2]

Because methane hydrate is so widespread, many people over the years have championed it as the fuel of the future. The trouble is, the ice cage is so unstable that mining it can trigger disaster. If you accidentally set a batch of methane hydrate decomposing when you're trying to extract it, the methane that bubbles out turns seawater into foam. Because foam is much less dense than liquid water, your drilling ship promptly sinks. In fact, since there's plenty of gas hydrate on the continental rise off the southeastern United States—the western portion of the so-called

Bermuda Triangle—this mechanism has frequently been invoked to explain the mysterious maritime disappearances that are supposed to have taken place there. The idea is that a sudden underwater landslide could reduce the pressure on the hydrate deposit, decomposing it and sending deadly bubbles of gas up to the surface. This is plausible enough geologically, but, sadly, the shipping end doesn't hold up. The Bermuda Triangle simply hasn't swallowed an abnormal number of ships—ask Lloyds of London.

Still, methane hydrates have apparently been responsible for real-life dramas in the past. In the Barents Sea, just off Norway's northeastern tip, hydrate deposits seem to have exploded thousands of years ago, leaving behind giant craters that pockmark the seafloor. That probably happened at the end of the last ice age, when warmer seas destabilized the hydrates there until they erupted like a volcano. And some researchers believe that destabilized hydrates released more than a million cubic kilometres of methane at the end of the Palaeocene, about 55 million years ago.

Martin Kennedy wants to explain the carbonates that blanketed the Earth immediately after the Snowball in the same way. Methane hydrates, he thinks, might have been more widespread than they are today. After all, everyone agrees that conditions were colder then. When warmer temperatures returned, those methane hydrates would have released their gas, to be quickly oxidized and precipitated into the ocean as a blanket of carbonate.[3]

Martin felt that he had found clear evidence for this idea in the Noonday dolomite in Death Valley. That's why he was so excited. The chevrons filled with cement, the rocks that looked as if they had been simultaneously squeezed and stretched, the "worm burrow" structures—all of these could have been caused by decomposing methane hydrates. As the light methane travelled upwards through the heavier mud, it would have created tube-like vertical passageways, just like the dark tubes in the rocks. And if the gas hit an obstruction—a microbial mat, say, which is dense and rubbery—it would have lifted the mat up into a dome, creating both the chevron shape and the cavity that cement would eventually move in to fill. The strange structures in the Noonday dolomite were Martin's missing evidence. "They're stunning," he breathed. "They do exactly what you'd predict. It's better than I could ever have imagined."

Martin wasn't just trying to disprove Paul and Dan. He had other reasons for preferring his explanation of the cap carbonates. Martin, like Nick, is deeply rooted in the world around him. He too thinks that seeing is believing. And he likes his methane theory precisely because it uses processes that we see happening today. There's plenty of methane hydrate in the world right now. This explanation for the caps doesn't require insanely high weathering rates, an extreme hothouse after the ice, a frozen ocean, all the things that make Paul and Dan's Snowball so radically different. "Paul and Dan's Snowball is *really* non-uniformitarian," Martin told me once. "It really worries me when you suddenly evoke a Martian-like world."

But there are some things that Martin is willing to admit, and that Paul and Dan immediately seized on. His methane idea doesn't explain nearly as much as Paul and Dan's theory can. It can't explain the iron formations or the ubiquitous ice. It tells you nothing about why the Snowball might ultimately have ended. And it doesn't necessarily oppose Dan's explanation for the cap carbonates. Even if Martin was right, the hydrates could have been decomposing *at the same time* that acid rain was lashing on to ground-up rocks. There's nothing to stop the two effects working in tandem. Martin's methane-hydrate theory, it turns out, doesn't disprove the Snowball at all.

Martin, however, had another challenge to make. All of Paul's information about the Snowball ocean came from the isotopes in his carbonate rocks, but he only had carbonates from before and after. Martin, on the other hand, had found something extraordinary. He had carbonates from *during* the Snowball. And they seemed to show that Paul's Snowball theory had a fatal flaw.

Just over 600 million years ago, colonies of bacteria floated invisibly in the sea that would one day become northwest Namibia. They hung in the water perhaps a hundred feet below the surface. Undisturbed by wind or waves, they busied themselves with the endless operation of their internal chemical factories. Make food. Consume food. Make food. Consume food. The sea around them turned hazy with the accumulation of their minuscule efforts. As the chemical balance of the water changed in response to their factory effluents, tiny flakes of carbonate sprang out of solution and floated softly down to the seafloor.

Though the Snowball had already gripped the outside world, there was little sign, this deep in the water, of the icebergs passing overhead. Just the occasional dropstone, a boulder released from the ice that would appear suddenly from above, pass gently by the clouds of bacteria, and sink into the flakes of mud that were slowly, steadily accumulating on the seafloor beneath.

Today this carbonate mud has transformed into layers of rock perhaps six inches thick. Its layers stand out clearly among the ice rocks of the Namibian outcrops. They're the colour of rancid butter, occasionally enlivened with a white or tan dropstone. Above and below them the rocks are grey with shattered carbonates, silicates and sand, landslides of debris brought suddenly in from the shore. But the serene yellow layers speak of gentler moments in the life of the Snowball ocean. Made of microscopic carbonate flakes, balled up into tiny spheres called peloids, they are extremely delicate. Intact peloids are a signal that the rock hasn't moved since they formed. If it had tumbled, the peloids would have been smashed to pieces.

The peloidal mud also has occasional cracks, filled with crystals of carbonate cement that jut outwards from the walls of the cracks like the spears of a tiny white picket fence. They too must have formed in the Snowball ocean.

Martin Kennedy had collected samples of these rocks from Namibia back in 1996. He'd also collected similar carbonates from his beloved Australian Outback in the early 1990s. The Australian samples were stromatolites, those strange domed structures created during the Slimeworld, when bacterial mats sat on the growing surface of a rocky edifice, and sediment accumulated beneath them. They were tall, thin columns of carbonate rock, perhaps six inches wide, embedded in yellow dolomite, and they provided another direct window into the Snowball ocean.

Now Martin looked back at these rocks and wondered. Could they help him test the Snowball idea? In early 2001 he returned to Death Valley to collect samples from yet another set of Snowball carbonates, which he'd spotted mixed in with the ice rocks there. These carbonates were called oolites, and were strange rock forms with a caviar-like texture. You find them today in places like the Bahamas. They grow as grains that roll backwards and forwards in the waves, coating themselves with carbonate that precipitates from the seawater. Like the peloids, they're extremely delicate. If you find them intact, they can't have come from some other place, or some other time slice. Since they are mixed in with Death Valley's ice rocks, they too must have formed in the Snowball ocean.

Martin realized that all these carbonate samples gave him a direct window into the Snowball ocean and, with it, a way to test the Snowball hypothesis. Remember that according to Paul and Dan, the Snowball ocean was essentially lifeless. That was Paul's first idea, the one that set the Snowball rolling. Paul had worked it out by looking at the ratio of light and heavy isotopes in carbonate rocks from just before the Snowball time. Heavy carbon equals life. Light carbon equals no life. And immediately before the Snowball, Paul had found unusually light carbon in his carbonate rocks. He concluded that many living things must have died off as the ice advanced, leaving only a few small groups to huddle together and wait for the thaw.

But because Paul had never managed to find carbonates from the Snowball ocean itself, he had no direct evidence for how much life there was then. If he was right, and the Snowball was an intensely cold, barren time, there was scarcely any life around. In that case, carbonates formed chemically from the Snowball ocean itself should have been light. If he was wrong, if the ocean was full of life, the carbonates from the seawater of the time should be heavy.

All right, then, Martin thought to himself. There's a hypothesis. Let's test it. He set about measuring the isotopes in all of these rock samples. And every time, he found the same thing.

They were all heavy.

Life in the Snowball ocean was apparently flourishing. There couldn't have been the extreme freeze-over that Paul demanded. Hot news—the Snowball wasn't that cold after all. This looked like the fatal flaw in Paul and Dan's hypothesis. Martin hastily wrote up the results and submitted them to a journal called *Geology*.

SNOWBRAWLS

Academic journals decide what to publish on the basis of "peer review", where anonymous scientists say what they think of the work. And one of the scientists to whom the editors at *Geology* sent the paper was Paul Hoffman. Paul's review was savage. He waived his anonymity ("I always sign my reviews") and firmly recommended that the paper be rejected. It contained, he said, a basic geological error.

According to Paul, Martin's samples had nothing to do with the Snowball ocean. They were, Paul believed, simply broken-off pieces of older rock. That would destroy Martin's argument. If the carbonate cements were from a much older time, their isotopes would be perfectly acceptable. Of course the ocean that existed long before the Snowball was full of life! The problems for life came only with the ice. If Paul was right, Martin was embarrassingly wrong.

Paul's review was still in the post, on its way to *Geology* and thence to Martin, when the two of them arrived in Edinburgh in June 2001 for a workshop about the Snowball. Paul was in pugnacious mood. A few days earlier he had stabbed his finger on the offending picture in the draft of Martin's *Geology* paper. "I'm hoping he'll show this photograph," he'd told me. "I'd like it to be demonstrated in public that this guy is incompetent in geology."

But Dan Schrag, Paul's smoother of relationships, was also in Edinburgh, and had decided to try bringing Martin on board. He took Martin to a pub with a crowd of other conference participants, and soon they were chatting with great good humour. They were discussing science, Snowballs, Dan's recent flight over Edinburgh in a friend's microlight aircraft. Martin was saying that he didn't want to be part of an "anti-Snowball team", that he didn't consider himself on anybody's "side", and that Nick should never have done the "snow job" talk, because it had just polarized everyone's opinions to little effect. What Martin and Dan both wanted, they agreed, was to keep talking, test the waters, come to the right answer.

Paul was in the pub too, struggling to keep quiet, doing his best not to spoil Dan's efforts. "This business of Martin Kennedy trying to kill the Snowball with his cements," he said to me later. "Half of me wanted to expose him and discredit him, so that nobody will believe his 'facts'. But half of me is appalled that I'd want to do that. It's not that I have any warm feelings for Martin. He's been a thorn in many of our sides for years. But to try to humiliate him in public would be cruel. Honestly, I'm not a malicious person. I'm certainly capable of being malicious, but it tends to be when I haven't thought it through."

Still, Paul couldn't quite contain himself. When Martin stood up to leave, Paul stood up too, determined to assert his conviction that the contentious cements were from older rocks. "You showed those cements and said they'd formed in place!" he said loudly, looking directly at Martin. "But I know what they are." All conversation around the table ceased. Everyone was staring. "I feel like I always did when my parents were fighting," interjected one of the students, *sotto voce.* "Poor Martin," mouthed another, silently. Dan sighed. He looked at Martin and said carefully: "Paul has sent you a signed review making that comment . . ."

"Yes," Paul interrupted, louder still. "And I had to tell you about it today in private or tomorrow in public." Martin by now was looking aghast. Dan jumped up and headed towards him. "Look, it's OK", he said soothingly. "You'll talk about it tomorrow." He put his arm under Martin's elbow and guided him out of the pub.

The next day, at the Snowball workshop, Paul was doing his best to be friendly again. Every so often during the presentations,

Paul would turn and mutter something conspiratorially to Martin, who would grin and whisper back. And at the end of the day, as the assembled researchers were leaving the lecture theatre, Paul walked over to Martin and shook his hand firmly.

"All the best, Martin," he said. "And I'm glad you're back . . . in the publishing world." Martin had spent some time working at Exxon, where commercial research is conducted behind closed doors. He'd only recently rejoined the academic world and started being allowed to publish his research again. Paul was trying to be polite. But as soon as he said it, he and Martin both thought of the cements manuscript, the one Paul had reviewed unfavourably, the one that might well not be published now because of Paul's comments.

"I guess I could have put that better," Paul said uncomfortably. Martin shrugged. "Well, you know . . ." he said, and turned to leave.

Paul started to follow him up the stairs. "But it could still be accepted, right? Mine was only one review."

"I think you have more influence than you realize," Martin replied.

"I was trying to spare you embarrassment."

"Oh, I don't think you have," said Martin. "In fact, you're going to spare me tenure if this goes on." Unlike Paul, Martin was not yet a tenured professor. Whether his position at the University of California at Riverside was made permanent would ultimately depend on a careful assessment of what papers he had published and how successful his research had been.

Paul was still trying to justify his review. "I wanted to show you these pictures," he said. "Listen, have you got time now?" Martin hesitated, then shrugged again. "OK," he said.

The lecture theatre was empty now. Paul climbed quickly to the top of the stairs, and began fumbling with his slides. Click. He showed a grainy image of a Namibian carbonate rock. This, he said, was the older rock formation, the one he believed Martin's samples originally came from. Click. There was another. This was the younger Snowball formation, with a chunk of the older rock embedded in it. Martin stared at the screen. He looked appalled. "Paul," he said, "that's not what I collected."

There was an awkward silence. Paul was standing stock-still at the top of the stairs. "What I wanted to say was . . ." he began, but Martin interrupted. He was clearly upset now, fighting to keep control of his voice. The images had convinced him that Paul's damning review was based on rocks that were entirely different from his own. "The cements that you showed are not the same as the ones I collected, Paul," he said formally. "But thank you for showing me the photographs." And then he turned abruptly and left the room.

A few weeks later, Martin's paper was accepted after all.[4] Even if his data hadn't impressed Paul, the other reviewers found his analysis sufficiently compelling. Paul, it turned out, had concluded his damning review with one of his favorite quotes: "False facts are highly injurious to the progress of science for they often endure long: but false views, if supported by some evidence, do little harm, for everyone takes a salutary pleasure in proving their falseness."

This comes from Charles Darwin, the originator of the theory of evolution, and it makes an important point. When people argue about ideas—"views", in Darwin's words—all the arguments have to be based on the available "facts". If one of the facts is wrong, the whole edifice can crumble—taking *all* the ideas with it.

But whose "facts" were wrong—Paul's or Martin's? Had the

rocks come from the Snowball ocean or not? Paul might be right about the "picket fence" crystals from Namibia and their related cements. It is possible that the crack they grew in was from some earlier time, and that the whole lump of rock had then broken off and been amalgamated into the younger Snowball formation. In the end, Martin left those samples out of his paper, just in case. But the peloids are harder to argue away, and the caviar-like oolites harder still. To break those off from older rock and transport them and mix them into a new formation would surely have smashed their delicate structures. They really did seem to come directly from the Snowball ocean, just as Martin said.

So was this indeed a fatal flaw in the Snowball argument? After all, the carbonates leading up to the Snowball had been light, and Paul had assumed that the ones that formed during the Snowball would be the same. But he hadn't made any direct assertions about this in his papers. Unlike Martin, Paul had no carbonate rocks from the Snowball period; with no data to interpret, neither he nor Dan had thought particularly hard about what the isotopes would be like. Now, though, they had an incentive. Spurred on by Martin's findings, Paul and Dan focused on this issue. And for two different though complementary reasons, they realized that you'd actually *expect* the Snowball ocean to be heavy—just as Martin had found. Ironically, Martin's heavy rocks didn't conflict with Paul and Dan's model at all.

Paul had realized that the oceans would have been heavy because they contained old material, dissolved from rocks on the seafloor. The floor of the Snowball ocean was lined with carbonate rocks that had been created when life was still abundant. And the acidic seawater would have dissolved this old carbonate, just as acidic lime juice can dissolve a marble cutting board. The effect, says Paul, was to change the signature of the ocean. If you pour hot milk onto cocoa powder, the milk turns brown because the dissolved cocoa swamps the milk's original colour. Similarly, the "heavy" signals of plentiful life dissolved from the old seafloor carbonate would have swamped the "light" signal from the largely lifeless Snowball ocean.

Dan has discovered another effect that reinforces this one. His answer involves a more arcane issue, beloved of geochemists and understood by few others. The carbon isotopes in the ocean don't depend only on the activity of living things; they are also affected by the way carbon dioxide gas migrates from the atmosphere into the ocean. And this in turn depends on what proportion of carbon the atmosphere already contains.

Nowadays the atmosphere has only a tiny proportion of carbon, less than 5 per cent. But the Snowball was very different. According to Paul and Dan's model, carbon dioxide gas had been building up in the atmosphere for millions of years. The Snowball atmosphere contained a much higher proportion of carbon, and that would have made all the difference.

Dan did the sums. He crunched through all the equations that predict how this change would affect the ocean isotopes. And he came up with a number that matched—precisely—the heavy values Martin had found. Even if the Snowball ocean was totally lifeless, the carbonate cements would be just as Martin had measured.[5] This was no fatal flaw. Paul and Dan concluded that Martin's evidence *confirmed* the Snowball theory.

When Nick Christie-Blick, Martin's co-author on the cement paper, heard Paul and Dan's new arguments, he immediately cried foul. Paul and Dan, he said, were simply shifting their goalposts.

How could he and his fellow critics test their theory if they kept changing what it said? Paul and Dan responded that they were naturally modifying a young theory, making it richer and fuller.

Who's right? Well, science works at its best when somebody puts forward a theory and everyone else tries to pull it down. Sacrosanct scientific philosophy holds that no theory can ever be *proved.* A theory can only be *disproved,* and the longer it survives the attacks against it, the more confidence you can place in it—while never knowing for certain if it is right. Following the philosopher of science Thomas Kuhn, many see science as a procession of revolutions, where a prevailing paradigm holds sway in researchers' minds until it is finally disproved and a new one takes its place.

The trouble comes in this process of disproving. Scientists often—perhaps usually—find it hard to let go of a theory that they care about. When some devastating new finding shows it to be wrong, that's hard for them to accept. There are many theories whose proponents have clung on to them for too long, rendering them more and more elaborate in a desperate attempt to accommodate the findings that disprove them and ward off the inevitable end.

But science moves much more in fits and starts than a simple reading of Kuhn's paradigm shifts would suggest. And it can be hard to decide whether a theory has truly been disproved. Often counter-arguments can be incorporated into the theory itself until it becomes richer for having adapted and allowed itself to grow. If a theory comes under attack when it's too young and raw to defend itself, it can also be destroyed prematurely. Wegener's theory of continental drift is one example of this. And Alvarez's asteroid hypothesis about the death of the dinosaurs could have encountered the same fate if someone hadn't found the crater—the "smoking gun". Even that was lucky. The crater could have long since been swallowed back up into the Earth's interior, as much of the rest of the Earth's crust has been since the time of the dinosaurs. Alvarez's theory might have been right, and yet could still have been killed.

Though Martin's cement paper may not have tripped up the Snowball idea, as he had thought it would, it had taught Paul one lesson at least: when under attack, try to spread out the target area. The next time I saw Paul at the beginning of a new lecture tour, the Snowball had become "Kirschvink's theory". Every time he mentioned the idea, Paul was reminding his audience that it had come, in much of its modern incarnation, from the insights of Joe Kirschvink, the Caltech professor who had dreamt up the volcanic aftermath, and coined the name Snowball Earth.

Paul even described himself as "Kirschvink's bulldog". That was neat, a direct reference to Thomas Henry Huxley, who earned the nickname "Darwin's bulldog" for his blunt and ferocious defence of Darwin's ideas on evolution. Darwin shrank from crusading against the disapproving Anglican establishment on behalf of his worrisome new theory. Huxley, however, had no such qualms. At a debate in 1860, when Bishop Samuel Wilberforce asked sarcastically whether Huxley would prefer to be descended from an ape on his grandfather's or grandmother's side, Huxley reportedly replied thus:

> If the question is put to me, would I rather have a miserable ape for a grandfather, or a man highly endowed by nature and possessed of great means and influence, and yet who employs these faculties and that influence for the

mere purpose of introducing ridicule into a grave scientific discussion, I unhesitatingly affirm my preference for the ape.[6]

Meanwhile, Martin's paper on the cements had done something else for the Snowball idea. He had finally focused attention back on to the question of what exactly was alive in the Snowball ocean. This was something that was beginning to worry many biologists. They knew that certain creatures must have survived the Snowball: bacteria, of course, in their enveloping mats of slime; slightly more sophisticated—but still single-celled—creatures, with their internal chemicals neatly packaged rather than floating freely in a soup; simple algae, brown and green and red. All these creatures left their faint fossil traces in rocks from both before and after the ice, so they, at least, must have lived through it. But the scale of Paul's freeze-over was troubling. If it was as severe as Paul insisted, how could *anything* have remained alive at all? This was to become the Snowball's next test.

Ice is an extraordinary substance. Subtle shifts in its structure can render it white or green or blue, translucent or opaque. It can shatter like glass, or creep like treacle. Ice is a tough building material, as strong as concrete. In the Second World War, plans were even developed to create giant aircraft carriers called "bergships" out of ice. They might have been built, too, if the range of aircraft hadn't increased enough to render them obsolete. Russian empresses used ice to build vast, glittering palaces: "The delightful material gave a new, fantastic beauty to every feature, sometimes white and sometimes clear green—dark and opaque where the shadows fell, and almost transparent in the sun. No dream castle of jasper or beryl . . . could be more beautiful than these wonderful buildings of ice."[7]

Ice is alien to life. Part of the attraction of the Antarctic ice cap is that the essentials of life—food, water, fuel and shelter—have all been stripped away. Explorers have talked for decades about the sublimity and purity of this landscape. "During the long hours of steady tramping across the trackless snow-fields, one's thoughts flow in a clear . . . stream," Antarctic explorer Sir Douglas Mawson wrote, while trying to explain his urge to return. "The mind is unruffled and composed and the passion of a great venture springing suddenly before the imagination is sobered by the calmness of pure reason."[8]

But there is danger as well as purity in this escape from life. Remember Wegener in Greenland, Scott in Antarctica, and Hornby in the bitter winter of the Canadian Arctic. Ice also kills. Every cell in your body is a squashy bag of water, with just a few other chemicals thrown in. If this water freezes, jagged crystals of ice appear, and they slash and tear at the cell's fragile walls. These membranes also spring leaks when their molecules begin to congeal together into clumps, like fat cooling in a frying pan. Within the cell, proteins unwind their complicated loops and become flaccid. With great care, and clever technology, certain cells can be preserved on ice—sperm, eggs or bone marrow. But for the most part, life depends on water; and ice brings death.[9]

That's why the biologists were so worried by Paul Hoffman's Snowball. If ice covered the world, how could even single-celled life have survived?

Water wasn't the whole problem. The oceans in Paul's world wouldn't have frozen solid, largely thanks to another strange property of ice. Most solids don't float. They become denser when they freeze, and in a bath of their own liquid, they'll sink. But ice is the exact opposite. When water turns to ice, its molecules become more loosely bound, forming a lacy network that's full of space. That's why ice floats, and why the Snowball oceans didn't freeze completely. If icebergs sank, lakes and oceans would freeze from the bottom up, instead of just growing an ice skin on their surface. So, even as Paul envisioned it, there would still have been plenty of liquid water *within* the Snowball oceans.

But living things also need sunlight. Because Paul argued that the entire surface of the ocean was frozen over, all the water would be beneath that ice layer, blocked from the sun. And for most of the simple denizens of the Snowball, darkness would mean death. Even the creatures that didn't make their living using sunlight depended on the ones that did. Living things needed both liquid water and sunlight, and for that, the ice had to have holes.

At first, Paul wouldn't hear of this. The oceans were fully covered with ice, and that was final. But once again he was forced to change his mind. And the impetus for this came from a rival ice world, a pretender to the Snowball crown, which began to tug at everyone's attention. This was a newer, gentler snowball, with a moniker of its own: "Slushball Earth". Climate modellers created it. They use computer programs to do what geologists can't—rerun the Earth's experiment and see what happens. As soon as they heard about the Snowball, they fired up their machines and tried to make one.

They couldn't.

However much the modellers wanted to generate an ice-covered world, their computers wouldn't oblige. Modelling had developed into a much more sophisticated affair since Mikhail Budyko's primitive attempts first turned up the "ice catastrophe" back in the 1960s. And the modern models stuck at a sort of halfway house, where ice advanced to somewhere near the tropics, but no further. A few models could generate ice *on land* near the equator—which would explain the ice rocks from Australia. But the equatorial *oceans* remained stubbornly ice-free.[10]

So the modelers began to talk of an alternative to Paul's "hard" Snowball, a new, softer variant. Nick Christie-Blick thought this Slushball was a wonderfully moderate solution to the Snowball conundrum; neither one extreme nor the other, it was a comfortable answer. It could also explain some of the evidence that had consistently bothered Nick. For instance, in many parts of the world the ice rocks are hundreds of feet thick. To make those deposits, icebergs had to be free to wander offshore and melt and drop their loads on the seafloor, and they certainly couldn't do this if the oceans were fully frozen. Paul argued that the ice rocks formed at the beginning and the end of the Snowball, when there was still a little open water. But Nick felt that to make such thick deposits, the process would have to continue throughout the Snowball. Open oceans at the equator, he felt, provided the perfect answer. For the biologists, too, the Slushball was just right. This, they felt, was exactly what life needed.

Paul, however, hated the Slushball. He called it "Loophole Earth", and said the models needed a few reality checks of their own. The Earth's climate is insanely complicated, and nobody claims that its every nuance can be encapsulated inside a computer. Modellers are good at reproducing today's climate mainly because they can compare their model output with records of temperature, wind and weather. But the only Precambrian weather reports are the ones written in rocks. And according to Paul, the Slushball came nowhere near explaining this geological evidence.

It couldn't account for the ironstones, the cap carbonates, or the strange chemical signatures in the rocks. Most important of all, it couldn't explain the extremely long *duration* of the ice.

Paul pointed in particular to the findings of Nick Christie-Blick's graduate student, Linda Sohl. Her magnetic work had shown that the glaciations must have lasted at least hundreds of thousands, if not millions, of years. The Slushball, said Paul, simply couldn't last that long. It was precarious, like a pencil balanced on its tip. Nudge the model world one way or another, and you would quickly force it to choose: Snowball or no-ball.

If you cooled the Slushball a little, Paul said, ice would quickly take over. White ice reflects sunlight, which cools the Earth, which breeds more ice in a runaway cycle, which, Paul said, would freeze over the tropical ocean. Warm the Slushball a little, on the other hand, and its ice would soon vanish. Warming melts ice, which opens up dark patches of ocean, which absorb more sunlight until all the remaining ice races back to the poles.

What, then, was Paul's explanation for how life survived the ice? Well, living things are extraordinarily resilient, especially simple ones. Bacteria survive—somehow—at the South Pole. Other bacteria have shown up beneath glaciers, and even inside solid rock. Unknown to the authorities, a small colony of *Streptococcus mitis* hitched a ride to the Moon in 1967 inside an Apollo TV camera, and the bacteria were still alive three years later when the camera was brought back to Earth. They had managed to survive without food, water or even air. Hot springs are often brilliant with living colour. The steaming, acidic pools of Yellowstone National Park, for instance, contain vivid bacterial patches of orange, red and green despite their boiling temperatures. Life has a habit of finding its way, no matter what.

The biologists pointed out, however, that many of these resilient creatures are weirdly adapted to their extreme conditions, whereas most of the ones that survived the Snowball were apparently more normal in their requirements, particularly in their need for sunlight. There had to be sunlight. There had, the biologists said firmly, to be holes.

So Paul and Dan changed tack. They obviously needed to provide some refuges for life within the frozen seas. What kind of openings might there have been? Well, any hot spring or volcano on a shallow enough ocean floor would have created at least a small hole in the ice above it. Also, the Snowball was not uniformly cold. Though global temperatures would initially have plunged to around minus 40 degrees C, they would gradually have risen as carbon dioxide built up in the air. And the equator would always be warmer than this bleak global average. Soon the ice at the equator would grow thinner, perhaps even thin enough to crack periodically.

Thinking the question through further, Paul and Dan also realized that the Snowball's stronger seasons would also have helped living things cope. Even if winter temperatures near the equator were 30 degrees below zero, summers could have crept above freezing for a few days each year. In melted puddles and ice cracks, living things could then have grabbed their chance to make and store food, as they do in Antarctica today. And even in winter there could well have been other patches of open water among the ice. In today's frozen oceans, odd currents keep certain places—called polynyas—ice-free throughout the year. Whales trapped in the pack ice use these open patches as breathing holes while they wait for spring to return and release them.

For the biologists, this line of thinking was much more encouraging. But were there *enough* refuges? Could each individual species huddle together in a big enough group to survive until the Snowball finally melted?

To find out, Dan Schrag called a friend, another hot young scientist, Doug Erwin, from the National Museum of Natural History in Washington, D.C. Doug is an expert on ancient life, and he also knows a fair amount about ecology in the modern world. To protect an endangered species, Doug said, you have to maintain its genetic variety. The genetic material that passes from one generation to the next is constantly changing—and not always for the better. In an isolated group—a herd of elephants in a national park, say—dangerous mutations can spread quickly. For the species as a whole to survive, any one group must contain enough individuals, enough variety, to dilute this danger. And there must be enough separate groups that if a few of them fail, the rest will still pull through.

Doug realized that the same would apply to the Snowball's inhabitants. He made a list of all the different species that needed to make it through the Snowball. Then he used conservation models to calculate two numbers: how many individuals of each Snowball species you'd need in a given refuge, and how many refuges you'd need overall.

The answer astonished both Doug and Dan. It was far easier than they'd expected. To get virtually all of the species through the Snowball, you only needed something like one thousand different refuges. And each refuge only needed to house around one thousand individuals. What's more, the Snowball creatures were no elephants. "Do you know how much open water you'd have needed to support one thousand of these individuals?" Dan demanded of me as we sat in a café. "This much." He spread his hands apart until they outlined a region of air the size and shape of a dinner plate.

For Doug and Dan, at least, this solved the problem of

survival. Of course you could poke one thousand small holes into the Snowball ocean. You could probably make tens of thousands without compromising the model, and many of them could be much bigger than a dinner plate. With this many refuges, says Doug, every species that needed to make it through the Snowball could do so easily. There's no need for a Slushball. Life could survive an all-out, full-on Snowball with no problem at all.[11]

So far the Snowball theory has survived every challenge that's been thrown at it. And thanks to Paul and his proselytizing, there has now been a radical transformation in scientific attitudes to the ice rocks. In a few short years Paul has achieved the feat that eluded Brian Harland, Joe Kirschvink, and anyone else who became intrigued by the Earth's Precambrian ice rocks. Virtually everyone now says that this was a time of extraordinary ice, cold and catastrophe. Even critics like Nick Christie-Blick, who still believe in the Slushball, admit that ice went almost all the way. Paul has taken an idea that was once too shocking to be considered, and brought it into the scientific limelight.

He could have stopped there. But there's another part of the Snowball story that he'd dearly love to be true. While Paul continues to strengthen his geological case, he's also fascinated by the biological implications. Was the Snowball the creative spark for the new life that followed?

Paul has believed from the beginning that the ice and its aftermath somehow triggered life's biggest evolutionary moment since it first appeared on Earth: the switch from simple to complex. Without the Snowball, he thinks, there would have been no animals, no richly diverse Earth, and no people to argue over it. Paul, though, is not a biologist. What do the experts say?

NINE

CREATION

Billions of years had passed, nearly nine-tenths of Earth history, when life finally made its vital leap into complexity. Now at last it could move on from dull primordial slime, and begin inventing the fabulous life-forms that we see today.

Of all the innovations conjured up by evolution, this was the most dramatic. It was the world's first industrial revolution. Before then, each individual cell had to be master of all trades: eat, digest; excrete; reproduce; perform all the essentials of life within one small squashy sac. Afterward, mighty corporations of cells sprang up to share the load. Specialization became the rule. Thanks to structural cells, bodies could grow large and adopt inventive new architectures. Muscle cells could move these bodies to new grazing grounds. Sensory cells could warn of danger, appendage cells could rake in supplies. Cells evolved to regulate temperature, transport information, innovate and consolidate.

And this specialization opened up a world of possibilities. Suddenly, in the Earth's late middle age, life began frantically procreating, evolving and developing new forms. First came trilobites and ammonites, then dinosaurs and octopuses, dromedaries, whales and wallabies, as the new complex creatures competed to find ever more imaginative ways of exploiting the world's resources. Life as big business was wildly successful.

Then why did it take so long? Though the history of life is ambiguous, traced through an imperfect record of fossils and rocks, most researchers believe that complexity was invented somewhere between 550 and 590 million years ago. That's after more than 3 billion years of simple, single-celled slime.

Biologists have been trying for decades to understand why complex life appeared on Earth at that particular moment. And then, along came Paul Hoffman, talking of global catastrophe. Paul's evidence suggested that at least two, and possibly as many as five, successive Snowballs had rocked the Earth starting around 750 million years ago. Most significantly, this series of Snowballs ended 590 million years ago, just around the time complex life was beginning to emerge. The news of Paul's ice sent biologists racing back to their fossils. "What did this?" they started asking themselves. "Was it the Snowball?"

These are early days for the biological part of the Snowball theory. Some biologists are automatically dismissive of *any* idea that makes biology subordinate to geology. "Genes don't care about the weather," one researcher told me. "Adding ice cubes doesn't give any explanatory insight," e-mailed another.

But others are intrigued by Paul's findings. There are now

signs that complexity really did appear soon after the ice receded. And though the picture is still far from clear, many biologists are beginning to think the Snowball could indeed have been the trigger.

To FIND the cause of a historical event, you first need to know when to look. And until recently, most biologists have assumed that complexity arose with an event called the Cambrian explosion. This episode has grabbed all the early-life attention for decades.

EVOLUTION'S BIG BANG! screamed the front cover of *Time* magazine on 4 December 1995. "New discoveries show that life as we know it began in an amazing biological frenzy that changed our planet overnight." The animals that reared up on its cover were from the beginning of the Cambrian period, around 545 million years ago. At first sight, that seems like a serious problem for Paul. The Cambrian explosion can't possibly have been triggered by his Snowballs. They ended around 590 million years ago,[1] and 45 million years is far too long to sit around with a lighted fuse waiting for the bang. Even Paul admits this.

But he also says that the Cambrian doesn't deserve quite as much attention as it receives. The beginning of the Cambrian was certainly a burgeoning, inventive time for life. During this rapid burst of new evolutionary shapes and strategies, the foundations were set for every modern family of animals. The Cambrian fossils have been known for centuries; they mark the end of the Dark Ages without fossils and the beginning of geological and biological enlightenment. They are Stephen Jay Gould's "Wonderful Life".[2] But all this fame has come to them mainly because they were *easy to preserve*. They show up everywhere. At the beginning of the Cambrian, life invented skeletons: scales, shells, spines, all the sorts of bodily supports that stick around long enough after death to turn into clear, unambiguous fossils.

So the Cambrian fossils weren't the first complex animals, any more than language began with the printing press, or with papyrus. Complex life could easily have been around for millions of years before then, and just not left such a clear record in the rocks.

The invention of multicellularity was certainly a *prerequisite* for the Cambrian explosion. Some biologists even say that it made the Cambrian explosion inevitable. Of course, life began experimenting with its new toy, exploring the many new possibilities it now had for shapes and functions, tissues and organs. With complexity already in place, the Cambrian explosion was just regular evolution in action.[3]

So forget the brash fossils of the Cambrian. To find the real moment that life learned to use many cells instead of one, biologists need to seek out creatures that are much more mysterious. If Paul Hoffman is right, and the Snowball truly triggered the invention of complexity, the world's first complex creations must have appeared shortly after the ice receded. The question is, did they?

CRUNCH! JIM Gehling's foot lands on a coke can and squashes it into the ground. He picks up the can, points to the misshapen circle its trace has left in the mud, and grins. "Go on," he says. "Look at that, and tell me what shape the can was originally, or what it was used for."

This is one of Jim's favourite metaphors for the work he does: reconstructing some of the first new creatures to emerge after the end of the Snowball. His task is extraordinarily hard. At least people studying more recent fossils such as dinosaurs have something concrete to dig up—bones, scales or shells. But the creatures that Jim studies had none of these attributes. They lived before skeletons had been invented. Their bodies were soft, like jellyfish. And the fossils they left behind are like the smudgy circle from the coke can—indistinct impressions, squashed into ancient mud. From these scant clues Jim and his colleagues have been trying to figure out whether these jelly-creatures were the world's first complex animals.

Jim is from Adelaide, Australia. He is in his mid-fifties, tall and slim, with white spiky hair, a long, thin face and a centurion's hooked nose. His eyes are deep blue, and—consciously or not—he tends to wear shirts that highlight them perfectly. His smile is engaging, his manner easy. He delights in his fossils and—most unusually, in the jealous field of palaeontology—he loves to share them. He is a prime candidate for the world's nicest man.

Everybody likes Jim. They all say so repeatedly, even Paul Hoffman, though Jim recalls that he and Paul fought heatedly about something or other when they first met. Jim has the knack. He can disagree with, and even correct, the most egotistical brains in the business without causing any apparent offence. He's not like Paul. You don't want to impress him, or struggle to win his approval. But within a few hours of meeting Jim, you find yourself confiding in him. That's why so many people claim him as their best friend.

Jim first encountered the jelly-fossils as an undergraduate at the University of Adelaide, working for a pioneering palaeontologist, Mary Wade. Mary was an eccentric but enthusiastic teacher. She and Jim camped among the fossils, and—this was the

1960s—she brought her eighty-year-old mother along as chaperon. From Mary, Jim caught the bug. He became addicted to hunting for new fossils, and trying to make sense of the ones he'd already found. What shape was it? How did it live? How did it die? He stayed on for a master's degree, but he quickly realized that job prospects in the field were poor. Jim didn't want to leave Adelaide—his family was settled there—and the only slot for a palaeontologist at the university was already taken.

So he started work teaching general science at a local teacher training college. Though he loved teaching and his students loved him, he couldn't stop thinking about fossils. He'd go to the library and find himself drifting toward the palaeontology journals rather than the ones he was supposed to be looking up. He read about the latest fossil research each night, and spent every holiday out in the field or attending fossil conferences. He produced so many academic papers that many palaeontologists were astonished to discover he was a part-timer.

Finally, when the kids had grown up and left home, Jim quit his day job to concentrate on fossils full-time. His friends were delighted. Bruce Runnegar—a brittle Australian professor at UCLA, with a wicked grin and an acerbic sense of humour—immediately invited Jim to come to California and complete the formality of a Ph.D. (He had already published more research than all the other graduate students put together.) Another friend, a bright-eyed Canadian palaeontologist named Guy Narbonne, arranged for Jim to spend time studying fossils in Newfoundland. Everyone liked Jim in Newfoundland, too. When they heard he was in town, people would appear with gifts for him: a cake, or a bag of berries. He was embarrassed by this.

Now Jim has a precarious adjunct position at the South

Australian Museum in Adelaide. It brings him an office but little money, and his duties as an exhibition organizer take precious time away from his research. There are still no openings at the University of Adelaide, and Jim is struggling to go on with his fossil work. He is neither angry nor bitter about this. He is the most well-adjusted person I have ever met.

Jim's fossils come from a place that has figured many times already in the Snowball story: the Flinders Ranges of South Australia. In 1947, geologist Reg Sprigg spotted what looked like squashed, petrified jellyfish in the rocks of an abandoned mine near Ediacara Hills, on the western edge of the Flinders. Similar creatures have since been found in rocks around the world,[4] but they are still collectively called Ediacarans in honour of Sprigg's discovery.

There's little point in visiting Ediacara today. These fossils are worth tidy money on the open market, and the site has been thoroughly despoiled. Any samples that weren't removed in daylight by palaeontologists were stolen in the night by people using crowbars and mechanical diggers. There's nothing left to see.

But if you swear not to reveal its whereabouts, Jim can take you to a secret location where the fossils are still intact, and in place. You go in the early morning or late afternoon to catch the slanting rays of what Jim calls "fossil light". First you take a paved road north into the Flinders, then a bush track that's strewn perilously with rocks. (Don't offer to drive; Jim doesn't like being driven. But he's gracious about it, and he's also such a good driver in the bush that you probably won't mind.) The landscape is muted, as if faded by the sun to shades of pale terracotta and drab olive grey. On the left, a line of straggly gum trees marks the bed of a dried creek. The ground around is stony, a desert pavement scattered with squat, round saltbushes. Apart from the wedgetailed eagle sheltering in one of the gums, and the ubiquitous, irritating Aussie flies around your face, there is no life to be seen.

The track swings around a corner and stops at the foot of a gentle hillside, covered with slabs of pale stone. They are irregularly shaped, an inch or two thick and several feet across like broken, prophetic tablets. Jim climbs to one of them, turns it over, and begins to scrub off the dirt with a yellow brush that he has pulled out of his pack. ("Here's the main instrument for this sort of work. A nylon dishwashing brush. Two dollars.") And then he holds the slab out for inspection.

The reason for going early in the morning is now clear. At first you see only the stippled red underside of the rock. But then the slanting light casts shadows that resolve into an oval indented image like a giant thumbprint, perhaps six inches long. The creature that left this imprint is called *Dickinsonia,* one of the icons of the Ediacaran world. Its body is segmented like a worm's, and split by a groove running down its centre. Perhaps this was a stiffening rod for its soft body. Perhaps it is the trace of a gut. At one end, the strange parabolic segments are slightly thicker and wider than at the other. Unlike the inhabitants of Slimeworld, this creature knew the difference between head and tail.

Now Jim is turning over more slabs, and finding more fossils. There's a squashed, sponge-like creature called *Palaeophragnodictya,* revealed as a small disc set slightly off-centre within a larger one. There's another disc with a set of grooves inside it, like the outline of a cartoon arrowhead. *"Aspidella,"* Jim says, and then moves on. Some Ediacaran fossils have flouncy, frilled edges like a Victorian petticoat. Others have discs and stems and branches. One looks like a Roman coin, another like a sheriff's badge—a tiny, five-pointed star inside a ridged circle. Some are truly enormous. One *Dickinsonia* found at this site, Jim says, was more than three feet long.

There's something extraordinary about seeing these ancient ancestors lying in front of you, everywhere you look. Perhaps one of these slabs bears a single indented shape that will shock the biological world. There might be a new species, one that you could name after yourself. The spirit of the hunt catches you, and you start to lift slab after slab. You find more ghostly *Dickinsonia* shapes; then a *Spriggina,* with a long, ridged body and blunt head. But then, suddenly, the sun is too high in the sky, and the images vanish.

The Ediacarans imprinted on these rocks lived—and died—in a shallow, sandy seafloor close to shore. They were the first large creatures to appear on Earth after the long epoch of microscopic slime. Theirs was an innocent age. Predators had not yet been invented, and big, defenceless sheets of flesh like *Dickinsonia* could lie around on the seafloor with impunity. "If you want to have a Garden of Eden from some time in the history of life, this was it," Jim says.

But death still came to these particular unfortunates, in the shape of a storm that stirred up the peaceful sea and brought sand cascading down to smother them where they lay. Each thin slab of rock on the hillside was created during one such underwater sandstorm. Even then, the Ediacarans' soft bodies would have rotted away to nothing, if Slimeworld hadn't intervened to preserve them. Most of the fossil slabs have a rough, stippled texture like elephant skin, the remnant of the slimy bacterial mats that were draped over the Ediacaran seafloor. A few years ago Jim realized that these mats also helped preserve a record of the Ediacaran fossils.

The idea came from an old image. As a child, Jim was flicking through an encyclopedia when he saw a picture of the death mask taken from the corpse of the notorious outlaw of the Australian Outback, Ned Kelly. Jim can still remember the imprint that Kelly's eyelashes had left behind, and the shape of his chin. And when Jim was studying the fossil slabs, he realized that the slimy mats would have provided each Ediacaran with a death mask just like Kelly's.

As soon as the Ediacarans died, bacteria would have rushed to cover them, greedily absorbing their nutrients, and extruding chemicals that bound the sand above into a tough, yellow mineral of iron called pyrite. Even when the soft body rotted away, this pyrite shell would have stood firm. Now, hundreds of millions of years later, that same iron crust survives on the underside of each sandstone slab; it has rusted now to red iron oxide, but still provides a faithful mold of the creature that once lay beneath.[5]

This means of preservation captured not just individuals, but a slice of life, a snapshot of the Ediacaran seafloor. Unfortunately the sand also squashed many of its victims before their death masks formed. This has left Jim and his colleagues with a quandary. Some Ediacarans were born flat, some became flat; how do you tell the difference? Interpreting the impressions of these crushed bodies is certainly an art—hence the business earlier with the coke can. ("Arm-wrestle for it!" one exasperated onlooker told the researchers who had spent most of a field trip wrangling pointlessly about the appearance—when alive—of a particularly smudgy fossil.) But there are clues in the rocks, if you know how to read them. Are the fossils ever folded over? Then they must have been flat when they were alive. Are they all lying in the same orientation? Then perhaps they were bound into the seafloor, and all swaying in the same current, when the deadly sandstorm hit.

From this kind of analysis, some things have become clear. The Ediacarans were definitely alive—no geological process can make shapes like theirs. They were also much larger than the microscopic creatures of Slimeworld. And though some look like nothing on Earth, others are uncannily like more modern animals—starfish, jellyfish, sponges and sea pens. These resemblances have led many researchers—Jim among them—to believe that at least some of the Ediacarans were direct ancestors of the complex animals we see today. But everyone admits that shape-matching isn't enough. Though the fossils look complicated, they could still be some strange aggregation of simpler creatures. Over the years people have speculated that each Ediacaran could have been one giant single cell, quilted into many fluid-filled compartments, like an air mattress; or perhaps even some exceptionally coordinated colonies of bacteria, banding together into deceptively complex shapes.[6]

As it turns out, they are neither of these. We now know that Ediacarans truly were the first complex, multicellular animals. The proof has emerged only in the past couple of years, and it comes not from the shape of the fossils, but from their *trails.*

In his collection at the Institute of Palaeontology in Moscow, Misha Fedonkin has some of the best Ediacaran fossils in the world. Misha is a dapper man in his early fifties, with a short, tidy moustache and jet-black hair. His English, like his manner, is fluent. He is charming and suave, often animated but never, ever ruffled. After decades of doing science in Russia, he is also infinitely resourceful. That doesn't just apply to fieldwork. Put him in the residential part of an unfamiliar city, and he will immediately find you a delightful little bar, club or restaurant, just around the corner. When Misha was young, he loved hunting and fishing, but now his heart lies elsewhere. He hasn't touched a gun since he found his first Ediacaran fossil, nearly thirty years ago. "Fossil hunting," he says, "captures your soul."

Misha's fossils come from the sea cliffs of Russia's White Sea coast, near the remote northern port of Arkhangel'sk. The train journey from Moscow takes twenty-two hours, crammed in a cabin with everything for the season: army-surplus tents, ropes and climbing gear, food tins, slabs of butter and cheese. From the port there's another ten-hour journey by boat to the campsite, squeezed on a beach between steep, sloping cliffs of clay and the bleak White Sea.

Occasionally a river has cut a canyon through the cliffs on its way to the sea, and Misha usually tries to camp near one of these for the fresh water it provides. Though the White Sea is a branch of the Arctic Ocean, there is no ice on it in the summer. But the weather can still be grim. When it rains, the soft clay of the cliffs turns to pale, glutinous mud that yanks at your boots, and coats everything it touches. Sometimes there is a storm out to sea, and the water rises up on to the beach in a foaming mass, and rips through the camp.

Good weather, on the other hand, brings that other famous Arctic hazard: the flies. At first sight the taiga (Arctic forest) along the rivers looks impenetrable. Then you realize that the stunted trees are in fact widely spaced, and that the gaps between them are dark with dancing clouds of mosquitoes and black flies. On any fine beach day, these blood-hungry beasts will come for you. Like Paul Hoffman in Canada, Misha and his co-workers have gradually become inured to this menace. But one American researcher who joined Misha in the White Sea a few years ago was so badly bitten that his face quickly swelled to twice its normal size. Still, he wasn't particularly troubled. That same year he discovered a new species of Ediacaran that is now named after him. What are a few insect bites, when you can achieve immortality with the neatest of twists—passing your name on not to a descendant, but to an ancestor?[7]

The White Sea cliffs are packed with Ediacaran fossils. Like their Australian cousins, the creatures preserved here lived in a warm, shallow sea, and were suffocated with periodic blankets of sand. Now the crumbly clay of the cliffs is interleaved with layers of sandstone bearing the familiar Ediacaran death masks. Each spring new landslides send sandstone slabs tumbling down on to the beach. Each summer Misha returns to see what spectacular new finds have been loosened by the rains.

And some of Misha's more recent discoveries baffled him. He found four *Dickinsonia,* all exactly the same size, grouped together on a single slab. Puzzlingly, three were raised up proud from the sandstone in positive relief, and only the fourth was the usual indented mould. He also found a slab bearing *Yorgia,* another oval creature, which had internal riblike structures and strange squashed shapes that could have been some kind of organs; there, once again, Misha saw four fossils together, three of them in positive relief, one negative. And there was *Kimberella,* a creature the shape of a teardrop, with a flouncy frill around its edges that looked to Misha like the undulating foot of a slug or snail. At the pointed end of one fossil, Misha found grooves in the rock, as if something had been raking the seafloor just before the sandstorm hit. At the ends of others he noticed long, dark traces, many times longer than the *Kimberella* itself.[8]

Trails. These were all trails. Misha realized that the four *Dickinsonia* fossils had all come from one individual. Three times this creature had rested on the slimy microbial mat that coated the seafloor, and left an imprint of its belly there. The first three death masks stood up from the sandstone slab because the sand had reached out to fill hollows in the seafloor. Only the fourth was indented—a true mould of an Ediacaran's body. And the same thing applied to the *Yorgia*. Misha's sandstone slabs had captured the three previous belly-prints of the organism as well as its own corpse.

When Jim Gehling heard about these finds, he raced off to look at his own collections. Sure enough, he found an Australian *Dickinsonia* doing just the same thing: three belly-prints and one final fossil. And that meant one thing: these creatures could clearly move.

Perhaps, Jim thinks, the *Dickinsonia* and *Yorgia* were using their belly-flops as a way of feeding, since neither had the benefit of teeth. They would lie on the slimy mat covering the seafloor and gradually consume the bacteria there. "If you lay on the lawn long enough, you'd rot the grass underneath you," says Jim. "And that's a source of food." Obviously, then, the creatures would have to move on when the food was exhausted. Misha agrees, and thinks that the *Kimberella* might also have been moving to eat. Those scrape marks could have been places where it used a proboscis to rake in food from the seafloor. And the travelling *Kimberella* left a long trail in its wake, just like a slug or a snail.

The ability to move sends an immediate message to biologists everywhere. These creatures had to be complex. To make trails, you need tissues that behave like muscles; you have to be a cooperative organism made of multiple specialized cells. Quilted air mattresses can't do it, nor can groups of bacteria. *Kimberella* didn't just look like a snail. It *moved* like a snail. The extraordinary traces from the White Sea prove beyond doubt that Ediacarans really were complex, multicellular animals.

So now we know what Ediacarans were. But *when* were they? The fossils at Ediacara and in the White Sea lived around 555 million years ago. That's an improvement on the Cambrian explosion, but it's still some forty million years after the Snowball. The next step would be to find Ediacarans from nearer the end of the ice.

MISTAKEN POINT is a windswept, godforsaken promontory on the southern tip of Newfoundland. It is surrounded by the barrens: blasted, treeless heaths covered with mosses, lichens and tart crown berries. Nobody could love these barren lands, not even their mother. They are dreary and damp, their plants the colour of overcooked spinach and rusty nails; when the wind is not buffeting them or rain beating them down, they are shrouded in fog. The pale, thin caribou wander over them like lost souls.

The seas hereabouts were once rich with fish, but now that cod stocks have crashed and cod fishing is banned, depression has descended on the area like one of its famous fogs. (This is, officially, the foggiest place in the world.) Nobody has lived at Mistaken Point for decades, and only a few people remain in the village of Trepassey, an hour's drive to the west. The locals there are friendly, their accents a fossilized form of Irish, their grammatical constructions archaic. "There you be. You likes that, doesn't you?" they will say as they hand you a plate of salt beef. Or more probably cod, flown in from somewhere, since old dietary habits die harder than most.

Trepassey means "the dead souls" in Basque, and was named by fishermen of the sixteenth century for the many ships wrecked on this craggy coast. These waters hold the remains of thousands of people who were betrayed over the centuries by fog and Arctic ice and high winds, especially at Mistaken Point, which was "mistaken" by many unlucky sailors for the next finger of land along the coast. Seeking the safe harbour at Cape Race, they would turn too soon, and founder. The lighthouse at Cape Race received distress signals from the stricken *Titanic,* and is the closest place on land to the ship's Atlantic grave.

The rocks of Mistaken Point are also grave sites, but for much more ancient creatures. To see these, on one particularly grim day at the beginning of June, I have joined a troupe of soggy, sodden, steaming geologists squelching over the saturated mosses and the black, muddy streams near the cliff top. The rain is relentless. There is no perceptible distinction between sea and sky—both are the colour of granite. Disembowelled sea urchins lie where they were dropped on the rocks and then eviscerated by seagulls or yellow-headed gannets.

The fossil surfaces extend out sideways from the cliffside like a toppled stack of books. We climb down on to one of them, and huddle miserably in the wind, praying for a gap in the clouds. The rock is dull in the flat grey light, and the surface appears blank. Though we are crouched some twenty feet above the sea, occasional waves crash over the edges of the rock layer and people leap out of the way, squealing. Everyone has reverted to geology talk; they are speaking of clasts and volcanic bombs, gravity flows and forearc basins. Someone is talking about Paul Hoffman, though he isn't on the field trip. Bragging about him. "Paul worked for me once. He was my junior assistant in the field." "Did you hear that?" my companion says. "Everyone wants to claim a piece of Paul."

The afternoon is drawing on, but it seems the rain may be easing. Now the wind is a blessing, as it dries off the rock surface. And then suddenly, miraculously, slanting rays of sunlight appear through the clouds. Fossil light! The surface of the rock is suddenly crammed with strange shapes: fronds and spindles and discs and branches. "Look at that!" Jim Gehling whoops. "Someone switched on a projector in the sky!"

Despite the damp rocks, everybody immediately takes off their boots. These fossil forms may have survived thousands of years of pounding by wind and waves, but nobody wants to risk scratching them. As geologists pad over the rock surface in their thick hiking socks, the scene seems oddly biblical. And yes, there's even a "burning bush" fossil, the size of a spread hand, with fronds curling upwards like tongues of flame. Jim is standing to one side, staring raptly at the shapes that have appeared in the rocks. "Just imagine if you could have this as the floor of your house," he says. "Imagine if you could walk on it *every day.*"

The fossils of Mistaken Point are different in many ways from those at Ediacara or the White Sea. These creatures were nowhere near a sandy shore when they died. Instead they rested on the deep, dark floor of an underwater canyon. And they were overwhelmed not by sand, but by ash.

Northwest of here, a range of volcanoes once poked through land that would later become parts of Central America and Brazil.

Now and then, there would be a rumble, a roar, and an explosive eruption that sent dark clouds of ash flying through the air and into the ocean. The Ediacarans had no ears to hear any warning sounds. A thick, swirling cloud would simply have appeared from nowhere and filled their world, blanketing the canyon floor and everything that lived there.

The ancient Roman town of Pompeii was smothered by just such an ash cloud when Vesuvius erupted in 79 A.D. Fleeing, crouching or writhing, the bodies of Pompeii's residents were preserved in death by the same ash that ended their lives. First this ash smothered them, then it hardened around their rotting corpses. Centuries later, the bodies themselves had disappeared. But archaeologists injected plaster into the voids left behind, and re-created the shapes of the dying residents in extraordinary, sometimes horrible, detail.

In just the same way, volcanic ash hardened over the rotting bodies of the Ediacarans. Then, like the plaster at Pompeii, mud from the former seafloor forced its way up into the hollows they left behind, making faithful images that turned slowly into rock.

Preservation by ash is unusual. Most Ediacaran fossils were smothered by sand, and their images are preserved only as indented death masks in an overhead layer of sandstone. But in Newfoundland the image comes from the underlying mud. The ash layers have mostly weathered away, and solid casts of the Ediacarans stick up from the rock surface. Mudstone on mudstone, they are still only visible when the sun is low enough to outline them with their own shadows. But then they are revealed exactly as they once lay—Newfoundland is the only place in the world where you can walk around on the Ediacaran seafloor.

This rare preservation method has another benefit. Each layer of ash provides the Ediacarans it killed with a handy age marker. Volcanic ash is a great way to date rocks, because it contains traces of radioactive elements; uranium, for instance, which decays into lead at a precise, well-defined rate. When a volcano explodes and its ash forms, this uranium clock starts ticking. With each tick, the rock loses uranium and gains lead, and the ratio of these two elements gradually changes over time. In an ash layer, geologists can measure the ratio of uranium to lead today, and can tell how much time has passed since the ash formed.

Here at Mistaken Point, the ash that smothered these jellied fronds and spindles gives an age of 565 million years.[9] The creatures here are much older than the ones in Australia or the White Sea. They lived just 25 million years or so after the Snowball.

They're not even the oldest fossils in Newfoundland. The next day we find ourselves on another part of the coast, this time close to a beach where each receding wave sucks the pebbles backwards with a sound like a crackling flame. The rock layers here are giant black slabs, tilted sideways like collapsed dominoes. We head for one that has some kind of large discs etched into it. "Pizza to go, anyone?" says Guy Narbonne, the Canadian researcher who is leading the trip. He's right. These fossils look exactly like pepperoni pizzas. Or at least they are the perfect size and shape; but the pepperoni pieces are mud-coloured and the region around them is apple green, like slightly mouldy dough. A careful hike over the slippery rocks takes us to a red surface, which bears the faintest trace of what looks like a long-stemmed cocktail glass. Near it are two thin fronds, several feet long. At least these don't remind anyone of food; they look more like the prints from a bicycle tyre.[10]

Where the rock layer disappears into the ground, we can see, edge-on, the ash layer that preserved these faint fossils. I'd half expected it to be dark and crumbly, but instead it's as solid as concrete, and the same pale green colour as the pizza "dough". This ash has been dated. The result bears the status of a "rumourchron"—geologist slang for a date that has been measured but is not officially out in the world. It comes from the lab of Sam Bowring at MIT, one of the most reliable geochronologists in the world. Sam has told the date to many people. He's presented it at conferences.[11] But he hasn't quite published it yet. It's 575 million years. If his date holds, these fossils are almost as old as the final days of the Snowball.

STEP BY step, the date for the invention of complexity is approaching the end of the Snowball. For many people this is beginning to look like more than a coincidence. But one serious problem still remains. A few researchers have turned up what they think could be signs of complexity *before* the Snowball. If multicellular life really did emerge before there were even glimmers of global ice, that could scotch the whole biological part of the Snowball theory.

Some of the evidence is still highly controversial. An outcrop in northern Canada bears perhaps a thousand discs, somewhere between the size of a dime and a quarter, impressed into the bottom of sandstone rocks. The rocks date from around 100 million years before the end of the Snowballs, and their discoverer, Guy Narbonne, insists that they are complex creatures.[12] But they left no trails, and have virtually no structure. Other biologists say they could well have been simple blobs of jelly or colonies of bacteria.

And etched on billion-year-old rocks from India are pencil-thin branching tubes, which their discoverer—a most respected researcher named Dolf Seilacher—believes were made by some kind of early worm.[13] But most of his colleagues sigh and point out that there's no sign of the creatures themselves among the "trails", which makes his argument much harder to swallow. Other researchers have just reported the discovery of blobby grooves, like worm casts, in 1.2-billion-year-old sandstones from southwestern Australia.[14] But once again, there's no sign of any animal, nor any clear evidence that complex creatures really created these "trails", and few biologists think they pose problems for the Snowball theory.

More troubling, though, are the algae. Algae are marine plants that live throughout the modern ocean. Some are small, hairy blobs floating through the water or clinging to rocks. Others are huge. Kelp is a form of algae, and the kelp forests off the coast of California contain plants that are hundreds of feet tall. Algae certainly existed before the Snowball. There was nothing like kelp; the biggest creatures were just a fraction of an inch across. But they were almost certainly multicellular.

For instance, Nick Butterfield, a Canadian biologist now at Cambridge University, has found fabulously preserved red algae in a lump of chert that he collected from Somerset Island in the Canadian Arctic. The rock is 1,200 million years old, and the fossils it contains are tiny, hairy things, scarcely visible to the naked eye. But when Nick put his samples under a microscope, he realized that the fossil images were dead ringers for a modern red alga called *Bangia,* which you can scrape off rocks on many seashores today. He saw the classic rows of disc-shaped cells that make up the *Bangia*'s filaments, and the wedge-shaped cells that adult *Bangia* possess, having divided their discs into eight, twelve and sixteen pieces. He also saw separate cells that were orientated vertically, and appeared to be making up a "holdfast", a kind of anchor that could bind the alga into the seafloor and enable it to grow upwards rather than merely sideways like the primitive flat mats of Slimeworld.[15]

Then there's fancy filamentous algae from Spitzbergen, which look much like green algae does today. And a strange beast named *Valkyria,* with appendages that look almost like legs (but aren't). And a Siberian fossil found by Andy Knoll, a colleague of Paul Hoffman's at Harvard, which looks just like a modern green alga called *Voucharia.* Many of these are not just collections of cells. They really look as if they've already learned to specialize.

These finds may seem to topple the biological part of the Snowball argument, but they leave open one big mystery. Algae apparently learned to be multicellular by 1.2 billion years ago. If they then passed on the secret to the rest of the world, why did it take another 600 million years before animals did the same? If this was the crucial step that changed the world for ever, why did the rest of the planet stay mired in simple slime for so long afterwards? *Nobody* believes that one evolutionary event can trigger another occurring hundreds of millions of years later. Even Nick Butterfield says so. "There's still this huge delay before things got rolling," he acknowledges, rather sadly. "Biology moves faster than that."

There are, then, two remaining possibilities. Either algae invented complexity separately, and kept the secret to themselves—which would pose no particular problem for the Snowball idea—or there were plenty of complex animals around before the Snowballs, but they left no trace in the rocks. That would definitely be a problem, since the Snowball couldn't trigger something that already existed long before. But how would you test this without fossils? There might just be a way. The evidence would come not in the form of fossils, but from applications of a more oblique approach known as a "molecular clock".

The genetic material—the molecule called DNA—inside every living cell contains information about its ancestry. In principle, with a sample of my DNA and some of yours, we could work out how closely you and I are related. Although we are both humans, your DNA differs slightly from mine. That's why our faces are perhaps different shapes, or our eyes a different colour. These changes in DNA have happened over many generations, as genetic material passed from parent to child, sometimes with small mistakes introduced, sometimes just through the natural mutations that appear over time.

When my DNA was last identical to yours, it resided in the cells of the person who was our last common ancestor, our mutual great-great-ever-so-great-grandparent. So if we wanted to work out when this ancestor lived, we wouldn't necessarily have to consult a genealogist. Instead we could simply measure the differences between my DNA and yours, and estimate how quickly the DNA clock ticks.

In practice, DNA changes aren't fast enough to help with recent family trees, though researchers have used this technique to show that we are all descended from one modern human "Eve," who lived a little more than 200,000 years ago. Molecular clocks can also identify the timing of more-distant ancestors—between a human and an orangutan, say, or a human and a fruit fly. Each tick of the clock, each slight change in the exact composition of two creatures' DNA, takes them further away from their mutual ancestor. If you measure how much their DNA has changed, and you know how fast the clock was ticking, you can track evolution backwards even without the help of fossils.

Many different research groups had already used this approach to try to find out when complex animals first emerged. They examined the DNA in different animal species alive today, and then worked backwards to try to date the appearance of their last common ancestor. The first of these studies, back in 1982, said this unique animal ancestor lived around 900 million years ago. Though others thought it might have been a little younger, the most recent studies have pushed the date much further back. In the past couple of years, several different molecular clocks have suggested that the animal ancestor lived some 1.2 billion years ago or more, long before the Snowballs had even begun.[16]

Kevin Peterson is a fervent young biologist from Dartmouth College in New Hampshire. He doesn't like the Snowball idea. He doesn't really like *any* big idea. What he cares about, he says, are hypotheses. Unlike ideas, hypotheses are *testable.* Kevin is very big on testability. If you told him you thought tomorrow was Saturday, he'd probably ask if your hypothesis was testable. Unless he can test something, he doesn't want to know.

To some extent, this is true of all scientists. Speculating is fun, but if you can't decide whether one speculation fits reality better than another, then why bother? "However beautiful we may find the constructions of our imagination," wrote the physicist Lee Smolin, "if they are meant to be representations of the natural world, we must take those constructions humbly to nature and seek its consent."[17]

There's nothing particularly humble about Kevin. He's a

young Turk, supremely confident that his own findings will shoot down those of his scientific seniors. But he certainly believes in consulting nature. And as soon as he saw Paul and Dan's papers about the Snowball, he was revolted. Yes, yes, lovely idea, but where's the *proof*? Kevin wasn't concerned with the geological side of the argument. That's not his speciality. But when he saw the part about biology, he was infuriated. Paul and Dan had drawn a diagram showing the first complex animals appearing after the Snowball. "I thought it was one of the silliest figures I'd ever seen in my life," Kevin says. He knew all about the molecular clocks. And every single one of them said that complex animals formed hundreds of millions of years before the Snowballs even began.

The problem was that these previous attempts produced a disturbingly wide spread of dates. Kevin decided that was because they hadn't been done correctly. There were flaws, he felt, in each of them. So he resolved to do a new, improved study. He would design the best-ever molecular clock. He would use it to pin down the timing of the animal ancestor once and for all, and—he presumed—to demonstrate that the biological part of Paul's idea was hopelessly wrong.

Kevin chose the creatures for his molecular clock carefully. He decided on echinoderms—the family that contains urchins and sea stars. For one thing, there was a complete, well-dated set of fossils for the ancestors of these creatures, so he could check more carefully than previous studies how well his clock was doing as he went backwards in time. For another, these creatures had similar body size, metabolic rate, and amount of time between one generation and the next. Kevin felt that previous studies had erred by picking creatures that were too different in all these respects. The more alike the creatures that he started with, Kevin realized, the more accurate the clock tracing their mutual ancestry should be.

So he set about uncovering the genetic sequence behind seven different creatures. He calculated how fast the DNA must have changed. He worked steadily backwards, checking his dates as he went. Whenever he had well-dated evidence from fossils for the timing of a particular ancestor, he checked whether the clock agreed. Each time, the clock looked good. Encouraged, Kevin projected his clock further back in time. Now there were no fossils to check against. Now he was getting closer and closer to the animal ancestor. And then, finally, he had his answer. The last common ancestor of all complex animals lived . . . somewhere around . . . 700 million years ago.[18]

Kevin was stunned. "I couldn't refute that diagram of Paul and Dan's," he told me several months later, still sounding dazed. Why not? After all, his clock didn't throw up that magic date of 590 million years—which marks both the ending of the ice and the beginning of the first complex fossils. But neither did it prove that animals had existed more than a billion years ago, as Kevin had expected. Instead, its date agreed almost exactly with the end of the very first Snowball.

Remember that Paul was dealing with a series of events, not just one. And the first Snowball ended around 700 million years ago, exactly the date Kevin's clock produces for the animal ancestor. Perhaps complex animals were triggered by the first Snowball, and then survived through the remaining episodes of ice. If each subsequent Snowball wiped out all but a few of those new animals, that might explain why widespread fossils didn't appear until the ice finally receded.

It's also possible that even Kevin's careful clock overestimated the age of complex animals. Genetic studies like these assume that their clocks have always ticked at the same rate. But some biologists think that genetic changes happened more quickly in the past, and that all the clocks give older times than they should. Kevin believes he solved many of the problems with the earlier clocks, but he says himself that he may not have solved them all.

THERE WILL be many more attempts to find traces of the earliest animals. Researchers are collecting genetic material, designing newer, better molecular clocks, and scouring the world's rocks for trails blazed by ancient life. But the more biologists try to pin down the timing of widespread complexity, the more their results seem to point towards the Snowball.

Was this just a coincidence? "Most people I know think they're connected," says Jim Gehling. And Kevin has changed his mind about the timing at least, though he still wants a testable hypothesis to explain the connection. Nick Butterfield, the algae man, says the same. He is annoyed about Paul's airy biological assumptions. He says that Paul's paper was "cringingly awful in its biology". He says it's up to Paul to explain exactly how ice could have triggered life's industrial revolution.

But forget Paul for a moment. What does Nick think about the biological evidence, the new fossils and the molecular clocks and all the other developments that are steadily moving the date of burgeoning complexity closer and closer to the ice? Nick pauses. Then says this. "I think it's *fascinating.*"

Biologists aren't so very different from geologists, under the skin. Some of them have jumped eagerly on to Paul's Snowball bandwagon, and some have declared furiously that it must be stopped. And some are still waiting to see what will happen.

Though the world of ancient fossils seems pretty well explored, it's still possible that someone, somewhere, will find a vast stash of complex animals from long before the Snowball. But in the absence of this, the evidence for some connection between ice and new life is looking more persuasive by the day.

So biologists are beginning to think of ways that the Snowball might have triggered complexity. Everyone agrees the capacity to be complex must already have existed in the creatures' genes, but nobody knows for sure what spurred those dormant genes into action. The theories are not yet fully formed; they're speculations in corridors rather than neat, tidy theories. But there are several intriguing ways in which ice, that most inimical of substances, might ultimately have given rise to this new life.

The Snowball itself could have encouraged life to diversify and experiment. New species often arise when a single population of creatures is separated from its fellows in an isolated refuge, for something upwards of a million years. Or perhaps the opportunity for complexity arose after ice wiped large areas of the planet clean of life. All living things need certain resources to survive—food, water and shelter—and as long as enveloping mats of slime were hogging all the resources, there would be no space left to innovate. Removing the extant occupants of Earth's ecological niches might have made room for life to experiment. We already know that worldwide extinctions make space for new species to emerge. When a meteorite wiped out the dinosaurs, for instance, the previously tiny mammals suddenly had free licence to grow, change shape, and consume the resources once reserved for the likes of *Diplodocus* and *T. rex.* Though there's no direct sign that the ice made any of the slime-creatures *extinct,* it may have killed off enough of each species to create the breathing space that evolution needed.

Another suggestion has come mainly from Jim Gehling. He wonders whether complex life was a response to the sheer changeability of life in the Snowball's aftermath. First the world endured its longest and most severe ice age, and then came a violent hothouse lashed with acid rain. With conditions changing as drastically as this, life had a natural incentive to spawn creatures that could protect themselves from external buffeting. Single-celled slime balls are at the mercy of current and weather, but large, multicelled animals have much more control. They can dig into the ground and hold tight in fierce currents. They can control their internal temperature, store food more effectively against lean times, and grow covers to protect themselves.

But the most popular idea for a trigger point involves oxygen. Large creatures need efficient ways of mobilizing their food into energy, and oxygen is one of the best. When we breathe, the oxygen we inhale is used to "burn" food, like burning petrol in a car engine, and that's what generates the energy that supports our vigorous lifestyles. Oxygen is also necessary to make collagen—the connecting tissue that binds muscles to bones and helps keep cells together, and that is found somewhere or other in every complex animal.

There are some signs in the rocks that atmospheric oxygen was increasing around the time of the ice. Perhaps whatever triggered the Snowball also created this excess of oxygen. Or perhaps there was a sudden pulse of oxygen immediately after the freeze ended. For millions of years, life would have been restricted to a few small refuges, and unused nutrients would have built the ocean up into a tasty chemical soup. As soon as the ice was over, the few remaining creatures would have seized on these nutrients and blossomed. The white planet would have become green with massive colonies of bacteria and algae stretching over the surface of the ocean. And those same colonies would have soaked up sunlight, made food, and belched out oxygen as a waste product of their endeavours. That sudden pulse of oxygen may have been exactly what complex life was waiting for.

Biologists are now trying to figure out how to test these ideas. But there's something they all agree on. Whatever creative role the Snowball may have played in shaping a new world order, it would also have been devastating for many of the life-forms it first encountered. And this raises a disturbing question: could another Snowball happen today? If the ice returned to haunt us, the consequences would be horrific. Earth has come a long way since the simple days of Slimeworld, and life is now a complex web of interdependent creatures. If another Snowball engulfed the Earth, many—perhaps most—of these creatures would perish.

TEN

EVER AGAIN

To figure out whether the ice will ever return, we first need to know why it appeared in the first place. What was so special about the Snowball times? Though the clues are scant, some evidence has emerged from another, even older part of Earth's geological history. Joe Kirschvink, the sparky, inventive Caltech professor who set much of the early Snowball rolling, has discovered that Paul's Snowball period wasn't the only one.

South Africa, September 2000

You'd expect the Kalahari Desert to be dry and hot, and so it usually is. Even the place names around here evoke its baking, insufferable summers—Hotazel, for instance, a remote mining outpost an hour or two north of here. (The land surveyor who proposed the name back in 1917 had to use this phonetic spelling because the authorities objected to his original suggestion: "Hot as Hell".)

But now, at the tail end of a southern winter, the desert is both cold and very wet. The rain began yesterday evening with a roar that eventually settled down into a night-long drumroll on the tin roof of our tiny motel. It has turned the dirt road into a skating rink of rich, red mud. I'm here with Joe Kirschvink, who has brought a phalanx of students to tour the geological sights of South Africa. As our five vehicles lurch in convoy through the puddles, fountains of red water fly into the air and separate out into thick, cartoonlike drops.

We turn, thankfully, on to a paved road again, and the rain begins to ease. The landscape in the southern Kalahari is just like Paul Hoffman's Namibian field sites: open and almost featureless, scattered with camel thorns and golden grasses and those towering red termite mounds, as tall as the trees. There's a striped gemsbok, sheltering among the thornbushes. And there, hanging from a telegraph pole, is a familiar weaverbird's nest, an impressive sack of entwined grasses nearly five feet long and almost as wide. The road begins to gain altitude, and the temperature drops further. A dense white mist descends around us, hovering off-road just beyond the barbed-wire fence. At the designated road cut, we climb out of the vehicles and wince as flecks of ice fly by in the freezing fog. But Joe is cheerful. "What did you expect?" he says with a grin. "This is Snowball country."

He's right. All around us, among thornbushes and orange clumps of grass, are the now-familiar signs of ancient ice. The background flame-coloured rock is studded with pebbles and stones that were once bulldozed along the ground by glaciers, and then tipped offshore into a shallow, ice-covered sea. Not all of these have remained embedded. The sides of the road are scattered with loose pebbles and we spread out, red-cheeked in the wind, seeking stones with tell-tale glacial scrapes. Here's one, a small rounded lump, indented with parallel grooves where it was scoured along the ground. These deposits bear all the hallmarks of a Snowball. Not just the jumbled, scratched ice rocks that we see here; back at Hotazel there are also thick layers of iron that rusted out of the Snowball ocean, and above them, the classic cap carbonates that brought Dan and Paul their eureka moment.

But this Snowball differs from Paul's in one crucial respect: it happened nearly two billion years earlier. These are the remnants from a Snowball that gripped the Earth not 600 million years ago, but a full 2.4 billion years ago. Rocks this old tend to have suffered greatly from the tectonic twisting and weathering that billions of years inevitably bring; intact outcrops are rare from the Earth's early history, and even the few that have survived are tremendously hard to interpret. So although geologists have known about these truly ancient signs of ice for decades, nobody has paid them much attention. Unlike the ones from Paul's more recent Snowball, these ice rocks don't appear on every continent, and not many bear the clear tropical hallmarks that gave Snowballers their first clues. Nobody would have guessed that these few deposits marked another, earlier global ice age. Until, that is, Joe came along and proved it, using yet another of his magic magnetic measurements.

Back in the warm trucks, we head for the site of these measurements: a road cut that's about an hour's drive away. There the route has sliced through slate-grey remnants of ancient volcanic eruptions. The deposits are immense, a thousand feet thick, and they once covered much of the Kalahari. They are "flood basalts", so called because the lava gushed from the ground in torrents for perhaps a million years, and created a smouldering new surface. These volcanics, Joe tells us, lie right in the middle of the Snowball rocks; they emerged from the ground into a world that was already gripped by ice. "It must," Joe says, "have been wild."

Above our heads was once a shallow, frozen sea, its shoreline just a little way over to the east. Beneath our feet, lava was issuing from cracks in the seafloor, heating the water, melting the overhead ice, and filling the air with hissing clouds of steam. The sea's icy surface was pocked with hot pools, which were green with grateful clumps of slime. As the lava flooded out into the cold seawater, it cooled immediately into gloops of rock like toothpaste writhing from a tube. These rocks are classic signs of an underwater eruption. Geologists call them pillow lavas, but they look bulbous and blubbery—more like elephant seals lolling on a beach. Here, they are collectively named the Ongeluk Formation; *ongeluk* means "misfortune" in Afrikaans, but they brought good luck for Joe. Using these rocks, he discovered that when this part of the Kalahari was coated in ice and fire, it rested within a few degrees of the equator.

Remember that many rocks carry a magnetic birth certificate. When they are young and soft, any magnetic minerals they contain will line up like tiny compass needles along the Earth's local magnetic field: vertical near the poles, and horizontal near the equator. And when the minerals harden into rock, this pattern is frozen in. Wherever these rocks wander, they take their birth field along with them.

So, simply measuring the magnetic field in the rocks should tell you where in the world they formed. There is, though, a catch. Remember, too, that subsequent events—heating, mangling or importing of new magnetic minerals—might have overwritten the original pattern, issuing the rocks with a fake birth certificate. To find out where the rocks were born, Joe didn't just have to measure their magnetic field; he also needed to prove that the field was original.

Joe realized that the pillow lavas could help him do this. The field test he planned was a little like the fold test he used to check whether the Australian rocks had a genuine magnetic memory. But in this case he wasn't seeking a fold in the rocks. Instead he wanted to find shattered volcanic shards. As red-hot molten rock spills into cold water to make pillow lavas, the outer surface immediately freezes in shock and becomes a brown, glassy coating called a "chill margin". The inner parts of the rock take a little longer to catch up. When they finally do cool, they shrink, and this process often shatters the brittle outer skin.

The chill margin takes on the local magnetic field the moment it cools. If it then shatters and the shards point in random directions, their magnetic arrows should be just as random. But if the magnetic field of these rocks has been overwritten after they formed, cooled and shattered, the fields in pillows, shards and everything else will all point neatly in one overall direction. Joe knew that if he measured the fields of the chill fragments, he would be able to tell if the magnetic birth certificate was original.

Here at the Ongeluk, we can still see the small cylindrical holes where Joe took his samples. Joe calls them "palaeomaggot holes", but he made them himself with a rock drill almost a decade ago. He eventually analyzed the results years later, at the prompting of a particularly insistent graduate student. And he then discovered that the volcanics, and the ice rocks that bracketed them, contained a field that was almost flat. What's more, the shards of shattered chill margin had fields that pointed in every direction. This meant the measurement was genuine, and that 2.4 billion years ago, ice lay within a few degrees of the equator. These ice rocks, in other words, are just like Paul's. They are the remnants of another, earlier Snowball.[1]

So now we know that Snowballs have happened twice. At least one occurred a little over 2 billion years ago, and then a series of perhaps four engulfed the Earth between 750 and 590 million years ago. There have apparently been no others. What, then, did these Snowball periods have in common, and what made them different from every other time period in Earth's long history? Was there anything unusual about them that could have triggered the ice onslaught?

Perhaps. There are intriguing magnetic hints that both of these time periods had a peculiar continental alignment. As the world's tectonic plates drift over its surface, the continents sometimes bunch and sometimes scatter. When they spread out, they can end up anywhere. But on a few rare occasions, they can find themselves in a band around the Earth's equator. And this might be exactly what happened during the Snowball periods.

Though magnetic measurements are difficult, and many of the sites have had their magnetic memories rewritten in the intervening time, decent data exist from about half the continents that were around during the later Snowballs. And every one of these lay near the equator. So, too, did the half-dozen sets of ice rocks that have now been measured. For the earlier Snowball, the task is harder and the measurements are fewer. But still, all of them point to low-latitude continents.

If the continents truly were arranged around the equator during these two Snowball periods, that could be just what the ice needed. One reason is that the tropics soak up most of the heat that arrives on Earth from the sun. Because land is more reflective than ocean, putting all available land in the tropics could reflect more of the incoming sunlight, and help the planet to cool. Joe Kirschvink suggested this in his short paper back in 1992.

Dan Schrag, the ideas man, has come up with another reason why equatorial continents could be the key. When continents spread out to the far north and south, he says, they act as an important brake on overenthusiastic polar ice caps.

Ice naturally wants to spread: white ice reflects sunlight, which causes cooling, which breeds more ice—and if this were left unchecked, Earth would spend its entire life as a Snowball. Fortunately for us, high-latitude continents stop that from happening by helping to warm the Earth back up again whenever polar ice becomes rampant.

Normally, rocks do the opposite. They help prevent the Earth from overheating by soaking up the greenhouse gases like carbon dioxide that are pumped out by volcanoes. But if polar ice starts to spread, any high-latitude continents will switch loyalties. Because their rocks become covered with ice, they can no longer soak up carbon dioxide. Instead it stays in the atmosphere to do its greenhouse thing, warming up the Earth and melting the excess ice. So if ever the polar caps start to grow, high-latitude continents will force them to shrink again.

Now imagine what would happen if all the continents were arranged in a band around the Earth's equator. In that case the polar ice caps could spread with impunity. There would be no high-latitude continents to cover, and hence nothing to stop the ice going all the way. By the time ice reached the equatorial continents, it would be too late to prevent a Snowball.[2]

This idea also neatly explains why, at least in Paul's period, there was a series of Snowballs rather than just one. Between 750 and 590 million years ago, the continents could simply have stayed near the equator. A Snowball would begin when some trivial cooling trigger set the ice moving. With no high-latitude continents to stop it, the ice would continue until the Earth was encased. Over the next 10 million years or so, carbon dioxide gas pouring out of volcanoes would build the atmosphere into a furnace, until it became so hot that the ice melted back. Gradually, then, the carbon dioxide levels in the atmosphere would drop, until the whole process started again. As long as the continents stayed near the equator, another cooling trigger would set another Snowball rolling. And another. And another. Until, eventually, the continents moved on and the world was spared.

There aren't enough outcrops from Joe's earlier Snowball to know whether it was a series of events or just one. But researchers believe that the sun was much feebler then, and that an individual Snowball would have lasted much longer. Perhaps Joe's single early Snowball lasted so long that the continents had begun to move away from the equator again by the time the ice receded.

So equatorial continents could provide the rare but reasonable recipe for a Snowball. If this explanation is right, that's encouraging news for our own future. Right now we have plenty of continents at high latitudes. Most of the world's landmasses are way up in the north—think of Canada, Europe and Russia. Presumably, these far northern lands are protecting us all from the ice. Well, not necessarily. It turns out that in spite of this reassuring continental arrangement, the Earth may even now be preparing for another descent into ice.

DAVE EVANS used to be a graduate student of Joe's. He's the one who prodded Joe into measuring the South African samples, and proving that the older ice rocks had been close to the equator. (Dave found the samples collecting dust, and resurrected them.) Now, in his early thirties, he's a professor at Yale University. He is thin and gangly, with thick, wavy hair and a pleasant smile, and looks younger than most of his students. Though he is organized and careful, from working in Joe's lab he also has this legacy: the capacity to consider crazy ideas that might just be true.

While he was at Caltech, Dave didn't just work with Joe on the ancient Snowball. He also investigated another of Joe's "nutty" ideas. As the Earth's tectonic plates creep over its surface, they usually travel at a sluggish few inches a year—the same speed that your fingernails grow. But Joe and Dave believe that at certain times in the past, the continents let rip, travelling at what for them was the breakneck speed of several feet a year. They did so, according to this theory, because they had an inexorable urge to reach the equator.

Spinning objects always prefer to have most of their weight around their middles. Think of a child's spinning top: tall, thin ones are much easier to knock over than short, fat ones, because they're more unstable. If the instability is too much to handle, the object will try to readjust. Suppose you dropped a large lump of clay on to the top of a spinning basketball. If the lump was heavy enough, the basketball would tip over until the excess weight was spinning around its waist, and the system was safely back in balance.

Joe and Dave believe the same thing applies to our spinning planet. They think that if the shifting continental lumps on its surface throw it off balance, the Earth will try to move them equatorwards. This doesn't happen all the time, they say. There has to be a big enough imbalance before the Earth will notice. But occasionally the random jitterbug of the continents brings them crashing together into one massive "supercontinent".

Even that's still not quite enough. All the continents in the world don't weigh much compared to the Earth's massive innards: its thick mantle of plastic flowing rock and its hefty iron core. But Dave believes that the supercontinent would act as an insulating cap over the mantle that lies beneath. Gradually the mantle would heat up, and a great plume of rock would rise up like lava beneath the supercontinent, lifting it up like a giant pustule. Now, with continents and mantle together, the lump would tip the balance and the Earth would respond. Both supercontinent and underlying mantle would go flying off to the equator, until the world became stable again.[3]

This idea is every bit as controversial as the Snowball, and Dave—one of the few people in the world to work on both—has now thought of an ingenious potential connection between the two theories. What if slipping supercontinents make Snowballs? First the continents would collect together into one enormous mass; then this mass would skid to the equator; then the hot plume of rock that still lay beneath the supercontinent would blast it apart in a frenzy of volcanic activity that left fragments scattered around the equator and tropics. And while this was going on, the polar caps could proceed, unchecked, to cover the Earth.

At least some of the available evidence fits this idea. A supercontinent that geologists have named Rodinia finally broke up around 750 million years ago, exactly when Paul's Snowball episodes began. Nobody knows whether there was a supercontinent before the earlier Snowball, the one whose remnants Dave and Joe measured in South Africa. But those same bulbous pillow lavas that provided their samples might also contain clues about the state of the continents then. Those massive volcanic floods didn't cover only South Africa; they also poured out on to many other parts of the world. And that's exactly what you'd expect if a supercontinent was breaking up, and huge amounts of lava were spilling through the cracks.

Were there any supercontinents that didn't produce Snowballs? Well, one called Pangaea existed around 225 million years ago, without generating any notable ice. But Dave points out that Pangaea broke up again relatively quickly. He suspects that it simply wasn't around long enough for that crucial plume of hot rock to form underneath, and unbalance the Earth.

Now we're at the outer reaches of the Snowball idea, with speculation heaped on speculation. But this idea of Dave's is intriguing. And if he's right, the corollary is also chilling. You see, the Earth is making another supercontinent right now.

SIXTY MILLION years ago, not long after an asteroid slammed into the Earth to end the rule of the dinosaurs, India began to sense the presence of Asia. The bulk of what is now the Indian subcontinent had been drifting, footloose, ever since it broke away from Antarctica during the shattering of Pangaea. Now it was moving steadily northwards at the rate of a few inches per year, and Asia was in its way. There was only one possible outcome: a continental pile-up. When ocean basins collide, one or other of the crusts tends to be forced downwards, back into the Earth's interior. But continents are not nearly dense enough to sink. When two continents crash, the only way is up.[4]

So India crashed into Asia, and the land began to rise. First the crust of Asia squeezed around the sides of the thrusting arriviste. Then, as India wedged itself like a chisel further beneath Asia, the surface crust crumpled and folded into a range of mountains more than two thousand miles long. These were the beginnings of the Himalayas. And the land around the mountains was forced up into a vast plateau, the "roof of the world", whose average height is greater than the highest mountain in America. India is still pushing. The Himalayas grow by nearly half an inch a year, and Everest and its kin would be even taller if their fresh young rocks weren't eroding away as they rose.

Meanwhile, partway round the world, Africa was aggressively reacquainting itself with its old Pangaean neighbour, Europe. The first part to hit was a peninsula, sticking out from the northern part of the African plate and bearing what is now Italy and Greece and the countries of former Yugoslavia. This collision threw up the beginning of the Alps. Spain crammed into France, and henceforth there were Pyrenees. And though Africa and Eurasia may seem as if they are only joined at their Arabian hip, the Mediterranean is slowly closing. When Africa itself collides with the European continent, a mighty new range of mountains will be born.

Arabia is now shoving into Iran. Europe and Asia have never been parted since Pangaea, and Australia is heading northwards to join in. In a few tens of millions of years, Australia's left shoulder will probably catch on the southernmost islands of Southeast Asia. It will twist and jerk upwards, to slam into Borneo and the southern parts of China.

Predicting the future of continental movements is an inexact science. But supercontinents come and go over time, and most of the world's landmasses are already crammed into this one gigantic block. Only the Americas and Antarctica remain aloof. As the Atlantic Ocean widens, America is moving steadily further from Europe, and the most dramatic drift within the North American continent is the one taking Los Angeles and Baja northwards. (In about 10 million years, L.A. will pass San Francisco, and by 60 million years from now, it will be heading down a trench into the Earth's interior, just south of Alaska.) But some researchers predict that the Americas, too, will be reunited with the rest of the world's landmasses. According to one attempt at constructing the future, over the next few hundred million years the Atlantic will begin to close again, bringing North and South America back into the fold, and Antarctica will head north to join India.[5]

If so, in 250 million years, the new Pangaea could form. Then it would need to survive intact for another hundred million years or so, while a plume of hot mantle built up beneath. It would shift to the equator in a geological eye blink, just a million years or so, to right the balance of the spinning world; it would break up, and scatter its pieces around the equator and tropics. And then the ice would return.

Gradually the frozen polar oceans begin to reach out with tentative feelers of ice. Finding nothing to stop them, they continue their spread. The whiteness advances like a disease that gradually covers the planet's blue surface. The oceans turn greasy, first, with smashed ice crystals. Then the pancakes of ice are back, and the frost flowers, and the transparent young coating of sea ice that bends with the swell. The ice thickens, and spreads, and thickens some more. By the time it reaches the tropics, it's unstoppable, and in just a few centuries it goes all the way. Global temperatures plummet; rain stops; clouds no longer form. Wisps of ice ripped from the frozen ocean are spread by the wind until they begin to build up on the world's highest mountains. Slowly, steadily, the ice forms glaciers that spill down on to the lowlands. And then the white-out is complete.

WHAT WILL our descendants do? Perhaps they will be so unimaginably advanced that they'll be able to prevent a Snowball. They might be routinely tapping additional energy from the sun, or stopping continents in their tracks. But the Earth is a powerful and stubborn force. She limits our resources, and her geological will is extremely hard to check.

If distant descendants of the human lineage can't stop the Snowball, can they weather it? That's hard to imagine, too. Getting a few simple marine creatures through the ice is one thing, but the complex creatures that inhabit our planet today would be another matter. Antarctica is the most hostile place on Earth. Unless you take your own life-support system of food and fuel and shelter there with you, you will die. And in a Snowball, Antarctica takes over the world. For any truly complex creatures, the result would surely be disastrous. Norse mythology has a word for it. After the catastrophe of Fimbulwinter comes Ragnarok, the end of the world.

But a new Snowball wouldn't be the end for *all* life on Earth, any more than the previous ones were. The destructive power of the last Snowball was followed by an extraordinary new beginning. Who knows what direction a post-Snowball Earth might encourage its living things to take?

Our planet is, after all, a master of invention. Through geological time, Earth has constantly sought out new forms and taken on remarkable new identities. Plumes of hot rock ascending from the interior drive continual reshaping of the continental surface. A mountain range rises; another falls. Oceans open here and close there. Earthquakes and eruptions and tidal waves that seem so catastrophic to us are all just part of Earth's irresistible transforming urge. Even the flimsy atmosphere plays its part in adapting, then reinforcing, our planet's shifting moods. Change doesn't alarm the Earth; it is a fundamental part of its nature. We humans, and the other creatures that share our geological slice of life, are the fragile ones.

EPILOGUE

He walks and walks and walks until
he's reached the summit of the hill.
There he rolls a ball of snow
and aims it at his friends below,
But then he slips, so now poor Paul
becomes a part of his own snowball![1]

When Paul Hoffman saw the children's cartoon book containing this poem, he couldn't believe it. Now he has both cartoon and poem proudly taped to his office door. The Paul of the story ends up careering down a hill, trapped inside a snowball, and wearing a look of horror along with his muffler and galoshes. But Paul Hoffman is delighted to be bound up in his Snowball theory. Remember how in 1991, before he had even gone to Namibia, his old university had asked what he would like to be remembered for. And how he'd promptly replied, "Something I haven't done yet." If you present him with the same question today, he hesitates. "I suppose I should say the same thing—something I haven't done yet," he says at last. "But the Snowball's going to be pretty hard to top."

Paul has won the medals he sought. He's particularly proud of the Alfred Wegener medal, awarded by the European Union of Geosciences for research that successfully brings together many different fields "in the spirit of Wegener", that other charismatic, vilified champion of a theory that rocked the world of science. Paul received the medal in April 2001, and gave a rousing rendition of the Snowball story to the assembled ceremonial crowd.

Some of Paul's critics still complain that he is resting too much of his reputation on this one idea. "Of all the people who need to prove themselves, I'd have thought Paul was the last," a researcher said to me one night, over the dregs of what had been a fine bottle of wine. "Whatever criterion you want to look at—he's a Harvard professor, he's fit, he's a member of the National Academy—apart from becoming Lord Hoffman, there's not much to go for, unless it's the big P. Posterity."

This isn't the whole story. Paul may love the attention that his work brings him, but his passion for the rocks themselves also runs deep. Working in the field, unearthing clues, and piecing them together into a picture that changes his understanding of the way the world works, that's when he feels fully alive. On the last day of a field season he often has tears in his eyes, though nobody has ever dared mention them. Paul goes to the rocks because he has to.

That combination of passion and aggrandizement has its inevitable effect. Paul sparks reactions in those around him. Some people are thrilled, others determined to cut him down to size. All are pulled into his story. This capacity to attract attention is crucial for really big scientific ideas. Theories like the Snowball often languish for decades without being properly probed. They need champions to drag them into the scientific limelight and expose them to scrutiny. They need people like Paul.

NOTES AND SUGGESTIONS FOR FURTHER READING

Most of the material in this book comes from interviews with the researchers involved, from their students or former students, and from visits I made to their field sites. For some of the historical information, good general books or articles are available, and I've listed them in the notes that follow. But most of the research is so recent that published academic papers provide the only available accounts. I've included references to these papers, for the truly dedicated reader. Some of the research has not yet been published; in those cases information came directly from the researchers themselves.

ONE: FIRST FUMBLINGS

1. There's a wonderful account of the life and times of stromatolites in Ken McNamara's slim volume *Stromatolites,* 2nd edition (Perth: W.A. Museum, 1997); also see the vivid description of the creatures of Slimeworld in Jan Zalasiewicz and Kim Freedman, "The Dawn of Slime", *New Scientist,* 11 March 2000, 30.
2. Using 4.55 billion years for the origin of the Earth, 3.85 billion years for the earliest life, and .54 billion years for the Cambrian explosion.
3. John McPhee, *Basin and Range* (New York: Farrar, Straus & Giroux, 1981), 126.
4. Stephen Jay Gould, *Time's Arrow, Time's Cycle* (Cambridge, Mass.: Harvard University Press, 1998), 1–2.

5. Stephen Pyne is a true ice-lover, and his book *The Ice* (Seattle: University of Washington Press, 1998) pays detailed homage to all things frozen. Also see the wistful, wonderful *Arctic Dreams* by Barry Lopez (New York: Scribner's, 1986).

TWO: THE SHELTERING DESERT

1. Information about the expedition comes from George Whalley, *The Legend of John Hornby* (London: John Murray, 1962), and from Edgar Christian's diary, published by his parents after his death with the title *Unflinching: A Diary of Tragic Adventure* (London: John Murray, 1937).
2. Whalley, *The Legend of John Hornby,* 282.
3. R. F. Scott, *Scott's Last Expedition,* 5th edition, vol. 1 (London: Smith, Elder, 1914), 542.
4. Of the many, many words that have been written about why Scott's expedition went so wrong, Susan Solomon's book *The Coldest March* (New Haven and London: Yale University Press, 2001) is among the best. A world-renowned atmospheric scientist, Solomon analysed the meteorological data from Scott's trip and concluded that the adventurers encountered exceptionally bad weather. The best contemporary description of the events leading up to the tragedy is in Apsley Cherry-Garrard, *The Worst Journey in the World* (London: Picador, 1994).
5. See Simon Winchester, *The Map That Changed the World* (London: Viking, 2001).
6. P. H. Hoffman, "United Plates of America, the birth of a craton: Early Proterozoic assembly and growth of Laurentia", in *Annual Review of Earth and Planetary Sciences,* vol. 16 (1988), 543–603.
7. *The Ottawa Citizen,* 14 July 1989.
8. Paul borrowed this quote from John Kenneth Galbraith's novel *The Tenured Professor* (Boston: Houghton Mifflin, 1990). The character who made the remark was a former president of the University of California, Berkeley, who had been forced out after colliding with the governor of California, Ronald Reagan. Another character in the

book, hearing the remark, says, "He lost big because he won big. That's my idea of life." Paul told me he wished he had read the book earlier, so that he could have used the quote when he was actually leaving the Survey.

THREE: IN THE BEGINNING

1. Apsley Cherry-Garrard, *The Worst Journey in the World* (London: Picador, 1994).
2. Ibid., 369.
3. Here's an example of Brian's scrupulousness. When he was still doing research long after the age of retirement, he would carefully claim his own travel costs at the reduced rate for pensioners.
4. W. B. Harland, "The Cambridge Spitsbergen Expedition, 1949", *Geographical Journal* 118 (1952), 309–31.
5. Brian put forth this argument in a paper in the *Geological Magazine* 93, no. 4 (1956), 22.
6. See Edmund Blair Bolles, *The Ice Finders* (Washington D.C.: Counterpoint, 1999). Another good account of the development of ice age theories is John and Mary Gribbin's *Ice Age* (London: Penguin, 2001).
7. W. B. Harland, "Evidence of late Precambrian glaciation and its significance", in *Problems in Palaeoclimatology: Proceedings of the NATO Palaeoclimates Conference held at the University of Newcastle-upon-Tyne, January 7–12, 1963*, edited by A. E. M. Nairn (London: Interscience Publishers, 1964), 119.
8. J. C. Crowell, "Climate significance of sedimentary deposits containing dispersed megaclasts", in *Problems in Palaeoclimatology* (see above), 86.
9. Mike Hambrey, a student of Brian's who is now a geology professor at the University of Aberystwyth in Wales, performed the most detailed analysis of ice rocks in the late 1970s, and published the results in *Earth's Pre-Pleistocene Glacial Record,* edited by M. J. Hambrey and W. B. Harland (Cambridge, England: Cambridge University Press, 1984).

10. Alfred Wegener, *Annals of Meteorology* 4 (1951), 1–13.
11. H. W. Menard, *The Ocean of Truth: A Personal History of Global Tectonics* (Princeton, N.J.: Princeton University Press, 1986), 20–21.
12. "Continental Drift and Plate Tectonics: A Revolution in Science", in J. Bernard Cohen, *Revolution in Science* (Cambridge, Mass.: Harvard University Press, 1985), 446–66.
13. More on this story can be found in the rather uneasy account given by one of the researchers who survived through the winter at the station: Johannes Giorgi, *Mid-Ice: The Story of the Wegener Expedition to Greenland,* translated by F. H. Lyon (London: Kegan Paul, Trench, Trubner & Co., 1934); see also Martin Schwarzbach, *Alfred Wegener: The Father of Continental Drift* (Madison, Wisc.: Science Tech Publishers, 1986).
14. Mott T. Greene, "Alfred Wegener", *Social Research* 51, no. 3 (1984), 747.
15. In 1964 he laid out his main arguments in a paper titled "Critical evidence for a great infra-Cambrian glaciation", in *Geologische Rundschau* 54: 45–61; a more accessible version of the idea appears in a marvellous article that Brian wrote with M. J. S. Rudwick and published in *Scientific American* that same year, "The Great Infra-Cambrian Glaciation" (August 1964), 28.
16. W. B. Harland, "The Geology of Svalbard", *Geological Society Memoir* no. 17 (London: Geological Society, 1997).

FOUR: MAGNETIC MOMENTS

1. Joe later published this finding in "South-seeking magnetic bacteria", *Journal of Experimental Biology* 86 (1980), 345–47.
2. J. L. Gould, J. L. Kirschvink, and K. S. Deffeyes, "Bees have magnetic remanence", *Science* 201 (1978), 1026–28; C. Walcott, J. L. Gould, and J. L. Kirschvink, "Pigeons have magnets", *Science* 205 (1979), 1027–29.
3. J. L. Kirschvink, A. Kobayashi-Kirschvink, and B. J. Woodford, "Magnetite biomineralization in the human brain", *Proceedings of the National Academy of Sciences* 89 (1992), 7683–87.

4. The two researchers then incorporated additional information to make a somewhat more persuasive case, and Joe eventually recommended publication. The paper was published under the title "Low palaeolatitude of deposition for late Precambrian periglacial varvites in South Australia", *Earth and Planetary Science Letters* 79 (1986), 419–30. It did not, however, include the crucial field test that Joe went on to perform.
5. M. I. Budyko, "The effect of solar radiation variations on the climate of the Earth", *Tellus* 21 (1969), 611–19.
6. George E. Williams, "Precambrian tidal and glacial clastic deposits: Implications for Precambrian Earth-Moon dynamics and palaeoclimate", *Sedimentary Geology* 120 (1998), 55–74.
7. Dawn Sumner presented the results at the autumn meeting of the American Geophysical Union in 1987.
8. J. L. Kirschvink, "Late Proterozoic low-latitude glaciation: The Snowball Earth", section 2.3, in J. W. Schopf, C. Klein, and D. Des Maris, eds., *The Proterozoic Biosphere: A Multidisciplinary Study* (Cambridge, England: Cambridge University Press, 1992), 51–52.

FIVE: EUREKA

1. Paul went on to publish this paper, as well as the one with Dan Schrag, in *Science*. Paul F. Hoffman, Alan J. Kaufman, and Galen P. Halverson, "Comings and goings of global glaciations on a neoproterozoic carbonate platform in Namibia", *GSA Today* 8 (1998), 1–9.
2. Dan has made some marvellous discoveries about past climate in his coral work. See, for example, K. A. Hughen, D. P. Schrag, S. B. Jacobsen, and W. Hantoro, "El Niño during the last Interglacial recorded by fossil corals from Indonesia", *Geophysical Research Letters* 26 (1999), 3129–32. This tale is written up in more accessible form in "Weather warning", *New Scientist* 164 (9 October 1999), 36.

SIX: ON THE ROAD

1. P. F. Hoffman, A. J. Kaufman, G. P. Halverson, and D. P. Schrag, "A Neoproterozoic snowball Earth", *Science* 281 (1998), 1342–46. Paul

and Dan went on to write a more popular rendition of their ideas: "Snowball Earth", *Scientific American,* January 2000, 68–75.

2. Alfred Wegener, *The Origin of Continents and Oceans,* translated from the third German edition by J. G. A. Skerl (London: Methuen & Co., 1924), 5.
3. E. W. Berry comments on the Wegener hypothesis in *The Theory of Continental Drift: A Symposium,* edited by W. A. J. M. van Waterschoot van der Gracht (London: John Murray, 1928), 124.
4. B. Willis, *American Journal of Science* 242 (1944), 510–13.
5. For a more detailed discussion of these disagreements, see Naomi Oreskes's extremely thorough analysis, *The Rejection of Continental Drift* (Oxford, England: Oxford University Press, 1999).
6. From Mott T. Greene, "Alfred Wegener", *Social Research* 51, no. 3 (1984), 753.
7. Oreskes gives an excellent description of uniformitarianism in *The Rejection of Continental Drift* (see above). Also, Stephen Jay Gould has considered the principle in many fine essays. See, for example, his discussion of uniformitarianism and catastrophism, "Lyell's Pillars of Wisdom", in *The Lying Stones of Marrakech* (London: Vintage, 2000), 147–68; or the discussion in his *Time's Arrow, Time's Cycle* (Cambridge, Mass.: Harvard University Press, 1998).
8. Walter Alvarez wrote an entertaining book about this process, *T. rex and the Crater of Doom* (Princeton, N.J.: Princeton University Press, 1997).
9. W. M. Davis, "The value of outrageous geological hypotheses", *Science* 63 (1926), 464.
10. H. W. Menard, *The Ocean of Truth: A Personal History of Global Tectonics* (Princeton, N.J.: Princeton University Press, 1986).
11. Yes, Pippa is a man, and he has no idea why his parents burdened him with what appears to be a woman's name. In publications he goes by his first name, Galen, which he also finds baffling.
12. John Playfair, *Illustrations of the Huttonian Theory of the Earth* (Edinburgh: William Creech, 1802).
13. Gould, *Time's Arrow, Time's Cycle,* 64.

14. ———, "James Hutton's Theory of the Earth", in *Time's Arrow, Time's Cycle,* 61–98.

SEVEN: DOWN UNDER

1. Linda Sohl, Nicholas Christie-Blick, and Dennis Kent, "Paleomagnetic polarity reversals in Marinoan glacial deposits of Australia", *GSA Bulletin* 111 (1999), 1120–39.
2. George Williams has described his idea about the tilting of the Earth in a series of academic papers. The broadest and best for nonspecialists is probably his chapter entitled "The enigmatic Late Proterozoic glacial climate: An Australian perspective", in *Earth's Glacial Record* (Cambridge, England: Cambridge University Press, 1994), 146–64.
3. *The Australian Geologist* 117 (31 December 2000), 21.
4. After some heated e-mail exchanges with the editor of *The Australian Geologist,* Paul abandoned his attempt to publish a ten-page rebuttal of George Williams's arguments in the magazine. Instead he posted his rebuttal on his website, http://www.eps.harvard.edu/people/faculty/hoffman/TAG.html
5. There is even some evidence that the presence of the moon *prevented* the Earth's tilt from further fluctuations. See, for example, J. Laskar, F. Joutel, and P. Robutel, "Stabilization of the Earth's obliquity by the Moon", *Nature* 361 (1993), 615–17. Recently one group of researchers did try to come up with a possible mechanism for righting the Earth: See Darren Williams, James Kasting, and Lawrence Frakes, "Low-latitude glaciation and rapid changes in the Earth's obliquity explained by obliquity-oblateness feedback", *Nature* 396 (1998), 453. But the authors say that their paper serves to demonstrate how difficult the task would be.
6. Jim Walker's paper is now in press at the *Proceedings of the National Academy of Sciences.*

NOTES AND SUGGESTIONS FOR FURTHER READING

EIGHT: SNOWBRAWLS

1. Once I was interviewing Martin Kennedy in an Italian restaurant, using a minidisk recorder and a microphone. When Martin left for the bathroom, the proprietor sidled up to me and asked, "Who is he? Why are you interviewing him? Is he someone famous?" When Martin returned, I related this story and he grinned. "Surely he's heard of the *Kennedys,*" he said in a stage whisper. Then, "Did he ask what kind of book you're writing? You did tell him, didn't you, that it's a thriller?"
2. For more on the remarkable properties of methane hydrates, see Erwin Suess, Gerhard Bohrmann, Jens Greinert, and Erwin Lausch, "Flammable ice", *Scientific American,* November 1999, 76–83. There's also a vivid essay by Nicola Jones: "Fire and Ice", *Chemistry and Industry* 26 (June 2000), 398–99.
3. Martin Kennedy, Nicholas Christie-Blick, and Linda Sohl, "Are Proterozoic cap carbonates and isotopic excursions a record of gas hydrate destabilization following Earth's coldest intervals?" *Geology* 29, no. 5 (2001), 443–46.
4. Martin Kennedy, Nicholas Christie-Blick, and Anthony Prave, "Carbon isotopic composition of Neoproterozoic glacial carbonates as a test of paleoceanographic models for snowball Earth phenomena", *Geology* 29, no. 12 (2001), 1135–38.
5. "The Aftermath of a Snowball Earth", by John Higgins and Daniel Schrag, submitted to the electronic journal *Geochemistry, Geophysics, Geosystems.*
6. Though Huxley's reply is widely quoted, the precise wording varies between versions, and sadly no verbatim account of the debate exists. See, for example, *The Columbia World of Quotations,* edited by R. Andrews, M. Biggs, and M. Seidel (New York: Columbia University Press, 1996).
7. This quote comes from the marvellous *Ice Palaces* by Fred Anderes and Ann Agranoff (New York: Abbeville Press, 1983). Sadly, the book is out of print, but it is well worth hunting around for a used copy.
8. Douglas Mawson, "The Home of the Blizzard", (New York: St.

Martin's Press, 1998), xvii. One of the best books ever written about Antarctic exploration, and yet little known outside Australia, this is a must for anyone who cares about ice.

9. See Philip Ball's vivid descriptions of this in *Life's Matrix: A Biography of Water* (Berkeley: University of California Press, 2001).
10. For example, the paper by William Hyde, Thomas Crowley, Steven Baum and Richard Peltier, "Neoproterozoic 'snowball Earth' simulations with a coupled climate/ice-sheet model", *Nature* 405 (2000), 425–29; also Bruce Runnegar's commentary, "Loophole for snowball Earth", on page 403 of the same issue; and Mark Chandler and Linda Sohl, "Climate forcings and the initiation of low-latitude ice sheets during the Neoproterozoic Varanger glacial interval", *Journal of Geophysical Research* 105 (2000), 20, 737–20, 756.
11. Doug and Dan are now writing up this work for publication.

NINE: CREATION

1. The dating is controversial, with a few Snowball deposits dated as early as 575 million years.
2. A wonderful book! Stephen Jay Gould, *Wonderful Life* (New York: Vintage, 2000).
3. Attempts to explain the *form* of the Cambrian explosion have been many and various. The fact that such an explosion ever occurred, though, can be traced back to the previous leap into multicellularity. For more on this, see Carl Zimmer, *Evolution: The Triumph of an Idea* (London: William Heinemann, 2002); and Bill Schopf, *Cradle of Life* (Princeton, N.J.: Princeton University Press, 1999).
4. A few Ediacarans had been found before Sprigg's discovery, for example from Charnwood Forest, in England. But they were not grouped together or given this collective name until the major finds in South Australia.
5. James Gehling, "Microbial Mats in terminal Proterozoic siliciclastics: Ediacaran death masks", *Palaios* 14 (1999), 40–57.
6. Researcher Mark McMenamin has written a book titled *The Garden of Ediacara* (New York: Columbia University Press, 1998), in which

he argues that the Ediacarans were a failed experiment, which then became extinct. (Be warned, though, this book was received with little enthusiasm by the rest of the Ediacaran community, and one researcher said in a review that it "falls as flat as a week-old *Dickinsonia* roadkill".) For more on this debate, see also Bennett Daviss, "Cast out of Eden", *New Scientist* 158 (16 May 1998), 26; and Richard Monastersky, "Life grows up", *National Geographic,* April 1998, 100–15. In an academic treatise, Jim Gehling sets out his arguments that some at least of the Ediacarans evolved into more familiar animals: "The case for Ediacaran fossil roots to the metazoan tree", *Geological Society of India Memoir No.* 20 (1991), 181–224.

7. The researcher is Ben Waggoner, now at the University of Central Arkansas, and the fossil he is so proud of is called *Yorgia waggoneri.* He says he finds the fact that nobody knows the exact nature of this creature "somehow deeply appropriate and satisfying".
8. These finds are all extremely recent. Misha is now preparing his descriptions and conclusions for publication.
9. A. P. Benus, *Bulletin of New York State Museum* 463 (1988), 8.
10. There is no sign of any animal trails among the Ediacarans at Mistaken Point. Since these creatures are much older than the ones found in South Australia or the White Sea, they are presumably at an earlier state of evolution. Many researchers believe that their large size and the intricate fronds and spindles mean that these, too, are complex, differentiated creatures.
11. S. A. Bowring et al., "Geochronological constraints on the duration of the Neoproterozoic-Cambrian transition", *Geological Society of America,* 1998 annual meeting, Abstract A147; S. A. Bowring and D. H. Erwin, "Progress in Calibrating the Tree of Life and Metazoan Phylogeny", *Eos Trans. AGU* 81 (48), Fall Meeting Supplement, Abstract B62A–04, 2000.
12. H. J. Hofmann, G. M. Narbonne, and J. D. Aitken, "Ediacaran Remains from Intertillite Beds in Northwestern Canada", *Geology* 18 (1990), 1199–1202.

13. A. Seilacher, P. K. Bose, and F. Pflüger, "Triploblastic animals more than 1 billion years ago: Trace fossil evidence from India", *Science* 282 (1998), 80.
14. B. Rasmussen, S. Bengston, I. Fletcher, and N. McNaughton, "Discoidal impressions and trace-like fossils more than 1200 million years old", *Science* 296 (2002), 1112–15.
15. N. J. Butterfield, "*Bangiomorpha pubescens:* Implications for the evolution of sex, multicellularity, and the Mesoproterozoic/Neoproterozoic radiation of eukaryotes", *Paleobiology* 26 (2000), 386–404.
16. There's a good, though somewhat technical, overview of this technique in Andrew Smith and Kevin Peterson, "Dating the time of origin of major clades: Molecular clocks and the fossil record", *Annual Reviews of Earth and Planetary Science* 30 (2002), 65–89.
17. Lee Smolin, "Art, science and democracy", written for a catalogue of an exhibit of sculpture by Elizabeth Turk at the Santa Barbara Contemporary Arts Forum, 24 February–14 April 2001.
18. Kevin is currently preparing his paper for publication.

TEN: EVER AGAIN

1. Joe published these results in D. A. Evans, N. J. Beukes, and J. L. Kirschvink, "Low-latitude glaciation in the Palaeoproterozoic era", *Nature* 386 (1997), 262–65. His primitive Snowball didn't seem to trigger the same dramatic evolutionary changes that accompanied the later one. It may simply have been too early. For any evolutionary leap, living things don't just need the environmental opportunity, they also need the genetic wherewithal. Genetic material changes through time and chance, and the creatures living during this early Snowball may not yet have had long enough to string together the genes they'd eventually need.
2. Dan Schrag also points out that rainfall in the tropics is much more intense than at higher latitudes. The more intense the rainfall, the more effective rocks are at trapping carbon dioxide. If all the world's continents were near the equator, their collective ability to suck up carbon dioxide would go into overdrive; both advancing ice

and continental rocks would be working in tandem to cool the Earth. He put all these arguments in a paper in which he also puts forward an intriguing idea for what specifically may have triggered the cooling that led to one of Paul's suite of Snowballs. G. P. Halverson, P. F. Hoffman, D. P. Schrag, and A. J. Kaufman, "A major perturbation of the carbon cycle before the Ghaun glaciation in Namibia: prelude to snowball Earth", *Geochemistry, Geophysics, Geosystems,* 27 June 2002.

3. David Evans, "True polar wander, a supercontinental legacy", *Earth and Planetary Science Letters* 157 (1998), 1–8; and D. A. Evans, "True polar wander and supercontinents", *Tectonophysics,* 2002, in press. A more accessible description of this idea is Robert Irion, "Slip-sliding away", *New Scientist* 18 (August 2001), 34.
4. For a good general description of the motions of continents, see David M. Harland *The Earth in Context* (Chichester, England: Springer-Praxis, 2001). There is a fuller and more technical description in Donald Turcotte and Gerald Schubert *Geodynamics,* 2nd edition (Cambridge, England: Cambridge University Press, 2002).
5. Chris Scotese's construction of the possible new supercontinent, which he calls "Pangaea Ultima", is part of his Paleomap Project. More information about this is on his website, www.scotese.com.

EPILOGUE

1. *The Biggest Snowball Ever.* Copyright © 1988 by John Rogan. Reproduced by permission of the publisher, Candlewick Press, Inc., Cambridge, MA, on behalf of Walker Books Ltd., London.

INDEX

INDEX

INDEX

INDEX

INDEX

INDEX

INDEX

INDEX

INDEX

A NOTE ON THE AUTHOR

Gabrielle Walker has travelled in search of science stories to all seven continents - including a stint at the South Pole. She has climbed trees in the Amazon rainforest, used a geological hammer to pull fresh lava from a volcano in Hawaii and dodged icebergs while sailing across Drake's passage and around Cape Horn. She has a PhD in Natural Science from Cambridge University and has been Editor at *Nature* and Features Editor at *New Scientist* for whom she now acts as consultant. An award-winning writer of more than sixty pieces for *New Scientist*, she has also written for *The Economist*, the *Independent*, the *Guardian* and the *Daily Telegraph*, and is a frequent studio guest and presenter for the BBC. She lives in London. This is her first book.